Large-Scale System Analysis under Uncertainty

Discover a comprehensive set of tools and techniques for analyzing the impact of uncertainty on large-scale engineered systems. Providing accessible yet rigorous coverage, this book showcases the theory through detailed case studies drawn from electric power application problems, including the impact of integration of renewable-based power generation in bulk power systems, the impact of corrupted measurement and communication devices in microgrid closed-loop controls, and the impact of components failures on the reliability of power supply systems. The case studies also serve as a guide on how to tackle similar problems that appear in other engineering application domains, including automotive and aerospace engineering.

This is essential reading for academic researchers and graduate students in power systems engineering, and dynamic systems and control engineering.

Alejandro D. Domínguez-García is a Professor in the Department of Electrical and Computer Engineering at the University of Illinois at Urbana-Champaign, where he also holds Research Professor appointments with the Coordinated Science Laboratory and the Information Trust Institute.

"Uncertainty and reliability are of higher and higher priority as marginal costs go down and the number of components goes up. This book is comprehensive and rigorous, full of derivations and nuanced examples – an excellent reference for anyone seeking solid foundations."

Josh Taylor, University of Toronto

"Professor Domínguez-García's book should be required reading for all graduate students in electric power systems. It fills a critical gap by focusing on the dynamical systems and control engineering foundations that underpin power system analysis and control, rather than focusing on the application domain problems themselves."

Johanna Mathieu, University of Michigan

"With climate change disruptions being a new normal, electric vehicles taking over our roads, and renewable energies being an existential matter for all of us, there is a considerable need for a reference book on power system analysis under uncertainties. This book is advantageous to both students and practitioners working in the planning and operation of the electric grid."

Reza Argandeh, Western Norway University of Applied Sciences

Large-Scale System Analysis under Uncertainty

With Electric Power Applications

ALEJANDRO D. DOMÍNGUEZ-GARCÍA
University of Illinois at Urbana-Champaign

CAMBRIDGE
UNIVERSITY PRESS

University Printing House, Cambridge CB2 8BS, United Kingdom

One Liberty Plaza, 20th Floor, New York, NY 10006, USA

477 Williamstown Road, Port Melbourne, VIC 3207, Australia

314–321, 3rd Floor, Plot 3, Splendor Forum, Jasola District Centre, New Delhi – 110025, India

103 Penang Road, #05–06/07, Visioncrest Commercial, Singapore 238467

Cambridge University Press is part of the University of Cambridge.

It furthers the University's mission by disseminating knowledge in the pursuit of education, learning, and research at the highest international levels of excellence.

www.cambridge.org
Information on this title: www.cambridge.org/9781107192089
DOI: 10.1017/9781108123853

© Cambridge University Press 2022

This publication is in copyright. Subject to statutory exception and to the provisions of relevant collective licensing agreements, no reproduction of any part may take place without the written permission of Cambridge University Press.

First published 2022

A catalogue record for this publication is available from the British Library.

Library of Congress Cataloging-in-Publication Data
Names: Domínguez–García, Alejandro D., 1977– author.
Title: Large-scale system analysis under uncertainty : with electric power applications / Alejandro D. Domínguez–García, Department of Electrical and Computer Engineering, University of Illinois at Urbana-Champaign.
Description: Cambridge, United Kingdom ; New York, NY, USA : Cambridge University Press, 2022. | Includes bibliographical references and index.
Identifiers: LCCN 2021024851 (print) | LCCN 2021024852 (ebook) | ISBN 9781107192089 (hardback) | ISBN 9781108123853 (epub)
Subjects: LCSH: Systems engineering–Risk assessment. | Electric power systems–Risk assessment. | Uncertainty.
Classification: LCC TA169.55.R57 D66 2022 (print) | LCC TA169.55.R57 (ebook) | DDC 621.31–dc23
LC record available at https://lccn.loc.gov/2021024851
LC ebook record available at https://lccn.loc.gov/2021024852

ISBN 978-1-107-19208-9 Hardback

Cambridge University Press has no responsibility for the persistence or accuracy of URLs for external or third-party internet websites referred to in this publication and does not guarantee that any content on such websites is, or will remain, accurate or appropriate.

In memory of my father,
Ángel Domínguez Casás (1930 – 2020),
who instilled in me a passion for mathematics

Contents

Preface and Acknowledgments *page* xi
Notation xiii

1 **Introduction** 1
1.1 Motivation 1
1.2 System Models 3
1.3 Uncertainty Models 7
1.3.1 Static Systems 7
1.3.2 Dynamical Systems 8
1.4 Application Examples 9
1.4.1 Power Flow Analysis under Active Power Injection Uncertainty 9
1.4.2 Analysis of Inertia-less AC Microgrids under Power Injection Uncertainty 10
1.4.3 Reliability Analysis of Static Systems 11
1.5 Book Road Map 12
1.6 Notes and References 15

2 **Preliminaries** 16
2.1 Probability and Stochastic Processes 16
2.1.1 Probability Spaces 16
2.1.2 Random Variables 17
2.1.3 Jointly Distributed Random Variables 21
2.1.4 Random Vectors 27
2.1.5 Stochastic Processes 29
2.2 Set Theory 35
2.2.1 Basic Notions and Notation 35
2.2.2 Sets in Euclidean Space 36
2.3 Linear Dynamical Systems 46
2.3.1 Discrete-Time Systems 46
2.3.2 Continuous-Time Systems 48
2.4 Notes and References 53

3	**Static Systems: Probabilistic Input Uncertainty**		54
	3.1	Introduction	54
	3.2	Moment Characterization	56
		3.2.1 Linear Setting	56
		3.2.2 Nonlinear Setting	58
	3.3	Distribution Characterization	62
		3.3.1 Linear Setting	62
		3.3.2 Nonlinear Setting	68
	3.4	Performance Characterization	74
		3.4.1 Known Input Moments	74
		3.4.2 Known Input Probability Density Function	76
	3.5	Application to Power Flow Analysis	77
		3.5.1 Power Flow Model	78
		3.5.2 Power Flow Vector Distribution	79
	3.6	Notes and References	90
4	**Static Systems: Probabilistic Structural Uncertainty**		91
	4.1	Introduction	91
	4.2	System Stochastic Model	92
	4.3	Markov Process Characterization	93
		4.3.1 Discrete-Time Case	94
		4.3.2 Continuous-Time Case	97
	4.4	Performance Characterization	100
	4.5	Application to Reliability and Availability Analysis	101
		4.5.1 Multi-Component System Input-to-State Characterization	101
		4.5.2 Systems with Non-Repairable Components	109
		4.5.3 Systems with Repairable Components	118
		4.5.4 Reduced-Order Models	124
	4.6	Notes and References	129
5	**Discrete-Time Systems: Probabilistic Input Uncertainty**		130
	5.1	Introduction	130
	5.2	Discrete-Time Linear Systems	131
		5.2.1 Characterization of First and Second Moments	131
		5.2.2 Probability Distribution	138
	5.3	Discrete-Time Nonlinear Systems	144
		5.3.1 Characterization of First and Second Moments	145
		5.3.2 Probability Distribution	150
	5.4	Analysis of Microgrids under Power Injection Uncertainty	151
		5.4.1 System Model	151
		5.4.2 Average Frequency Error Statistical Characterization	156
		5.4.3 Phase Angle Statistical Characterization	161
	5.5	Notes and References	165

6 **Continuous-Time Systems: Probabilistic Input Uncertainty** 166
6.1 Introduction 166
6.2 Continuous-Time Linear Systems 168
6.2.1 Characterization of First and Second Moments 171
6.2.2 Gaussian Systems 181
6.3 Continuous-Time Nonlinear Systems 187
6.3.1 Moments 188
6.3.2 Probability Distribution 189
6.4 Analysis of Microgrids under Sensor Measurement Uncertainty 192
6.4.1 System Model 192
6.4.2 Average Frequency Error Statistical Characterization 197
6.5 Notes and References 201

7 **Static Systems: Set-Theoretic Input Uncertainty** 202
7.1 Introduction 202
7.2 Ellipsoid-Based Input Set Description 204
7.2.1 Linear Setting 204
7.2.2 Nonlinear Setting 212
7.3 Zonotope-Based Input Set Description 218
7.3.1 Linear Setting 219
7.3.2 Nonlinear Setting 221
7.4 Performance Requirements Verification 225
7.5 Application to Power Flow Analysis 226
7.5.1 Power Flow Model 227
7.5.2 Ellipsoidal-Based Description of Possible Extraneous Power Injection Values 228
7.5.3 Zonotope-Based Description of Possible Extraneous Power Injection Values 233
7.6 Notes and References 236

8 **Discrete-Time Systems: Set-Theoretic Input Uncertainty** 237
8.1 Introduction 237
8.2 Discrete-Time Linear Systems 238
8.2.1 Ellipsoidal-Based Input Description 239
8.2.2 Choice of Parameter γ_k 242
8.2.3 Deterministic Inputs 258
8.3 Discrete-Time Nonlinear Systems 259
8.4 Performance Requirements Verification 261
8.5 Analysis of Microgrids under Power Injection Uncertainty 263
8.5.1 System Model 263
8.5.2 Characterization of the Set Containing the Average Frequency Error 265
8.5.3 Characterization of Set Containing the Bus Phase Angles 270
8.6 Notes and References 272

9 Continuous-Time Systems: Set-Theoretic Input Uncertainty 273
9.1 Introduction 273
9.2 Continuous-Time Linear Systems 275
9.2.1 Choice of Parameter $\beta(t)$ 278
9.2.2 Deterministic Inputs 289
9.3 Continuous-Time Nonlinear Systems 290
9.4 Performance Requirements Verification 293
9.5 Case Studies 297
9.5.1 Buck Converter 297
9.5.2 Three-Bus Power System 300
9.6 Notes and References 306

Appendix A Mathematical Background 307

Appendix B Power Flow Modeling 320

References 330
Index 334

Preface and Acknowledgments

This book presents a collection of uncertainty analysis techniques for systems whose behavior can be mathematically represented by a set of algebraic or differential equations describing the relation between certain variables of interest. The techniques included revolve around probabilistic and set-theoretic descriptions of some uncertain phenomena that drive the system response, for example, random load variations in an electric power system, or manifest themselves as changes in the system structure, such as a power line outage caused by a storm. The case studies used throughout the book draw heavily from electric power system applications; however, the techniques presented are general and can be used in other applications, such as aerospace and automotive.

Many of the techniques presented in the book were developed in the area of systems theory and control. These techniques are very powerful and universally applicable; however, it requires a certain level of mathematical sophistication to understand the theory behind them. The goal of the book is to make these techniques accessible to applied researchers and engineers across multiple domains while maintaining a certain level of rigor in the exposition. In doing so, I have tried to make the book as self-contained as possible by including a preliminaries chapter and a mathematical background appendix that reviews most fundamental concepts used throughout the book, namely probability, stochastic processes, set theory, and linear dynamical systems theory. Except for the introductory and preliminaries chapters, the structure of subsequent chapters is always the same. Specifically, the first part of each chapter presents the general theory for a particular analysis technique interspersed with small examples that illustrate its use. The second part of each chapter illustrates the application of the particular technique to the analysis of problems encountered in electric power applications.

The inspiration for this book came from the book by Fred C. Schweppe entitled *Uncertain Dynamic Systems*, which published in 1973. I first became familiar with the book while I was working on my PhD when Professor George Verghese, who was a member of my thesis committee, pointed it out to me because of the material it contained on set-theoretic techniques for uncertainty analysis of linear dynamical systems. This material ended up being very relevant for my graduate research work on reliability and performance analysis of fault-tolerant systems, and it formed the core of one of the chapters of my PhD thesis; thus, I am greatly indebted to Professor Verghese.

Shortly after joining the University of Illinois, I developed a graduate-level course entitled Dynamic System Reliability. Early on in this course, I adopted Schweppe's aforementioned book as a reference. The book had been long out of print, but I was fortunate enough to get permission from the publisher to have the book reprinted locally for exclusive use in the course. As the course material evolved, over numerous offerings, it became apparent that Schweppe's book, while a fantastic reference, was not a perfect match for the course syllabus. As a result, I decided to develop a set of lecture notes that would align better with the course material; those notes eventually became the core material for this book.

Since my arrival at Illinois, I have worked with many undergraduate and graduate students – too many to name them all here – and I am grateful for having had the opportunity to work with a group of such talented individuals. Special thanks go to my former graduate students Stanton Cady, Sairaj Dhople, Christine Chen, Xichen Jiang, Eric Hope, and Madi Zholbaryssov as several of the application examples featured in the book are drawn from our joint research. I would also like to thank the students who have taken my graduate course and provided feedback on early versions of some of the book chapters. My former PhD student, Madi Zholbaryssov, read the whole manuscript very thoroughly and helped in fixing several issues; the end result is better because of him and I am very thankful for it. I am eternally grateful to my friend and long-term collaborator Christoforos Hadjicostis, who read early versions of the manuscript, providing encouragement and critical feedback early in the writing process, and also gave a thorough read to the final manuscript. Finally, I would like to thank my colleagues at the Department of Electrical Engineering at Illinois for providing a stimulating intellectual environment – George Gross, Daniel Liberzon, Pete Sauer, and Venu Veeravalli deserve a special mention for all the mentoring and encouragement they have provided over the years.

This book is dedicated to my father, Ángel, who unexpectedly passed away as I was applying the final touches to the manuscript. He and my mother, Vicenta, provided a nurturing environment when I was growing up and prioritized the education of their six children over all material things. They instilled in me a curiosity for learning and the importance of hard work that led me to pursue an academic career and ultimately resulted in the writing of this book. The final words are for the three loves of my life, my two daughters, Maia and Lia, and my wife, Cristina. Maia and Lia were both born in the six-year span that it took me to complete the book, and they brightened some difficult periods in the writing process. Cristina's appetite for learning, work ethic, and determination were a continuous source of inspiration during the writing process, and a constant reminder of values I hold dear.

Notation

Set Theory

$\mathbb{N}$	Set of natural numbers
$\mathbb{Z}$	Set of integer numbers
$\mathbb{R}$	Set of real numbers
$\mathbb{C}$	Set of complex numbers
$\mathbb{R}^n$	Set of n-dimensional real vectors
$\mathbb{R}^{n \times m}$	Set of $(n \times m)$-dimensional real matrices
$\mathbf{int}(\mathcal{X})$	Interior of the set $\mathcal{X} \in \mathbb{R}^n$
$\mathbf{cl}(\mathcal{X})$	Closure of the set $\mathcal{X} \in \mathbb{R}^n$
$\mathbf{bd}(\mathcal{X}),\ \partial\mathcal{X}$	Boundary of the set $\mathcal{X} \in \mathbb{R}^n$
$S_{\mathcal{X}}(\cdot)$	Support function of the set $\mathcal{X} \in \mathbb{R}^n$

Vectors and Matrices

$x^\top y$, $<x, y>$	Inner product of two vectors, x, y
$\|x\|_2$	Euclidean norm, or 2-norm, of a vector x
$\mathbf{0}_n$	All-zeros vector of dimension n
$\mathbf{1}_n$	All-ones vector of dimension n
$\mathbf{0}_{n \times m}$	All-zeros matrix of dimensions $n \times m$
I_n	Identity matrix of dimensions $n \times n$
$A^\top$	Transpose of a real matrix A
A^{-1}	Inverse of a nonsingular matrix A
$\mathbf{diag}(x_1, x_2, \ldots, x_n)$	$(n \times n)$-dimensional diagonal matrix whose (i, i) entry is equal to x_i
$\mathbf{rank}(A)$	Rank of a matrix A
$\mathbf{tr}(A)$	Trace of a square matrix A
$\mathbf{det}(A)$	Determinant of a square matrix A
$\mathbf{vec}(A)$	Column vector obtained by stacking the columns of a matrix A
$\mathbf{real}(\lambda)$	Real part of eigenvector λ
$\sigma(A)$	Spectrum of a square matrix $A \in \mathbb{R}^{n \times n}$, i.e., set of eigenvalues of A
$A \otimes B$	Kronecker product of matrices A and B.

Functions

$f\colon \mathcal{X} \to \mathcal{Y}$	A function mapping elements in $\mathcal{X}$ into elements in $\mathcal{Y}$
$f^{-1}\colon\ \mathcal{Y} \to \mathcal{X}$	Inverse function of $f\colon\ \mathcal{X} \to \mathcal{Y}$
$f, f(\cdot)$	Shorthand notation for $f\colon\ \mathcal{X} \to \mathcal{Y}$
$f\colon \mathbb{R}^n \to \mathbb{R}$	A function mapping n-dimensional real vectors into real numbers
$f\colon \mathbb{R}^n \to \mathbb{R}^m$	A function mapping n-dimensional real vectors into m-dimensional real vectors
$\nabla f(x)$	Gradient of a real-valued function $f(\cdot)$
$\frac{\partial f(x)}{\partial x}$, $J_f(x)$	Jacobian of a vector-valued function $f(\cdot)$ at x
$\left.\frac{\partial f(x)}{\partial x}\right\vert_{x=x_0}$	Jacobian of a vector-valued function $f(\cdot)$ at $x = x_0$

Probability and Stochastic Processes

$\Pr(A)$	Probability that event A has occurred
$F_X(\cdot)$	Cumulative distribution function (cdf) of random variable X
$p_X(\cdot)$	Probability mass function (pmf) of a discrete random variable X
$f_X(\cdot)$	Probability density function (pdf) of continuous random variable X
$F_{X,Y}(\cdot,\cdot)$	Joint cdf of random variables X and Y
$p_{X,Y}(\cdot,\cdot)$	Joint pmf of discrete random variables X and Y
$f_{X,Y}(\cdot,\cdot)$	Joint pdf of continuous random variables X and Y
$f_{X\,\vert\,Y}(\cdot\,\vert\,y)$	Conditional pdf of random variable X given $Y = y$
$\mathrm{E}[\cdot]$	Expectation operator
μ_X	Mean of random variable X
$c_{X,Y}$	Covariance of random variables X and Y
$r_{X,Y}$	Correlation of random variables X and Y
σ_X^2	Variance of random variable X
m_X	Mean of random vector X
Σ_X	Covariance matrix of random vector X
S_X	Correlation matrix of random vector X
$C_{X,Y}$	Covariance matrix of random vectors X and Y
$R_{X,Y}$	Correlation matrix of random vectors X and Y
$m_X[\cdot]$	Mean function of discrete-time stochastic vector process X
$C_X[\cdot,\cdot]$	Covariance function of discrete-time stochastic vector process X
$R_X[\cdot,\cdot]$	Correlation function of discrete-time stochastic vector process X
$m_X(\cdot)$	Mean function of continuous-time stochastic vector process X
$C_X(\cdot,\cdot)$	Covariance function of continuous-time stochastic vector process X
$R_X(\cdot,\cdot)$	Correlation function of continuous-time stochastic vector process X
$\delta(\cdot)$	Dirac delta function

Linear Dynamical Systems

$\Phi_{k,\ell}$	State-transition matrix of a discrete-time state-space model
$\Phi(t,\tau)$	State-transition matrix of a continuous-time state-space model

Power Networks

θ_i	Phase angle of the phasor associated with bus i's sinusoidal voltage
V_i	Magnitude of the phasor associated with bus i's sinusoidal voltage
p_i	Active power injected into a power network at bus i
q_i	Reactive power injected into a power network at bus i
$\overline{Y}_{ik}$	Shunt admittance of a transmission line linking bus i and bus j
$\overline{Z}_{ik}$	Series impedance of a transmission line linking bus i and bus j
X_{ik}	Imaginary part of $\overline{Z}_{ik}$
$\overline{Y}$	Power network admittance matrix
G	Real part of $\overline{Y}$
B	Imaginary part of $\overline{Y}$
$\overline{y}_{i,k}$	(i,k) entry of the power network admittance matrix
$g_{i,k}$	Real part of $\overline{y}_{i,k}$
$b_{i,k}$	Imaginary part of $\overline{y}_{i,k}$
M	Incidence matrix of the graph associated with a power network
p.u.	Per unit

1 Introduction

In this chapter, we first provide some motivation for the type of modeling problems we address in this book. Then we provide an overview of the types of mathematical model used to describe the behavior of the classes of systems of interest. We also describe the types of uncertainty model adopted and how they fit into the mathematical models that describe system behavior. In addition, we provide a preview of the applications discussed throughout the book, mostly centered around electric power systems. We conclude the chapter by providing a brief summary of the content of subsequent chapters.

1.1 Motivation

Loosely speaking, an engineered system is a collection of hardware and software components assembled and interacting in a particular way so that they collectively fulfill some function. The interaction between components can be of physical nature, i.e., components can be electrically, mechanically, or thermally coupled, and thus may involve some exchange of energy. Components can also be coupled in the sense that they exchange information with each other. Because of phenomena external to the system, there is some uncertainty as to how the system will perform. This phenomenon can materialize as an external (time-varying) input that drives the system response, or as a change in the system structure. In both cases, these external phenomena will alter the system nominal response and might cause the system to fail to perform its function. While most systems are typically designed to withstand some structural and operational uncertainty, it is important to verify that this is the case before the system is deployed.

To illustrate these ideas, consider the power supply system in Fig. 1.1, whose function is to reliably provide electric power to a mission-critical computer load, labeled as IT load, at a certain voltage level. To this end, there are three sources of power: two utility feeders, labeled as feeder 1 and feeder 2, and a backup generator, labeled as genset. Having such a redundant arrangement ensures delivery of power to the IT load with high assurance. While not depicted in the figure, there is a computer-based control system in charge of monitoring and controlling the power sources and switchgear, which plays an important role in the analysis of the system.

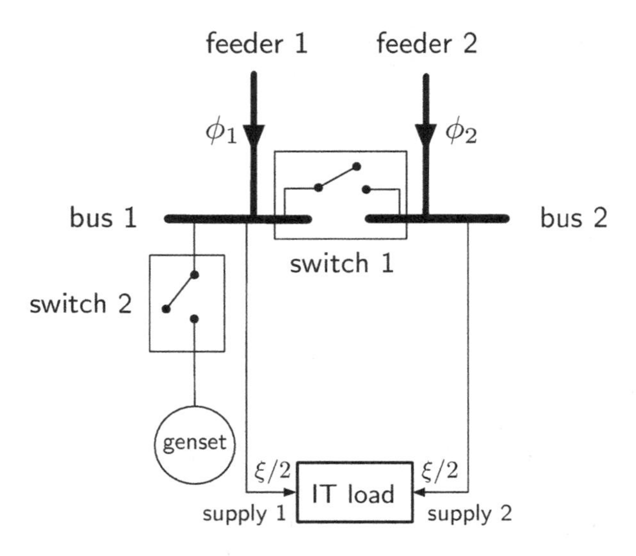

Figure 1.1 Schematic of power supply system for mission-critical load, where ϕ_1 and ϕ_2 denote the active power flowing through feeder 1 and feeder 2, respectively, and ξ denotes the total power delivered to the IT load.

Under normal operating conditions, both switch 1 and switch 2 are open (therefore genset is not initially used to supply power to the IT load). Then, feeder 1 supplies power to bus 1, which, in turn, serves half of the power demanded by the IT load via supply 1. Similarly, feeder 2 supplies bus 2, which in turn serves the other half of the power demanded by the IT load via supply 2. The total power demanded by the IT load is determined by the computer workload, which can vary according to external requests received by the computer and is a priori unknown. Thus, from the point of view of the power supply system, the computer workload is an external input that drives the power supply system response. Since the workload evolves over time and is unknown a priori, there is some uncertainty on how the power supply system will perform its function.

In the event that there is an outage in either feeder 1 or feeder 2, switch 1 will close and all power to the IT load will be supplied by the available feeder. In the event that there is an outage in both feeders (either sequential in time or simultaneous), switch 2 is closed and the genset will supply all the power to the IT load until either (i) one or both utility feeders are restored back to operation, at which point in time, switch 2 is open and the power to the IT load is once again supplied by one or both utility feeders, or (ii) there is an event that causes the genset to fail, at which point the IT load is shut down offline. Thus, the phenomena causing feeder outages and genset failure result in a change in the system structure in the sense of how power is routed from the available sources to the IT load.

In the context of the system above, one might be interested in quantifying the impact of workload variability on certain variables of interest, e.g., the flows of power through the wires connecting the buses and the IT load, and the magnitude of the voltage at bus 1 and bus 2 (when not connected together). This analysis is necessary to ensure that wires are sized correctly and protection equipment, e.g., under- and over-voltage protection relays, is calibrated appropriately. It is also necessary to ensure that after outages in one or both feeders occur, subsequent switching actions are correctly executed. In addition, one could be interested in quantifying the impact of equipment failure on the system ability to perform its function over some period of time.

The goal of this book is to develop analysis tools to perform the types of analysis described above. The applications and examples throughout the book draw heavily from electric power applications, including bulk power systems and microgrids, and linear and switched linear circuits encountered in power electronics applications. However, the modeling framework and techniques presented are general and can be applied to other engineering domains, including automotive and aerospace applications. For example, they can be used to assess the dynamic performance of an automotive steer-by-wire system and the lateral-directional control system of a fighter aircraft.

1.2 System Models

In a broad sense, one can think of an engineered system as an entity imposing constraints on certain *variables* associated with the system energy and information content. With this point of view, we can represent the behavior of the system by a set of mathematical relations between the aforementioned variables; this is what we refer to as the model of the system. These relations can be a result of physical laws, e.g., Kirchhoff's laws, energy conservation law, or moment conservation law. They can also arise from algorithms implemented in a digital computer, for modifying (controlling) the physical behavior of the system, e.g., the proportional-integral control scheme used in a bulk power system to automatically regulate frequency across the system. These mathematical relations will, in general, also include numerous *parameters*, i.e., quantities defining physical or information properties of the components comprising the system. Such system parameters can be constant or vary with time, and their value can be a priori unknown or uncertain. When modeling a system, the distinction between parameters and variables is typically clear because the values taken by the system parameters should not be affected by the values taken by the system variables or other parameters. Indeed, if the value of some parameter p is actually affected by the values taken by the system variables or other parameters, then the model should reflect this dependence and instead of being treated as a parameter in the model, p should be considered as an additional system variable and treated in the model as such. Next, we illustrate the ideas above via some examples.

Example 1.1 (Power supply system) Consider the system in Fig. 1.1 and assume there are no losses in any of its components. Let ϕ_1 and ϕ_2 denote the active power flowing through feeder 1 into bus 1 and the active power flowing through feeder 2 into bus 2, respectively. Let p_g denote the active power supplied by the genset. Let ξ denote the active power demanded by the IT load. Recall that under normal operating conditions, half of the power to the IT load is supplied by feeder 1, while the other half is supplied by feeder 2 (the genset does not supply any power). Then, since the active power flowing in and out of both bus 1 and bus 2 needs to be balanced, we have that

$$\phi_1 = \frac{\xi}{2}, \quad \phi_2 = \frac{\xi}{2}, \quad p_g = 0. \tag{1.1}$$

Now, recall that if there is an outage in feeder 1 (feeder 2), switch 1 will close and all the power to the IT load will be delivered by feeder 2 (feeder 1); thus,

$$\begin{aligned} &\phi_1 = 0, \quad \phi_2 = \xi, \quad p_g = 0, \quad \text{if outage in feeder 1 and feeder 2 in service,} \\ &\phi_1 = \xi, \quad \phi_2 = 0, \quad p_g = 0, \quad \text{if outage in feeder 2 and feeder 1 in service.} \end{aligned} \tag{1.2}$$

Also, recall that if there is an outage in both feeders, the genset will supply all the power to the load; thus,

$$\phi_1 = 0, \quad \phi_2 = 0, \quad p_g = \xi. \tag{1.3}$$

Finally, for the case when there is an outage in both feeders and a failure in the genset, we have that

$$\phi_1 = 0, \quad \phi_2 = 0, \quad p_g = 0, \tag{1.4}$$

and the IT load is shut down offline; thus, $\xi = 0$. Note that the relations in (1.1–1.4) only involve ξ, ϕ_1, ϕ_2, and p_g, which in this case are the system variables, i.e., the model here does not involve any parameters.

Example 1.2 (Linear circuit) Consider the circuit in Fig. 1.2 and assume that $c(t) = c$, $t \geq 0$, where c is a positive scalar. First, note that

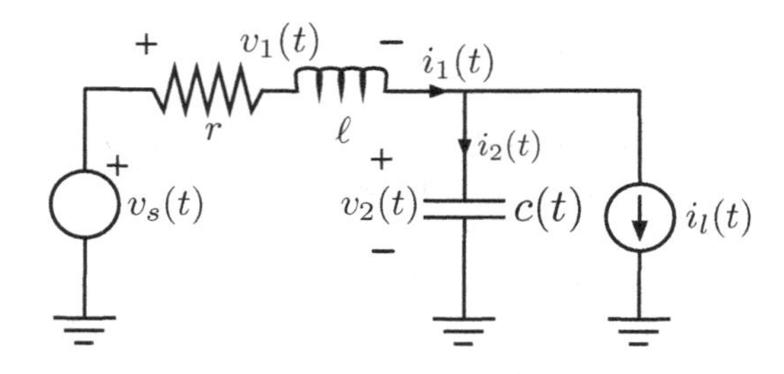

Figure 1.2 Linear circuit.

$$v_1(t) = r\, i_1(t) + \ell \frac{di_1(t)}{dt},$$

$$i_2(t) = c\frac{dv_2(t)}{dt}. \tag{1.5}$$

Then, by using Kirchhoff's laws, we can obtain the following model describing the relation between the currents $i_1(t)$ and $i_l(t)$, and the voltages $v_s(t)$ and $v_2(t)$:

$$0 = i_1(t) - c\frac{dv_2(t)}{dt} - i_l(t),$$

$$0 = v_s(t) - r\, i_1(t) - \ell\frac{di_1(t)}{dt} - v_2(t), \tag{1.6}$$

where r, ℓ, and c are positive constants. Here, $i_1(t)$, $i_2(t)$, $v_s(t)$, and $v_2(t)$ are variables, whereas r, ℓ, and c are parameters. Now, assume that $c(t)$ is known to evolve according to

$$\frac{dc(t)}{dt} = -c(t) + \alpha u(t),$$

where α is a positive scalar; thus,

$$i_2(t) = c(t)\frac{dv_2(t)}{dt} + v_2(t)\big(- c(t) + \alpha u(t)\big).$$

Then, the model describing the circuit behavior is as follows:

$$0 = i_1(t) - c(t)\frac{dv_2(t)}{dt} - v_2(t)\big(- c(t) + \alpha u(t)\big) - i_l(t),$$

$$0 = v_s(t) - r i_1(t) - \ell\frac{di_1(t)}{dt} - v_2(t),$$

$$0 = \frac{dc(t)}{dt} + c(t) - \alpha u(t); \tag{1.7}$$

thus, in this model, the capacitance, $c(t)$, is no longer a parameter but a variable.

There are some fundamental differences between the models in (1.1–1.4), (1.6), and (1.7). First, in the models in (1.1–1.4) and (1.6), the relation between the variables is linear, whereas in the model in (1.7), the relation between the variables is nonlinear. Second, in the model in (1.1–1.4), the constraints imposed on the system variables are in the form of a system of algebraic equations, whereas in the models in (1.6) and (1.7), the constraints imposed on the system variables are in the form of a set of ordinary differential equations (ODEs). In this book, we refer to systems whose behavior can be described by a set of algebraic equations as *static systems*, whereas systems whose behavior can be described by a set of ODEs are referred to as *continuous-time dynamical systems*. Furthermore, we refer to systems whose behavior can be described iteratively by a set of recurrent relations as *discrete-time dynamical systems*.

So far, we have not discussed the nature of the variables describing the energy and information state of a system; in general, we will categorize them as either inputs or states. By inputs, we refer to variables that are set and can be varied extraneously, whereas by states, we refer to variables that result from the constraints describing the system behavior and the values the inputs take. With this categorization, we can rewrite the static system in (1.1–1.4) as follows:

$$x = H_i \xi, \quad i = 1, 2, 3, \tag{1.8}$$

where $x = [\phi_1, \phi_2, p_g]^\top$, $\xi \geq 0$, and

$$H_1 = \begin{bmatrix} \frac{1}{2} \\ \frac{1}{2} \\ 0 \end{bmatrix}, \quad H_2 = \begin{bmatrix} 0 \\ 1 \\ 0 \end{bmatrix}, \quad H_3 = \begin{bmatrix} 1 \\ 0 \\ 0 \end{bmatrix}, \tag{1.9}$$

and

$$x = H_4 \xi, \tag{1.10}$$

where $\xi = 0$, and

$$H_4 = \begin{bmatrix} 0 \\ 0 \\ 0 \end{bmatrix}. \tag{1.11}$$

More generally, in this book we will consider static systems of the form

$$x = h_i(w), \quad i \in \mathcal{Q}, \tag{1.12}$$

where $x \in \mathbb{R}^n$ is referred to as the state vector, $w \in \mathbb{R}^m$ is referred to as the input vector, $\mathcal{Q}$ takes values in some finite set, and $h_i \colon \mathbb{R}^m \to \mathbb{R}^n$, which we refer to as the system input-to-state mapping, is defined by the relations between the state variables (i.e., the entries of the state vector), the inputs (i.e., the entries of the input vector), and the system parameters.

By using the same categorization of variables as inputs or states, we can rewrite the model in (1.6) in state-space form as follows:

$$\frac{d}{dt} x(t) = A x(t) + B w(t), \quad t \geq 0, \tag{1.13}$$

where $x(t) = \big[v_2(t), i_1(t)\big]^\top$ is referred to as the state vector, $w(t) = \big[v_s(t), i_l(t)\big]^\top$ is referred to as the input vector, and

$$A = \begin{bmatrix} 0 & \frac{1}{c} \\ -\frac{1}{\ell} & -\frac{r}{\ell} \end{bmatrix}, \quad B = \begin{bmatrix} 0 & -\frac{1}{c} \\ \frac{1}{\ell} & 0 \end{bmatrix}. \tag{1.14}$$

More generally, we will consider continuous-time dynamical systems of the form

$$\frac{d}{dt} x(t) = f\big(t, x(t), w(t)\big), \quad t \geq 0, \tag{1.15}$$

where $f \colon [0, \infty) \times \mathbb{R}^n \times \mathbb{R}^m \to \mathbb{R}^n$, is defined by the relations between the state variables (i.e., the entries of the state vector, $x \in \mathbb{R}^n$); and the inputs (i.e., the entries of the input vector, $w(t) \in \mathbb{R}^m$). For example, the system in (1.7) can be written as

$$\frac{d}{dt}x(t) = f\big(x(t), w(t)\big), \tag{1.16}$$

where $x(t) = \big[v_2(t), i_1(t), c(t)\big]^\top$, $w(t) = \big[v_s(t), i_l(t), u(t)\big]^\top$, and

$$f\big(x(t), w(t)\big) = \begin{bmatrix} \frac{1}{c(t)}\big(v_2(t)c(t) - \alpha v_2(t)u(t) + i_1(t) - i_l(t)\big) \\ \frac{1}{\ell}\big(- v_2(t) - r\, i_1(t) + v_s(t)\big) \\ -c(t) + \alpha u(t) \end{bmatrix}. \tag{1.17}$$

Finally, in this book we also consider discrete-time dynamical systems of the form

$$x_{k+1} = h_k(x_k, w_k), \quad k = 0, 1, 2, \dots, \tag{1.18}$$

where $x_k \in \mathbb{R}^n$, $w_k \in \mathbb{R}^m$, and $h_k \colon \mathbb{R}^n \times \mathbb{R}^m \to \mathbb{R}^n$, $k = 0, 1, 2, \dots,$ is defined by the relations between the state variables and inputs of the particular system under consideration.

1.3 Uncertainty Models

Here, we provide an overview of the different uncertainty models considered in subsequent chapters for both static and dynamical systems. In the process, we also state the analysis objectives for both classes of systems under each uncertainty model considered.

1.3.1 Static Systems

Consider a static system of the form:

$$x = h_i(w), \quad i \in \mathcal{Q}. \tag{1.19}$$

If the value that w takes is uncertain, we say the system is subject to input uncertainty, whereas if the value that i takes is uncertain, we say the system is subject to structural uncertainty. In terms of formally describing input uncertainty, we will consider both probabilistic models and set-theoretic models. In terms of formally describing structural uncertainty, we will only consider a probabilistic model. Each of these models is briefly described next.

Probabilistic Input Uncertainty Model: We assume the values that w can take are described by a random vector with known first and second moments or known probability density function (pdf). Then, for each $h_i(\cdot)$, $i \in \mathcal{Q}$, the values that $x = h_i(w)$ can take will be also random and described by some random vector, and the objective is to characterize the first and second moments (or the pdf) of this random vector.

Set-Theoretic Input Uncertainty Model: We assume the values that the input vector, w, can take belong to some convex set. While in general this set can have any shape, in this book we restrict our analysis to two particular classes of closed, convex sets, namely ellipsoids and zonotopes. Then, for each $h_i(\cdot)$, $i \in \mathcal{Q}$, the possible values that the state, $x = h_i(w)$, can take will belong to some set, and the objective is to characterize such a set.

Structural Uncertainty Model: We assume that the input-to-state mapping, $h_i(\cdot)$, evolves with time according to a Markov chain with known transition probabilities. Thus, for a given w, the value that $x = h_i(w)$ takes will evolve according to the aforementioned Markov process. Here, the value that w can take is either a known constant or can be described by a probabilistic model like the one above. Then, the objective is to characterize the state vector statistics.

1.3.2 Dynamical Systems

Consider continuous-time systems of the form

$$\frac{d}{dt}x(t) = f\big(t, x(t), w(t)\big), \tag{1.20}$$

and discrete-time systems of the form

$$x_{k+1} = h_k(x_k, w_k). \tag{1.21}$$

Here we will only analyze the system behavior under input uncertainty, i.e., the values that $w(t)$ and w_k can take are not a priori known, and will use both probabilistic and set-theoretic models to describe them.

Probabilistic Input Uncertainty Model: For continuous-time dynamical systems, we consider the case when the function $f(\cdot,\cdot,\cdot)$ is defined as follows:

$$f(t, x, w) = \alpha\big(t, x\big) + \beta(t, x)w, \tag{1.22}$$

whereas, for discrete-time dynamical systems, we do not impose any restrictions on $h_k(\cdot,\cdot)$. We assume that the values that $w(t)$ and w_k take are random and governed by a "white noise" process; thus, the values that $x(t)$ and x_k take are also random and governed by some stochastic process. We first consider the case when we only know the mean and covariance functions of the "white noise" input process and characterize the mean and covariance functions of the stochastic process describing the evolution of $x(t)$ and x_k. Then, we further assume that the "white noise" input process is Gaussian and provide the complete probabilistic characterization of the stochastic process describing the evolution of $x(t)$ and x_k.

Set-Theoretic Input Uncertainty Model: We consider general functions $f(\cdot,\cdot,\cdot)$ and $h_k(\cdot,\cdot)$, and assume the values that $w(t)$ and w_k can take are known to belong to a convex set, namely an ellipsoid. Then, the values that $x(t)$ and x_k can take also belong to some set (not necessarily an ellipsoid), and the objective is to characterize such a set. Providing an exact characterization of such a set is difficult in general (even if $f(\cdot,\cdot,\cdot)$ and $h_k(\cdot,\cdot)$ are affine functions); thus, we settle for obtaining ellipsoidal upper bounds.

1.4 Application Examples

Most of the techniques presented in the book are illustrated by using examples from circuit theory, electric power systems, and power electronics. For example, we utilize a simplified formulation of the power flow problem in AC power systems to illustrate the techniques developed for analyzing static systems subject to input uncertainty. Also, in order to illustrate the analysis techniques for dynamical systems subject to input uncertainty, we utilize a simplified model of the dynamics of an inertia-less AC microgrid, i.e., a small AC power system whose generators and loads are interfaced with the network via power electronics.

1.4.1 Power Flow Analysis under Active Power Injection Uncertainty

Consider a three-phase power system comprising n buses ($n > 1$) indexed by the elements in $\mathcal{V} = \{1, 2, \ldots, n\}$, and l transmission lines $\big(n - 1 \leq l \leq n(n-1)/2\big)$ indexed by the elements in the set $\mathcal{L} = \{1, 2, \ldots, l\}$, and assume the following hold:

A1. The system is balanced and operating in sinusoidal regime.
A2. There is at most one transmission line connecting each pair of buses.
A3. Each transmission line is short and lossless.
A4. The voltage magnitude at each bus is fixed by some control mechanism.

Let p_i denote the active power injected into the system network via bus i, and let ϕ_e denote the active power flowing on transmission line e, $e = 1, 2, \ldots, l$. Assume that

$$p_i = \xi_i, \quad i = 1, 2, \ldots, n-1,$$

where ξ_i is extraneously set and a priori unknown. Then, since the system is lossless, the injection into bus n must be such that $\sum_{j=1}^{n} p_j = 0$; thus,

$$p_n = -\sum_{j=1}^{n-1} \xi_j.$$

Define $\xi = [\xi_1, \xi_2, \ldots, \xi_{n-1}]^\top$ and $\phi = [\phi_1, \phi_2, \ldots, \phi_l]^\top$. Then, by imposing some conditions on the values that ξ can take, there exists a function $f \colon \mathbb{R}^l \to \mathbb{R}^l$ encapsulating the system network topology and transmission line parameters such that

$$w = f(\phi), \tag{1.23}$$

where $w = \left[\xi^\top, \mathbf{0}_{l-n+1}^\top\right]^\top$. The formulation above can be generalized to the case when there are m, $1 \leq m \leq n-1$, power injections being extraneously set with the remaining power injections being adjusted so that $\sum_{j=1}^{n} p_j = 0$, which is a necessary condition that the power injections need to satisfy because of the assumption on the transmission lines in the system being lossless.

If the network is a tree, then we have that $l = n - 1$ and

$$\xi = \widetilde{M}\phi,$$

where $\widetilde{M} \in \mathbb{R}^{(n-1)\times(n-1)}$ is invertible; thus, we can write

$$\phi = \widetilde{M}^{-1}\xi. \tag{1.24}$$

For the case when the network is not a tree, because of the assumptions made on the values that ξ can take, we can ensure that there exists $f^{-1}\colon \mathbb{R}^l \to \mathbb{R}^l$ such that

$$\phi = f^{-1}(w). \tag{1.25}$$

Then, given either a probabilistic model or a set-theoretic model describing the values that the vector of extraneous power injections, ξ, can take, the problem is to characterize the values that the vector of line flows, ϕ, can take. We explore such settings in detail in Chapter 3 and Chapter 7.

1.4.2 Analysis of Inertia-less AC Microgrids under Power Injection Uncertainty

Consider a three-phase microgrid comprising n buses ($n > 1$) indexed by the elements in $\mathcal{V} = \{1, 2, \ldots, n\}$, and l transmission lines $\big(n - 1 \leq l \leq n(n-1)/2\big)$ and assume the following hold:

B1. The microgrid is balanced and operating in sinusoidal regime.

B2. There is at most one transmission line connecting each pair of buses.

B3. Each transmission line is short and lossless.

B4. Connected to each bus there is either a generating- or a load-type resource interfaced via a three-phase voltage source inverter.

B5. The reactance of each voltage source inverter output filter is small when compared to the reactance values of the network transmission lines.

B6. The phase angle of the inverter connected to each bus is regulated via a droop control scheme updated at discrete time instants indexed by $k = 0, 1, 2, \ldots$.

B7. The inverter outer voltage and inner current control loops hold the inverter output voltage magnitude constant throughout time.

Consider the case when the frequency-droop control setpoints of the inverters at buses $1, 2, \ldots, m$ change randomly (this is typically the case in photovoltaic installations endowed with maximum power point tracking). Let ξ_k denote an m-dimensional vector whose entries correspond to the values taken at instant k of said setpoints. Assume that in order to mitigate the effect of these random fluctuations, the frequency-droop control setpoints of the inverters at buses $m + 1, m + 2, \ldots, n$ are regulated via a closed-loop integral control scheme so that the frequency across the microgrid network remains close to some nominal value. Let

z_k denote an $(n-m)$-dimensional vector whose entries correspond to the internal states at instant k of said closed-loop integral control scheme. Let θ_k denote an n-dimensional vector whose entries correspond to the bus voltage phase angles at time instant k. Then, the microgrid closed-loop dynamics can be written as follows:

$$\begin{aligned} \theta_{k+1} &= h_1(\theta_k, z_k, \xi_k), \\ z_{k+1} &= h_2(z_k, \xi_k), \end{aligned} \tag{1.26}$$

where the functions $h_1 \colon \mathbb{R}^n \times \mathbb{R}^{n-m} \times \mathbb{R}^m \to \mathbb{R}^n$ and $h_2 \colon \mathbb{R}^{n-m} \times \mathbb{R}^m \to \mathbb{R}^{n-m}$ encapsulate the microgrid network topology and transmission line parameters, and the control gains of the frequency regulation scheme. Then, given either a probabilistic model or a set-theoretic model describing the values that ξ_k, $k \geq 0$, can take, the problem is to characterize the values that θ_k, $k \geq 0$, and z_k, $k \geq 0$, can take. This information can then be used to characterize the system frequency. We explore such settings in detail in Chapters 5 and 8. In addition, in Chapter 6, we formulate the continuous-time counterpart of the model in (1.26) and study the system performance when the measurements utilized by the frequency regulation scheme are corrupted.

1.4.3 Reliability Analysis of Static Systems

Consider a static system comprising r components indexed by the elements in $\mathcal{C} = \{c_1, c_2, \dots, c_r\}$. Assume that each component c_i can operate in one of two modes: nominal and off-nominal (failed). Further assume that initially all the components are operating in their nominal mode, and as time evolves, transitions from the nominal to the failed mode may occur, and once a transition occurs, the component remains in its failed mode. Such static systems are called non-repairable and their input-to-state behavior can be described by

$$x = h_i(w), \tag{1.27}$$

where $x \in \mathbb{R}^n$, $w \in \mathbb{R}^m$ is constant, and $h_i \colon \mathbb{R}^m \to \mathbb{R}^n$, $i \in \mathcal{Q} = \{1, 2, \dots, N\}$, is determined by (i) the components that are operating in their nominal mode, and (ii) the components that have transitioned from operating in their nominal to their failed mode, and (iii) the chronological order in which these transitions occurred.

Assume that the time it takes for each component c_i to transition from its nominal mode to its failed mode is random and can be described by a continuous random variable, T_{c_i}, with continuously differentiable cumulative distribution function, $F_{T_{c_i}}(\cdot)$. Further, assume that these random variables are pairwise independent. Then, transitions among the elements in $\mathcal{Q}$ are governed by a continuous-time Markov chain, $Q = \{Q(t) \colon t \geq 0\}$, $Q(t) \in \mathcal{Q}$, whose transition rates can be obtained from the individual component failure rates, denoted by $\lambda_{c_i}(t),\ t \geq 0$, and defined as follows:

$$\lambda_{c_i}(t) = \frac{f_{T_{c_i}}(t)}{1 - F_{T_{c_i}}(t)}, \tag{1.28}$$

where $f_{T_{c_i}}(t) = \frac{dF_{T_{c_i}}(t)}{dt}$.

Let $\mathcal{R} = \mathcal{R}_1 \times \mathcal{R}_2 \times \cdots \times \mathcal{R}_n$, with $\mathcal{R}_i = (x_i^m, x_i^M)$, where $x_i^m, x_i^M \in \mathbb{R}$ respectively denote the minimum and maximum values that the ith entry of x can take so as to guarantee that the system is performing its intended function. Now, because w is assumed to be constant, the value that x takes for each i is also constant. Then, given $i \in \mathcal{Q}$, the system is said to be operational if $x = h_i(w) \in \mathcal{R}$, and nonoperational otherwise. Let $\pi_i(t)$ denote the probability that the aforementioned continuous-time Markov chain is in state $i \in \mathcal{Q}$. Then, the reliability of the system at time t, which we denote by $R(t)$, can be defined as the probability that the non-repairable system is operating at time t, and can be computed as follows:

$$R(t) = \sum_{i \in \mathcal{Q}_0} \pi_i(t), \tag{1.29}$$

where $\mathcal{Q}_0 = \{i \in \mathcal{Q} \colon x = h_i(w) \in \mathcal{R}\}$.

The ideas above can be generalized in several directions. First, it is possible to take into account input uncertainty by modeling w as a random vector with known first and second moments (or known pdf). In addition, it is possible to take into account events that may cause several components to simultaneously transition from operating in their nominal mode to their failed mode; such events are referred to as common cause failures. Finally, one can extend the framework to analyze static systems comprising components that can be repaired. All these ideas are explored in detail in Chapter 4.

1.5 Book Road Map

Chapter 2 starts by reviewing important concepts from probability theory and stochastic processes. Subsequent chapters on probabilistic input and structural uncertainty make heavy use of random vectors and vector-valued stochastic process, so the reader should be familiar with the material included on these concepts. Next, the chapter provides a review of set-theoretic notions. The material on sets in Euclidean space included in this part is key to understanding the set-theoretic approach to input uncertainty modeling. The chapter concludes with a review of several fundamental concepts from the theory of discrete- and continuous-time linear dynamical systems.

Chapter 3 covers the analysis of static systems under probabilistic input uncertainty. The first part of the chapter is devoted to analyzing both linear and nonlinear static systems when the first and second moments of the input vector are known, and provides techniques for characterizing the first and second moments of the state vector. For the linear case, the techniques provide the exact moment characterization, whereas for the nonlinear case, the characterization, which is based on a linearization of the system model, is approximate. The second part of the chapter provides techniques for the analysis of both linear and nonlinear static systems when the pdf of the input vector is known. The techniques included provide exact characterizations of the state pdf for both linear and nonlinear systems. In both cases, the inversion of the input-to-state mapping is required, which in the linear case involves the computation of the inverse of a matrix; however, for the nonlinear, it involves obtaining an analytical expression for the input-to-state mapping, which might be difficult in general. Thus, for the nonlinear case, we also provide a technique based on linearization that yields an approximation of the state pdf. The chapter concludes by utilizing the techniques developed to study the power flow problem under active power injection uncertainty.

Chapter 4 studies static systems under structural uncertainty. The first part of the chapter is devoted to the development of a model describing the system stochastic behavior. To this end, we assume that the system can only adopt a finite number of input-to-state mappings, and that transitions among these different mappings are random and governed by a Markov chain. We consider both discrete- and continuous-time settings and provide expressions governing the evolution of the probability distribution associated with the resulting Markov chains. The second part of the chapter tailors the techniques developed earlier to analyze multi-component systems subject to component failures and repairs. Techniques for constructing the system input-to-state model are extensively covered, as this is in general the most difficult part of the analysis when analyzing systems with a large number of components.

Chapter 5 provides techniques for analyzing discrete-time dynamical systems under probabilistic input uncertainty. Here, the relation between the input and the state is described by a discrete-time state-space model. The input vector is modeled as a vector-valued stochastic process with known first and second moments (or known pdf). The first part of the chapter is devoted to the analysis of linear systems and provides techniques for characterizing the first and second moments and the pdf of the state vector. The second part of the chapter is devoted to the analysis of nonlinear systems, where we use the techniques developed in Chapter 4 to exactly characterize the distribution of the state vector when the pdf of the input vector is given. In addition, we rely on linearization techniques to obtain expressions that approximately characterize the first and second moments and the pdf of the state vector. The third part of the chapter illustrates the application of the techniques developed to the analysis of inertia-less AC microgrids under random active power injections.

Chapter 6 is the continuous-time counterpart of Chapter 5, i.e., it studies continuous-time dynamical systems described by a continuous-time state-space model whose input is subject to probabilistic uncertainty. The first part of the chapter is devoted to the analysis of linear systems and provides techniques for computing the first and second moments of the state vector when the evolution of the input vector is governed by a "white noise" process with known mean and covariance functions. Then, by additionally imposing this "white noise" process to be Gaussian, we provide a partial differential equation whose solution yields the pdf of the state vector. The second part of the chapter extends these techniques to the analysis of nonlinear systems, with a special focus on the case when the "white noise" governing the evolution of the input vector is Gaussian. The third part of the chapter illustrates the application of the techniques developed to the analysis of inertia-less AC microgrids when the measurements utilized by the frequency control system are corrupted by additive disturbances.

Chapter 7 is the set-theoretic counterpart of Chapter 3, i.e., it covers the analysis of static systems under set-theoretic input uncertainty. In the first part of the chapter, we assume that the input belongs to an ellipsoid and analyze both linear and nonlinear systems. For the linear case, we provide techniques to exactly characterize the set containing all possible values that the state can take. For the nonlinear case, we again resort to linearization to approximately characterize the set containing all possible values that the state can take. The second part of the chapter considers linear and nonlinear systems when the input is known to belong to a zonotope. For the linear case, we are able to compute the exact set containing all possible values the state can take, whereas for the nonlinear case, we settle for an approximation thereof obtained via linearization. The techniques developed are utilized to analyze the power flow problem under uncertain active power injections.

Chapter 8 is the set-theoretic counterpart of Chapter 5, i.e., it analyzes linear and nonlinear discrete-time systems described by a discrete-time state-space model whose inputs are uncertain but known to belong to an ellipsoid. For the linear case, even if the input set is an ellipsoid, the set containing all possible values that the state can take is not an ellipsoid in general; however, it can be outer bounded by an ellipsoid. We develop techniques for recursively computing a family of such outer-bounding ellipsoids. Within this family, we then show how to choose ellipsoids that are optimal in some sense, e.g., they have minimum volume. For the nonlinear case, we will again resort to linearization techniques to approximately characterize the set containing all possible values that the state can take. Application of the techniques is illustrated using the same inertia-less AC microgrid model used in Chapter 5.

Chapter 9 is the continuous-time counterpart of Chapter 8, i.e., it covers the analysis of linear and nonlinear continuous-time dynamical systems described by a continuous-time state-space model whose input belongs to an ellipsoid. Similar to the linear discrete-time case, the set containing all possible values that the state can take is not an ellipsoid in general; however, it can be outer bounded by a family of ellipsoids whose evolution is governed by a differential equation that can be derived from the system state-space model. As in the discrete-time case, it is possible to choose ellipsoids within this family that are optimal in some sense. The nonlinear case is again handled using linearization. The techniques developed in the chapter are used to analyze the performance of a buck DC-DC power converter. In addition, we show how the techniques can be used to assess the effect of variability associated with renewable-based electricity generation on bulk power system dynamics, with a focus on time-scales involving electromechanical phenomena.

There are several ways in which the material in subsequent chapters can be studied depending on the end goal of the reader. For example, a reader familiar with the foundations on which the book builds – probability and stochastic processes, set theory, and linear dynamical system theory – can skip Chapter 2, although its reading is recommended so one becomes familiar with the notation adopted. A reader interested exclusively in the theory can skip the final section in each subsequent chapter as these provide detailed case studies drawn from electric power applications that showcase the potential of the theoretical tools for tackling real-world problems. A reader interested exclusively in probabilistic uncertainty analysis tools can focus on Chapters 3 to 6, whereas a reader interested exclusively in set-theoretic uncertainty analysis tools can focus on Chapters 7 to 9. A reader interested in the analysis of static systems can focus on Chapters 3, 4, and 7. A reader interested in the analysis of discrete-time dynamical systems can focus on Chapters 5 and 8, whereas a reader interested in continuous-time dynamical systems can focus on Chapters 6 and 9.

1.6 Notes and References

Formal definitions of a dynamical system, including the notions of input and state and the state-space formalism, can be found in [1, 2]. Early references on the use of state-space models to describe the behavior of dynamical systems include [3, 4]. The topic is now standard and covered in detail in many modern control theory textbooks (e.g., [5, 6, 7, 8, 9]). Several examples on how to describe particular dynamical systems, e.g., an inverted pendulum and a nonlinear circuit, are included in [7]. The use of probabilistic and set-theoretic uncertainty models in the context of static and dynamical systems is extensively covered in [10] and many of the results discussed in later chapters can be traced back to this seminal book.

2 Preliminaries

This chapter provides a succinct review of key concepts from probability and stochastic processes, set theory (with a focus on sets in Euclidean space), and linear dynamical systems that appear throughout the book. The treatment of each subject is far from being complete and is mostly included as a reference for the developments in later chapters. The reader is expected to have certain mathematical maturity and already be familiar with the material as the majority of the results are presented without deriving them. The material presented builds on several concepts from matrix and vector analysis, all of which are briefly reviewed in Appendix A.

2.1 Probability and Stochastic Processes

2.1.1 Probability Spaces

An experiment whose outcome is uncertain can be mathematically described by a so-called probability space, whose components are:

P1. the sample space, denoted by Ω, which is the set of all possible outcomes of the experiment;

P2. an event algebra, denoted by $\mathcal{F}$, which is a set whose elements, referred to as events, are subsets of the sample space chosen so that (i) $\Omega \in \mathcal{F}$, (ii) if $A \in \mathcal{F}$, then $A^c \in \mathcal{F}$ (clearly $\emptyset \in \mathcal{F}$ because $\Omega \in \mathcal{F}$), and (iii) if $A_1, A_2, \ldots \in \mathcal{F}$, then $\bigcup_i A_i \in \mathcal{F}$; and

P3. a probability measure, which assigns each event $A \in \mathcal{F}$ a number $\Pr(A)$, referred to as the probability of event A, so that: (i) $\Pr(A) \geq 0$ for any $A \in \mathcal{F}$, (ii) $\Pr\left(\bigcup_i A_i\right) = \sum_i \Pr(A_i)$, for any events $A_1, A_2, \ldots,$ such that $A_i \cap A_j = \emptyset,\ i \neq j$, and (iii) $\Pr(\Omega) = 1$.

Conditional Probability and Independence

We say an event A has occurred if the outcome of the experiment is within the set A. Given events A, B, the conditional probability of event A given that event B has occurred is defined as follows:

$$\Pr(A \mid B) = \begin{cases} \frac{\Pr(A \cap B)}{\Pr(B)}, & \text{if } \Pr(B) > 0, \\ \text{undefined}, & \text{if } \Pr(B) = 0. \end{cases} \tag{2.1}$$

A similar expression can be obtained for $\Pr(B \mid A)$, i.e., the conditional probability of event B given event A has occurred.

Events A, B are said to be independent if $\Pr(A \cap B) = \Pr(A)\Pr(B)$; thus, if $\Pr(B) > 0$, we have that $\Pr(A \mid B) = \Pr(A)$, i.e., knowing event B has occurred does not yield additional information about event A. Similarly, if events A and B are independent and $\Pr(A) > 0$, we have that $\Pr(B \mid A) = \Pr(B)$, which is to say that knowing event A has occurred does not give us any additional information about event B. Events A, B, C are said to be pairwise independent if $\Pr(A \cap B) = \Pr(A)\Pr(B)$, $\Pr(A \cap C) = \Pr(A)\Pr(C)$, and $\Pr(B \cap C) = \Pr(B)\Pr(C)$. Events A, B, C are said to be independent if they are pairwise independent and

$$\Pr(A \cap B \cap C) = \Pr(A)\Pr(B)\Pr(C).$$

Bayes' Formula and the Law of Total Probability

Given events A, B, if both $\Pr(A) > 0$ and $\Pr(B) > 0$, then $\Pr(A \cap B) = \Pr(A \mid B)\ \Pr(B) = \Pr(B \mid A)\Pr(A)$; this leads to the following expression

$$\Pr(B \mid A) = \frac{\Pr(A \mid B)\Pr(B)}{\Pr(A)},$$

which is known as Bayes' formula.

Events $E_1, E_2, \dots, E_n$ are said to be mutually exclusive if $E_i \cap E_j = \emptyset$, $i \neq j$. Events $E_1, E_2, \dots, E_n$ are said to form a partition of Ω if they are mutually exclusive and $\bigcup_{i=1}^{n} E_i = \Omega$. For any event A and any partition of the sample space, $E_1, E_2, \dots E_n$, satisfying $\Pr(E_i) > 0$, $i = 1, 2, \dots, n$, we have that

$$\Pr(A) = \Pr(A \mid E_1)\Pr(E_1) + \Pr(A \mid E_2)\Pr(E_2) + \cdots + \Pr(A \mid E_n)\Pr(E_n);$$

this is known as the law of total probability.

By using Bayes' formula and the law of total probability one can check that, for any event A and any partition of the sample space, $E_1, E_2, \dots, E_n$, satisfying $\Pr(E_i) > 0, i = 1, 2, \dots, n$, the following holds true:

$$\Pr(E_i \mid A) = \frac{\Pr(A \mid E_i)\Pr(E_i)}{\Pr(A \mid E_1)\Pr(E_1) + \Pr(A \mid E_2)\Pr(E_2) + \cdots + \Pr(A \mid E_n)\Pr(E_n)},$$

$i = 1, 2, \dots, n$.

2.1.2 Random Variables

Given a probability space $(\Omega, \mathcal{F}, \Pr)$, a random variable is a function $X(\cdot)$ that maps each outcome $\omega \in \Omega$ to a real number, while satisfying that

$$\{\omega \in \Omega\colon X(\omega) \leq x\} \in \mathcal{F}$$

for every $x \in \mathbb{R}$. The values that $X(\cdot)$ takes are referred to as the realizations of the random variable. In order to simplify notation, we write $\{X \leq x\}$ as a shorthand for the event $\{\omega \in \Omega\colon X(\omega) \leq x\}$, and $\Pr(X \leq x)$ as a shorthand for its probability, $\Pr\left(\{\omega \in \Omega\colon X(\omega) \leq x\}\right)$.

Since $\{\omega \in \Omega : X(\omega) \leq x\}$ is an event for every $x \in \mathbb{R}$, we can define the following function:

$$F_X(x) = \Pr(X \leq x), \quad x \in \mathbb{R},$$

which is referred to as the cumulative distribution function (cdf) of X. The cdf of a random variable X, $F_X(\cdot)$, always satisfies the following properties:

C1. It is always nondecreasing.
C2. $\lim_{x \to -\infty} F_X(x) = 0$, and $\lim_{x \to \infty} F_X(x) = 1$.
C3. It is right continuous, i.e., $\lim\limits_{\substack{\epsilon \to 0 \\ \epsilon > 0}} F_X(x + \epsilon) = F_X(x)$.

Given events $\{X \leq a\}$, $\{X \leq b\}$, $a < b$, we have that

$$\{X \leq b\} = \{X \leq a\} \cup \{a < X \leq b\},$$

with events $\{X \leq a\}$ and $\{a < X \leq b\}$ being mutually exclusive; thus, we can write $\Pr(X \leq b) = \Pr(X \leq a) + \Pr(a < X \leq b)$, and by using the definition of cdf, we obtain that

$$\Pr(a < X \leq b) = F_X(b) - F_X(a).$$

Discrete Random Variables

A random variable X is said to be discrete if it takes values in a finite or countable set $\mathcal{X} = \{x_1, x_2, \ldots\}$, $x_i \in \mathbb{R}$, (i.e., there is one-to-one correspondence between each element in $\mathcal{X}$ and a natural number) such that

$$\sum_{x_i \in \mathcal{X}} \Pr(X = x_i) = 1. \tag{2.2}$$

For each $x_i \in \mathcal{X}$, let $p_X(x_i)$ denote the probability that X takes value x_i; $p_X(\cdot)$ is referred to as the probability mass function (pmf) of X and in light of (2.2), it must satisfy

$$\sum_{x_i \in \mathcal{X}} p_X(x_i) = 1.$$

The expectation, or first moment, of a discrete random variable X with pmf $p_X(x_i)$, $x_i \in \mathcal{X}$, which we denote by $\mathrm{E}[X]$ and sometimes by μ_X, is defined as

$$\mathrm{E}[X] = \sum_{x_i \in \mathcal{X}} x_i p_X(x_i). \tag{2.3}$$

Given a discrete random variable X with its pmf taking values $p_X(x_i)$, $x_i \in \mathcal{X}$, and a function $g \colon \mathbb{R} \to \mathbb{R}$, we have that

$$\mathrm{E}\big[g(X)\big] = \sum_{x_i \in \mathcal{X}} g(x_i) p_X(x_i). \tag{2.4}$$

The variance of a discrete random variable X with its pmf taking values $p_X(x_i), x_i \in \mathcal{X}$, which we denote by σ_X^2, is defined as

$$\sigma_X^2 = \mathrm{E}\big[(X - \mu_X)^2\big]$$
$$= \sum_{x_i \in \mathcal{X}} (x_i - \mu_X)^2 p_X(x_i). \tag{2.5}$$

By using simple manipulations, one can check that

$$\sigma_X^2 = \mathrm{E}\big[(X - \mu_X)^2\big] = \mathrm{E}\big[X^2\big] - \mu_X^2.$$

The term $\mathrm{E}\big[X^2\big]$ is referred to as the second moment of X. More generally, $\mathrm{E}[X^n], n = 1, 2, \ldots,$ is referred to as the nth moment of X. The square root of the variance, σ_X, is referred to as the standard deviation of X.

Example 2.1 (Bernoulli distribution) A discrete random variable X with pmf $p_X(\cdot)$ defined as follows

$$p_X(x) = \begin{cases} p, & \text{if } x = 1, \\ 1 - p, & \text{if } x = 0, \end{cases}$$

where $0 \le p \le 1$, is said to have a Bernoulli distribution with parameter p. By using (2.3) and (2.5), one can check that $\mu_X = p$ and $\sigma_X^2 = p(1 - p)$.

Example 2.2 (Binomial distribution) A discrete random variable X with pmf $p_X(\cdot)$ defined as follows

$$p_X(x) = \binom{n}{x} p^x (1 - p)^{n-x}, \quad x = 0, 1, 2, \ldots, n,$$

where $0 \le p \le 1$, is said to have a Binomial distribution with parameters n and p. By using (2.3) and (2.5), one can check that $\mu_X = np$ and $\sigma_X^2 = np(1 - p)$.

Example 2.3 (Geometric distribution) A discrete random variable X with pmf $p_X(\cdot)$ defined as follows

$$p_X(x) = (1 - p)^{x-1} p, \quad x = 1, 2, \ldots,$$

where $0 \le p \le 1$, is said to have a Geometric distribution with parameter p. By using (2.3) and (2.5), one can check that $\mu_X = \frac{1}{p}$ and $\sigma_X^2 = \frac{1-p}{p^2}$.

Example 2.4 (Poisson distribution) A discrete random variable X with pmf $p_X(\cdot)$ defined as follows

$$p_X(x) = \frac{\lambda^x e^{-\lambda}}{x!}, \quad x = 0, 1, \ldots,$$

where $\lambda \ge 0$, is said to have a Poisson distribution with parameter λ. By using (2.3) and (2.5), one can check that $\mu_X = \lambda$ and $\sigma_X^2 = \lambda$.

Continuous Random Variables

A random variable X is said to be continuous if there exists some function $f_X\colon \mathbb{R} \to [0, \infty)$, referred to as the probability density function (pdf) of X, such that

$$F_X(x) = \int_{-\infty}^{x} f_X(y)dy;$$

thus, because of Property C2 above, it follows that

$$\int_{-\infty}^{\infty} f_X(x)dx = 1.$$

If the pdf of a continuous random variable X, $f_X(\cdot)$, is continuous, then its cdf, $F_X(\cdot)$, is continuously differentiable and

$$f_X(x) = \frac{dF_X(x)}{dx}. \tag{2.6}$$

In this remainder, we only consider continuous random variables whose pdf is piecewise continuous with a finite or countable number of discontinuity points; thus, except for those points at which the pdf is discontinuous, (2.6) holds.

The expectation, or first moment, of a continuous random variable with pdf $f_X(\cdot)$ is defined as

$$\mu_X = \mathrm{E}[X] = \int_{-\infty}^{\infty} x f_X(x)dx. \tag{2.7}$$

Given a continuous random variable X with pdf $f_X(x)$, we have that

$$\mathrm{E}\big[g(X)\big] = \int_{-\infty}^{\infty} g(x)f_X(x)dx; \tag{2.8}$$

this result is analogous to that in (2.4).

The variance of a continuous random variable with pdf $f_X(\cdot)$ is defined as

$$\begin{aligned}\sigma_X^2 &= \mathrm{E}[(X - \mu_X)^2] \\ &= \int_{-\infty}^{\infty} (x - \mu_X)^2 f_X(x)dx. \end{aligned} \tag{2.9}$$

As in the discrete case, we have that

$$\sigma_X^2 = \mathrm{E}[(X - \mu_X)^2] = \mathrm{E}[X^2] - \mu_X^2.$$

Example 2.5 (Uniform distribution) A continuous random variable X with pdf

$$f_X(x) = \begin{cases} \frac{1}{b-a}, & \text{if } a < x \leq b, \\ 0, & \text{otherwise,} \end{cases}$$

$a, b \in \mathbb{R}$, $a < b$, is said to have a Uniform distribution on the interval $[a, b]$. One can check by using (2.7) and (2.9) that $\mu_X = \frac{a+b}{2}$ and $\sigma_X^2 = \frac{(b-a)^2}{12}$.

Example 2.6 (Exponential distribution) A continuous random variable X with pdf

$$f_X(x) = \begin{cases} \lambda e^{-\lambda x}, & \text{if } x \geq 0, \\ 0, & \text{otherwise,} \end{cases}$$

$\lambda > 0$, is said to have an Exponential distribution with parameter λ. One can check by using (2.7) and (2.9) that $\mu_X = \frac{1}{\lambda}$ and $\sigma_X^2 = \lambda^2$.

Example 2.7 (Gaussian distribution) A continuous random variable X with pdf

$$f_X(x) = \frac{1}{\sqrt{2\pi\sigma_X^2}} e^{-\frac{(x-\mu_X)^2}{2\sigma_X^2}}, \quad -\infty \leq x \leq \infty,$$

$\mu \in \mathbb{R}$, $\sigma > 0$, is said to have a Gaussian (or Normal) distribution with parameters μ_X and σ_X^2. One can check by using (2.7) and (2.9) that indeed μ_X and σ_X^2 are respectively the mean and variance of X.

2.1.3 Jointly Distributed Random Variables

Given two random variables X and Y defined on the same probability space $(\Omega, \mathcal{F}, \Pr)$ and any $x, y \in \mathbb{R}$, we can define an event of the form

$$\{\omega \in \Omega \colon X(\omega) \leq x, \ Y(\omega) \leq y\},$$

and as before, we will write $\Pr(X \leq x, Y \leq y)$ as a shorthand for its probability. Then, we can define the function

$$F_{X,Y}(x, y) = \Pr(X \leq x, Y \leq y),$$

$x, y \in \mathbb{R}$, which we refer to as the joint cdf of X and Y. In the remainder of this section, we mostly focus on continuous random variables; however, there are counterpart notions for discrete random variables to the majority of concepts introduced.

Joint and Marginal pdfs

Given two continuous random variables X, Y, there exists a function $f_{X,Y}(\cdot, \cdot)$, referred to as the joint probability density function of X and Y, such that

$$F_{X,Y}(x, y) = \int_{-\infty}^{x} \int_{-\infty}^{y} f_{X,Y}(u, v) du dv,$$

and, similar to the single continuous random variable case, we have that

$$\int_{-\infty}^{\infty} \int_{-\infty}^{\infty} f_{X,Y}(x, y) dx dy = 1.$$

Given two continuous random variables with joint pdf $f_{X,Y}(\cdot,\cdot)$, the marginal pdf $f_X(\cdot)$ is the pdf of X when considered by itself, and can be obtained as follows:

$$f_X(x) = \int_{-\infty}^{\infty} f_{X,Y}(x,y)dy.$$

One can similarly obtain an expression for the marginal pdf $f_Y(\cdot)$.

Example 2.8 Consider two continuous random variables X, Y with joint pdf as follows:

$$f_{X,Y}(x,y) = \begin{cases} \frac{1}{2}, & \text{if } |x| + |y| \leq 1, \\ 0, & \text{otherwise;} \end{cases} \tag{2.10}$$

see Fig. 2.1 (left) for a graphical representation of the support of $f_{X,Y}(\cdot,\cdot)$. The marginal pdf $f_X(x)$ can be computed as follows. For $0 \leq x \leq 1$, we have that $f_{X,Y}(x,y) = \frac{1}{2}$ if $-1+x \leq y \leq 1-x$ and $f_{X,Y}(x,y) = 0$ otherwise; thus

$$\begin{aligned} f_X(x) &= \int_{-1+x}^{1-x} \frac{1}{2} dy \\ &= 1 - x, \quad 0 \leq x \leq 1. \end{aligned}$$

A similar argument yields

$$f_X(x) = 1 + x, \qquad -1 \leq x < 0.$$

Finally, since $f_{X,Y}(x,y) = 0$ when $x > 1$ or $x < -1$, we have that $f_X(x) = 0$ for $x > 1$ or $x < -1$. Putting together the results above yields

$$f_X(x) = \begin{cases} 1 - |x|, & \text{if } -1 \leq x \leq 1, \\ 0, & \text{otherwise.} \end{cases} \tag{2.11}$$

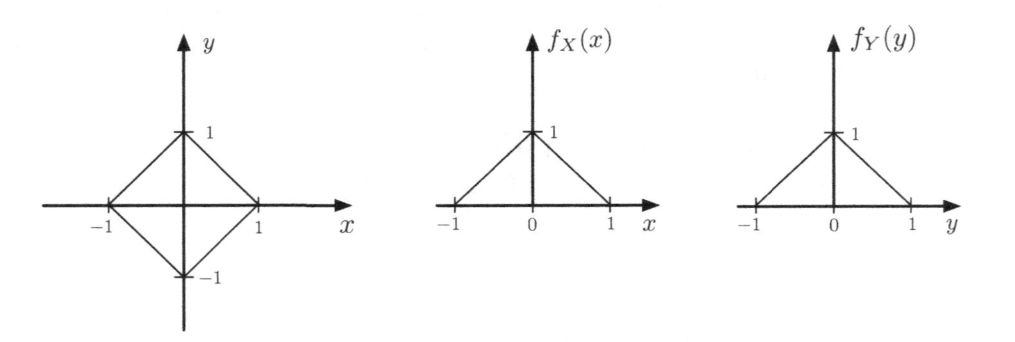

Figure 2.1 Example 2.8: support of joint pdf $f_{X,Y}(\cdot,\cdot)$ (left), marginal pdf $f_X(\cdot)$ (center), and marginal pdf $f_Y(\cdot)$ (right).

Because of symmetry, it is easy to see that

$$f_Y(y) = \begin{cases} 1 - |y|, & \text{if } -1 \le y \le 1, \\ 0, & \text{otherwise.} \end{cases} \tag{2.12}$$

The graphs of $f_X(\cdot)$ and $f_Y(\cdot)$ are displayed in Fig. 2.1 (center) and Fig. 2.1 (right), respectively.

Conditional pdfs and Independence

As we did with events, if X and Y are continuous random variables, we can define the conditional pdf of X given $Y = y$ such that $f_Y(y) > 0$, which we denote by $f_{X\,|\,Y}(\cdot\,|\,y)$, as follows:

$$f_{X\,|\,Y}(x\,|\,y) = \frac{f_{X,Y}(x,y)}{f_Y(y)}. \tag{2.13}$$

One can also write a similar expression for the conditional pdf of Y given $X = x$.

Two jointly distributed continuous random variables X and Y are said to be independent if we can factor their joint pdf as the product of the marginal pdfs, i.e.,

$$f_{X,Y}(x,y) = f_X(x) f_Y(y), \tag{2.14}$$

for all x and y. Then, by plugging (2.14) into (2.13), we obtain

$$f_{X\,|\,Y}(x\,|\,y) = f_X(x),$$

for all x and all y such that $f_Y(y) > 0$.

Example 2.9 We continue with Example 2.8 and compute the conditional pdf of X given Y. Recall from (2.12) that $f_Y(y) = 1 - |y| > 0$ if $-1 < y < 1$, and $f_Y(y) = 0$ otherwise; thus, $f_{X\,|\,Y}(x\,|\,y)$ is only defined if $y \in (-1, 1)$. Also, recall from (2.10) that for any $y \in (-1, 1)$, we have that $f_{X,Y}(x,y) = 1/2$ if $|x| \le 1 - |y|$, and $f_{X,Y}(x,y) = 0$ otherwise; thus,

$$f_{X\,|\,Y}(x\,|\,y) = \begin{cases} \frac{1}{2(1-|y|)}, & \text{if } -1 + |y| \le x \le 1 - |y|, \\ 0, & \text{otherwise,} \end{cases} \tag{2.15}$$

for any $y \in (-1, 1)$, i.e., given $Y = y$, $y \in (-1, 1)$, the distribution of X on the interval $\left[-1+|y|, 1-|y|\right]$ is uniform. The graph of $f_{X\,|\,Y}(\cdot\,|\,y_0)$ for some $y_0 \in (0, 1)$ is displayed in Fig. 2.2, where one can visualize the relation between the support of $f_{X\,|\,Y}(\cdot\,|\,y_0)$ and the shape of the support of the joint pdf $f_{X,Y}(x,y)$. By inspection of (2.10), (2.11), and (2.12), it is clear that $f_{X,Y}(x,y) \ne f_X(x) f_Y(y)$; thus, the random variables X and Y are not independent.

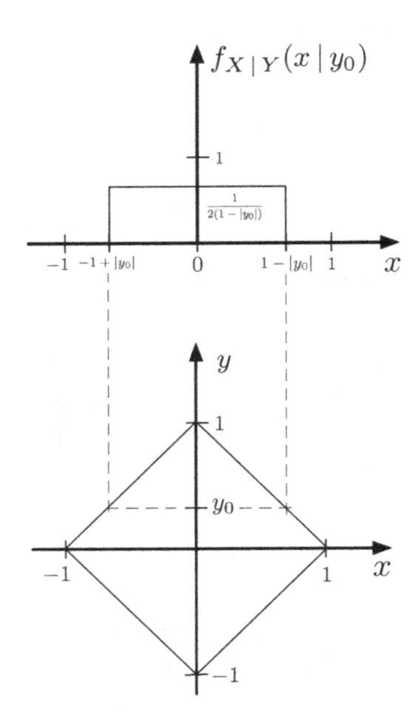

Figure 2.2 Example 2.9: conditional pdf $f_{X\,|\,Y}(\cdot\,|\,y_0)$ for some $y_0 \in (0,1)$ (top), and support of joint pdf $f_{X,Y}(\cdot,\cdot)$ (bottom).

Expectation, Covariance, and Correlation

Given two jointly distributed continuous random variables with joint pdf $f_{X,Y}(x,y)$, and a function $g\colon \mathbb{R}\times\mathbb{R}\to\mathbb{R}$, we have that

$$\mathrm{E}\big[g(X,Y)\big] = \int_{-\infty}^{\infty}\int_{-\infty}^{\infty} g(x,y)f_{X,Y}(x,y)dxdy; \tag{2.16}$$

this result is similar in spirit to the one in (2.8) for a single random variable and a real-valued function.

If X and Y are independent, then we have that

$$\mathrm{E}\big[XY\big] = \mathrm{E}\big[X\big]\mathrm{E}\big[Y\big].$$

To see this, we use (2.16) and (2.14) and proceed as follows:

$$\begin{aligned}\mathrm{E}\big[XY\big] &= \int_{-\infty}^{\infty}\int_{-\infty}^{\infty} xyf_{X,Y}(x,y)dxdy \\ &= \int_{-\infty}^{\infty}\int_{-\infty}^{\infty} xyf_X(x)f_Y(y)dxdy \\ &= \underbrace{\int_{-\infty}^{\infty} xf_X(x)dx}_{\mathrm{E}[X]}\underbrace{\int_{-\infty}^{\infty} yf_Y(y)dy}_{\mathrm{E}[Y]}.\end{aligned}$$

Given two jointly distributed continuous random variables X, Y, the conditional expectation of X given $Y = y$ is as follows:

$$\mathrm{E}\big[X \,|\, Y = y\big] = \int_{-\infty}^{\infty} x f_{X \,|\, Y}(x \,|\, y) dx, \tag{2.17}$$

which depends on y; thus, there exists some real-valued function $h(\cdot)$ such that

$$\begin{aligned} h(y) &:= \mathrm{E}\big[X \,|\, Y = y\big] \\ &= \int_{-\infty}^{\infty} x f_{X \,|\, Y}(x \,|\, y) dx. \end{aligned}$$

Then, $h(Y)$ is another random variable, denoted by $\mathrm{E}\big[X \,|\, Y\big]$, and referred to as the conditional expectation of X given Y. Now, by using (2.8), we have that

$$\begin{aligned} \mathrm{E}\Big[\mathrm{E}\big[X \,|\, Y\big]\Big] &= \mathrm{E}\big[h(Y)\big] \\ &= \int_{-\infty}^{\infty} h(y) f_Y(y) dy \\ &= \int_{-\infty}^{\infty} \left(\int_{-\infty}^{\infty} x f_{X \,|\, Y}(x \,|y) dx \right) f_Y(y) dy \\ &= \int_{-\infty}^{\infty} x \left(\int_{-\infty}^{\infty} \underbrace{f_{X \,|\, Y}(x \,|y) f_Y(y)}_{= f_{X,Y}(x,y)} dy \right) dx \\ &= \int_{-\infty}^{\infty} x f_X(x) dx \\ &= \mathrm{E}[X]. \end{aligned} \tag{2.18}$$

The covariance of two jointly distributed continuous random variables X, Y, which we denote by $c_{X,Y}$, is defined as follows:

$$\begin{aligned} c_{X,Y} &= \mathrm{E}\big[(X - \mu_X)(Y - \mu_Y)\big] \\ &= \int_{-\infty}^{\infty} \int_{-\infty}^{\infty} (x - \mu_X)(y - \mu_Y) f_{X,Y}(x,y) dx dy, \end{aligned}$$

and similarly to the variance of a single random variable, one can check that

$$\begin{aligned} c_{X,Y} &= \mathrm{E}[XY] - \mu_X \mu_Y \\ &= \underbrace{\int_{-\infty}^{\infty} \int_{-\infty}^{\infty} xy f_{X,Y}(x,y) dx dy}_{=: r_{X,Y}} - \mu_X \mu_Y. \end{aligned}$$

The term $r_{X,Y} := \mathrm{E}[XY]$ is referred to as the correlation of X and Y. Random variables X and Y are said to be uncorrelated when $c_{X,Y} = 0$. If this is the case, we have that

$$r_{X,Y} = \mu_X \mu_Y;$$

the converse is also true, i.e., if $r_{X,Y} = \mu_X \mu_Y$, then X and Y are uncorrelated. Recall that if two random variables X and Y are independent, then

$\mathrm{E}[XY] = \mu_X \mu_Y$; thus, if X and Y are independent, they are also uncorrelated. The converse is not true in general, i.e., if X and Y are uncorrelated, it does not imply that they are independent unless both X and Y are jointly distributed Gaussian random variables.

Example 2.10 We continue with Example 2.9. By plugging (2.15) into (2.17), we obtain

$$\begin{aligned} \mathrm{E}[X \mid Y = y] &= \int_{-1+|y|}^{1-|y|} x \frac{1}{2(1-|y|)} dx \\ &= 0 \\ &=: h(y), \quad -1 \leq y \leq 1. \end{aligned} \tag{2.19}$$

Then, by plugging (2.12) and (2.19) into (2.18), we obtain

$$\begin{aligned} \mathrm{E}[X] &= \mathrm{E}\big[\mathrm{E}[X \mid Y]\big] \\ &= \mathrm{E}\big[h(Y)\big] \\ &= 0, \end{aligned} \tag{2.20}$$

which matches the value of $\mathrm{E}[X]$ computed by using (2.11):

$$\begin{aligned} \mathrm{E}[X] &= \int_{-\infty}^{\infty} x f_X(x) dx \\ &= \int_{-1}^{1} x(1-|x|) dx \\ &= \int_{-1}^{0} x(1+x) dx + \int_{0}^{1} x(1-x) dx \\ &= 0, \end{aligned} \tag{2.21}$$

as expected. Similar calculation yields $\mathrm{E}[Y] = 0$. Thus,

$$\begin{aligned} c_{X,Y} &= r_{X,Y} \\ &= \int_{-\infty}^{\infty} x \left(\int_{-\infty}^{\infty} y f_{X,Y}(x,y) dy \right) dx \\ &= \frac{1}{2} \int_{-1}^{0} x \left(\int_{-(1+x)}^{1+x} y dy \right) dx + \frac{1}{2} \int_{0}^{1} x \left(\int_{-(1-x)}^{1-x} y dy \right) dx \\ &= 0, \end{aligned} \tag{2.22}$$

therefore, X and Y are uncorrelated; however, they are not independent as we established in Example 2.9.

2.1.4 Random Vectors

An n-dimensional random vector X is an n-tuple whose components, denoted by $X_1, X_2, \ldots, X_n$, are random variables all defined on the same probability space, $(\Omega, \mathcal{F}, \Pr)$. Because the X_i's take values in $\mathbb{R}$, the values that X takes, denoted by x, are real vectors in $\mathbb{R}^n$. (Unless otherwise stated, in this book we adopt the convention that the components of a vector are arranged in a column format.)

Now, for any $x = [x_1, x_2, \ldots, x_n]^\top \in \mathbb{R}^n$, we can define events of the form:

$$\{\omega \in \Omega \colon X_1(\omega) \leq x_1,\ X_2(\omega) \leq x_2, \ldots, X_n(\omega) \leq x_n\}, \tag{2.23}$$

and as before, we write $\Pr(X_1 \leq x_1,\ X_2 \leq x_2, \ldots, X_n \leq x_n)$ as a shorthand for their probability. Then, we can define a function $F_X \colon \mathbb{R}^n \to [0, 1]$ as follows:

$$F_X(x_1, x_2, \ldots, x_n) = \Pr(X_1 \leq x_1,\ X_2 \leq x_2, \ldots, X_n \leq x_n);$$

this function is referred to as the joint cdf of the components of X.

If the components of an n-dimensional random vector X are jointly distributed discrete random variables, then X will take values in a finite or countable set $\mathcal{X} \subset \mathbb{R}^n$, and we refer to X as a discrete random vector. Thus, as in the case of a single random variable, the pmf of the random vector X, denoted by $p_X(\cdot, \cdot, \ldots, \cdot)$, is defined as follows:

$$p_X(x_1, x_2, \ldots, x_n) = \Pr(X_1 = x_1,\ X_2 = x_2, \ldots, X_n = x_n),$$

$x = [x_1, x_2, \ldots, x_n]^\top \in \mathcal{X}$, which satisfies $\sum_{x \in \mathcal{X}} p_X(x_1, x_2, \ldots, x_n) = 1$.

We can write the set $\mathcal{X} \subset \mathbb{R}^n$ containing the values taken by a discrete random vector $X = [X_1, X_2, \ldots, X_n]^\top$ as follows:

$$\mathcal{X} = \mathcal{X}_1 \times \mathcal{X}_2 \times \cdots \times \mathcal{X}_n \subset \mathbb{R}^n,$$

where $\mathcal{X}_i \subset \mathbb{R},\ i = 1, 2, \ldots, n$, is a countable set containing the values that X_i can take. Then, we can define the marginal pmf of X_i, denoted by $p_{X_i}(\cdot)$, as follows:

$$p_{X_i}(x_i) = \sum_{x_j \in \mathcal{X}_j, j \neq i} p_X(x_1, x_2, \ldots, x_i, \ldots, x_n).$$

For the case when the components of X are jointly distributed continuous random variables, in which case we refer to X as a continuous random vector, there exists some function $f_X \colon \mathbb{R}^n \to [0, \infty)$, referred to as the joint pdf of the components of X, such that

$$F_X(x_1, x_2, \ldots, x_n) = \int_{-\infty}^{x_1} \int_{-\infty}^{x_2} \cdots \int_{-\infty}^{x_n} f_X(y_1, y_2, \ldots, y_n) dy_1 dy_2 \ldots dy_n,$$

and

$$\int_{-\infty}^{\infty}\int_{-\infty}^{\infty}\cdots\int_{-\infty}^{\infty} f_X(x_1, x_2, \ldots, x_n)dx_1dx_2\ldots dx_n = 1.$$

Given an n-dimensional continuous random vector X with pdf $f_X(\cdot)$, the marginal pdf of its ith component, denoted by $f_{X_i}(\cdot)$, is defined as follows:

$$f_{X_i}(x_i) = \int_{-\infty}^{\infty}\int_{-\infty}^{\infty}\cdots\int_{-\infty}^{\infty} f_X(x_1, x_2, \ldots, x_i, \ldots, x_n)dx_1dx_2\ldots dx_{i-1}dx_{i+1}\ldots dx_n.$$

Conditional pdfs and Independence

Given two random vectors $X = [X_1, X_2, \ldots, X_n]^\top$ and $Y = [Y_1, Y_2, \ldots, Y_m]^\top$ defined on the same probability space, $(\Omega, \mathcal{F}, \Pr)$, we can define their joint cdf, which we denote by $F_{X,Y}(\cdot,\cdot)$, by using a similar approach to the one above used for defining the joint pdf of the components of a random vector X. Similarly, whether X and Y are discrete or continuous random vectors, we can also define their joint pmf and pdf, which we denote by $p_{X,Y}(\cdot,\cdot)$ and $f_{X,Y}(\cdot,\cdot)$, respectively, and the marginal pmf and pdf of X (or Y), which we denote by $p_X(x)$ $\big($or $p_Y(y)\big)$ and $f_X(x)$ $\big($or $f_Y(y)\big)$, respectively.

As we did with events, if X and Y are discrete random vectors, we can define the conditional pmf of X given Y, denoted by $p_{X\,|\,Y}(\cdot\,|\,\cdot)$, as follows:

$$p_{X\,|\,Y}(x\,|\,y) = \frac{p_{X,Y}(x,y)}{p_Y(x)},$$

and if

$$p_{X,Y}(x,y) = p_X(x)p_Y(y),$$

we say they are independent. Similarly, if X and Y are continuous random vectors, we can define the conditional pdf of X given Y, denoted by $f_{X\,|\,Y}(\cdot\,|\,\cdot)$, as follows:

$$f_{X\,|\,Y}(x\,|\,y) = \frac{f_{X,Y}(x,y)}{f_Y(x)},$$

and if

$$f_{X,Y}(x,y) = f_X(x)f_Y(y),$$

we say they are independent.

Expectation, Covariance, and Correlation

The mean of a random vector $X = [X_1, X_2, \ldots, X_n]^\top$, denoted by $\mathrm{E}[X]$ or m_X, is a vector in $\mathbb{R}^n$ whose components are the expectations of the components of X, i.e.,

$$\mathrm{E}[X] = \Big[\mathrm{E}[X_1], \mathrm{E}[X_2], \ldots, \mathrm{E}[X_n]\Big]^\top. \tag{2.24}$$

The covariance matrix of a random vector $X = [X_1, X_2, \ldots, X_n]^\top$, denoted by $\mathrm{E}[(X - m_X)(X - m_X)^\top]$ or Σ_X, is defined as:

$$[\Sigma_X]_{i,j} = \mathrm{E}[(X_i - m_{X_i})(X_j - m_{X_j})],$$

$i = 1, 2, \ldots, n$, $j = 1, 2, \ldots, n$. Similarly, the correlation matrix of a random vector $X = [X_1, X_2, \ldots, X_n]^\top$, denoted by $\mathrm{E}[XX^\top]$ or S_X, is defined as:

$$[S_X]_{i,j} = \mathrm{E}[X_i X_j],$$

$i = 1, 2, \ldots, n$, $j = 1, 2, \ldots, n$. One can check that the relation between the covariance and correlation matrices of X is given by:

$$\Sigma_X = S_X - m_X m_X^\top.$$

Given two random vectors, $X = [X_1, X_2, \ldots, X_n]^\top$ and $Y = [Y_1, Y_2, \ldots, Y_n]^\top$, defined on the same probability space, we can define their covariance matrix, which we denote by $\mathrm{E}[(X - m_X)(Y - m_Y)^\top]$ or $C_{X,Y}$, as follows:

$$[C_{X,Y}]_{i,j} = \mathrm{E}\Big[\big(X_i - \mathrm{E}[X_i]\big)\big(Y_j - \mathrm{E}[Y_j]\big)\Big], \tag{2.25}$$

$i = 1, 2, \ldots, n$, $j = 1, 2, \ldots, n$. Similarly, we can define their correlation matrix, which we denote by $\mathrm{E}[XY^\top]$ or $R_{X,Y}$, as follows:

$$[R_{X,Y}]_{i,j} = \mathrm{E}[X_i Y_j], \tag{2.26}$$

$i = 1, 2, \ldots, n$, $j = 1, 2, \ldots, n$. For the case when $Y = X$, we clearly have $C_{X,X} = \Sigma_X$ and $R_{X,X} = S_X$. Given two random vectors X and Y, one can check that the relation between their covariance and correlation matrices is given by:

$$C_{X,Y} = R_{X,Y} - m_X m_Y^\top.$$

2.1.5 Stochastic Processes

An n-dimensional stochastic process X is a collection of n-dimensional random vectors indexed in some set $\mathcal{T}$, all of which are defined on the same probability space $(\Omega, \mathcal{F}, \Pr)$. When $\mathcal{T} = \{0, 1, 2, \ldots\}$, X is called a vector-valued discrete-time stochastic process, and we typically write $X = \{X_k : k \in \mathcal{T}\}$, $\mathcal{T} = \{0, 1, 2, \ldots\}$, or $X = \{X_k : k \geq 0\}$ to represent it. When $\mathcal{T} = [0, \infty)$, X is called a vector-valued continuous-time stochastic process, and we typically write $X = \{X(t) : t \in \mathcal{T}\}, \mathcal{T} = [0, \infty)$, or $X = \{X(t) : t \geq 0\}$ to represent it.

To completely characterize a vector-valued discrete-time stochastic process, $X = \{X_k : k \geq 0\}$, it is necessary to obtain the joint cdf of the random vectors $X_{k_1}, X_{k_2}, \ldots, X_{k_\ell}$,

$$F_{X_{k_1}, X_{k_2}, \ldots, X_{k_\ell}}(x_1, x_2, \ldots, x_\ell),$$

for all ℓ and any $k_1, k_2, \ldots, k_\ell$. Similarly, to fully characterize a vector-valued continuous-time stochastic process, $X = \{X(t) : t \geq 0\}$, it is necessary to obtain the joint cdf of the random vectors $X(t_1), X(t_2), \ldots, X(t_\ell)$,

$$F_{X(t_1),X(t_2),\ldots,X(t_\ell)}(x_1, x_2, \ldots, x_\ell),$$

for all ℓ and all $t_1, t_2, \ldots, t_\ell$.

Example 2.11 (Bernoulli process) Consider a discrete-time stochastic process $X = \{X_k \colon k \geq 0\}$, where the X_k's are Bernoulli independent and identically distributed (i.i.d.) random variables with parameter p, i.e., $\Pr(X_k = 1) = p$ and $\Pr(X_k = 0) = 1 - p$; such process is referred to as a Bernoulli process. In this case, instead of characterizing

$$F_{X_{k_1},X_{k_2},\ldots,X_{k_\ell}}(x_1, x_2, \ldots, x_\ell),$$

for all ℓ and any $k_1, k_2, \ldots, k_\ell$, we can equivalently provide a complete description of the stochastic process by providing the joint pmf of $X_{k_1}, X_{k_2}, \ldots, X_{k_\ell}$, $p_{X_{k_1},X_{k_2},\ldots,X_{k_\ell}}(x_1, x_2, \ldots, x_\ell)$, for all ℓ and any $k_1, k_2, \ldots, k_\ell$. First note that

$$\Pr(X_k = x_k) = p^{x_k}(1-p)^{1-x_k},$$

with $x_k \in \{0, 1\}$; then, we have that

$$\begin{aligned} p_{X_{k_1},X_{k_2},\ldots,X_{k_\ell}}(x_1, x_2, \ldots, x_\ell) &= \Pr\left(X_{k_1} = x_1, X_{k_2} = x_2, \ldots, X_{k_\ell} = x_\ell\right) \\ &= \Pr\left(X_{k_1} = x_1\right) \Pr\left(X_{k_2} = x_2\right) \cdots \Pr\left(X_{k_\ell} = x_\ell\right) \\ &= p^{x_1}(1-p)^{1-x_1} p^{x_2}(1-p)^{1-x_2} \ldots p^{x_\ell}(1-p)^{1-x_\ell} \\ &= p^{\sum_{k=1}^{\ell} x_k}(1-p)^{\ell - \sum_{k=1}^{\ell} x_k}, \end{aligned} \tag{2.27}$$

with $x_i \in \{0, 1\}$, $i = 1, 2, \ldots, \ell$.

Mean, Covariance, and Correlation Functions

Except for specific cases (as in the example above), it is hard in general to obtain a full characterization of a stochastic process. However, in many applications it suffices to characterize the first and second moments of the stochastic process, which we define next.

The first moment, or mean function, of an n-dimensional discrete-time stochastic process X, which we denote by $m_X[\cdot]$, is defined as follows:

$$m_X[k] = \mathrm{E}\left[X_k\right] \in \mathbb{R}^n, \quad k \geq 0. \tag{2.28}$$

Similarly, if X is an n-dimensional continuous-time stochastic process, its mean function, which we denote by $m_X(\cdot)$, is defined as follows:

$$m_X(t) = \mathrm{E}\big[X(t)\big] \in \mathbb{R}^n, \quad t \geq 0. \tag{2.29}$$

The covariance function of an n-dimensional discrete-time stochastic process, which we denote by $C_X[\cdot,\cdot]$, is defined as follows:

$$\begin{aligned} C_X[k_1,k_2] &:= C_{X_{k_1},X_{k_2}} \\ &= \mathrm{E}\Big[\big(X_{k_1} - m_X[k_1]\big)\big(X_{k_2} - m_X[k_2]\big)^\top\Big] \in \mathbb{R}^{n\times n}, \quad k_1,k_2 \geq 0. \end{aligned} \tag{2.30}$$

Similarly, if X is an n-dimensional continuous-time stochastic process, its covariance function, which we denote by $C_X(\cdot,\cdot)$, is defined as follows:

$$\begin{aligned} C_X(t_1,t_2) &:= C_{X(t_1),X(t_2)} \\ &= \mathrm{E}\Big[\big(X(t_1) - m_X(t_1)\big)\big(X(t_2) - m_X(t_2)\big)^\top\Big] \in \mathbb{R}^{n\times n}, \quad t_1,t_2 \geq 0. \end{aligned} \tag{2.31}$$

In words, given $k_1, k_2 \geq 0$ ($t_1, t_2 \geq 0$), the value of the covariance function of X, $C_X[k_1,k_2]$ $\big(C_X(t_1,t_2)\big)$, is equal to the covariance of the random vectors X_{k_1} and X_{k_2} $\big(X(t_1)$ and $X(t_2)\big)$.

The correlation function of an n-dimensional discrete-time stochastic process, which we denote by $R_X[\cdot,\cdot]$, is defined as follows:

$$\begin{aligned} R_X[k_1,k_2] &:= R_{X_{k_1},X_{k_2}} \\ &= \mathrm{E}\big[X_{k_1}X_{k_2}^\top\big] \in \mathbb{R}^{n\times n}, \quad k_1,k_2 \geq 0. \end{aligned} \tag{2.32}$$

Similarly, if X is an n-dimensional continuous-time stochastic process, its covariance function, which we denote by $R_X(\cdot,\cdot)$, is defined as follows:

$$\begin{aligned} R_X(t_1,t_2) &:= R_{X(t_1),X(t_2)} \\ &= \mathrm{E}\big[X(t_1)X^\top(t_2)\big] \in \mathbb{R}^{n\times n}, \quad t_1,t_2 \geq 0. \end{aligned} \tag{2.33}$$

In words, given $k_1, k_2 \geq 0$ ($t_1, t_2 \geq 0$), the value of the correlation function of X, $R_X[k_1,k_2]$ $\big(R_X(t_1,t_2)\big)$, is equal to the correlation of the random vectors X_{k_1} and X_{k_2} $\big(X(t_1)$ and $X(t_2)\big)$.

The correlation and covariance functions of a stochastic process X are referred to as the second moments of the process, and they are related as follows:

$$C_X[k_1,k_2] = R_X[k_1,k_2] - m_X[k_1]m_X^\top[k_2], \quad k_1,k_2 \geq 0, \tag{2.34}$$

if X is a discrete-time stochastic process, and

$$C_X(t_1,t_2) = R_X(t_1,t_2) - m_X(t_1)m_X^\top(t_2), \quad t_1,t_2 \geq 0, \tag{2.35}$$

if X is a continuous-time stochastic process.

Example 2.12 (Bernoulli process moments) Consider again the Bernoulli process in Example 2.11. In this case, the mean, covariance, and correlation functions of the process are:

$$\begin{aligned} m_X[k] &= \mathrm{E}[X_k] \\ &= p, \quad k = 0, 1, 2, \ldots, \end{aligned} \tag{2.36}$$

$$\begin{aligned} C_X[k_1, k_2] &= C_{X_{k_1}, X_{k_2}} \\ &= \mathrm{E}\Big[\big(X_{k_1} - m_X[k_1]\big)\big(X_{k_2} - m_X[k_2]\big)\Big] \\ &= \begin{cases} p(1-p), & \text{if } k_1 = k_2, \\ 0, & \text{if } k_1 \neq k_2, \end{cases} \end{aligned} \tag{2.37}$$

$$\begin{aligned} R_X[k_1, k_2] &= C_X[k_1, k_2] + m_X[k_1] m_X[k_2] \\ &= \begin{cases} p, & \text{if } k_1 = k_2, \\ p^2, & \text{if } k_1 \neq k_2. \end{cases} \end{aligned} \tag{2.38}$$

Stationarity

An n-dimensional discrete-time stochastic process $X = \{X_k : k \geq 0\}$ is said to be strict-sense stationary if

$$F_{X_{k_1}, X_{k_2}, \ldots, X_{k_\ell}}(x_1, x_2, \ldots, x_\ell) = F_{X_{k_1+k'}, X_{k_2+k'}, \ldots, X_{k_\ell+k'}}(x_1, x_2, \ldots, x_\ell)$$

for all ℓ and all $k_1, k_2, \ldots, k_\ell, k' \geq 0$. Similarly, an n-dimensional continuous-time stochastic process $X = \big\{X(t) : t \in [0, \infty)\big\}$ is said to be strict-sense stationary if

$$F_{X(t_1), X(t_2), \ldots, X(t_\ell)}(x_1, x_2, \ldots, x_\ell) = F_{X(t_1+s), X(t_2+s), \ldots, X(t_\ell+s)}(x_1, x_2, \ldots, x_\ell)$$

for all ℓ and all $t_1, t_2, \ldots, t_\ell, s \in [0, \infty)$.

A discrete-time stochastic process $X = \{X_k : k \geq 0\}$ is said to be wide-sense stationary if

$$m_X[k] = m_X[k + k'], \tag{2.39}$$

$$C_X[k_1, k_2] = C_X[k_1 + k', k_2 + k'], \tag{2.40}$$

for all $k, k', k_1, k_2 \geq 0$, i.e., its mean function does not depend on time, and the value of the covariance function for any k_1, k_2, $C_X(k_1, k_2)$, only depends on $k_1 - k_2$. Similarly, a continuous-time stochastic process $X = \{X(t) : t \in [0, \infty)\}$ is said to be wide-sense stationary if

$$m_X(t) = m_X(t + s), \tag{2.41}$$

$$C_X(t_1, t_2) = C_X(t_1 + s, t_2 + s), \tag{2.42}$$

for all $t, s, t_1, t_2 \geq 0$. Any strict-sense stationary process is also wide-sense stationary; the converse is not true in general.

Example 2.13 Consider again the Bernoulli process of Examples 2.11, 2.12. For any $k, k' \geq 0$, we have that

$$\begin{aligned} m_X[k+k'] &= \mathrm{E}[X_{k+k'}] \\ &= p; \end{aligned} \tag{2.43}$$

which matches the expression for $m_X[k]$ in (2.36). Similarly, for any $k_1, k_2, k' \geq 0$, direct calculation of $\mathrm{E}\big[\big(X_{k_1+k'} - m_X[k_1+k']\big)\big(X_{k_2+k'} - m_X[k_2+k']\big)\big]$ yields

$$C_X[k_1+k', k_2+k'] = \begin{cases} p(1-p), & \text{if } k_1 = k_2, \\ 0, & \text{if } k_1 \neq k_2, \end{cases} \tag{2.44}$$

which matches the expression for $C_X[k_1, k_2]$ in (2.37). Thus, we conclude that the Bernoulli process is wide-sense stationary.

By using a similar procedure to the one used in (2.27), we obtain that

$$p_{X_{k_1+k'}, X_{k_2+k'}, \ldots, X_{k_\ell+k'}}(x_1, x_2, \ldots, x_\ell) = p^{\sum_{k=1}^{\ell} x_k}(1-p)^{\ell - \sum_{k=1}^{\ell} x_k}, \tag{2.45}$$

for any $k_1, k_2, \ldots, k_\ell, k' \geq 0$, which matches the expression for

$$p_{X_{k_1}, X_{k_2}, \ldots, X_{k_\ell}}(x_1, x_2, \ldots, x_\ell)$$

in (2.27); thus, the Bernoulli process is also strict-sense stationary.

The Wiener Process

The Wiener process is a continuous-time stochastic process that will be heavily featured in Chapter 6. It can be used to formally describe the random motion of a tiny particle suspended in water, a phenomenon first observed by the botanist Robert Brown in 1827, and for this reason, a Wiener process is also referred to as Brownian motion process, or Brownian motion. However, it was not until the 1920s that the mathematician Norbert Wiener provided a rigorous mathematical description of the phenomenon (Albert Einstein also studied the problem in 1905). The formal definition of the Wiener process is as follows:

DEFINITION 2.1 Let $W = \{W(t) : t \in \mathcal{T} = [0, \infty)\}$ denote a real-valued stochastic process. We say W is a *standard Wiener process* if it satisfies the following properties:

B1. $W(0) = 0$.

B2. W has independent increments, i.e., the distribution of $W(t) - W(s)$ depends on $t - s$ alone, and the variables $W(t_j) - W(s_j)$, $j = 1, 2, \ldots, n$, are independent whenever the intervals $(s_j, t_j]$ are disjoint.

B3. $W(s+t) - W(s)$ is normally distributed with zero mean and variance t for all $s, t \geq 0$.

B4. The sample paths of W are continuous.

Mean, Covariance, and Correlation Function. A standard Wiener process W is a zero-mean process, i.e.,

$$m_W(t) = 0, \tag{2.46}$$

for all $t \geq 0$; this can be easily established as follows. First note that $\mathrm{E}\big[W(0)\big] = 0$ by Property B1; then we have

$$\begin{aligned} m_W(t) &= \mathrm{E}\big[W(t)\big] \\ &= \mathrm{E}\big[W(t)\big] - \mathrm{E}\big[W(0)\big] \\ &= \mathrm{E}\big[W(t) - W(0)\big] \\ &= 0, \end{aligned} \tag{2.47}$$

where the last equality follows from Property B3. In addition, the values taken by the covariance and correlation functions of a Wiener process W, $C_W(t,s),\ t,s \geq 0$, and $R(t,s),\ t,s \geq 0$, respectively, are equal and given by

$$\begin{aligned} C_W(t,s) &= R_W(t,s) \\ &= \min\{t,s\}; \end{aligned} \tag{2.48}$$

this formula can be established as follows. Clearly $C_W(t,s) = R_W(t,s)$ since $m_W(t) = 0$ for all $t \geq 0$. Now, assume $t \geq s$, then by noting Property B1 (i.e., $W(0) = 0$), we have that

$$\begin{aligned} \mathrm{E}\big[W(t)W(s)\big] &= \mathrm{E}\Big[\big(W(t) - W(0)\big)\big(W(s) - W(0)\big)\Big] \\ &= \mathrm{E}\Big[\big(W(s) - W(0)\big)^2 + \big(W(t) - W(s)\big)\big(W(s) - W(0)\big)\Big] \\ &= \mathrm{E}\Big[\big(W(s) - W(0)\big)^2\Big] + \underbrace{\mathrm{E}\Big[\big(W(t) - W(s)\big)\big(W(s) - W(0)\big)\Big]}_{= 0 \text{ by Properties B2 and B3}} \\ &= \mathrm{E}\Big[\big(W(s) - W(0)\big)^2\Big] \\ &= s, \end{aligned} \tag{2.49}$$

where the last equality follows from Property B3. A similar derivation for the case when $t \leq s$ results in $\mathrm{E}\big[W(t)W(s)\big] = t$; thus, $C_W(t,s) = R_W(t,s) = \min\{t,s\}$.

Probability Distribution. By Property B3, $\Delta W(t) = W(t) - W(0)$ is normally distributed with zero mean and variance t; thus, its pdf, which we denote by $f_{\Delta W}(t,\cdot)$, is given by

$$f_{\Delta W}(t,\Delta w) = \frac{1}{\sqrt{2\pi t}} e^{-\frac{\Delta w^2}{2t}}, \tag{2.50}$$

and since $W(0) = 0$ by Property B1, we have that the pdf of

$$W(t) = W(0) + \Delta W(t),$$

denoted by $f_W(t,\cdot)$, is given by

$$f_W(t,w) = \frac{1}{\sqrt{2\pi t}} e^{-\frac{w^2}{2t}}. \tag{2.51}$$

One can check that (2.51) satisfies the following partial differential equation

$$\frac{\partial f_W(t,w)}{\partial t} = \frac{1}{2}\frac{\partial^2 f_W(t,w)}{\partial w^2}, \tag{2.52}$$

with $f_W(0,w) = \delta(w)$.

Assume that $W(s) = x$, $s \geq 0$ and $x \in \mathbb{R}$; then, conditioned on this, we have that $W(t)$, $t \geq s$, is normally distributed with mean x and variance $t-s$, i.e., the pdf of $W(t)$ conditioned on $W(s) = x$, which we denote by $f_W(t,\cdot \mid s,x)$, is given by

$$f_W(t,w \mid s,x) = \frac{1}{\sqrt{2\pi(t-s)}} e^{-\frac{(w-x)^2}{2(t-s)}}. \tag{2.53}$$

One can check that $f_W(t,w \mid s,x)$ satisfies the following partial differential equation

$$\frac{\partial f_W(t,w \mid s,x)}{\partial t} = \frac{1}{2}\frac{\partial^2 f_W(t,w \mid s,x)}{\partial w^2}, \tag{2.54}$$

which has the same form as that in (2.52) governing the evolution of the unconditional pdf.

2.2 Set Theory

2.2.1 Basic Notions and Notation

A set $\mathcal{X}$ is a collection of distinct objects, referred to as the elements of $\mathcal{X}$. We write $x \in \mathcal{X}$ to denote that x is an element of $\mathcal{X}$ and $x \notin \mathcal{X}$ to denote that x is not an element of $\mathcal{X}$. The cardinality of a set $\mathcal{X}$, denoted by $|\mathcal{X}|$, is the number of elements in $\mathcal{X}$. A set $\mathcal{X}$ is called a singleton if it contains a single element, i.e., $|\mathcal{X}| = 1$. The empty set, denoted by $\emptyset$, is a set containing no elements; thus, $|\emptyset| = 0$. The sets of natural numbers (including 0), integer numbers, real numbers, and complex numbers are denoted by $\mathbb{N}$, $\mathbb{Z}$, $\mathbb{R}$, and $\mathbb{C}$, respectively.

A set can be described in terms of some relations that its elements satisfy. For example, let $\mathcal{X}$ denote a set whose elements, denoted by x, are real numbers satisfying the following relation:

$$x^2 - 1 \leq 0;$$

then we write

$$\mathcal{X} = \{x \in \mathbb{R} \colon x^2 - 1 \leq 0\}.$$

We say two sets $\mathcal{X}$ and $\mathcal{Y}$ are equal, and denote it by $\mathcal{X} = \mathcal{Y}$, if every element in $\mathcal{X}$ is also an element of $\mathcal{Y}$ and vice versa. We say the set $\mathcal{X}$ is a subset of another set $\mathcal{Y}$, and denote it by $\mathcal{X} \subseteq \mathcal{Y}$, if every element in $\mathcal{X}$ is also an element of $\mathcal{Y}$. We say the set $\mathcal{X}$ is a strict subset of another set $\mathcal{Y}$, and denote it by $\mathcal{X} \subset \mathcal{Y}$ if $\mathcal{X} \subseteq \mathcal{Y}$ and $\mathcal{X}$ and $\mathcal{Y}$ are not equal, i.e., $\mathcal{X} \neq \mathcal{Y}$. We say two sets $\mathcal{X}$ and $\mathcal{Y}$ are disjoint if none of the elements in the set $\mathcal{X}$ are contained in the set $\mathcal{Y}$ and vice versa.

The union of two sets $\mathcal{X}$ and $\mathcal{Y}$, denoted by $\mathcal{X} \cup \mathcal{Y}$, is another set containing all the elements that are in either $\mathcal{X}$ or $\mathcal{Y}$; this can be written as

$$\mathcal{X} \cup \mathcal{Y} = \{x \colon x \in \mathcal{X} \text{ or } x \in \mathcal{Y}\}.$$

The intersection of two sets $\mathcal{X}$ and $\mathcal{Y}$, denoted by $\mathcal{X} \cap \mathcal{Y}$, is another set containing all the elements that are in both $\mathcal{X}$ and $\mathcal{Y}$; this can be written as

$$\mathcal{X} \cap \mathcal{Y} = \{x \colon x \in \mathcal{X} \text{ and } x \in \mathcal{Y}\}.$$

The difference of two sets $\mathcal{X}$ and $\mathcal{Y}$, denoted by $\mathcal{X} \setminus \mathcal{Y}$, is another set containing the elements in $\mathcal{X}$ that are not in $\mathcal{Y}$; this can be written as

$$\mathcal{X} \setminus \mathcal{Y} = \{x \colon x \in \mathcal{X} \text{ and } x \notin \mathcal{Y}\}.$$

Let $\mathcal{X} \subseteq \mathcal{U}$, where $\mathcal{U}$ denotes the universal set, i.e., the set that contains all considered elements; then we can define the complement of $\mathcal{X}$, which we denote by $\overline{\mathcal{X}}$ or $\mathcal{X}^c$, as $\overline{\mathcal{X}} = \mathcal{U} \setminus \mathcal{X}$.

The Cartesian product of two sets $\mathcal{X}$ and $\mathcal{Y}$, denoted by $\mathcal{X} \times \mathcal{Y}$, is another set containing ordered pairs of the form (x, y); this can be written as:

$$\mathcal{X} \times \mathcal{Y} = \{(x, y) \colon x \in \mathcal{X},\ y \in \mathcal{Y}\}.$$

2.2.2 Sets in Euclidean Space

We write $\mathcal{X} \subseteq \mathbb{R}^n$ to denote a set whose elements are vectors in the n-dimensional Euclidean space. Given a set $\mathcal{X} \subseteq \mathbb{R}^n$, we say an element $x \in \mathcal{X}$ is an interior point of $\mathcal{X}$ if there exists some $\epsilon > 0$ so that the set

$$\{y \in \mathbb{R}^n \colon (y - x)^\top (y - x) \leq \epsilon\}$$

is a subset of $\mathcal{X}$. The set of all interior points of $\mathcal{X}$ is called the interior of $\mathcal{X}$ and is denoted by $\mathbf{int}(\mathcal{X})$. A set $\mathcal{X} \subseteq \mathbb{R}^n$ is said to be open if $\mathbf{int}(\mathcal{X}) = \mathcal{X}$. A set $\mathcal{X} \subseteq \mathbb{R}^n$ is said to be closed if its complement,

$$\overline{\mathcal{X}} = \{x \in \mathbb{R}^n \colon x \notin \mathcal{X}\},$$

is open. The closure of a set $\mathcal{X} \subseteq \mathbb{R}^n$, denoted by $\mathbf{cl}(\mathcal{X})$, is defined as

$$\mathbf{cl}(\mathcal{X}) = \mathbb{R}^n \setminus \mathbf{int}(\mathbb{R}^n \setminus \mathcal{X}).$$

The boundary of a set $\mathcal{X} \subseteq \mathbb{R}^n$, denoted by $\mathbf{bd}(\mathcal{X})$ or $\partial\mathcal{X}$, is defined as

$$\mathbf{bd}(\mathcal{X}) = \mathbf{cl}(\mathcal{X}) \setminus \mathbf{int}(\mathcal{X}).$$

Operations with Sets in $\mathbb{R}^n$

The (geometric) Minkowski sum of two sets $\mathcal{X} \in \mathbb{R}^n$ and $\mathcal{Y} \in \mathbb{R}^n$, denoted by $\mathcal{X} + \mathcal{Y}$, is another set whose elements result from adding each element in $\mathcal{X}$ to each element in $\mathcal{Y}$; this can be written as

$$\mathcal{X} + \mathcal{Y} = \{z \in \mathbb{R}^n : z = x + y,\ x \in \mathcal{X},\ y \in \mathcal{Y}\}.$$

The (geometric) Minkowski difference of two sets, $\mathcal{X} \in \mathbb{R}^n$ and $\mathcal{Y} \in \mathbb{R}^n$, denoted by $\mathcal{X} - \mathcal{Y}$, is another set defined as follows:

$$\mathcal{X} - \mathcal{Y} = \{x \in \mathbb{R}^n : \{x\} + \mathcal{Y} \subseteq \mathcal{X}\}.$$

For a given $H \in \mathbb{R}^{p \times n}$, the linear transformation of a set $\mathcal{X} \subseteq \mathbb{R}^n$ is another set $H\mathcal{X} \subseteq \mathbb{R}^p$ defined as follows:

$$H\mathcal{X} = \{y \in \mathbb{R}^p : y = Hx,\ x \in \mathcal{X}\}.$$

The Support Function

Given a closed set $\mathcal{X} \subseteq \mathbb{R}^n$, its support function, denoted by $S_{\mathcal{X}}(\cdot)$, is defined as

$$S_{\mathcal{X}}(\eta) = \max_{x \in \mathcal{X}} \eta^\top x, \quad \eta \in \mathbb{R}^n. \tag{2.55}$$

Let

$$x^*(\eta) = \arg \underbrace{\max_{x \in \mathcal{X}} \eta^\top x}_{S_{\mathcal{X}}(\eta)}; \tag{2.56}$$

then, clearly $S_{\mathcal{X}}(\eta) = \eta^\top x^*(\eta)$ and $x^*(\eta) \in \mathbf{bd}(\mathcal{X})$; to see this, assume that $S_{\mathcal{X}}(\eta) > 0$. Then, if $x^*(\eta) \notin \mathbf{bd}(\mathcal{X})$, there exists some $\tilde{x}^* \in \mathcal{X}$ satisfying

$$\tilde{x}^*(\eta) = x^*(\eta) + \varepsilon\eta$$

for some $\varepsilon > 0$. Then we have that

$$\begin{aligned} \eta^\top \tilde{x}^* &= \eta^\top \left(x^*(\eta) + \varepsilon\eta\right) \\ &= S_{\mathcal{X}}(\eta) + \varepsilon\eta^\top\eta \\ &> S_{\mathcal{X}}(\eta), \end{aligned}$$

but this contradicts the fact that $S_{\mathcal{X}}(\eta) = \max_{x \in \mathcal{X}} \eta^\top x$. A similar argument can be made when $S_{\mathcal{X}}(\eta) < 0$.

The support function of the Minkowski sum of two sets $\mathcal{X} \subseteq \mathbb{R}^n$ and $\mathcal{Y} \subseteq \mathbb{R}^n$, denoted by $S_{\mathcal{X}+\mathcal{Y}}(\cdot)$, is given by

$$S_{\mathcal{X}+\mathcal{Y}}(\eta) = S_{\mathcal{X}}(\eta) + S_{\mathcal{Y}}(\eta), \quad \eta \in \mathbb{R}^n. \tag{2.57}$$

To see this, define

$$\mathcal{Z} := \mathcal{X} + \mathcal{Y} = \{z \in \mathbb{R}^n : z = x + y,\ x \in \mathcal{X},\ y \in \mathcal{Y}\};$$

then, by using (2.55), we have

$$\begin{aligned} S_{\mathcal{X}+\mathcal{Y}}(\eta) &= S_{\mathcal{Z}}(\eta) \\ &= \max_{z\in\mathcal{Z}} \eta^\top z \\ &= \max_{x\in\mathcal{X},\ y\in\mathcal{Y}} \eta^\top (x+y) \\ &= \max_{x\in\mathcal{X}} \eta^\top x + \max_{y\in\mathcal{Y}} \eta^\top y \\ &= S_{\mathcal{X}}(\eta) + S_{\mathcal{Y}}(\eta), \end{aligned} \tag{2.58}$$

as claimed in (2.57).

The support function of $H\mathcal{X}$, where $H \in \mathbb{R}^{p\times n}$ and $\mathcal{X} \subseteq \mathbb{R}^n$, is given by

$$S_{H\mathcal{X}}(\eta) = S_{\mathcal{X}}(H^\top \eta); \tag{2.59}$$

this can be established as follows. Let $\mathcal{Y} := H\mathcal{X}$; then, by using (2.55), we have

$$\begin{aligned} S_{\mathcal{H}\mathcal{X}}(\eta) &= S_{\mathcal{Y}}(\eta) \\ &= \max_{y\in\mathcal{Y}} \eta^\top y \\ &= \max_{x\in\mathcal{X}} \eta^\top (Hx) \\ &= \max_{x\in\mathcal{X}} \left(H^\top \eta\right)^\top x \\ &= S_{\mathcal{X}}(H^\top \eta), \end{aligned} \tag{2.60}$$

as claimed in (2.59).

Convex Sets

A set $\mathcal{X} \in \mathbb{R}^n$ is convex if the line segment between any two points in $\mathcal{X}$ is contained in $\mathcal{X}$, i.e., for any $x_1, x_2 \in \mathcal{X}$, and any $\theta \in [0,1]$, we have that

$$\theta x_1 + (1-\theta)x_2 \in \mathcal{X}.$$

Let $\mathcal{B} \subseteq \mathbb{R}^n$ denote the unit ball, i.e.,

$$\mathcal{B} = \left\{\eta \in \mathbb{R}^n \colon \eta^\top \eta = 1\right\}.$$

A closed convex set $\mathcal{X} \subseteq \mathbb{R}^n$ can then be described via its support function as follows:

$$\mathcal{X} = \left\{x \in \mathbb{R}^n \colon \eta^\top x \leq S_{\mathcal{X}}(\eta),\ \eta \in \mathcal{B}\right\}. \tag{2.61}$$

To establish this, we need the following result.

THEOREM 2.2 (Supporting Hyperplane Theorem) *If $\mathcal{X} \subseteq \mathbb{R}^n$ is a nonempty convex set, then for any $\overline{x}$ in the boundary of $\mathcal{X}$, there exists a vector $a \in \mathbb{R}^n$ such that*

$$a^\top x \leq a^\top \overline{x} \tag{2.62}$$

for all $x \in \mathcal{X}$.

The geometric interpretation of the result in the theorem above is that the hyperplane,

$$\mathcal{H}(a) = \{x \in \mathbb{R}^n : a^\top x = a^\top \overline{x}\},$$

is tangent to $\mathcal{X}$ at $\overline{x}$; thus, $\mathcal{H}(a)$ is called a supporting hyperplane to $\mathcal{X}$ at $\overline{x}$, hence the name of the theorem.

From the supporting hyperplane theorem, associated to each $\overline{x} \in \mathbf{bd}(\mathcal{X})$, there exists some a so that $a^\top x \leq a^\top \overline{x}$ for all $x \in \mathcal{X}$, which is equivalent to saying that

$$\eta^\top x \leq \eta^\top \overline{x}, \quad \eta = \frac{a}{\|a\|},$$

for all $x \in \mathcal{X}$. Now, recall from (2.56) that $S_\mathcal{X}(\eta) = \eta^\top x^*(\eta)$, with

$$x^*(\eta) = \arg\max_{x \in \mathcal{X}} \eta^\top x$$

in the boundary of $\mathcal{X}$; thus,

$$\eta^\top x \leq \eta^\top x^*(\eta) \tag{2.63}$$

for all $x \in \mathcal{X}$. Therefore, each $\overline{x} \in \mathbf{bd}(\mathcal{X})$ can be written as

$$\overline{x} = x^*(\eta), \quad \eta \in \mathcal{B},$$

thus,

$$\mathcal{X} = \{x \in \mathbb{R}^n : \eta^\top x \leq S_\mathcal{X}(\eta),\ \eta \in \mathcal{B}\}, \tag{2.64}$$

with

$$\mathcal{H}(\eta) = \{x \in \mathbb{R}^n : \eta^\top x = S_\mathcal{X}(\eta)\} \tag{2.65}$$

defining a supporting hyperplane to $\mathcal{X}$ at $x^*(\eta)$, i.e., $\mathcal{H}(\eta)$ is tangent to $\mathcal{X}$ at $x^*(\eta)$.

Consider two closed and convex sets $\mathcal{X}, \mathcal{Y} \in \mathbb{R}^n$. Then, as a consequence of the supporting hyperplane theorem and the result in (2.65), if $S_\mathcal{X}(\eta) = S_\mathcal{Y}(\eta)$ for some $\eta \in \mathbb{R}^n$, then the boundaries of $\mathcal{X}$ and $\mathcal{Y}$ touch each other at the following point:

$$\begin{aligned} x^*(\eta) &= \arg\underbrace{\max_{x \in \mathcal{X}} \eta^\top x}_{= S_\mathcal{X}(\eta)} \\ &= \arg\underbrace{\max_{x \in \mathcal{Y}} \eta^\top x}_{= S_\mathcal{Y}(\eta)}. \end{aligned} \tag{2.66}$$

Let $\mathcal{X}, \mathcal{Y} \in \mathbb{R}^n$ be closed and convex, then, $\mathcal{X} \subseteq \mathcal{Y}$ if and only if $S_\mathcal{X}(\eta) \leq S_\mathcal{Y}(\eta)$ for all $\eta \in \mathbb{R}^n$; this can be established as follows. Since $\mathcal{X}$ and $\mathcal{Y}$ are convex, we can describe them using (2.61) as follows:

$$\mathcal{X} = \{x \in \mathbb{R}^n : \eta^\top x \leq S_\mathcal{X}(\eta),\ \eta \in \mathcal{B}\}, \tag{2.67}$$

$$\mathcal{Y} = \{y \in \mathbb{R}^n : \eta^\top y \leq S_\mathcal{Y}(\eta),\ \eta \in \mathcal{B}\}. \tag{2.68}$$

If $S_{\mathcal{X}}(\eta) \leq S_{\mathcal{Y}}(\eta)$ for all $\eta \in \mathbb{R}^n$, then, by using (2.67), we have that every $x \in \mathcal{X}$ satisfies $\eta^\top x \leq S_{\mathcal{X}}(\eta) \leq S_{\mathcal{Y}}(\eta)$, $\eta \in \mathcal{B}$; thus, every $x \in \mathcal{X}$ is contained in $\mathcal{Y}$ by virtue of (2.68), therefore, $\mathcal{X} \subseteq \mathcal{Y}$. If $\mathcal{X} \subseteq \mathcal{Y}$, then for every $\eta \in \mathcal{B}$, and because $\mathcal{X}$ is closed, there exists some $x' \in \mathcal{X}$ such that $\eta^\top x' = S_{\mathcal{X}}(\eta)$ by virtue of (2.67), which also satisfies $\eta^\top x' \leq S_{\mathcal{Y}}(\eta)$ by virtue of (2.68); thus, $S_{\mathcal{X}}(\eta) \leq S_{\mathcal{Y}}(\eta)$ for all $\eta \in \mathcal{B}$. Now, if we multiply η by any $\alpha \in \mathbb{R}$, it follows from the definition of support function that $S_{\mathcal{X}}(\alpha\eta) = \alpha S_{\mathcal{X}}(\eta)$ and $S_{\mathcal{Y}}(\alpha\eta) = \alpha S_{\mathcal{Y}}(\eta)$, thus $S_{\mathcal{X}}(\alpha\eta) \leq S_{\mathcal{Y}}(\alpha\eta)$ for any $\alpha \in \mathbb{R}$ and all $\eta \in \mathcal{B}$, therefore, $S_{\mathcal{X}}(\eta) \leq S_{\mathcal{Y}}(\eta)$ for all $\eta \in \mathbb{R}^n$.

The Minkowski sum of two closed convex sets $\mathcal{X}, \mathcal{Y} \in \mathbb{R}^n$ is also a closed convex set. To see this, let w and z denote two elements of $\mathcal{X} + \mathcal{Y}$. Then, there exist $w_1, z_1 \in \mathcal{X}$ and $w_2, z_2 \in \mathcal{Y}$ so that $w = w_1 + w_2$ and $z = z_1 + z_2$. Now, for $\theta \in [0, 1]$, we have that

$$\theta w + (1-\theta)z = \underbrace{\theta w_1 + (1-\theta)z_1}_{=:\, x} + \underbrace{\theta w_2 + (1-\theta)z_2}_{=:\, y},$$

and by convexity of $\mathcal{X}$ and $\mathcal{Y}$, we have that $x \in \mathcal{X}$ and $y \in \mathcal{Y}$; thus, we conclude that $\theta w + (1-\theta)z \in \mathcal{X} + \mathcal{Y}$, therefore $\mathcal{X} + \mathcal{Y}$ is convex.

Ellipsoids. An ellipsoid is a closed convex set $\mathcal{E} \subseteq \mathbb{R}^n$ defined as follows:

$$\mathcal{E} = \left\{x \in \mathbb{R}^n \colon (x - x_0)^\top E^{-1}(x - x_0) \leq 1\right\}, \tag{2.69}$$

where $x_0 \in \mathbb{R}^n$ is the center of the ellipsoid, and $E \in \mathbb{R}^{n \times n}$ is a symmetric positive definite matrix referred to as the shape matrix. The directions of the the semi-axes of $\mathcal{E}$ are defined by the eigenvectors of the matrix E, whereas its eigenvalues are the squares of the semi-axis lengths.

Support function: Consider an ellipsoid $\mathcal{E} \subseteq \mathbb{R}^n$ with center $x_0 \in \mathbb{R}^n$ and shape matrix $E \in \mathbb{R}^{n \times n}$. Then, its support function, $S_{\mathcal{E}}(\cdot)$, is given by

$$S_{\mathcal{E}}(\eta) = \eta^\top x_0 + \sqrt{\eta^\top E \eta}, \quad \eta \in \mathbb{R}^n, \tag{2.70}$$

and

$$\mathcal{H}(\eta) = \left\{x \in \mathbb{R}^n \colon \eta^\top x = \underbrace{\eta^\top x_0 + \sqrt{\eta^\top E \eta}}_{= S_{\mathcal{E}}(\eta)}\right\}$$

is a supporting hyperplane to $\mathcal{E}$ at

$$x^*(\eta) = x_0 + \frac{1}{\sqrt{\eta^\top E \eta}} E\eta.$$

To see this, recall that $S_{\mathcal{E}}(\eta) = \eta^\top x^*(\eta)$, where $x^*(\eta) = \arg\max_{x \in \mathcal{E}} \eta^\top x$. Since $x^*(\eta) \in \mathbf{bd}(\mathcal{E})$, we can obtain $x^*(\eta)$ from local extremum points of the following optimization problem:

$$\begin{aligned} &\underset{x}{\text{maximize}} && \eta^\top x \\ &\text{subject to} && (x - x_0)^\top E^{-1}(x - x_0) = 1. \end{aligned} \tag{2.71}$$

By introducing a Lagrange multiplier, λ, we can reformulate (2.71) as an unconstrained optimization problem as follows:

$$\underset{x,\lambda}{\text{maximize}}\ f(x,\lambda), \tag{2.72}$$

where $f(x,\lambda) = \eta^\top x - \lambda\big((x-x_0)^\top E^{-1}(x-x_0) - 1\big)$. The values that maximize $f(x,\lambda)$ can be obtained by computing the values of x and λ for which the gradient of $f(\cdot,\cdot)$ is equal to zero. Thus, we have that

$$\begin{aligned}\frac{\partial f(x,\lambda)}{\partial x} &= \eta^\top - 2\lambda(x-x_0)^\top E^{-1} \\ &= 0, \end{aligned} \tag{2.73}$$

$$\begin{aligned}\frac{\partial f(x,\lambda)}{\partial \lambda} &= (x-x_0)^\top E^{-1}(x-x_0) - 1 \\ &= 0. \end{aligned} \tag{2.74}$$

Then, by using (2.73), we obtain

$$x = x_0 + \frac{1}{2\lambda}E\eta, \tag{2.75}$$

and by plugging this expression into (2.74), and solving for λ, we obtain

$$\lambda = \pm\frac{1}{2}\sqrt{\eta^\top E\eta}. \tag{2.76}$$

Now, one can easily see that

$$x = x_0 + \frac{1}{\sqrt{\eta^\top E\eta}}E\eta, \qquad \lambda = \frac{1}{2}\sqrt{\eta^\top E\eta},$$

maximize the value of $f(x,\lambda)$, while

$$x = x_0 - \frac{1}{\sqrt{\eta^\top E\eta}}E\eta, \qquad \lambda = -\frac{1}{2}\sqrt{\eta^\top E\eta},$$

minimize it. Thus,

$$x^*(\eta) = x_0 + \frac{1}{\sqrt{\eta^\top E\eta}}E\eta,$$

which, by plugging into $S_{\mathcal{E}}(\eta) = \eta^\top x^*(\eta)$, yields

$$S_{\mathcal{E}}(\eta) = \eta^\top x_0 + \sqrt{\eta^\top E\eta},$$

as claimed in (2.70). The fact that

$$\mathcal{H}(\eta) = \Big\{x \in \mathbb{R}^n \colon \eta^\top x = \eta^\top x_0 + \sqrt{\eta^\top E\eta}\Big\}$$

is a supporting hyperplane to $\mathcal{E}$ at

$$x^*(\eta) = x_0 + \frac{1}{\sqrt{\eta^\top E \eta}} E\eta$$

follows directly from (2.65).

Minkowski sum: Consider two ellipsoids $\mathcal{X} \subseteq \mathbb{R}^n$ and $\mathcal{Y} \subseteq \mathbb{R}^n$ defined as follows:

$$\begin{aligned}\mathcal{X} &= \{x \in \mathbb{R}^n : (x - x_0)^\top X^{-1}(x - x_0) \leq 1\},\\ \mathcal{Y} &= \{y \in \mathbb{R}^n : (y - y_0)^\top Y^{-1}(y - y_0) \leq 1\},\end{aligned}$$

where X and Y are positive definite matrices; thus, their support functions are

$$\begin{aligned}S_{\mathcal{X}}(\eta) &= \eta^\top x_0 + \sqrt{\eta^\top X \eta},\\ S_{\mathcal{Y}}(\eta) &= \eta^\top y_0 + \sqrt{\eta^\top Y \eta},\end{aligned}$$

respectively. Now, let $\mathcal{Z} \subseteq \mathbb{R}^n$ denote the set that results from the Minkowski sum of $\mathcal{X}$ and $\mathcal{Y}$. Then, by using (2.57), we have that

$$\begin{aligned}S_{\mathcal{Z}}(\eta) &= S_{\mathcal{X}}(\eta) + S_{\mathcal{Y}}(\eta)\\ &= \eta^\top (x_0 + y_0) + \sqrt{\eta^\top X \eta} + \sqrt{\eta^\top Y \eta};\end{aligned} \tag{2.77}$$

thus, in general, $\mathcal{Z}$ is not an ellipsoid because, except for specific cases, we can not find some positive definite matrix Z such that $\sqrt{\eta^\top Z \eta} = \sqrt{\eta^\top X \eta} + \sqrt{\eta^\top Y \eta}$.

Volume: Consider an ellipsoid $\mathcal{E} \subseteq R^n$ with center $x_0 \in \mathbb{R}^n$ and shape matrix $E \in \mathbb{R}^{n \times n}$. Then its volume, denoted by $\mathbf{vol}(\mathcal{E})$, is defined as follows:

$$\mathbf{vol}(\mathcal{E}) = \frac{\pi^{n/2}\sqrt{\mathbf{det}(E)}}{\Gamma(\frac{n}{2} + 1)}, \tag{2.78}$$

where $\Gamma(\cdot)$ is Euler's gamma function, which is defined as follows:

$$\Gamma(x) = \int_0^\infty u^{x-1} e^{-u} du. \tag{2.79}$$

For $n = 2$, we have that

$$\begin{aligned}\Gamma(2) &= \int_0^\infty u e^{-u} du\\ &= 1;\end{aligned} \tag{2.80}$$

thus, in this case, $\mathbf{vol}(\mathcal{X}) = \pi\sqrt{\mathbf{det}(E)}$. Now, if $E = r^2 I_2$, the resulting ellipsoid is just a disc of radius r, and since $\mathbf{det}(E) = r^4$, we have that $\mathbf{vol}(\mathcal{X}) = \pi r^2$, which is the familiar formula for the surface area of a disc of radius r. For $n = 3$, we have that

$$\begin{aligned}\Gamma(2.5) &= \int_0^\infty u^{1.5} e^{-u} du\\ &= \frac{3\sqrt{\pi}}{4};\end{aligned} \tag{2.81}$$

thus, in this case, $\mathbf{vol}(\mathcal{X}) = \frac{4\pi}{3}\sqrt{\mathbf{det}(E)}$. Now, if $E = r^2 I_3$, the resulting ellipsoid is just a sphere of radius r, and since $\mathbf{det}(E) = r^6$, we have that $\mathbf{vol}(\mathcal{X}) = \frac{4\pi r^3}{3}$, which is the familiar formula for the volume of a sphere of radius r.

Zonotopes. A zonotope is a closed convex set $\mathcal{Z} \subseteq \mathbb{R}^n$ defined as follows:

$$\mathcal{Z} = \left\{ x \in \mathbb{R}^n \colon x = x_0 + \sum_{i=1}^{s} \alpha_i e_i, \quad -1 \leq \alpha_i \leq 1 \right\}, \tag{2.82}$$

where $x_0 \in \mathbb{R}^n$ is the center of the zonotope, and $e_1, e_2, \ldots, e_s$ are vectors in $\mathbb{R}^n$ referred to as the generators of the zonotope.

Support function: Consider a zonotope $\mathcal{Z} \subseteq \mathbb{R}^n$ with center $x_0 \in \mathbb{R}^n$ and generators $e_1, e_2, \ldots, e_s$. Then, its support function, $S_{\mathcal{Z}}(\cdot)$, is defined as follows:

$$S_{\mathcal{Z}}(\eta) = \eta^\top x_0 + \sum_{i=1}^{s} |\eta^\top e_i|, \quad \eta \in \mathbb{R}^n. \tag{2.83}$$

To see this, define $\mathcal{X}_0 = \{x_0\}$ and

$$\mathcal{L}_i = \{x \in \mathbb{R}^n \colon x = \alpha_i e_i, \quad -1 \leq \alpha_i \leq 1\}, \quad i = 1, 2, \ldots, s;$$

then, clearly $\mathcal{Z}$ can be written as the Minkowski sum of $\mathcal{X}_0$, $\mathcal{L}_i$, $i = 1, 2, \ldots, s$; thus,

$$S_{\mathcal{Z}}(\eta) = S_{\mathcal{X}_0}(\eta) + \sum_{i=1}^{s} S_{\mathcal{L}_i}(\eta).$$

Now, from the definition of support function in (2.55), we have

$$\begin{aligned} S_{\mathcal{X}_0}(\eta) &= \max_{x \in \mathcal{X}_0} \eta^\top x \\ &= \eta^\top x_0, \end{aligned} \tag{2.84}$$

and

$$\begin{aligned} S_{\mathcal{L}_i}(\eta) &= \max_{x \in \mathcal{L}_i} \eta^\top x \\ &= \max_{-1 \leq \alpha_i \leq 1} \alpha_i \eta^\top e_i \\ &= |\eta^\top e_i|; \end{aligned} \tag{2.85}$$

thus,

$$S_{\mathcal{Z}}(\eta) = \eta^\top x_0 + \sum_{i=1}^{s} |\eta^\top e_i|,$$

as claimed in (2.83).

Minkowski sum: Consider two zonotopes $\mathcal{X} \subseteq \mathbb{R}^n$ and $\mathcal{Y} \subseteq \mathbb{R}^n$ defined as follows:

$$\mathcal{X} = \left\{ x \in \mathbb{R}^n \colon x = x_0 + \sum_{i=1}^{p} \alpha_i e_i, \quad -1 \leq \alpha_i \leq 1 \right\},$$

$$\mathcal{Y} = \left\{ y \in \mathbb{R}^n : y = y_0 + \sum_{i=1}^{q} \beta_i f_i, \ -1 \leq \beta_i \leq 1 \right\}; \tag{2.86}$$

thus, their support functions are

$$S_{\mathcal{X}}(\eta) = \eta^\top x_0 + \sum_{i=1}^{p} |\eta^\top e_i|,$$

$$S_{\mathcal{Y}}(\eta) = \eta^\top y_0 + \sum_{i=1}^{q} |\eta^\top f_i|, \tag{2.87}$$

respectively. Now, let $\mathcal{Z} \subseteq \mathbb{R}^n$ denote the set that results from the Minkowski sum of $\mathcal{X}$ and $\mathcal{Y}$. Then, by using (2.57), we have that

$$\begin{aligned} S_{\mathcal{Z}}(\eta) &= S_{\mathcal{X}}(\eta) + S_{\mathcal{Y}}(\eta) \\ &= \eta^\top (x_0 + y_0) + \sum_{i=1}^{p} |\eta^\top e_i| + \sum_{i=1}^{q} |\eta^\top f_i|; \end{aligned} \tag{2.88}$$

thus, $\mathcal{Z}$ is clearly a zonotope with center $z_0 = x_0 + y_0$ and generators

$$e_1, e_2, \ldots, e_p, f_1, f_2, \ldots, f_q.$$

Volume: Consider a zonotope $\mathcal{Z} \subseteq \mathbb{R}^n$ with center $x_0 \in \mathbb{R}^n$ and generators $e_1, e_2, \ldots, e_s$. Then, its volume, denoted by $\mathbf{vol}(\mathcal{Z})$, is defined as follows. For $n > s$, we have that

$$\mathbf{vol}(\mathcal{Z}) = 0. \tag{2.89}$$

For $n \leq s$, consider all n-combinations of the set of generators,

$$\{e_1, e_2, \ldots, e_s\},$$

i.e., all sets formed by taking n distinct elements of the set $\{e_1, e_2, \ldots, e_s\}$; there are $N = \binom{s}{n} = \frac{s!}{n!(s-n)!}$ such sets, which we denote by $\mathcal{E}_i$, $i = 1, 2, \ldots, N$. Let E_i denote the $(n \times n)$-dimensional matrix formed by horizontal concatenation of the elements in $\mathcal{E}_i$, $i = 1, 2, \ldots, N$, then,

$$\mathbf{vol}(\mathcal{Z}) = 2^n \sum_{i=1}^{N} \left|\mathbf{det}(E_i)\right|. \tag{2.90}$$

Consider the case when $\mathcal{Z} \subseteq R^3$ is a zonotope with center $x_0 = [0,0,0]^\top$ and generators $e_1 = [l/2, 0, 0]^\top$, $e_2 = [0, l/2, 0]^\top$, and $e_3 = [0, 0, l/2]^\top$; thus $\mathcal{Z}$ is a cube with sides of length l. In this case, $N = 1$ and $E_1 = \frac{l}{2} I_3$; thus,

$$\mathbf{vol}(\mathcal{Z}) = 2^3 \mathbf{det}\left(\frac{l}{2} I_3\right) = l^3,$$

which is the familiar formula for the volume of a cube of length l.

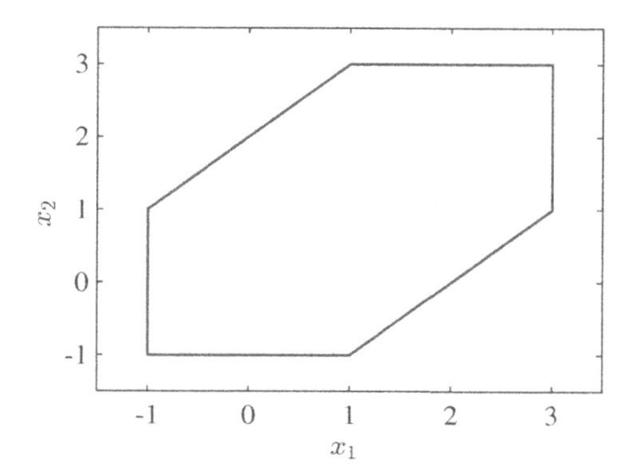

Figure 2.3 Zonotope in $\mathbb{R}^2$ with center $x_0 = [1,1]^\top$ and generators $e_1 = [1,0]^\top$, $e_2 = [0,1]^\top$, and $e_3 = [1,1]^\top$.

Example 2.14 (Zonotope in $\mathbb{R}^2$) Consider a zonotope $\mathcal{Z} \subseteq \mathbb{R}^2$ with center

$$x_0 = [1,1]^\top,$$

and generators

$$e_1 = [1,0]^\top, \quad e_2 = [0,1]^\top, \quad e_3 = [1,1]^\top; \tag{2.91}$$

thus,

$$\mathcal{Z} = \left\{ \begin{bmatrix} x_1 \\ x_2 \end{bmatrix} : \begin{bmatrix} x_1 \\ x_2 \end{bmatrix} = \begin{bmatrix} 1+\alpha_1+\alpha_3 \\ 1+\alpha_2+\alpha_3 \end{bmatrix} \right.$$
$$\left. -1 \leq \alpha_1 \leq 1,\ -1 \leq \alpha_2 \leq 1,\ -1 \leq \alpha_3 \leq 1 \right\}; \tag{2.92}$$

see Fig. 2.3 for a graphical depiction.

Now, by tailoring (2.90) to the setting here, we have that

$$E_1 = \begin{bmatrix} 1 & 0 \\ 0 & 1 \end{bmatrix}, \quad E_2 = \begin{bmatrix} 1 & 1 \\ 0 & 1 \end{bmatrix}, \quad E_3 = \begin{bmatrix} 0 & 1 \\ 1 & 1 \end{bmatrix}, \tag{2.93}$$

from where it follows that

$$\begin{aligned} \mathbf{vol}(\mathcal{Z}) &= 2^2\Big(\big|\mathbf{det}(E_1)\big| + \big|\mathbf{det}(E_2)\big| + \big|\mathbf{det}(E_3)\big|\Big) \\ &= 12. \end{aligned} \tag{2.94}$$

By inspection of Fig. 2.3, one can verify that the area enclosed by the boundary of the zonotope is 12, which is consistent with the result in (2.94).

2.3 Linear Dynamical Systems

In this section, we study discrete-time and continuous-time linear dynamical systems described by state-space models. We first introduce the notion of the state-transition matrix, which we subsequently use for trajectory characterization and stability analysis.

2.3.1 Discrete-Time Systems

Here we study linear time-varying dynamical systems described by a difference equation of the form:

$$x_{k+1} = G_k x_k + H_k w_k, \quad k = 0, 1, \ldots, \tag{2.95}$$

where $x_k \in \mathbb{R}^n$ is referred to as the system state, $w_k \in \mathbb{R}^m$ is referred to as the input, $G_k \in \mathbb{R}^{n \times n}$, and $H_k \in \mathbb{R}^{n \times m}$. Given x_0, and a sequence of inputs, $w_0, w_1, \ldots,$ we would like to find an expression describing the trajectory followed by the state, $x_k, k \geq 0$. In addition, we would like to provide a formal characterization of the stability of such systems.

Trajectory Characterization. Consider first the case when $w_k = \mathbf{0}_m$ for all $k \geq 0$; then, (2.95) reduces to

$$x_{k+1} = G_k x_k, \quad k \geq 0, \tag{2.96}$$

from where it follows that

$$x_k = G_{k-1} G_{k-2} \ldots G_1 G_0 x_0, \quad k \geq 0. \tag{2.97}$$

More generally, we can relate the value the system state takes at time instant k to the value taken at an earlier time instant ℓ as follows, By defining the following matrix,

$$\Phi_{k,\ell} = \begin{cases} G_{k-1} G_{k-2} \cdots G_{\ell+1} G_\ell, & \text{if } k > \ell \geq 0, \\ I_n, & \text{if } k = \ell, \end{cases} \tag{2.98}$$

referred to as the discrete-time state-transition matrix, we can write

$$x_k = \Phi_{k,\ell} x_\ell, \quad k \geq \ell \geq 0, \tag{2.99}$$

and in particular

$$x_k = \Phi_{k,0} x_0, \quad k \geq 0. \tag{2.100}$$

From (2.98), it is clear that discrete-time state-transition matrix, $\Phi_{k,\ell}$, can also be described recursively as follows:

$$\Phi_{k+1,\ell} = G_k \Phi_{k,\ell}, \quad k \geq \ell \geq 0. \tag{2.101}$$

If G_k is invertible for all k, we can relate the value that the state takes at time instant k to the value it would take at a later time instant ℓ as follows:

$$x_k = G_k^{-1} G_{k+1}^{-1} \ldots G_{\ell-2}^{-1} G_{\ell-1}^{-1} x_\ell; \tag{2.102}$$

one can see this by noting that

$$x_\ell = G_{\ell-1}G_{\ell-2}\dots G_{k+1}G_kx_k;$$

thus,

$$\begin{aligned} x_k &= (G_{\ell-1}G_{\ell-2}\dots G_{k+1}G_k)^{-1}x_\ell \\ &= G_k^{-1}G_{k+1}^{-1}\dots G_{\ell-2}^{-1}G_{\ell-1}^{-1}x_\ell. \end{aligned} \tag{2.103}$$

Thus, for such class of systems, we could generalize the definition of the discrete-time state-transition matrix in (2.98) as follows:

$$\Phi_{k,\ell} = \begin{cases} G_{k-1}G_{k-2}\cdots G_{\ell+1}G_\ell, & \text{if } k > \ell \geq 0, \\ I_n, & \text{if } k = \ell, \\ G_k^{-1}G_{k+1}^{-1}\dots G_{\ell-2}^{-1}G_{\ell-1}^{-1}, & \text{if } 0 \leq k < \ell, \end{cases} \tag{2.104}$$

which would allow us to generalize (2.99) as

$$x_k = \Phi_{k,\ell}x_\ell, \quad k, \ell \geq 0, \tag{2.105}$$

and (2.101) as

$$\Phi_{k+1,\ell} = G_k\Phi_{k,\ell}, \quad k, \ell \geq 0, \tag{2.106}$$

with $\Phi_{\ell,\ell} = I_n$. Unless otherwise stated, we will assume $\Phi_{k,\ell}$ is only defined for $k \geq \ell \geq 0$ as in (2.98).

Now, consider the general case when w_k can take any values in $\mathbb{R}^m$; then, by using (2.95) for $k = 0, 1$, we have

$$\begin{aligned} x_1 &= G_0x_0 + H_0w_0, \\ x_2 &= G_1x_1 + H_1w_1 \\ &= G_1\big(G_0x_0 + H_0w_0\big) + H_1w_1 \\ &= G_1G_0x_0 + G_1H_0w_0 + H_1w_1 \\ &= \Phi_{2,0}x_0 + \Phi_{2,1}H_0w_0 + \Phi_{2,2}H_1w_1, \end{aligned} \tag{2.107}$$

and more generally one can check that

$$x_k = \Phi_{k,0}x_0 + \sum_{\ell=0}^{k-1}\Phi_{k,\ell+1}H_\ell w_\ell, \quad k \geq 0. \tag{2.108}$$

When the system is time-invariant, i.e., $G_k = G$ and $H_k = H$ for all k, where G and H are some constant matrices, we have that

$$\Phi_{k,\ell} = G^{k-\ell}, \; k \geq \ell \geq 0;$$

thus, (2.108) reduces to

$$x_k = G^kx_0 + \sum_{\ell=0}^{k-1}G^{k-\ell-1}Hw_\ell, \quad k \geq 0. \tag{2.109}$$

Stability Characterization. Consider the homogeneous part of the system in (2.95), i.e.,

$$x_{k+1} = G_k x_k, \tag{2.110}$$

with $x_0 \in \mathbb{R}^n$ given. Recall the expression in (2.99):

$$x_k = \Phi_{k,\ell} x_\ell, \tag{2.111}$$

with $\Phi_{k,\ell}$ as defined in (2.98). If $x_\ell = \mathbf{0}_n$, it follows that $x_k = \mathbf{0}_n$, $k \geq \ell$; thus, we say $x^\circ = \mathbf{0}_n$ is an equilibrium point of the system in (2.110). Next, we introduce two important notions for characterizing the stability of the system in (2.110) around x°.

The system in (2.110) is said to be *stable in the sense of Lyapunov* around x° if the following property is satisfied:

S1. For a given $\epsilon > 0$, there exists some $\delta_1 > 0$ such that if $\|x_0\|_2 \leq \delta_1$, then $\|x_k\|_2 < \epsilon$ for all $k \geq 0$.

Furthermore, the system in (2.110) is said to be *asymptotically stable* around x° if it is stable in the sense of Lyapunov and the following property is additionally satisfied:

S2. There exists some $\delta_2 > 0$ such that if $\|x_0\|_2 \leq \delta_2$, then $\lim_{k\to\infty} x_k = \mathbf{0}_n$.

Since $x_k = \Phi_{k,0} x_0$, $k \geq 0$, the matrix $\Phi_{k,0}$ completely determines whether or not Properties S1 and S2 are satisfied.

For homogeneous linear time-invariant systems, i.e., $x_{k+1} = G x_k$, we have that $\Phi_{k,0} = G^k$. Then, Property S1 is satisfied if and only if (i) the magnitude of all the eigenvalues of the matrix G is smaller than or equal to one, and (ii) for each eigenvalue whose magnitude is equal to one, the associated algebraic and geometric multiplicities must be identical. For Property S2 to be satisfied, the magnitude of all the eigenvalues of the matrix G must be strictly smaller than one. Finally, consider time-invariant systems of the form

$$x_{k+1} = G x_k + H w_k,$$

where w_k, $k \geq 0$, is bounded, i.e., there exists some positive constant, K_w, such that $\|w_k\|_2 \leq K_w$ for all k. Then, the system state will also remain bounded, i.e., there exists some positive constant, K_x, such that $\|x_k\|_2 \leq K_x$ for all k, if the magnitude of all the eigenvalues of the matrix G is strictly smaller than one.

2.3.2 Continuous-Time Systems

Consider the continuous-time counterpart of the system in (2.95), which is described by a differential equation as follows:

$$\frac{d}{dt}x(t) = A(t)x(t) + B(t)w(t), \tag{2.112}$$

where $x(t) \in \mathbb{R}^n$ is the system state, $w(t) \in \mathbb{R}^m$ is the input, $A(t) \in \mathbb{R}^{n\times n}$, and $B(t) \in \mathbb{R}^{n\times m}$. Given $x(0)$, and $w(\tau)$, $0 \leq \tau \leq t$, where $w(\cdot)$ is some integrable function, and assuming the entries of $A(t)$ and $B(t)$ are sufficiently well behaved, we would like to find an expression describing the trajectory followed by the state, $x(t)$, $t \geq 0$.

In the discrete-time case discussed earlier, we saw that the solution can be expressed as a function of the state-transition matrix $\Phi_{k,\ell}$, which is completely characterized by the recursion in (2.101); here, we will follow a similar approach and describe the solution in terms of some matrix $\Phi(t,s)$, $t,s \geq 0$, referred to as the continuous-time state-transition matrix. To this end, we first consider a matrix $F(t) \in \mathbb{R}^{n\times n}$ that satisfies the following matrix differential equation:

$$\frac{d}{dt}F(t) = A(t)F(t), \quad t \geq 0, \tag{2.113}$$

with $F(0)$ given such that $\mathbf{det}(F(0)) \neq 0$; we refer to $F(t)$ as a fundamental matrix of the system in (2.112) for the special case when $B(t)w(t) = 0$ for all $t \geq 0$. The matrix $F(t)$ is invertible for all $t \geq 0$; one can see this by using Jacobi's formula (see (A.11)) as follows:

$$\begin{aligned}\frac{d}{dt}\mathbf{det}\Big(F(t)\Big) &= \mathbf{tr}\left(\mathbf{adj}(F(t))\frac{d}{dt}F(t)\right)\\ &= \mathbf{tr}\Big(\mathbf{adj}(F(t))A(t)F(t)\Big)\\ &= \mathbf{tr}\Big(\underbrace{F(t)\mathbf{adj}(F(t))}_{=\,\mathbf{det}(F(t))I_n}A(t)\Big)\\ &= \mathbf{det}(F(t))\mathbf{tr}(A(t)),\end{aligned} \tag{2.114}$$

which by integrating results in

$$\mathbf{det}(F(t)) = \mathbf{det}(F(0))e^{\int_0^t \mathbf{tr}(A(\tau))d\tau}; \tag{2.115}$$

thus, clearly $\mathbf{det}(F(t)) \neq 0$, therefore $F(t)$ is invertible. Now, we use $F(t)$ to define the continuous-time state-transition matrix $\Phi(t,\tau)$, $t,\tau \geq 0$, as follows:

$$\Phi(t,\tau) := F(t)F^{-1}(\tau), \quad t,\tau \geq 0. \tag{2.116}$$

The continuous-time state-transition matrix satisfies the following properties:

F1. $\Phi(t,t) = I_n$ for any $t \geq 0$. One can see this as follows:

$$\begin{aligned}\Phi(t,t) &= F(t)F^{-1}(t)\\ &= I_n.\end{aligned} \tag{2.117}$$

F2. $\Phi^{-1}(t,s) = \Phi(s,t)$ for any $t, s \geq 0$. One can see this as follows:

$$\begin{aligned}\Phi^{-1}(t,s) &= \left(F(t)F^{-1}(s)\right)^{-1} \\ &= F(s)F^{-1}(t) \\ &= \Phi(s,t). \end{aligned} \tag{2.118}$$

F3. $\Phi(t,s) = \Phi(t,\tau)\Phi(\tau,s)$ for any $t, \tau, s \geq 0$. One can see this as follows:

$$\begin{aligned}\Phi(t,s) &= F(t)F^{-1}(s) \\ &= F(t)F^{-1}(\tau)F(\tau)F^{-1}(s) \\ &= \Phi(t,\tau)\Phi(\tau,s). \end{aligned} \tag{2.119}$$

F4. $\frac{\partial}{\partial t}\Phi(t,\tau) = A(t)\Phi(t,\tau)$, for any $t, \tau \geq 0$. One can see this as follows:

$$\begin{aligned}\frac{\partial}{\partial t}\Phi(t,\tau) &= \frac{\partial}{\partial t}\left(F(t)F^{-1}(\tau)\right) \\ &= \frac{d}{dt}F(t)F^{-1}(\tau) = A(t)F(t)F^{-1}(\tau) \\ &= A(t)\Phi(t,\tau). \end{aligned} \tag{2.120}$$

Now, we use $\Phi(t,\tau)$ to construct the solution of (2.112). For the special case when $w(\tau) = 0$ for all $\tau \in [0,t]$, the expression in (2.112) reduces to

$$\frac{d}{dt}x(t) = A(t)x(t), \tag{2.121}$$

with $x(0)$ given, the solution of which is given by

$$x(t) = \Phi(t,0)x(0); \tag{2.122}$$

this can be seen by differentiating both sides of the expression above with respect to t and using Property F4:

$$\begin{aligned}\frac{d}{dt}x(t) &= \frac{d}{dt}\Phi(t,0)x(0) \\ &= A(t)\Phi(t,0)x(0) \\ &= A(t)x(t), \end{aligned} \tag{2.123}$$

which matches the expression in (2.121). More generally, we have

$$x(t) = \Phi(t,s)x(s), \tag{2.124}$$

for any $t, s \geq 0$. (This is unlike the discrete time case, where, unless the matrix G_k is invertible for all k, the state-transition matrix, $\Phi_{k,\ell}$, was only defined for $k \geq \ell \geq 0$.) To see this, note that

$$\begin{aligned}x(t) &= \Phi(t,0)x(0), \\ x(s) &= \Phi(s,0)x(0). \end{aligned}$$

Now, by using Property F2, we have that

$$\begin{aligned} x(0) &= \Phi^{-1}(s,0)x(s) \\ &= \Phi(0,s)x(s), \end{aligned}$$

thus,

$$\begin{aligned} x(t) &= \Phi(t,0)x(0) \\ &= \Phi(t,0)\Phi(0,s)x(s) \\ &= \Phi(t,s)x(s), \end{aligned}$$

where the last equality follows from Property F3.

For the general case when $w(\tau)$ is not equal to zero for all $\tau \in [0,t]$, we have that the solution of (2.112) is given by

$$x(t) = \Phi(t,0)x(0) + \int_0^t \Phi(t,\tau)B(\tau)w(\tau)d\tau. \tag{2.125}$$

To see this, we differentiate both sides of (2.125) with respect to t and check that the result matches (2.112). To this end, recall Leibniz's rule for differentiating an integral:

$$\begin{aligned} \frac{d}{dx}\left(\int_{a(x)}^{b(x)} f(x,y)dy\right) &= f\big(x,b(x)\big)\frac{d}{dx}b(x) - f\big(x,a(x)\big)\frac{d}{dx}a(x) \\ &\quad + \int_{a(x)}^{b(x)} \frac{\partial}{\partial x} f(x,y)dy; \end{aligned} \tag{2.126}$$

then, it follows that

$$\begin{aligned} \frac{d}{dt}x(t) &= \frac{d}{dt}\Phi(t,0)x(0) + \frac{d}{dt}\int_0^t \Phi(t,\tau)B(\tau)w(\tau)d\tau \\ &= \frac{d}{dt}\Phi(t,0)x(0) + \int_0^t \left(\frac{\partial}{\partial t}\Phi(t,\tau)\right)B(\tau)w(\tau)d\tau + \underbrace{\Phi(t,t)}_{I_n}B(t)w(t) \\ &= A(t)\Phi(t,0)x(0) + \int_0^t A(t)\Phi(t,\tau)B(\tau)w(\tau)d\tau + B(t)w(t) \\ &= \underbrace{A(t)\left[\Phi(t,0)x(0) + \int_0^t \Phi(t,\tau)B(\tau)w(\tau)d\tau\right]}_{=\,x(t) \text{ by } (2.125)} + B(t)w(t) \\ &= A(t)x(t) + B(t)w(t), \end{aligned} \tag{2.127}$$

which matches the expression in (2.112).

When the system is time-invariant, i.e., $A(t) = A$ and $B(t) = B$ for all t, where A and B are some constant matrices, it follows from (2.113) that

$$\frac{d}{dt}\Phi(t,0) = A\Phi(t,0), \quad t \geq 0, \tag{2.128}$$

with $\Phi(0,0) = I_n$, the solution of which is

$$\begin{aligned}\Phi(t,0) &= e^{tA} \\ &:= \sum_{k=0}^{\infty} \frac{1}{k!} t^k A^k; \end{aligned} \tag{2.129}$$

one can check this by noting that

$$\begin{aligned}\frac{d}{dt}\Phi(t,0) &= \frac{d}{dt}\left(\sum_{k=0}^{\infty} \frac{1}{k!} t^k A^k\right) \\ &= \sum_{k=1}^{\infty} \frac{1}{(k-1)!} t^{k-1} A^k \\ &= A \sum_{k=1}^{\infty} \frac{1}{(k-1)!} t^{k-1} A^{k-1} \\ &= A \sum_{m=0}^{\infty} \frac{1}{m!} t^m A^m \\ &= A\Phi(t,0). \end{aligned} \tag{2.130}$$

Then, by plugging $\Phi(t,0) = e^{tA}$ into (2.125), we obtain

$$x(t) = e^{tA} + \int_0^t e^{(t-\tau)A} Bw(\tau) d\tau. \tag{2.131}$$

Stability Characterization. Consider the homogeneous part of the system in (2.112), i.e.,

$$\frac{d}{dt}x(t) = A(t)x(t), \tag{2.132}$$

with $x(0) \in \mathbb{R}^n$. As in the discrete-time case, $x^\circ = \mathbf{0}_n$ is an equilibrium point of the system in (2.132). Similar to the discrete-time case, the system in (2.132) is said to be *stable in the sense of Lyapunov* around x°, if, for a given $\epsilon > 0$, there exists some $\delta_1 > 0$ such that if $\|x(0)\|_2 \leq \delta_1$; then $\|x(t)\|_2 < \epsilon$ for all $t \geq 0$. Furthermore, the system in (2.132) is said to be *asymptotically stable* around x°, if it is stable in the sense of Lyapunov and there exists some $\delta_2 > 0$ such that if $\|x(0)\|_2 \leq \delta_2$, then $\lim_{t\to\infty} x(t) = \mathbf{0}_n$.

For the time-invariant case, i.e., $\frac{d}{dt}x(t) = Ax(t)$, stability in the sense of Lyapunov around x° is guaranteed if and only if (i) the real part of all the eigenvalues of A is smaller than or equal to zero, and (ii) for those eigenvalues whose real part is equal to zero, the associated algebraic and geometric multiplicities are identical. Furthermore, asymptotic stability around x° is guaranteed if and only if each eigenvalue of the matrix A has real part strictly smaller than zero. Finally, consider the non-homogeneous case $\frac{d}{dt}x(t) = Ax(t) + Bw(t)$, where $w(t)$, $t \geq 0$, is bounded, i.e., $\|w(t)\|_2 \leq K_w$, where K_w is some positive constant.

Then, the $x(t)$, $t \geq 0$, will be bounded, i.e., $\|x(t)\|_2 \leq K_x$, where K_x is some positive constant if the real part of all the eigenvalues of A is strictly negative.

2.4 Notes and References

The material on probability theory is standard and follows the developments in [11, 12, 13, 14]. The basic material on stochastic processes can be found in [15]. The material on the Wiener process follows the developments in [11]. Basic set-theoretic notions are covered in [16]. The material on sets in Euclidean space can be found in [10, 17, 18, 19, 20]. The derivation of the formula for the support function of an ellipsoid follows ideas from [10]. The formula for the volume of an ellipsoid given in (2.78) can be obtained from that given in [21], where an n-dimensional ellipsoid is referred to as a hyperellipsoid and the term "content" is used instead of "volume." The formula for the volume of a zonotope given in (2.90) follows from that given in [22], where a zonotope $\mathcal{Z} \subseteq \mathbb{R}^n$ is parametrized as follows:

$$\mathcal{Z} = \left\{ x \in \mathbb{R}^n : x = \sum_{i=1}^{s} \alpha_i' e_i', \ 0 \leq \alpha_i' \leq 1 \right\}. \tag{2.133}$$

Let $\mathcal{E}_i'$, $i = 1, 2, \ldots, N$, where $N = \binom{s}{n} = \frac{(s)!}{n!(s-n)!}$, denote all n-combinations of the set $\{e_1', e_2', \ldots, e_s'\}$. Then, the formula for the volume of $\mathcal{Z}$ given in [22] is as follows:

$$\mathbf{vol}(\mathcal{Z}) = \sum_{i=1}^{N} \left|\mathbf{det}(E_i')\right|, \tag{2.134}$$

where E_i' denotes the $(n \times n)$-dimensional matrix formed by horizontal concatenation of the elements in $\mathcal{E}_i'$. Note that $\mathcal{Z}$ in (2.133) can be equivalently written as

$$\mathcal{Z} = \left\{ x = \mathbb{R}^n : x = x_0 + \sum_{i=1}^{s} \alpha_i e_i, \ -1 \leq \alpha_i \leq 1 \right\}, \tag{2.135}$$

where $e_i = \frac{1}{2} e_i'$ and $\alpha_i = -1 + 2\alpha_i'$ for all $i = 1, 2, \ldots, s$, and $x_0 = \frac{1}{2} \sum_{i=1}^{s} e_i'$ (this is the parametrization given in (2.82)). As above, consider all n-combinations of the set $\{e_1, e_2, \ldots, e_s\}$, which we denote by $\mathcal{E}_i$, $i = 1, 2, \ldots, N$, and associated to each $\mathcal{E}_i$, define a matrix E_i formed by horizontal concatenation of the elements in $\mathcal{E}_i$. Clearly $E_i' = 2E_i$, $i = 1, 2, \ldots, s$; thus

$$\begin{aligned}\mathbf{vol}(\mathcal{Z}) &= \sum_{i=1}^{N} \left|\mathbf{det}(E_i')\right| \\ &= 2^n \sum_{i=1}^{N} \left|\mathbf{det}(E_i)\right|,\end{aligned} \tag{2.136}$$

which coincides with the formula of a zonotope given in (2.90).

3 Static Systems: Probabilistic Input Uncertainty

In this chapter, we provide tools to model the impact on system performance, as determined by certain system variables referred to as *states*, of randomness in some other system variables, referred to as *inputs*. We focus on the static case, i.e., systems in which the relation between inputs and states is described by some mapping. In this case, the system inputs are modeled as a collection of random variables with known joint distribution (or alternatively, known first few moments), and the objective is to characterize the joint distribution of the states (or their moments). The reader is referred to Section 2.1 for a review of probability theory concepts used throughout this chapter.

3.1 Introduction

Consider the DC circuit in Fig. 3.1, where the function of the DC voltage source on the left (with v_s denoting the voltage across its terminals and i_1 denoting the value of the current it delivers) is to provide power to the current source on the right (with i_l denoting the current value it delivers). Assume that the values that i_l and v_s take are not a priori known, but can be described by some random variables, I_l, V_s, with known joint distribution or known first few moments. Further, assume that for the circuit to perform as intended, the voltage v_2 across the current source terminals must be kept within some interval, (v_2^m, v_2^M), whereas the current i_1 delivered by the voltage source must be kept within some interval, (i_1^m, i_1^M).

A question that naturally arises in this setting is whether or not v_2 and i_1 will indeed be guaranteed to lie within the intervals (v_2^m, v_2^M) and (i_1^m, i_1^M), respectively, for all possible values that i_l and v_s take. If not, one could attempt to compute the probability that this is the case; this would provide a measure of the degree to which the system is delivering its intended function. To this end, we need to characterize the statistics of v_2 and i_1, a task that can be accomplished by using the available statistical information about i_l, v_s together with the governing circuit equations (which can be obtained by using Kirchhoff's laws):

$$\begin{bmatrix} v_2 \\ i_1 \end{bmatrix} = \begin{bmatrix} \frac{r_2}{r_1+r_2} & -\frac{r_1 r_2}{r_1+r_2} \\ \frac{1}{r_1+r_2} & \frac{r_2}{r_1+r_2} \end{bmatrix} \begin{bmatrix} v_s \\ i_l \end{bmatrix}. \tag{3.1}$$

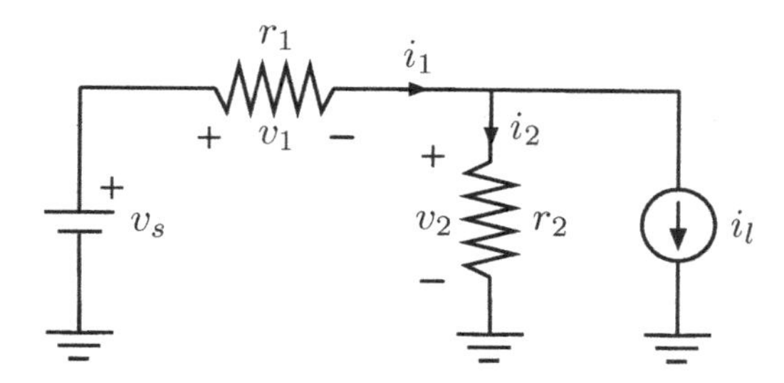

Figure 3.1 DC linear circuit.

The discussion above pertains to the case when the behavior of some system of interest can be described by a linear relation between a vector $x \in \mathbb{R}^n$, referred to as state vector (in the context of the example above, $x = [v_2, i_1]^\top$), the entries of which are important in determining whether or not the system is performing its function as intended, and some vector $w \in \mathbb{R}^m$, referred to as input vector (in the context of the example above, $w = [v_s, i_l]^\top$), the entries of which are uncertain, as follows:

$$x = Hw, \tag{3.2}$$

with $H \in \mathbb{R}^{n \times m}$ (in the context of the example above, the 2×2 matrix on the right-hand side of the expression in (3.1)). More generally, the relation between the input vector, w, and the state vector, x, could be nonlinear:

$$x = h(w), \tag{3.3}$$

where $h(\cdot)$ is some function mapping elements in $\mathbb{R}^m$ into elements in $\mathbb{R}^n$.

In this chapter, we consider systems of the form in (3.2) and (3.3), and analyze them separately in the following two scenarios:

S1. The values that w takes are described by a random vector, W, the entries of which are jointly distributed with known first and second moments.

S2. The values that w takes are described by a random vector, W, the entries of which are jointly distributed with known joint probability density function.

The objective in Scenario S1 is to characterize the first and second moments of X – the random vector describing the values that the state vector, x, takes, whereas in Scenario S2, the objective is to characterize the pdf of X. Once this information is obtained, we will utilize it to compute some performance metrics that measure the degree to which a system described by either (3.2) or (3.3) is delivering its intended function. Specifically, we will be able to compute the probability that X belongs to some set $\mathcal{R} \subseteq \mathbb{R}^n$ defined by the values that x must take so as to satisfy some performance requirements.

3.2 Moment Characterization

In this section, we consider Scenario S1, i.e., the first and second moments of the input vector are known, and the objective is to compute the first and second moments of the state vector. We first address the case in which the mapping between input and state is linear and provide an exact characterization of the state's first and second moments. Then we address the case in which the input-to-state mapping is nonlinear and resort to linearization techniques to obtain approximations of the actual value taken by the first and second moments of the state vector.

3.2.1 Linear Setting

Consider a linear static system of the form

$$X = HW, \tag{3.4}$$

where $H \in \mathbb{R}^{n \times m}$ is full rank, and W is an m-dimensional random vector with known mean,

$$m_W = \mathrm{E}[W] \in \mathbb{R}^m,$$

and known covariance matrix,

$$\Sigma_W = \mathrm{E}\Big[(W - m_W)(W - m_W)^\top\Big] \in \mathbb{R}^{m \times m}.$$

The goal here is to obtain the mean, the covariance matrix, and the correlation matrix of X, denoted by m_X, Σ_X, and S_X, respectively.

First, we can obtain the mean of X as follows:

$$\begin{aligned} m_X &= \mathrm{E}[X] \\ &= \mathrm{E}[HW] \\ &= H\mathrm{E}[W] \\ &= Hm_W. \end{aligned} \tag{3.5}$$

Then, we can characterize the covariance matrix of X as follows:

$$\begin{aligned} \Sigma_X &= \mathrm{E}\Big[(X - m_X)(X - m_X)^\top\Big] \\ &= \mathrm{E}\Big[H(W - m_W)(W - m_W)^\top H^\top\Big] \\ &= H\mathrm{E}\Big[(W - m_W)(W - m_W)^\top\Big]H^\top \\ &= H\Sigma_W H^\top. \end{aligned} \tag{3.6}$$

Finally, we can characterize the correlation matrix of X as follows. First note that

$$\Sigma_X = \mathrm{E}\Big[(X - m_X)(X - m_X)^\top\Big]$$
$$= \mathrm{E}[XX^\top] - m_X m_X^\top$$
$$= S_X - m_X m_X^\top; \tag{3.7}$$

then, by using (3.5) and (3.6), it follows that

$$S_X = \Sigma_X + m_X m_X^\top$$
$$= H\Sigma_W H^\top + H m_W m_W^\top H^\top$$
$$= H(\Sigma_W + m_W m_W^\top)H^\top. \tag{3.8}$$

Example 3.1 (Moment calculation for linear DC circuit) Consider the circuit in Fig. 3.1 and assume that $r_1 = r_2 = 1\ \Omega$; then, it follows from (3.1) that

$$\begin{bmatrix} v_2 \\ i_1 \end{bmatrix} = \begin{bmatrix} \frac{1}{2} & -\frac{1}{2} \\ \frac{1}{2} & \frac{1}{2} \end{bmatrix} \begin{bmatrix} v_s \\ i_l \end{bmatrix}. \tag{3.9}$$

Assume that the values the input vector, $w = [v_s, i_l]^\top$, takes can be described by a random vector, W, with known mean, m_W, and covariance matrix, Σ_W, given by

$$m_W = \begin{bmatrix} 10 \\ 5 \end{bmatrix},$$
$$\Sigma_W = \begin{bmatrix} 1 & 0 \\ 0 & 1 \end{bmatrix}. \tag{3.10}$$

Then, by using (3.5) and (3.6), we obtain that

$$m_X = \begin{bmatrix} \frac{1}{2} & -\frac{1}{2} \\ \frac{1}{2} & \frac{1}{2} \end{bmatrix} \begin{bmatrix} 10 \\ 5 \end{bmatrix}$$
$$= \begin{bmatrix} 2.5 \\ 7.5 \end{bmatrix},$$
$$\Sigma_X = \begin{bmatrix} \frac{1}{2} & -\frac{1}{2} \\ \frac{1}{2} & \frac{1}{2} \end{bmatrix} \begin{bmatrix} 1 & 0 \\ 0 & 1 \end{bmatrix} \begin{bmatrix} \frac{1}{2} & \frac{1}{2} \\ -\frac{1}{2} & \frac{1}{2} \end{bmatrix}$$
$$= \begin{bmatrix} \frac{1}{2} & 0 \\ 0 & \frac{1}{2} \end{bmatrix}. \tag{3.11}$$

Finally, by using (3.7), we obtain that

$$S_X = \begin{bmatrix} \frac{1}{2} & 0 \\ 0 & \frac{1}{2} \end{bmatrix} + \begin{bmatrix} 2.5 \\ 7.5 \end{bmatrix} \begin{bmatrix} 2.5 & 7.5 \end{bmatrix}$$
$$= \begin{bmatrix} 6.75 & 18.75 \\ 18.75 & 56.75 \end{bmatrix}. \tag{3.12}$$

Deterministic Inputs

The model in (3.4) can be augmented by including a deterministic input $u \in \mathbb{R}^l$ as follows:

$$X = HW + Lu, \tag{3.13}$$

with $L \in \mathbb{R}^{n \times l}$. The mean and covariance matrix of X can then be computed as follows:

$$\begin{aligned} m_X &= \mathrm{E}[HW + Lu] \\ &= Hm_W + Lu, \\ \Sigma_X &= \mathrm{E}\big[(X - m_X)(X - m_X)^\top\big] \\ &= \mathrm{E}\Big[\big(HW + Lu - (Hm_W + Lu)\big)\big(HW + Lu - (Hm_W + Lu)\big)^\top\Big] \\ &= \mathrm{E}\Big[\big(HW - Hm_W\big)\big(HW - Hm_W\big)^\top\Big] \\ &= H\Sigma_W H^\top, \end{aligned} \tag{3.14}$$

i.e., the deterministic input results in a shift of the mean of X by Lu, but leaves the covariance matrix unchanged. Because of the mean shift, the correlation matrix will also change as follows:

$$\begin{aligned} S_X &= \Sigma_X + m_X m_X^\top \\ &= H\Sigma_W H^\top + (Hm_W + Lu)(Hm_W + Lu)^\top. \end{aligned} \tag{3.15}$$

3.2.2 Nonlinear Setting

Consider a nonlinear static system of the form

$$X = h(W), \tag{3.16}$$

where $h\colon \mathbb{R}^m \to \mathbb{R}^n$ is continuously differentiable, and W is an m-dimensional random vector with known mean, $m_W \in \mathbb{R}^m$, and known covariance matrix, $\Sigma_W \in \mathbb{R}^{m \times m}$. As in the linear setting, the goal is to characterize the mean of X, m_X, the covariance matrix of X, Σ_X, and the correlation matrix of X, S_X. However, in general, it is not easy to obtain the exact expression for m_X, Σ_X, and S_X by direct manipulation of (3.16) together with m_W, Σ_W, and S_W. Thus, instead, we will settle for approximate expressions of m_X, S_X, and Σ_X obtained via linearization of $h(\cdot)$.

Let the pair w^*, x^* satisfy $x^* = h(w^*)$, and let Δx denote a small variation in x around x^* that arises from a small variation in w around w^*, which we similarly denote by Δw; then, we have that

$$x^* + \Delta x = h(w^* + \Delta w). \tag{3.17}$$

Now, by using the Taylor series expansion of $h(\cdot)$ around w^* and only keeping the first-order term, we obtain that

$$\Delta x \approx H(w^*)\Delta w, \tag{3.18}$$

where $H(w^*) = \frac{\partial h(w)}{\partial w}\Big|_{w=w^*} \in \mathbb{R}^{n\times m}$.

Define $\Delta W = W - m_W$ and $\Delta X = X - h(m_W)$; then, by using the linearization result in (3.18) with $w^* = m_W$, we have that

$$\Delta X \approx H(m_W)\Delta W, \tag{3.19}$$

where the mean, the covariance matrix, and the correlation matrix of ΔW can be obtained from the mean and covariance matrix of W as follows:

$$\begin{aligned}
m_{\Delta W} &= \mathrm{E}[\Delta W] \\
&= \mathrm{E}[W - m_W] \\
&= 0, \\
\Sigma_{\Delta W} &= \mathrm{E}\Big[(\Delta W - m_{\Delta W})(\Delta W - m_{\Delta W})^\top\Big] \\
&= \mathrm{E}\Big[(W - m_W)(W - m_W)^\top\Big] \\
&= \Sigma_W, \\
S_{\Delta W} &= \mathrm{E}\Big[\Delta W \Delta W^\top\Big] \\
&= \mathrm{E}\Big[(W - m_W)(W - m_W)^\top\Big] \\
&= \Sigma_W.
\end{aligned} \tag{3.20}$$

Define $\widetilde{\Delta X} := H(m_W)\Delta W$; then, we can use the results in (3.5), (3.6), and (3.8) to compute the mean, covariance matrix, and correlation matrix of X, yielding

$$\begin{aligned}
m_{\widetilde{\Delta X}} &= \mathbf{0}_n, \\
\Sigma_{\widetilde{\Delta X}} &= H(m_W)\Sigma_W H^\top(m_W), \\
S_{\widetilde{\Delta X}} &= H(m_W)\Sigma_W H^\top(m_W).
\end{aligned} \tag{3.21}$$

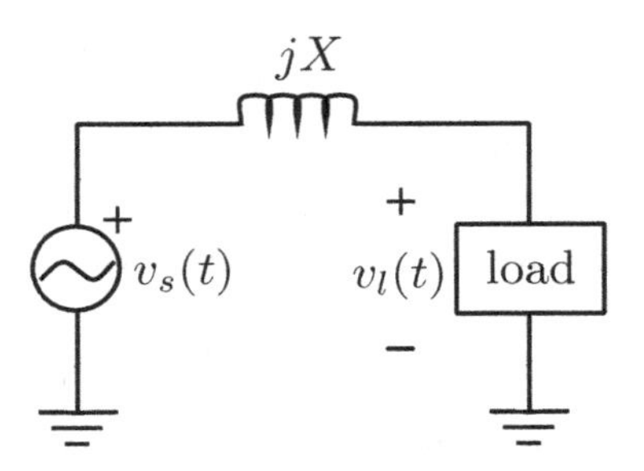

Figure 3.2 AC circuit with nonlinear load.

Then, since $X = h(m_W) + \Delta X \approx h(m_W) + \widetilde{\Delta X}$, we can approximate the mean, the covariance matrix, and the correlation matrix of X as follows:

$$\begin{aligned} m_X &= \mathrm{E}[h(m_W) + \Delta X] \\ &= h(m_W) + m_{\Delta X} \\ &\approx h(m_W) + m_{\widetilde{\Delta X}}, \\ &= h(m_W), \end{aligned} \tag{3.22}$$

$$\begin{aligned} \Sigma_X &= \mathrm{E}\Big[(X - m_X)(X - m_X)^\top\Big] \\ &= \mathrm{E}\Big[\Big(h(m_W) + \Delta X - \big(h(m_W) + m_{\Delta X}\big)\Big) \\ &\qquad \times \Big(h(m_W) + \Delta X - \big(h(m_W) + m_{\Delta X}\big)\Big)^\top\Big] \\ &= \mathrm{E}\Big[(\Delta X - m_{\Delta X})(\Delta X - m_{\Delta X})^\top\Big] \\ &= \underbrace{\mathrm{E}\big[\Delta X \Delta X^\top\big]}_{= S_{\Delta X}} - m_{\Delta X} m_{\Delta X}^\top \\ &\approx S_{\widetilde{\Delta X}} - m_{\widetilde{\Delta X}} m_{\widetilde{\Delta X}}^\top \\ &= H(m_W)\Sigma_W H^\top(m_W), \end{aligned} \tag{3.23}$$

$$\begin{aligned} S_X &= \Sigma_X + m_X m_X^\top \\ &\approx H(m_W)\Sigma_W H^\top(m_W) + h(m_W)h^\top(m_W), \end{aligned} \tag{3.24}$$

where the last line above follows from (3.22) and (3.23).

Example 3.2 (Moment calculation for AC circuit) Consider the AC circuit in Fig 3.2, where the function of the AC voltage source on the left (whose terminal voltage is $v_s(t) = \sqrt{2}\cos(\omega t)$) is to provide the active power and reactive power demanded by the load, which are denoted by p_l and q_l respectively, at some voltage $v_l(t) = \sqrt{2}\overline{v}_l \cos(\omega t + \theta_l)$, via an inductance with reactance value, X, being one p.u. (see Appendix B.5 for an overview of the per-unit normalization system used in power system analysis). The relation between all the variables defining the circuit behavior is as follows:

$$-p_l = \overline{v}_l \sin(\theta_l), \tag{3.25}$$

$$-q_l = \overline{v}_l^2 - \overline{v}_l \cos(\theta_l). \tag{3.26}$$

The value that p_l takes is random and can be described by a random variable, P_l, taking values in $[0, 1]$, with mean $\frac{\sqrt{3}}{2}$ p.u. and variance 0.01 p.u., whereas the value that q_l takes is controlled as follows:

$$q_l = -1 + \sqrt{1 - p_l^2}, \tag{3.27}$$

which in turn enforces the value of the magnitude of the voltage across the load, $\overline{v}_l$, to be unity for any value of p_l. Thus, we can write

$$-P_l = \sin(\Theta_l), \tag{3.28}$$

where Θ_l denotes the random variable describing the values that θ_l can take. We are now interested in finding the mean and the variance of Θ_l. To this end, since P_l can take values in $[0, 1]$, we can invert the mapping in (3.28) as follows:

$$\Theta_l = \arcsin(-P_l). \tag{3.29}$$

Then, by defining $\Delta P_l = P_l - p_l^*$, where $p_l^* = \mathrm{E}[P_l] = \frac{\sqrt{3}}{2}$, and $\Delta\Theta_l = \Theta_l - \theta_l^*$, where $\theta_l^* = \arcsin(-p_l^*) = -\frac{\pi}{3}$, and using (3.19), we obtain that

$$\begin{aligned} \Delta\Theta_l &\approx -\frac{1}{\sqrt{1-(-p_l^*)^2}}\Delta P_l \\ &= -\frac{1}{\cos(\theta_l^*)}\Delta P_l \\ &=: \widetilde{\Delta\Theta}_l. \end{aligned} \tag{3.30}$$

[Linearizing (3.28) around $P_l = p_l^*$ and $\Theta_l = \theta_l^*$ yields $\Delta P_l \approx -\cos(\theta_l^*)\Delta\Theta_l$. Then, by inverting this relation, we arrive at the same result in (3.30).] After plugging $\theta_l^* = -\frac{\pi}{3}$ in (3.30), we arrive at:

$$\Delta\Theta_l \approx \widetilde{\Delta\Theta}_l := -2\Delta P_l; \tag{3.31}$$

thus $\mathrm{E}[\widetilde{\Delta\Theta}_l] = 0$ and $\mathrm{E}[\widetilde{\Delta\Theta}_l^2] = 0.04$. Then, by using (3.22–3.24), we have that

$$\begin{aligned} \mathrm{E}[\Theta_l] &\approx \theta^* + \mathrm{E}[\widetilde{\Delta\Theta}_l] \\ &= -\frac{\pi}{3}, \\ \mathrm{E}[(\Theta_l - m_{\Theta_l})^2] &\approx \mathrm{E}[\widetilde{\Delta\Theta}_l^2] \\ &= 0.04, \\ \mathrm{E}[\Theta_l^2] &\approx (\theta_l^*)^2 + \mathrm{E}[\widetilde{\Delta\Theta}_l^2] \\ &= 1.1366. \end{aligned} \tag{3.32}$$

3.3 Distribution Characterization

In this section, we consider Scenario S2, i.e., the pdf of the input vector is known, and the objective is to compute the pdf of the state vector. We first address the case in which the mapping between input and state is linear and provide an exact characterization of the pdf of the state vector by using standard results pertaining to transformations of random vectors. Then, we address the case in which the input-to-state mapping is nonlinear and provide two methods; the first of these uses the same random vector transformation techniques used in the linear case and provides an exact characterization of the state pdf, whereas the second method resorts to linearization techniques and provides an approximation of the pdf of the state vector.

3.3.1 Linear Setting

Consider a linear static system of the form

$$X = HW, \tag{3.33}$$

where $H \in \mathbb{R}^{n\times m}$ is full rank, i.e., $\mathbf{rank}(H) = \min(n, m)$, and $W \in \mathbb{R}^m$ is random with known pdf, $f_W(w)$. Then, the goal is to obtain the pdf of X. We need to consider three different cases:

C1. H is a square matrix, i.e., $m = n$;
C2. H is a fat matrix, i.e., $m > n$; and
C3. H is a tall matrix, i.e., $m < n$.

Case C1. We will first consider the case when $m = n = 1$, i.e., X and W are random variables and $H = a$ is scalar such that $a \neq 0$. Now, let $F_W(w)$ and $F_X(x)$ denote the cdf of W and X respectively; then,

$$\begin{aligned} F_X(x) &= \Pr\left(X \leq x\right) = \Pr\left(aW \leq x\right) \\ &= \begin{cases} \Pr\left(W \leq \frac{x}{a}\right), & \text{if } a > 0, \\ \Pr\left(W \geq \frac{x}{a}\right), & \text{if } a < 0, \end{cases} \\ &= \begin{cases} F_W\left(\frac{x}{a}\right), & \text{if } a > 0, \\ 1 - F_W\left(\frac{x}{a}\right), & \text{if } a < 0. \end{cases} \end{aligned} \tag{3.34}$$

Then, we can obtain $f_X(x)$ by differentiating $F_X(x)$ as follows:

$$\begin{aligned} f_X(x) &= \begin{cases} \frac{1}{a} f_W\left(\frac{x}{a}\right), & \text{if } a > 0 \text{ and } f_W\left(\frac{x}{a}\right) \neq 0, \\ -\frac{1}{a} f_W\left(\frac{x}{a}\right), & \text{if } a < 0 \text{ and } f_W\left(\frac{x}{a}\right) \neq 0, \\ 0, & \text{otherwise,} \end{cases} \\ &= \begin{cases} \frac{1}{|a|} f_W\left(\frac{x}{a}\right), & \text{if } f_W\left(\frac{x}{a}\right) \neq 0, \\ 0, & \text{otherwise.} \end{cases} \end{aligned} \tag{3.35}$$

In the multi-dimensional case, i.e, $m = n > 1$, the expression for the pdf of X is as follows:

$$f_X(x) = \begin{cases} \left|\mathbf{det}(H^{-1})\right| f_W(H^{-1}x), & \text{if } x \in \{x\colon f_W(H^{-1}x) \neq 0\}, \\ 0, & \text{otherwise}, \end{cases} \tag{3.36}$$

where H^{-1} is guaranteed to exist because of the assumption on H being full rank. To establish this result, consider a closed and bounded set $\mathcal{W} \in \mathbb{R}^m$ and define

$$\mathcal{X} = \left\{x \in \mathbb{R}^m \colon x = Hw, \ w \in \mathcal{W}\right\}.$$

Then, since $X = HW$, we have that $\Pr(W \in \mathcal{W}) = \Pr(X \in \mathcal{X})$. Now, if $\mathcal{W}$ is contained in the support of $f_W(\cdot)$, on the one hand, we have that

$$\int_{\mathcal{W}} f_W(w)dw = \int_{\mathcal{X}} f_X(x)dx, \tag{3.37}$$

where $dw = dw_1 dw_2 \ldots dw_m$ and $dx = dx_1 dx_2 \ldots dx_m$. On the other hand, since the matrix H is invertible, we can write $w = H^{-1}x$, and by using the integration by substitution formula in (A.27), it follows that

$$\int_{\mathcal{W}} f_W(w)dw = \int_{\mathcal{X}} f_W(H^{-1}x)\left|\mathbf{det}(H^{-1})\right|dx. \tag{3.38}$$

Then, the right-hand side of (3.37) must be identical to the right-hand side of (3.38); therefore,

$$f_X(x) = \left|\mathbf{det}(H^{-1})\right| f_W(H^{-1}x), \tag{3.39}$$

as claimed in the first line of (3.36). If the intersection of $\mathcal{W}$ and the support of $f_W(\cdot)$ is empty, we have that $\Pr(W \in \mathcal{W}) = \Pr(X \in \mathcal{X}) = 0$; thus, since $\mathcal{X}$ is nonempty, it follows that $f_X(x) = 0$, as claimed in the second line of (3.36).

Example 3.3 (Distribution calculation for DC circuit when $m = n = 1$) Consider the DC circuit in Fig. 3.1, with $r_1 = r_2 = 1\ \Omega$ and $i_l = 0$ (i.e., the current source becomes an open circuit), and assume that v_s is random and can be described by a random variable, V_s, uniformly distributed in $[9.5, 10.5]$ V; thus,

$$f_{V_s}(v_s) = \begin{cases} 1, & \text{if } 9.5 \leq v_s \leq 10.5, \\ 0, & \text{otherwise}. \end{cases} \tag{3.40}$$

Then, by applying (3.35), we can obtain the pdf of I_2 – the random variable describing the value that the current through the resistor with resistance r_2 takes:

$$f_{I_2}(i_2) = \begin{cases} 2, & \text{if } 4.75 \leq i_2 \leq 5.25, \\ 0, & \text{otherwise}. \end{cases} \tag{3.41}$$

Similarly, one could obtain the pdf of any other variable in the circuit, e.g., the voltage across r_1, denoted by v_1.

Example 3.4 (Distribution calculation for DC circuit when $m = n = 2$) Consider again the DC circuit in Fig. 3.1 with the same values for r_1 and r_2 used in Example 3.1; thus, the relation between $x = [v_2, i_1]^\top$ and $w = [v_s, i_l]^\top$ is given by (3.9). Now assume that $W = [V_s,\ I_l]^\top$ is uniformly distributed in

$$\mathcal{W} = \{w = [v_s,\ i_l]^\top : 9 \le v_s \le 11,\ 4 \le i_l \le 6\};$$

then, the pdf of W is given by

$$f_W(w) = \begin{cases} \frac{1}{4}, & \text{if } w \in \mathcal{W}, \\ 0, & \text{otherwise.} \end{cases} \tag{3.42}$$

In order to obtain the pdf of $X = [V_2,\ I_1]^\top$, we first need to find its support, i.e., the set

$$\mathcal{X} = \{x = [v_2,\ i_1]^\top : f_X(x) \neq 0\}.$$

By inspection of (3.36), we see that

$$\begin{aligned} \mathcal{X} &= \{x = [v_2,\ i_1]^\top : f_X(x) \neq 0\} \\ &= \{x = [v_2,\ i_1]^\top : f_W(H^{-1}x) \neq 0\} \\ &= \{x = [v_2,\ i_1]^\top : x = Hw,\ w \in \mathcal{W}\}, \end{aligned} \tag{3.43}$$

where

$$H = \begin{bmatrix} \frac{1}{2} & -\frac{1}{2} \\ \frac{1}{2} & \frac{1}{2} \end{bmatrix}. \tag{3.44}$$

Since $\mathcal{W}$ is a polytope and $\mathcal{X}$ is obtained from $\mathcal{W}$ via the linear map H in (3.44), then, $\mathcal{X}$ will also be a polytope. Thus, in order to characterize $\mathcal{X}$, we just need to characterize its vertices, which can be obtained by mapping the vertices of $\mathcal{W}$ via H. One can see that since the set of vertices of $\mathcal{W}$ is

$$\mathcal{V}_{\mathcal{W}} = \{[9,4]^\top,\ [9,6]^\top,\ [11,4]^\top,\ [11,6]^\top\},$$

then, that of $\mathcal{X}$ is

$$\begin{aligned} \mathcal{V}_{\mathcal{X}} &= \{H[9,4]^\top,\ H[9,6]^\top,\ H[11,4]^\top,\ H[11,6]^\top\} \\ &= \{[2.5,6.5]^\top,\ [1.5,7.5]^\top,\ [3.5,7.5]^\top,\ [2.5,8.5]^\top\}; \end{aligned}$$

Fig. 3.3 depicts both sets $\mathcal{W}$ and $\mathcal{X}$. Finally, one can check that $\mathbf{det}(H) = 1/2$; then, by using (3.36), it follows that

$$f_X(x) = \begin{cases} \frac{1}{2}, & \text{if } x \in \mathcal{X}, \\ 0, & \text{otherwise.} \end{cases} \tag{3.45}$$

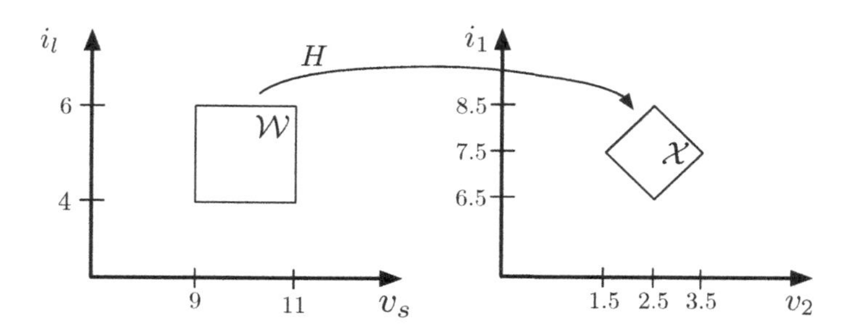

Figure 3.3 Support of $f_W(w)$ and $f_X(x)$ in Example 3.4.

Case C2. In this case, $m > n$, i.e., the dimension of the state vector, X, is smaller than the dimension of the input vector, W. The way to deal with this case is to form an "augmented" state vector $\widetilde{X} = [X^\top, X_{n+1}, X_{n+2}, \ldots, X_m]^\top \in \mathbb{R}^m$, where $X_{n+1}, X_{n+2}, \ldots, X_m$ are obtained by linearly combining entries of W; thus, $\widetilde{X}$ now has the same dimension as W. In essence, this process results in the addition of rows to the matrix H so as to make it square; this process needs to be done in such a way that the resulting matrix, $\widetilde{H}$, is invertible, which can always be accomplished because H is full rank. In summary, after the "augmentation" process described above is completed, we end up with a linear static system of the form

$$\widetilde{X} = \widetilde{H}W, \tag{3.46}$$

where $\widetilde{H} \in \mathbb{R}^{m\times m}$, and $W \in \mathbb{R}^m$ is random with known pdf $f_W(w)$.

Then, in order to obtain the pdf of X, $f_X(x)$, we first obtain the pdf of $\widetilde{X}$, $f_{\widetilde{X}}(\widetilde{x})$, by using the same technique as in Case C1; thus, we have that

$$f_{\widetilde{X}}(\widetilde{x}) = \begin{cases} |\mathbf{det}(\widetilde{H}^{-1})| f_W(\widetilde{H}^{-1}\widetilde{x}), & \widetilde{x} \in \{\widetilde{x}\colon f_W(\widetilde{H}^{-1}\widetilde{x}) \neq 0\}, \\ 0, & \text{otherwise.} \end{cases} \tag{3.47}$$

Finally, we need to integrate $f_{\widetilde{X}}(\widetilde{x})$ over the values that $x_{n+1}, x_{n+2}, \ldots, x_m$ take to obtain $f_X(x)$:

$$f_X(x) = \int_{-\infty}^{\infty}\int_{-\infty}^{\infty}\cdots\int_{-\infty}^{\infty} f_{\widetilde{X}}(\widetilde{x})dx_{n+1}dx_{n+2}\ldots dx_m. \tag{3.48}$$

Example 3.5 (Distribution calculation for DC circuit when $m = 2$, $n = 1$) Consider again the DC circuit in Fig. 3.1, and assume that $r_1 = 1\ \Omega$ and $r_2 = \infty\ \Omega$. Assume that the distribution of $W = [V_s,\ I_l]^\top$ is the same as that of Example 3.4, i.e., uniform with pdf given in (3.42). Here we are interested in finding the pdf of V_2, the random variable describing the values that the voltage across the current source, v_2, can take. To this end, we use Kirchhoff's voltage law to obtain the following relation between V_2, and V_s and I_l:

$$V_2 = \begin{bmatrix} 1 & -1 \end{bmatrix} \begin{bmatrix} V_s \\ I_l \end{bmatrix}. \tag{3.49}$$

In this case, the number of states is just one, whereas the number of inputs is two; thus, following the ideas described above, we need to define an augmented state vector $\widetilde{X}$ by introducing an additional state variable X_2 obtained as a linear combination of V_s and I_l. One such simple choice is $X_2 = I_l$. Then, by following the notation in (3.46), we have that

$$\widetilde{X} = \widetilde{H}W, \tag{3.50}$$

where $W = [V_s,\ I_l]^\top$, $\widetilde{X} = [V_2,\ X_2]^\top$, and

$$\widetilde{H} = \begin{bmatrix} 1 & -1 \\ 0 & 1 \end{bmatrix}. \tag{3.51}$$

Now, since $\mathbf{det}(\widetilde{H}) = 1$, it follows from (3.47) that

$$f_{\widetilde{X}}(\widetilde{x}) = \begin{cases} \frac{1}{4}, & \text{if } \widetilde{x} \in \mathcal{X}, \\ 0, & \text{otherwise}, \end{cases} \tag{3.52}$$

where $\widetilde{\mathcal{X}} = \{\widetilde{x} = [v_2,\ x_2]^\top : \widetilde{x} = \widetilde{H}w,\ w \in \mathcal{W}\}$, with

$$\mathcal{W} = \big\{w = [v_s,\ i_l]^\top : 9 \le v_s \le 11,\ 4 \le i_l \le 6\big\};$$

Fig. 3.4 (left) depicts the set $\widetilde{\mathcal{X}}$.

Finally, in order to obtain $f_{V_2}(v_2)$, we need to integrate $f_{\widetilde{X}}(\widetilde{x})$ over the values that x_2 takes. By inspection of the set $\widetilde{X}$ in Fig 3.4 (left), and using (3.48), we obtain the following:

$$\begin{aligned} f_{V_2}(v_2) &= \begin{cases} \displaystyle\int_{-v_2+9}^{6} \frac{1}{4} dx_2, & \text{if } 3 \le v_2 < 5, \\ \displaystyle\int_{4}^{-v_2+11} \frac{1}{4} dx_2, & \text{if } 5 \le v_2 < 7, \\ 0, & \text{otherwise}, \end{cases} \\ &= \begin{cases} \frac{1}{4}(v_2 - 3), & \text{if } 3 \le v_2 < 5, \\ \frac{1}{4}(-v_2 + 7), & \text{if } 5 \le v_2 < 7, \\ 0, & \text{otherwise}; \end{cases} \end{aligned} \tag{3.53}$$

the graph of which is displayed in Fig. 3.4 (right).

Case C3. In this case, $m < n$, i.e., the dimension of the state vector, X, is greater than the dimension of the input vector W and therefore, H has more

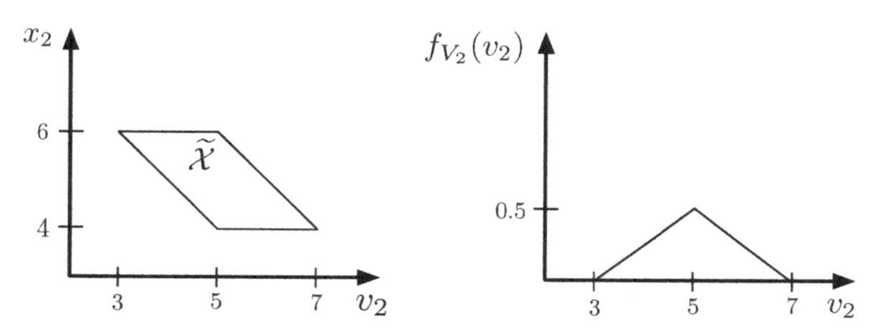

Figure 3.4 Support of $f_{\widetilde{X}}(\widetilde{x})$ and pdf of V_2 in Example 3.5.

rows than columns. Since H is full rank, there are m rows of H that are linearly independent; without loss of generality, assume that those are the first m ones. Then, we can rewrite the expression in (3.33) as

$$\begin{bmatrix} X_1 \\ X_2 \end{bmatrix} = \begin{bmatrix} H_1 \\ H_2 \end{bmatrix} W, \tag{3.54}$$

where $X_1 \in \mathbb{R}^m$, $X_2 \in \mathbb{R}^{n-m}$, $H_1 \in \mathbb{R}^{m \times m}$, and $H_2 \in \mathbb{R}^{(n-m)\times m}$. Now, since the rows of H_1 are linearly independent, we have that $W = H_1^{-1}X_1$, and therefore, we can express X_2 as a function of X_1 as follows:

$$X_2 = H_2 H_1^{-1} X_1. \tag{3.55}$$

Thus, a complete probabilistic description of X is obtained through the pdf of X_1, which can be computed by using the same results in Case C1, since H_1 is a full rank square matrix, i.e.,

$$f_{X_1}(x_1) = \begin{cases} |\mathbf{det}(H_1^{-1})| f_W(H_1^{-1}x_1), & \text{if } x_1 \in \{x_1\colon f_W(H_1^{-1}x_1) \neq 0\}, \\ 0, & \text{otherwise,} \end{cases} \tag{3.56}$$

together with the relation between X_1 and X_2 in (3.55).

Alternatively, it is possible to obtain an expression for the pdf of X by using (3.55) and Dirac delta functions as follows:

$$f_X(x) = f_{X_1}(x_1)\delta(z_1)\delta(z_2)\ldots\delta(z_{n-m}), \tag{3.57}$$

where $\delta(\cdot)$ is the Dirac delta function, and where z_i is the ith entry of the vector $z := x_2 - H_2H_1^{-1}x_1$. However, we prefer to use the characterization given by (3.55) and (3.56) as it is more practical for performing computations.

Example 3.6 (Distribution calculation for DC circuit when $m = 1$, $n = 2$) Consider again the circuit in Fig. 3.1 and assume that $i_l = 0$ and $r_1 = r_2 = 1\ \Omega$. Also, assume that V_s is uniformly distributed over the interval $[9.5, 10.5]$ V. Here, we are interested in finding the joint pdf of the random variables V_2 and I_1, respectively describing the values that the voltage across r_2 can take, and the current flowing through this resistor. In this case, we have that $X = [V_2, I_1]^\top$ relates to V_s as follows:

$$\begin{bmatrix} V_2 \\ I_1 \end{bmatrix} = \begin{bmatrix} \frac{1}{2} \\ \frac{1}{2} \end{bmatrix} V_s. \tag{3.58}$$

Then, a complete probabilistic description of $X = [V_2,\ I_1]^\top$ is given by the pdf of V_2 and the relation between I_1 and V_2, i.e., $I_1 = V_2$. Thus, since $V_2 = \frac{1}{2}V_s$, and V_s is uniformly distributed over $[9.5, 10.5]$ V, it follows that V_2 is also uniformly distributed over the interval $[4.75, 5.25]$ V; therefore,

$$f_{V_2}(v_2) = \begin{cases} 2, & \text{if } 4.75 \leq v_2 \leq 5.25, \\ 0, & \text{otherwise.} \end{cases} \tag{3.59}$$

Deterministic Inputs

Augment the system in (3.33) with a deterministic input $u \in \mathbb{R}^l$ as follows:

$$X = HW + Lu, \tag{3.60}$$

where $L \in \mathbb{R}^{n \times l}$ is known, H is as in (3.33) and W is an m-dimensional random vector with known pdf, $f_W(w)$. Define $Z = X - Lu$, then we have that

$$Z = HW, \tag{3.61}$$

which is of the form in (3.33); thus, we can apply the techniques discussed in Cases C1–C3 to obtain the pdf of Z, $f_Z(z)$. Then, since $X = Z + Lu$, it follows that

$$f_X(x) = f_Z(x - Lu). \tag{3.62}$$

3.3.2 Nonlinear Setting

We first provide a method to compute the exact pdf of X. This method is just an extension of the ideas in Section 3.3.1; however, it requires obtaining an analytical expression for the Jacobian of the input-to-state inverse mapping. Thus, since obtaining such an analytical expression is in general a hard problem, we also provide an alternative method to approximate the pdf of X. This alternative method relies on applying the linear setting techniques discussed in Section 3.3.1 to the linearized model obtained in Section 3.2.2.

Exact Distribution Characterization

Consider a system of the form

$$X = h(W), \tag{3.63}$$

where $h\colon \mathbb{R}^m \to \mathbb{R}^n$ is continuously differentiable, $W \in \mathbb{R}^m$ is random with known pdf $f_W(w)$, and $X \in \mathbb{R}^n$. Assume the Jacobian of $h(\cdot)$ is full rank for all the values of w in the support of $f_W(\cdot)$; then, the goal is to obtain the pdf of X, $f_X(x)$. Similar to the linear case, we need to consider three cases:

D1. the dimension of X is the same as the dimension of W, i.e., $m = n$;
D2. the dimension of X is smaller than the dimension of W, i.e., $m > n$; and
D3. the dimension of X is greater than the dimension of W, i.e., $m < n$.

Case D1. The pdf of $X \in \mathbb{R}^m$ is given by

$$f_X(x) = \begin{cases} \left|\mathbf{det}\left(J_{\widetilde{h}^{-1}}(\widetilde{x})\right)\right| f_W\left(h^{-1}(x)\right), & \text{if } x \in \{x\colon f_W\left(h^{-1}(x)\right) \neq 0\}, \\ 0, & \text{otherwise,} \end{cases} \tag{3.64}$$

where $J_{h^{-1}}(x)$ denotes the Jacobian of $h^{-1}(\cdot)$ evaluated at x. To establish this result, we follow a procedure similar to that used in the linear case. First, consider a closed and bounded set $\mathcal{W} \in \mathbb{R}^m$ and define

$$\mathcal{X} = \left\{x \in \mathbb{R}^m\colon x = h(w),\ w \in \mathcal{W}\right\}.$$

Then, since $X = h(W)$, we have that $\Pr(W \in \mathcal{W}) = \Pr(X \in \mathcal{X})$. Now, if $\mathcal{W}$ is contained in the support of $f_W(\cdot)$, on the one hand, we have that

$$\int_{\mathcal{W}} f_W(w)dw = \int_{\mathcal{X}} f_X(x)dx, \tag{3.65}$$

where $dw = dw_1 dw_2 \ldots dw_m$ and $dx = dx_1 dx_2 \ldots dx_m$. Now, recall the assumption on the function $h(\cdot)$ being continuously differentiable with its Jacobian being full rank for all the values of w in the support of $f_W(\cdot)$. Then, by the inverse function theorem (see Appendix A.3), there exists a continuously differentiable function $h^{-1}(\cdot)$ so that $w = h^{-1}(x)$ for all x such that $f_W\left(h^{-1}(x)\right) \neq 0$. Then, by using the integration by substitution formula in (A.27), it follows that

$$\int_{\mathcal{W}} f_W(w)dx = \int_{\mathcal{X}} f_W\left(h^{-1}(x)\right)\left|\mathbf{det}\left(J_{h^{-1}}(x)\right)\right|dx. \tag{3.66}$$

Thus, since the right-hand side of (3.65) must be identical to the right-hand side of (3.66), it follows that

$$f_X(x) = \left|\mathbf{det}\left(J_{h^{-1}}(x)\right)\right| f_W\left(h^{-1}(x)\right), \tag{3.67}$$

as claimed in the first line of (3.64). Finally, if the intersection of $\mathcal{W}$ and the support of $f_W(\cdot)$ is empty, it follows that $\Pr(W \in \mathcal{W}) = \Pr(X \in \mathcal{X}) = 0$, but since the set $\mathcal{X}$ is nonempty, it follows that $f_X(x) = 0$, as claimed in the second line of (3.64).

Example 3.7 (Exact distribution calculation for AC circuit) Consider again the AC circuit in Fig 3.2, with the same setting as in Example 3.2; then the relation between Θ_l and P_l is as given in (3.29), i.e., $\Theta_l = \arcsin(-P_l)$. Assume that P_l is uniformly distributed over the interval $\mathcal{P}_l = \left[\frac{\sqrt{3}}{2} - 0.1, \frac{\sqrt{3}}{2} + 0.1\right]$; thus

$$f_{P_l}(p_l) = \begin{cases} 5, & \text{if } p_l \in \mathcal{P}_l = \left[\frac{\sqrt{3}}{2} - 0.1, \frac{\sqrt{3}}{2} + 0.1\right], \\ 0, & \text{otherwise.} \end{cases} \tag{3.68}$$

In this case, $w = p_l$, and $h(p_l) = \arcsin(-p_l)$; thus, clearly we have that

$$\frac{dh(p_l)}{dp_l} = -\frac{1}{\sqrt{1-p_l^2}} \neq 0, \ p_l \in \mathcal{P}_l = \left[\frac{\sqrt{3}}{2} - 0.1, \frac{\sqrt{3}}{2} + 0.1\right]. \tag{3.69}$$

Thus there exists $h^{-1} \colon \mathbb{R} \to \mathcal{P}_l$ such that $p_l = h^{-1}(\theta_l)$, and it is clear that $h^{-1}(\theta_l) = -\sin(\theta_l)$. Then, if we use (3.64) together with (3.68), we obtain that

$$f_{\Theta_l}(\theta_l) = \begin{cases} 5\cos(\theta_l), & \text{if } \theta_l \in \left[\arcsin\left(-\frac{\sqrt{3}}{2} - 0.1\right), \arcsin\left(-\frac{\sqrt{3}}{2} + 0.1\right)\right], \\ 0, & \text{otherwise.} \end{cases} \tag{3.70}$$

Case D2. The procedure to handle this case is very similar to that for the linear case. The first step is to construct $\widetilde{x} = [x^\top, x_{n+1}, x_{n+2}, \ldots, x_m]^\top \in \mathbb{R}^m$, where $x_{n+1}, x_{n+2}, \ldots, x_m$ are obtained as functions of the entries of w, i.e., $x_i = h_i(w), i = n+1, n+2, \ldots, m$, for some $h_i \colon \mathbb{R}^m \to \mathbb{R}$ (e.g., the x_i's can be chosen to be linear combinations of the w_i's). Then, similar to the linear case, we can stack the entries of $h(\cdot)$ with $h_{n+1}(\cdot), h_{n+2}(\cdot), \ldots, h_m(\cdot)$ to form a function $\widetilde{h} \colon \mathbb{R}^m \to \mathbb{R}^m$. Here it is important to note that the choice of the $h_i(\cdot)$'s is not arbitrary but they need to be chosen so that the Jacobian of $\widetilde{h}(\cdot)$ is full rank for all the values of w in the support of $f_W(w)$. Then we can write

$$\widetilde{X} = \widetilde{h}(W), \tag{3.71}$$

and by using the result in (3.64), we obtain that

$$f_{\widetilde{X}}(\widetilde{x}) = \begin{cases} \left|\mathbf{det}\left(J_{\widetilde{h}^{-1}}(\widetilde{x})\right)\right| f_W\left(\widetilde{h}^{-1}(\widetilde{x})\right), & \text{if } \widetilde{x} \in \left\{\widetilde{x} \colon f_W\left(\widetilde{h}^{-1}(\widetilde{x})\right) \neq 0\right\}, \\ 0, & \text{otherwise,} \end{cases} \tag{3.72}$$

where $\widetilde{h}^{-1}(\cdot)$ is continuously differentiable for all $\widetilde{x}$ such that $f_W\left(\widetilde{h}^{-1}(\widetilde{x})\right) \neq 0$, and $J_{\widetilde{h}^{-1}}(\widetilde{x})$ denotes the Jacobian of $\widetilde{h}^{-1}(\cdot)$ evaluated at $\widetilde{x}$. Finally, as in the linear case, in order to obtain $f_X(x)$, we need to integrate $f_{\widetilde{X}}(\widetilde{x})$ over $x_{n+1}, x_{n+2}, \ldots, x_m$.

Case D3. Again, the procedure to handle this case is somewhat similar to that used for the linear case. Since the Jacobian of $h(\cdot)$ is full rank for all the values of w in the support of $f_W(\cdot)$, m of its rows are linearly independent. Without loss of generality, assume that those correspond to the first m entries of $h(\cdot)$; then, we can rewrite (3.63) as follows:

$$\begin{bmatrix} X_1 \\ X_2 \end{bmatrix} = \begin{bmatrix} h_1(W) \\ h_2(W) \end{bmatrix}, \tag{3.73}$$

where $h_1 \colon \mathbb{R}^m \to \mathbb{R}^m$, $h_2 \colon \mathbb{R}^m \to \mathbb{R}^{n-m}$, $W \in \mathbb{R}^m$, $X_1 \in \mathbb{R}^m$, and $X_2 \in \mathbb{R}^{n-m}$. Now, since the function $h(\cdot)$ is continuously differentiable, so is the function $h_1(\cdot)$. Also, since the entries of $h_1(\cdot)$ correspond to those rows of the Jacobian of $h(\cdot)$ that are linearly independent, the Jacobian of $h_1(\cdot)$, is also full rank (and therefore invertible) for the values of w in the support of $f_W(\cdot)$. Thus, by the inverse function theorem, there exists a continuously differentiable function $h_1^{-1}(\cdot)$ so that $w = h_1^{-1}(x_1)$ for all x_1 such that $f_W\big(h_1^{-1}(x_1)\big) \neq 0$, and thus X_2 can be expressed as a function of X_1 as follows:

$$X_2 = h_2\big(h_1^{-1}(X_1)\big). \tag{3.74}$$

Therefore, as in the linear case, a complete probabilistic description of X can be obtained through the pdf of X_1 (which can be computed using the results in Case D1):

$$f_{X_1}(x_1) = \begin{cases} \left|\mathbf{det}\big(J_{h_1^{-1}}(x_1)\big)\right| f_W\big(h_1^{-1}(x_1)\big), & \text{if } x_1 \in \{x_1 \colon f_W\big(h_1^{-1}(x_1)\big) \neq 0\}, \\ 0, & \text{otherwise,} \end{cases} \tag{3.75}$$

where $J_{h_1^{-1}}(x_1)$ denotes the Jacobian of $h_1^{-1}(\cdot)$ at x_1, and the mapping between X_1 and X_2 in (3.74).

Approximate Distribution Characterization

In the developments above, we provided an analytical expression to derive the pdf of X given the pdf of W; however, in practice, obtaining these closed-form solutions might be difficult as they involve the computation of inverses of nonlinear functions. In the remainder of this section, we will provide a method to obtain approximate expressions for the pdf of X. As in the method provided in Section 3.2 to obtain approximate expressions for the moments of X, here we also rely on the Taylor series expansion of $h(\cdot)$ around the mean of W, m_W.

Let the pair w^*, x^* satisfy $x^* = h(w^*)$, and let Δx denote a small variation in x around x^* that arises from a small variation in w around w^*, which we similarly denote by Δw. Then, by using the Taylor series expansion of $h(\cdot)$ around w^* and only keeping the first-order term, we obtain that

$$\Delta x \approx H(w^*)\Delta w, \tag{3.76}$$

where $H(w^*) = \frac{\partial h(w)}{\partial w}\Big|_{w=w^*} \in \mathbb{R}^{n \times m}$.

Now, let $w^* = m_W$ and define $\Delta W = W - w^*$ and $\Delta X = X - x^*$; then, by using (3.76), we can write

$$\Delta X \approx H(m_W)\Delta W, \tag{3.77}$$

where $\Delta W \in \mathbb{R}^m$ is random with pdf $f_W(\Delta w + m_W)$. Now, assume that the Jacobian of $h(\cdot)$ is full rank at $w = m_W$. Then, obtaining a complete probabilistic description of $\widetilde{\Delta X} := H(m_W)\Delta W$ reduces to Cases C1–C3.

For Cases C1 and C2, i.e., $m \geq n$, let $f_{\widetilde{\Delta X}}(\Delta\widetilde{x})$ denote the pdf of $\widetilde{\Delta X}$. Then, since $X = x^* + \Delta X$, where $x^* = h(m_W)$, we have that

$$\begin{aligned} f_X(x) &= f_{\Delta X}(x - h(m_W)) \\ &\approx f_{\widetilde{\Delta X}}(x - h(m_W)). \end{aligned} \tag{3.78}$$

For Case C3, i.e., $m < n$, without loss of generality, assume that the first rows of $\widetilde{\Delta X} := H(m_W)\Delta W$ are linearly independent, and write:

$$\begin{bmatrix} \widetilde{\Delta X}_1 \\ \widetilde{\Delta X}_2 \end{bmatrix} = \begin{bmatrix} H_1(m_W) \\ H_2(m_W) \end{bmatrix} \Delta W, \tag{3.79}$$

where $\widetilde{\Delta X}_1 \in \mathbb{R}^m$, $\widetilde{\Delta X}_2 \in \mathbb{R}^{n-m}$, $H_1(m_W) \in \mathbb{R}^{m\times m}$, and $H_2(m_W) \in \mathbb{R}^{(n-m)\times m}$. Then, following the same development in (3.54–3.56), a complete probabilistic characterization of $\widetilde{\Delta X}$ is given by the pdf of $\widetilde{\Delta X}_1$, $f_{\widetilde{\Delta X}_1}(\Delta\widetilde{x}_1)$, which can be obtained by using (3.56), and the mapping between $\widetilde{\Delta X}_1$ and $\widetilde{\Delta X}_2$:

$$\widetilde{\Delta X}_2 = H_2(m_W)H_1^{-1}(m_W)\widetilde{\Delta X}_1. \tag{3.80}$$

Finally, since $X_1 = h_1(m_W) + \Delta X_1$, and $X_2 = h_2(m_W) + \Delta X_2$, where $h_1(\cdot)$ and $h_2(\cdot)$ are functions obtained by removing components of $h(\cdot)$, so that their Jacobians evaluated at $w = m_W$ are $H_1(m_W)$ and $H_2(m_W)$, respectively, we can approximate the complete probabilistic description of X via an approximation to the pdf of X_1 given by:

$$\begin{aligned} f_{X_1}(x_1) &= f_{\Delta X_1}(x_1 - h_1(m_W)) \\ &\approx f_{\widetilde{\Delta X}_1}(x_1 - h_1(m_W)), \end{aligned} \tag{3.81}$$

and an approximate mapping between X_1 and X_2 given by:

$$X_2 \approx H_2(m_W)H_1^{-1}(m_W)X_1 - H_2(m_W)H_1^{-1}(m_W)h_1(m_W) + h_2(m_W). \tag{3.82}$$

Example 3.8 (Approximate distribution calculation for AC circuit) Consider again the AC circuit in Fig 3.2, with the same setting as in Example 3.7; thus, we have that $\Theta_l = \arcsin(-P_l)$, where P_l is uniformly distributed over the interval $\mathcal{P}_l = \left[\frac{\sqrt{3}}{2} - 0.1, \frac{\sqrt{3}}{2} + 0.1\right]$. It is easy to see that $m_{P_l} = \frac{\sqrt{3}}{2}$; thus, we have that $\Delta P_l = P_l - m_{P_l}$ is uniformly distributed over the interval $[-0.1, 0.1]$. Now, define

$$\Delta\Theta_l = \Theta_l - \arcsin(-\sqrt{3}/2); \tag{3.83}$$

then, it follows that

$$\Delta\Theta_l \approx -\frac{1}{\sqrt{1-m_{P_l}^2}}\Delta P_l = \\ = -2\Delta P_l; \tag{3.84}$$

thus, we have that $\widetilde{\Delta\Theta_l} := -2\Delta P_l$ is uniformly distributed over the interval $[-0.2, 0.2]$. Finally, by using the result in (3.78), we obtain that

$$f_{\Theta_l}(\theta_l) \approx f_{\widetilde{\Delta\Theta_l}}(\theta_l + \pi/3) \\ = \begin{cases} 2.5, & \text{if } \theta_l \in \big[-\pi/3 - 0.2, -\pi/3 + 0.2\big] \approx [-1.25, -0.85], \\ 0, & \text{otherwise.} \end{cases} \tag{3.85}$$

Let us compare (3.85) with the exact pdf of Θ_l obtained in Example 3.7:

$$f_{\Theta_l}(\theta_l) = \begin{cases} 5\cos(\theta_l), & \text{if } \theta_l \in \left[\arcsin\Big(-\frac{\sqrt{3}}{2} - 0.1\Big), \arcsin\Big(-\frac{\sqrt{3}}{2} + 0.1\Big)\right], \\ 0, & \text{otherwise,} \end{cases} \tag{3.86}$$

where $\left[\arcsin\big(-\frac{\sqrt{3}}{2} - 0.1\big), \arcsin\big(-\frac{\sqrt{3}}{2} + 0.1\big)\right] \approx [-1.31, -0.87]$. Obviously, the match between the expressions in (3.85) and (3.86) is not perfect. One can easily check that for smaller values of m_{P_l} the matching improves; this intuitively makes sense, as those values would correspond to angles that are closer to zero, where the approximation of the sine function by its argument is closer. For example, for $m_{P_l} = 1/2$, we have that $\widetilde{\Delta\Theta_l} = -\frac{2}{\sqrt{3}}\Delta P_l$, from where it follows that

$$f_{\Theta_l}(\theta_l) \approx f_{\widetilde{\Delta\Theta_l}}\Big(\theta_l + \frac{\pi}{6}\Big) \\ = \begin{cases} 2.5\sqrt{3}, & \text{if } \theta_l \in \Big[-\frac{\pi}{6} - \frac{0.2}{\sqrt{3}}, -\frac{\pi}{6} + \frac{0.2}{\sqrt{3}}\Big] \approx [-0.64, -0.41], \\ 0, & \text{otherwise,} \end{cases} \tag{3.87}$$

which matches closely the expression of the exact pdf, which is given by:

$$f_{\Theta_l}(\theta_l) = \begin{cases} 5\cos(\theta_l), & \text{if } \theta_l \in \left[\arcsin\Big(-\frac{1}{2} - 0.1\Big), \arcsin\Big(-\frac{1}{2} + 0.1\Big)\right], \\ 0, & \text{otherwise,} \end{cases} \tag{3.88}$$

where $\left[\arcsin\Big(-\frac{1}{2} - 0.1\Big), \arcsin\Big(-\frac{1}{2} + 0.1\Big)\right] \approx [-0.64, -0.41]$. Thus, it is important to understand the limitations of the approximate method.

3.4 Performance Characterization

Next, we provide a way to quantify the performance of linear and nonlinear static systems. Consider a system of the form

$$X = h(W), \tag{3.89}$$

where $X \in \mathbb{R}^n$, and $W \in \mathbb{R}^m$ is random with either (i) known mean, m_W, and covariance matrix, Σ_W, or (ii) known pdf $f_W(w)$. As in earlier developments, assume that the Jacobian of $h\colon \mathbb{R}^m \to \mathbb{R}^n$ is full rank for all the values of w in the support of $f_W(w)$. [Note that this formulation includes the linear setting as a particular case by making $h(W) = HW$, with $H \in \mathbb{R}^{n\times m}$ full rank.]

Let $\mathcal{R}_i = (x_i^m, x_i^M)$, where $x_i^m, x_i^M \in \mathbb{R}$ respectively denote the minimum and maximum values that the ith entry of x can take so as to guarantee that the system is performing its function as intended. Let $\mathcal{R} = \mathcal{R}_1 \times \mathcal{R}_2 \times \cdots \times \mathcal{R}_n$ and define the indicator of the event $\{X = h(W) \in \mathcal{R}\}$:

$$\varphi(W) = \begin{cases} 1, & \text{if } X = h(W) \in \mathcal{R}, \\ 0, & \text{if } X = h(W) \notin \mathcal{R}. \end{cases} \tag{3.90}$$

Then, we can measure the degree to which the system in (3.89) will perform its intended function as the expected value of $\varphi(W)$:

$$\begin{aligned} \rho &= \mathrm{E}[\varphi(W)] \\ &= \Pr\big(\varphi(W) = 1\big) \\ &= \Pr\big(X \in \mathcal{R}\big). \end{aligned} \tag{3.91}$$

Next, we show how to use the techniques presented in earlier sections to compute the exact value of ρ, or an approximation.

3.4.1 Known Input Moments

In this case, since m_W and Σ_W are known, we can use the techniques in Section 3.2 to compute the exact values of m_X, Σ_X, and S_X, or an approximation thereof. First, by using the union bound, we have that

$$\begin{aligned} \Pr\big(X \notin \mathcal{R}\big) &= \Pr\big(X_1 \notin \mathcal{R}_1 \cup X_2 \notin \mathcal{R}_2 \cup \cdots \cup X_n \notin \mathcal{R}_n\big) \\ &\leq \sum_{i=1}^{n} \Pr\big(X_i \notin \mathcal{R}_i\big). \end{aligned} \tag{3.92}$$

Now, each element in the summation in (3.92) can be rewritten as follows:

$$\begin{aligned}
\Pr\left(X_i \notin \mathcal{R}_i\right) &= 1 - \Pr\left(X_i \in \mathcal{R}_i\right) \\
&= 1 - \Pr\left(x_i^m < X_i < x_i^M\right) \\
&= 1 - \Pr\left(-\frac{x_i^M - x_i^m}{2} < X_i - \frac{x_i^M + x_i^m}{2} < \frac{x_i^M - x_i^m}{2}\right) \\
&= 1 - \Pr\left(\left|X_i - \frac{x_i^M + x_i^m}{2}\right| < \frac{x_i^M - x_i^m}{2}\right) \\
&= \Pr\left(\left|X_i - \frac{x_i^M + x_i^m}{2}\right| \geq \frac{x_i^M - x_i^m}{2}\right). \qquad (3.93)
\end{aligned}$$

By using Chebyshev's inequality, we obtain that

$$\begin{aligned}
\Pr\left(\left|X_i - \frac{x_i^M + x_i^m}{2}\right| \geq \frac{x_i^M - x_i^m}{2}\right) &\leq \frac{4\mathrm{E}\left[\left(X_i - \frac{x_i^M + x_i^m}{2}\right)^2\right]}{(x_i^M - x_i^m)^2} \\
&= a_i \mathrm{E}[X_i^2] + b_i \mathrm{E}[X_i] + c_i, \qquad (3.94)
\end{aligned}$$

where $a_i = 4/(x_i^M - x_i^m)^2$, $b_i = -4(x_i^M + x_i^m)/(x_i^M - x_i^m)^2$, and $c_i = \left(\frac{x_i^M + x_i^m}{x_i^M - x_i^m}\right)^2$; and $\mathrm{E}[X_i^2]$ is the (i, i) entry of the matrix $S_X = \Sigma_X + m_X m_X^\top$, and $\mathrm{E}[X_i]$ is the ith entry of m_X, both of which are assumed to be known either exactly or approximately. Then, we obtain that

$$\Pr\left(X \notin \mathcal{R}\right) \leq a^\top s_X + b^\top m_X + c^\top \mathbf{1}_n, \qquad (3.95)$$

where s_X denotes a column vector, the entries of which correspond to the diagonal entries of S_X. Finally, we can use the result in (3.95) to obtain a lower bound on the value of ρ as follows:

$$\begin{aligned}
\rho &= 1 - \Pr\left(X \notin \mathcal{R}\right) \\
&\geq 1 - (a^\top s_X + b^\top m_X + c^\top \mathbf{1}_n). \qquad (3.96)
\end{aligned}$$

(Note that in the case when instead of the exact values of m_X and S_X, we only have some approximate values, we can use these in (3.96), although of course there is no guarantee a priori whether or not the inequality will still hold.)

Example 3.9 (DC circuit performance when input moments are known) Consider the circuit in Fig. 3.1 with the model parameters as in Example 3.1, i.e., $r_1 = r_2 = 1\ \Omega$, and where the values taken by $w = [v_s, i_l]^\top$ are described by a random vector with known mean, m_W, and covariance matrix, Σ_W, as given in (3.10), which, as shown in Example 3.1, result in the mean and correlation matrix of $X = [V_2, I_1]^\top$ being:

$$m_X = \begin{bmatrix} 2.5 \\ 7.5 \end{bmatrix}, \quad S_X = \begin{bmatrix} 6.75 & 18.75 \\ 18.75 & 56.75 \end{bmatrix}. \tag{3.97}$$

Assume that for the circuit to deliver its intended function satisfactorily, the values that $x = [v_2, i_1]^\top$ takes must be within $\mathcal{R} = (0, 10) \times (0, 20)$. Thus, in this case, we have that $a = [1/25, 1/100]^\top$, $b = [-2/5, -1/5]^\top$, $c = [1, 1]^\top$, and $s_X = [6.75, 56.75]^\top$. Then, by applying (3.96), we obtain that

$$\rho \geq 0.6625. \tag{3.98}$$

3.4.2 Known Input Probability Density Function

Assume that we can use the techniques introduced in Section 3.3 to obtain the complete probabilistic characterization of X. When the dimension of the input vector is greater than or equal to the dimension of the state vector, i.e., $m \geq n$, we have an expression for the pdf of X; thus, we can compute the value of ρ in (3.91) as follows:

$$\begin{aligned} \rho &= \Pr\left(X \in \mathcal{R}\right) \\ &= \int_{\mathcal{R}} f_X(x) dx. \end{aligned} \tag{3.99}$$

When $m < n$, we do not have an expression for the pdf of X, but an expression for the pdf of some X_1 (a vector comprising a subset of entries of X), as given in (3.75), and the mapping between X_1 and X_2 (a vector comprising the entries of X that are not in X_1), as given in (3.74). Let $\mathcal{R}^{(X_1)} = \mathcal{R}_1 \times \mathcal{R}_2 \times \cdots \times \mathcal{R}_m$, and $\mathcal{R}^{(X_2)} = \mathcal{R}_{m+1} \times \mathcal{R}_{m+2} \times \cdots \times \mathcal{R}_n$. Then,

$$\begin{aligned} \rho &= \Pr\left(X \in \mathcal{R}\right) \\ &= \Pr\left(X_1 \in \mathcal{R}^{(X_1)} \cap X_2 \in \mathcal{R}^{(X_2)}\right) \\ &= \Pr\left(X_1 \in \mathcal{R}^{(X_1)} \cap h_2(h_1^{-1}(X_1)) \in \mathcal{R}^{(X_2)}\right) \\ &= \Pr\left(X_1 \in \mathcal{R}_r\right) \\ &= \int_{\mathcal{R}_r} f_{X_1}(x_1) dx_1, \end{aligned} \tag{3.100}$$

where

$$\mathcal{R}_r = \left\{x_1 \in \mathcal{R}^{(X_1)} \colon h_2(h_1^{-1}(x_1)) \in \mathcal{R}^{(X_2)}\right\}. \tag{3.101}$$

Alternatively, when $m \geq n$, if instead of having the exact expression for $f_X(x)$, we have an approximate expression, $\widehat{f}_X(x)$ (obtained using the techniques discussed in Section 3.3), then, we can compute an estimate of ρ, denoted by $\widehat{\rho}$, as follows:

$$\widehat{\rho} = \int_{\mathcal{R}} \widehat{f}_X(x)dx. \tag{3.102}$$

Similarly, when $m < n$, we can use the expressions in (3.81–3.82) to obtain

$$\widehat{\rho} = \int_{\widehat{\mathcal{R}}_r} \widehat{f}_{X_1}(x_1)dx_1, \tag{3.103}$$

where

$$\widehat{\mathcal{R}}_r = \Big\{x_1 : x_1 \in \mathcal{R}^{(X_1)}, \quad H_2(m_W)H_1(m_W)^{-1}x_1 - H_2(m_W)H_1(m_W)^{-1}h_1(m_W) + h_2(m_W) \in \mathcal{R}^{(X_2)}\Big\}. \tag{3.104}$$

Example 3.10 (AC circuit performance when input pdf is known) For the AC circuit in Fig. 3.2, consider the exact and approximate pdf characterization of Θ_l given in (3.86) and (3.85), respectively. Assume that for the circuit to deliver its function, the angle θ_l must belong to the interval $\mathcal{R} = \{\theta_l : \ -1.1 \leq \theta_l \leq -0.9\}$. Then, by using (3.86) in (3.99), we have that

$$\rho = \int_{-1.1}^{-0.9} 5\cos(\theta_l)\, d\theta_l = 0.53. \tag{3.105}$$

Similarly, by using (3.85) in (3.102), we have that

$$\widehat{\rho} = \int_{-1.1}^{-0.9} 2.5\, d\theta_2 = 0.5. \tag{3.106}$$

The error introduced by using $\widehat{\rho}$ instead of ρ is $e = \frac{\rho - \widehat{\rho}}{\rho} = 0.0566$ (i.e., 5.66%), and the discrepancy can be attributed to the fact that $\widehat{f}_{\Theta_l}(\theta_l)$ does not closely match $f_{\Theta_l}(\theta_l)$. On the other hand, if we let $\mathcal{R} = \{\theta_l : \ -0.6 \leq \theta_l \leq -0.45\}$, and use the expressions in (3.88) and (3.87), we obtain that $\rho = 0.648$ and $\widehat{\rho} = 0.649$, which should not come as a surprise, given that the match between (3.88) and (3.87) was much closer than that between (3.86) and (3.85).

3.5 Application to Power Flow Analysis

In this section, we apply the ideas introduced thus far to analyze the impact on the quasi-steady-state behavior of a power system, as described by a power flow model, of random fluctuations in active power injections that may arise from load variability and renewable-based generation. The reader is referred to Appendix B for the detailed derivation of the power flow model adopted here.

3.5.1 Power Flow Model

Consider a three-phase power system comprising n buses ($n > 1$) and l transmission lines $\big(n-1 \le l \le n(n-1)/2\big)$. Assume that there is at most one transmission line connecting each pair of buses. Assign an arbitrary direction for the positive flow of power between any two buses connected by a transmission line. Then, the power system network topology together with the chosen orientation can be described by a directed graph, $\mathcal{G} = \{\mathcal{V}, \mathcal{E}\}$, where each element in the set $\mathcal{V} = \{1, 2, \ldots, n\}$ corresponds to a bus in the system, and $\mathcal{E} \subset \mathcal{V} \times \mathcal{V} \backslash \{(i,i) \colon i \in \mathcal{V}\}$ so that $(i,j) \in \mathcal{E}$ if there is a transmission line connecting bus i and bus j with the flow of power on this line assigned to be positive from bus i to bus j. Let $\mathbb{I}$ denote a one-to-one mapping that assigns each element $(i,j) \in \mathcal{E}$ to an element $e \in \mathcal{L} = \{1, 2, \ldots, l\}$. Then, we can define the node-to-edge incidence matrix of $\mathcal{G}$, denoted by $M = [m_{i,e}] \in \mathbb{R}^{n \times l}$, as follows:

$$m_{i,e} = \begin{cases} 1, & \text{if } e = \mathbb{I}\big((i,j)\big),\ (i,j) \in \mathcal{E}, \\ -1, & \text{if } e = \mathbb{I}\big((l,i)\big),\ (l,i) \in \mathcal{E}, \\ 0, & \text{otherwise}. \end{cases} \tag{3.107}$$

Assume the system is balanced and operating in sinusoidal steady-state regime. Also, assume that all transmission lines in the network are short and lossless. Thus, the lumped-parameter circuit describing the terminal behavior of the line connecting bus i and bus j reduces to the series element of the Π-equivalent circuit model (see Appendix B), which in this case is a purely reactive impedance with a reactance value $X_{ij} > 0$. Let v_i denote the magnitude of the phasor associated with the (sinusoidal) voltage at bus i, and assume there is some control mechanism that ensures, $v_i = V_i$, $i \in \mathcal{V}$, where V_i is a positive scalar. Define $\Gamma := \mathbf{diag}(\gamma_1, \gamma_2, \ldots, \gamma_l)$, where

$$\gamma_e = \frac{V_i V_j}{X_{ij}},\ e = \mathbb{I}\big((i,j)\big),\ (i,j) \in \mathcal{E}.$$

Let p_i denote the active power injected into the power system network via bus i, and let ϕ_e denote the power flowing on transmission line $e = \mathbb{I}\big((i,j)\big)$. Define $p = [p_1, p_2, \ldots, p_n]^\top$ and $\phi = [\phi_1, \phi_2, \ldots, \phi_l]^\top$. Then, under some assumptions on the values that p takes and the resulting operating conditions of the system (see Appendix B for the details), we have that

$$p = M\phi, \tag{3.108}$$

$$\mathbf{0}_{l-n+1} = N\,\mathbf{arcsin}\,(\Gamma^{-1}\phi), \tag{3.109}$$

where the rows of $N \in \mathbb{R}^{(l-n+1)\times l}$ form a basis for the null space of M, i.e., N is a full rank matrix satisfying $MN^\top = \mathbf{0}_{n,l-n+1}$; and for $y = [y_1, y_2, \ldots, y_l]^\top$, $\mathbf{arcsin}\,(y)$, is defined as:

$$\mathbf{arcsin}\,(y) = \big[\arcsin(y_1), \arcsin(y_2), \ldots, \arcsin(y_l)\big]^\top. \tag{3.110}$$

Note that since the system is lossless, there is an additional constraint in that the sum of the active power injections needs to be equal to zero, i.e.,

$$\sum_{i=1}^{n} p_i = 0; \tag{3.111}$$

this can be easily checked by adding up the equations in (3.108).

Assume that each bus i, $i = 1, 2, \dots, m$, $1 \leq m \leq n-1$, corresponds to a generator or a load with the active power it injects into the network, denoted by ξ_i, being extraneously set (taking a positive value if bus i corresponds to a generator and taking a negative value if bus i corresponds to a load), and a priori uncertain (random), i.e.,

$$p_i = \xi_i, \quad i = 1, 2, \dots, m. \tag{3.112}$$

Similarly, assume that each bus $i = m+1, m+2, \dots, n$, corresponds to a generator or a load with the active power it injects into the network defined as follows:

$$p_i = r_i + \alpha_i \left(\sum_{j=1}^{m} \xi_j + \sum_{j=m+1}^{n} r_j \right), \quad i = m+1, m+2, \dots, n, \tag{3.113}$$

where r_i denotes some nominal setpoint, and α_i is a nonpositive scalar such that $\sum_{i=m+1}^{n} \alpha_i = -1$. One can check that (3.112) and (3.113) satisfy (3.111). Define $\xi = [\xi_1, \xi_2, \dots, \xi_m]^\top$, $r = [r_{m+1}, r_{m+2}, \dots, r_n]^\top$, and $\alpha = [\alpha_{m+1}, \alpha_{m+2}, \dots, \alpha_n]^\top$; then, by using (3.108–3.109) and (3.112–3.113), we obtain that

$$\begin{bmatrix} I_m \\ \alpha \mathbf{1}_m^\top \end{bmatrix} \xi + \begin{bmatrix} \mathbf{0}_{m \times (n-m)} \\ I_{n-m} + \alpha \mathbf{1}_{n-m}^\top \end{bmatrix} r = M\phi, \tag{3.114}$$

$$\mathbf{0}_{l-n+1} = N \mathbf{arcsin}\,(\Gamma^{-1}\phi). \tag{3.115}$$

Note that for a given ξ, there are $l+1$ equations in (3.114–3.115), while the dimension of the (unknown) line flow vector, ϕ, is l. Thus, when solving for ϕ, one can remove one of the equations in (3.114), which is linearly dependent on the others as the rank of M is $n-1$ when $\mathcal{G}$ is connected. In this regard, while any equation can be removed, in the remainder, we always remove the last one.

3.5.2 Power Flow Vector Distribution

Let Ξ denote a random vector describing the values that the vector of extraneous power injections, ξ, can take. Assume that the entries of Ξ are independent and uniformly distributed in the interval $\mathcal{Z}_i = [m_{\xi_i} + \underline{\xi}_i, m_{\xi_i} + \overline{\xi}_i]$, $\overline{\xi}_i = -\underline{\xi}_i \geq 0$, i.e.,

$$f_{\Xi}(\xi) = \begin{cases} \dfrac{1}{\text{vol}(\mathcal{Z})}, & \xi \in \mathcal{Z}, \\ 0, & \text{otherwise}, \end{cases} \tag{3.116}$$

where $\mathcal{Z} = \mathcal{Z}_1 \times \mathcal{Z}_2 \times \cdots \times \mathcal{Z}_m$, and

$$\text{vol}(\mathcal{Z}) = \Pi_{i=1}^{m}(\overline{\xi}_i - \underline{\xi}_i).$$

Also, assume that for every $\xi \in \mathcal{Z}$, there exists a ϕ that satisfies (3.114–3.115). Then, the goal is to characterize the statistical distribution of Φ – the random vector describing the values that the flow vector, ϕ, can take.

Tree Networks

In this case, $N = 0$; thus, (3.114–3.115) reduces to

$$\begin{bmatrix} I_m \\ \alpha\mathbf{1}_m^\top \end{bmatrix} \xi + \begin{bmatrix} \mathbf{0}_{m\times(n-m)} \\ I_{n-m} + \alpha\mathbf{1}_{n-m}^\top \end{bmatrix} r = M\phi, \tag{3.117}$$

where $M \in \mathbb{R}^{n\times(n-1)}$. As mentioned earlier, one of the equations in (3.117) is linearly dependent on the others; thus, it can be removed when solving for ϕ. To this end, let $\widetilde{M} \in \mathbb{R}^{(n-1)\times(n-1)}$ denote the matrix that results from removing the last row in M; this matrix is clearly invertible. Similarly, let $\widetilde{I}_{n-m} \in \mathbb{R}^{(n-m-1)\times(n-m)}$ denote the matrix that results from removing the last row in I_{n-m}. Also, let $\widetilde{\alpha} \in \mathbb{R}^{n-m-1}$ denote the vector that results from removing the last entry in α. Then, we can write

$$\phi = H\xi + Lr, \tag{3.118}$$

where

$$\begin{aligned} H &= \widetilde{M}^{-1} \begin{bmatrix} I_m \\ \widetilde{\alpha}\mathbf{1}_m^\top \end{bmatrix} \in \mathbb{R}^{(n-1)\times m}, \\ L &= \widetilde{M}^{-1} \begin{bmatrix} \mathbf{0}_{m\times(n-m)} \\ \widetilde{I}_{n-m} + \widetilde{\alpha}\mathbf{1}_{n-m}^\top \end{bmatrix} \in \mathbb{R}^{(n-1)\times(n-m)}. \end{aligned} \tag{3.119}$$

Then, by defining $\widetilde{\Phi} = \Phi - Lr$, we have that

$$\widetilde{\Phi} = H\Xi, \tag{3.120}$$

with $\Xi = [\Xi_1, \Xi_2, \ldots, \Xi_m]^\top$, where the Ξ_i's are independent random variables uniformly distributed over the interval $\mathcal{Z}_i$. Now, since the rank of $\widetilde{M}^{-1}$ is $n-1$, and the rank of

$$\begin{bmatrix} I_m \\ \widetilde{\alpha}\mathbf{1}_m^\top \end{bmatrix} \tag{3.121}$$

is $m \leq n-1$; then, the rank of H is m, which makes it full rank. Thus, (3.120) is of the form in (3.33), therefore, we can apply the techniques discussed in

Section 3.3.1 to obtain the pdf of $\widetilde{\Phi}$, $f_{\widetilde{\Phi}}(\widetilde{\phi})$, from where we can obtain the pdf of Φ as follows:

$$f_{\Phi}(\phi) = f_{\widetilde{\Phi}}(\phi - Lr).$$

Here, we need to differentiate between the case when $m = n - 1$ and the case when $1 \leq m < n - 1$.

When $m = n - 1$, it follows trivially that

$$H = \widetilde{M}^{-1} \in \mathbb{R}^{(n-1)\times(n-1)}, \tag{3.122}$$

and this scenario reduces to the setting in Case C1 in Section 3.3.1. Therefore, we can find the pdf of $\widetilde{\Phi}$ by using the expression in (3.36):

$$f_{\widetilde{\Phi}}(\widetilde{\phi}) = \begin{cases} \left|\mathbf{det}(\widetilde{M})\right| \dfrac{1}{\mathrm{vol}(\mathcal{Z})}, & \{\widetilde{\phi}\colon \widetilde{\phi} = \widetilde{M}^{-1}\xi,\ \xi \in \mathcal{Z}\}, \\ 0, & \text{otherwise}; \end{cases} \tag{3.123}$$

from where it follows that

$$f_{\Phi}(\phi) = \begin{cases} \left|\mathbf{det}(\widetilde{M})\right| \dfrac{1}{\mathrm{vol}(\mathcal{Z})}, & \phi \in \mathcal{F}, \\ 0, & \text{otherwise}, \end{cases} \tag{3.124}$$

where

$$\mathcal{F} = \{\phi\colon \phi = \widetilde{M}^{-1}\xi + Lr,\ \xi \in \mathcal{Z}\}. \tag{3.125}$$

When $1 \leq m < n - 1$, we first need to find m rows in H that are linearly independent; without loss of generality, we assume these are the first m ones (otherwise, we reorder the rows of the matrix H so that this is the case and relabel the entries of H). Then, we can write

$$\begin{bmatrix} \widetilde{\Phi}_1 \\ \widetilde{\Phi}_2 \end{bmatrix} = \begin{bmatrix} H_1 \\ H_2 \end{bmatrix} \Xi, \tag{3.126}$$

where $\widetilde{\Phi}_1 \in \mathbb{R}^m$, $\widetilde{\Phi}_2 \in \mathbb{R}^{n-m-1}$, $H_1 \in \mathbb{R}^{m\times m}$, and $H_2 \in \mathbb{R}^{(n-m-1)\times m}$. Next, as in the developments for Case C3 earlier, we have that

$$\widetilde{\Phi}_2 = H_2 H_1^{-1} \widetilde{\Phi}_1; \tag{3.127}$$

thus, a complete characterization of $\widetilde{\Phi}$ is given by the pdf of $\widetilde{\Phi}_1$, which can be obtained by using (3.56):

$$f_{\widetilde{\Phi}_1}(\widetilde{\phi}_1) = \begin{cases} \left|\mathbf{det}(H_1^{-1})\right| \dfrac{1}{\mathrm{vol}(\mathcal{Z})}, & \{\widetilde{\phi}_1\colon \widetilde{\phi}_1 = H_1\xi,\ \xi \in \mathcal{Z}\}, \\ 0, & \text{otherwise}, \end{cases} \tag{3.128}$$

and the relation between $\widetilde{\Phi}_1$ and $\widetilde{\Phi}_2$ in (3.127). Finally, by partitioning L as

$$\begin{bmatrix} L_1 \\ L_2 \end{bmatrix}, \tag{3.129}$$

with $L_1 \in \mathbb{R}^{m \times (n-m)}$, $L_2 \in \mathbb{R}^{(n-m-1)\times(n-m)}$, the complete probabilistic characterization of $\Phi = [\Phi_1^\top, \Phi_2^\top]^\top$, where $\Phi_1 \in \mathbb{R}^m$ and $\Phi_2 \in \mathbb{R}^{n-m-1}$, is given by

$$f_{\Phi_1}(\phi_1) = \begin{cases} \left|\mathbf{det}(H_1^{-1})\right| \dfrac{1}{\mathrm{vol}(\mathcal{Z})}, & \{\phi_1 \colon \phi_1 = H_1\xi + L_1 r,\ \xi \in \mathcal{Z}\}, \\ 0, & \text{otherwise}, \end{cases} \tag{3.130}$$

and

$$\Phi_2 = H_2 H_1^{-1} \Phi_1 + (L_2 - H_2 H_1^{-1} L_1) r. \tag{3.131}$$

Example 3.11 (Three-bus tree network with two random injections) Consider a three-bus lossless power network, the topology of which can be described by the graph in Fig. 3.5. By using the orientation indicated in the figure, the node-to-edge incidence matrix is given by:

$$M = \begin{bmatrix} 1 & 0 \\ -1 & 1 \\ 0 & -1 \end{bmatrix}. \tag{3.132}$$

Assume that the reactances of the transmission lines linking bus 1 and bus 2, and bus 2 and bus 3, denoted by X_{12} and X_{23}, respectively, are both equal to 1 p.u., and that the voltages at all three buses 1, 2, 3, denoted by v_1, v_2, and v_3, respectively, are the same and equal to 1 p.u.. Let ϕ_1 and ϕ_2 denote the active power flow along the transmission line linking bus 1 and bus 2, and bus 2 and bus 3, respectively, defined to be positive in the direction corresponding to the edge orientation in Fig. 3.5. Assume that the power injections into bus 1 and bus 2 are set extraneously, with the injection into bus 1, ξ_1, taking values in the interval $[0.1, 0.3]$, and with the injection into bus 2, ξ_2, taking values in the interval $[-0.6, -0.4]$. Thus, connected to bus 1, there is a generator with uncertain output, whereas connected to bus 2, there is a load with uncertain demand. Further, assume that ξ_1 and ξ_2 are jointly uniformly distributed; then, the joint pdf of $\Xi = [\Xi_1, \Xi_2]^\top$ that describes the value $\xi = [\xi_1, \xi_2]^\top$ takes is given by:

$$f_\Xi(\xi) = \begin{cases} 25, & 0.1 \le \xi_1 \le 0.3, \ -0.6 \le \xi_2 \le -0.4, \\ 0, & \text{otherwise.} \end{cases} \tag{3.133}$$

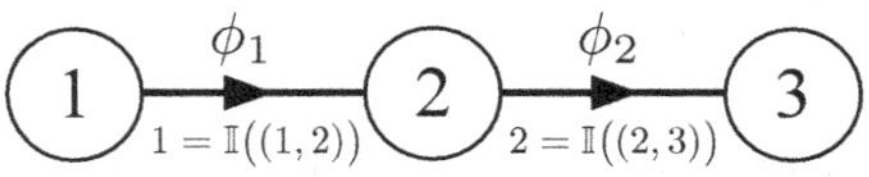

ϕ_1 ϕ_2

(1) → (2) → (3)

$1 = \mathbb{I}((1,2))$ $2 = \mathbb{I}((2,3))$

Figure 3.5 Graph describing the topology and flow assignment of a three-bus tree power network.

Assume that the power injection into bus 3 can be set as in (3.113) with $r_3 = 0$ and $\alpha_3 = -1$; thus,

$$p_3 = -(\xi_1 + \xi_2), \tag{3.134}$$

which makes (3.117) reduce to

$$\begin{bmatrix} 1 & 0 \\ 0 & 1 \\ -1 & -1 \end{bmatrix} \begin{bmatrix} \xi_1 \\ \xi_2 \end{bmatrix} = M \begin{bmatrix} \phi_1 \\ \phi_2 \end{bmatrix}, \tag{3.135}$$

with M as in (3.132). As in earlier developments, we eliminate the last equation in (3.135), as it is linearly dependent on the other two, to obtain:

$$\begin{bmatrix} 1 & 0 \\ 0 & 1 \end{bmatrix} \begin{bmatrix} \xi_1 \\ \xi_2 \end{bmatrix} = \widetilde{M} \begin{bmatrix} \phi_1 \\ \phi_2 \end{bmatrix}, \tag{3.136}$$

where

$$\widetilde{M} = \begin{bmatrix} 1 & 0 \\ -1 & 1 \end{bmatrix}, \tag{3.137}$$

from where it follows that:

$$\begin{aligned} \begin{bmatrix} \phi_1 \\ \phi_2 \end{bmatrix} &= \widetilde{M}^{-1} \begin{bmatrix} \xi_1 \\ \xi_2 \end{bmatrix} \\ &= \begin{bmatrix} 1 & 0 \\ 1 & 1 \end{bmatrix} \begin{bmatrix} \xi_1 \\ \xi_2 \end{bmatrix}. \end{aligned} \tag{3.138}$$

Then, by using (3.124) with $\widetilde{M}$ as in (3.137) and $f_\Xi(\xi)$ as in (3.133), we obtain

$$f_\Phi(\phi) = \begin{cases} 25, & 0.1 \le \phi_1 \le 0.3, \quad \phi_1 - 0.6 \le \phi_2 \le \phi_1 - 0.4, \\ 0, & \text{otherwise}. \end{cases} \tag{3.139}$$

Example 3.12 (Three-bus tree network with one random injection) Consider the same three-bus lossless power network as in Example 3.11 (see Fig. 3.5 for the topology and edge orientation used to obtain the corresponding node-to-edge incidence matrix given in (3.132)). Here, we make the same assumption about the line series reactances and voltage magnitudes, i.e., $X_{12} = X_{23} = 1$ p.u., and $v_1 = v_2 = v_3 = 1$ p.u.; however, in this case, the only extraneous power injection into the network is that of bus 1, with the value it takes, ξ_1, being random and uniformly distributed in the interval $[0.1, 0.3]$. Thus, the pdf of Ξ_1 that describes the value that ξ_1 takes is given by

$$f_{\Xi_1}(\xi_1) = \begin{cases} 5, & 0.1 \le \xi_1 \le 0.3, \\ 0, & \text{otherwise}. \end{cases} \tag{3.140}$$

Then, the controlled power injections into bus 2 and bus 3 are given by

$$\begin{aligned} p_2 &= r_2 + \alpha_2(\xi_1 + r_2 + r_3), \\ p_3 &= r_3 + \alpha_3(\xi_1 + r_2 + r_3), \end{aligned} \tag{3.141}$$

with $\alpha_2 + \alpha_3 = -1$. By making $r_2 = r_3 = 0$, and $\alpha_2 = \alpha_3 = -1/2$, the value that the random load connected to bus 1 takes is evenly split among the generators in bus 2 and bus 3; thus, (3.117) reduces to

$$\begin{bmatrix} 1 \\ -\frac{1}{2} \\ -\frac{1}{2} \end{bmatrix} \xi_1 = M \begin{bmatrix} \phi_1 \\ \phi_2 \end{bmatrix}, \tag{3.142}$$

with M as in (3.132). Then, by removing the last equation in (3.142), and solving for $\phi = [\phi_1, \phi_2]^\top$, we obtain

$$\begin{aligned} \begin{bmatrix} \phi_1 \\ \phi_2 \end{bmatrix} &= \widetilde{M}^{-1} \begin{bmatrix} 1 \\ -\frac{1}{2} \end{bmatrix} \xi_1 \\ &= \begin{bmatrix} 1 \\ \frac{1}{2} \end{bmatrix} \xi_1. \end{aligned} \tag{3.143}$$

Then, since $\phi_1 = \xi_1$ and $\phi_2 = \frac{1}{2}\phi_1$, by using (3.140) and following the developments in (3.126–3.131), Φ is characterized by

$$f_{\Phi_1}(\phi_1) = \begin{cases} 5, & 0.1 \leq \phi_1 \leq 0.3, \\ 0, & \text{otherwise}, \end{cases} \tag{3.144}$$

and

$$\Phi_2 = \frac{1}{2}\Phi_1. \tag{3.145}$$

Example 3.13 (Three-bus tree network with one deterministic injection) Assume the same setting as in Example 3.12, with $r_2 = 0.1$, $r_3 = 0$, $\alpha_2 = 0$, and $\alpha_3 = -1$; thus, (3.117) reduces to

$$\begin{bmatrix} 1 \\ 0 \\ -1 \end{bmatrix} \xi_1 + \begin{bmatrix} 0 \\ 0.1 \\ -0.1 \end{bmatrix} = M \begin{bmatrix} \phi_1 \\ \phi_2 \end{bmatrix}, \tag{3.146}$$

with M as in (3.132). Now, as in the two previous examples, by removing the last equation in (3.146), we obtain

$$\begin{aligned} \begin{bmatrix} \phi_1 \\ \phi_2 \end{bmatrix} &= \widetilde{M}^{-1} \begin{bmatrix} 1 \\ 0 \end{bmatrix} \xi_1 + \widetilde{M}^{-1} \begin{bmatrix} 0 \\ 0.1 \end{bmatrix} \\ &= \begin{bmatrix} 1 \\ 1 \end{bmatrix} \xi_1 + \begin{bmatrix} 0 \\ 0.1 \end{bmatrix}, \end{aligned} \tag{3.147}$$

from where it follows that

$$\begin{bmatrix} \widetilde{\phi}_1 \\ \widetilde{\phi}_2 \end{bmatrix} = \begin{bmatrix} 1 \\ 1 \end{bmatrix} \xi_1, \tag{3.148}$$

where $\widetilde{\phi}_1 = \phi_1$, and $\widetilde{\phi}_2 = \phi_2 - 0.1$. Then, similar to Example 3.12, we have that the complete probabilistic characterization of $\widetilde{\Phi}$ is given by

$$f_{\widetilde{\Phi}_1}(\widetilde{\phi}_1) = \begin{cases} 5, & 0.1 \leq \widetilde{\phi}_1 \leq 0.3, \\ 0, & \text{otherwise;} \end{cases} \tag{3.149}$$

and

$$\widetilde{\Phi}_2 = \widetilde{\Phi}_1; \tag{3.150}$$

thus, the complete probabilistic characterization of Φ is given by

$$f_{\Phi_1}(\phi_1) = \begin{cases} 5, & 0.1 \leq \phi_1 \leq 0.3, \\ 0, & \text{otherwise,} \end{cases} \tag{3.151}$$

and

$$\Phi_2 = \Phi_1 + 0.1. \tag{3.152}$$

Loopy Networks

In this case, $N \neq 0$; thus, the mapping between the extraneous power injection vector, ξ, and the flow vector, ϕ, is nonlinear as given by the expressions in (3.114–3.115). One approach to solve the problem at hand – which would provide the exact characterization of the statistical distribution of Φ – is to invert the mapping in (3.114–3.115). However, this approach does not yield a workable solution as it is not possible in general to obtain an analytical expression for such an inverse mapping. Thus, we settle for obtaining an approximation of the pdf of Φ by using the approach in Section 3.3.2.

Let ϕ^* denote the value of ϕ satisfying the relations in (3.114–3.115) for $\xi = m_\xi = [m_{\xi_1}, m_{\xi_2}, \ldots, m_{\xi_m}]^\top$, and let $\Delta\phi$ denote a variation in ϕ around ϕ^* that arises from variations in ξ around m_ξ, which is similarly denoted by $\Delta\xi$. Then, by using the Taylor series expansion of $N\mathbf{arcsin}(\cdot)$ around $\Gamma^{-1}\phi^*$ and only keeping the first-order term, we obtain that

$$\begin{bmatrix} I_m \\ \alpha \mathbf{1}_m^\top \end{bmatrix} \Delta\xi = M\Delta\phi, \tag{3.153}$$

$$0 \approx J(\phi^*)\Delta\phi, \tag{3.154}$$

with

$$J(\phi^*) = N \frac{\partial \mathbf{arcsin}\,(y)}{\partial y}\bigg|_{y=\Gamma^{-1}\phi^*} \Gamma^{-1} \in \mathbb{R}^{(l-n+1)\times l}, \tag{3.155}$$

where

$$\left.\frac{\partial \mathbf{arcsin}\,(y)}{\partial y}\right|_{y=\Gamma^{-1}\phi^*} = \mathbf{diag}\left(\frac{1}{\sqrt{1-y_1^2}}, \frac{1}{\sqrt{1-y_2^2}}, \dots, \frac{1}{\sqrt{1-y_l^2}}\right),$$

with $y_e = \phi_e^*/\gamma_e$.

Next, note that in (3.153–3.154), there are $l+1$ equations, whereas the dimensions of the unknown vector, $\Delta\phi$, is l. As in the tree case, one of the equations in (3.153) is linearly dependent on the others; therefore, it can be removed when solving for $\Delta\phi$. Let $\widetilde{M} \in \mathbb{R}^{(n-1)\times l}$ denote the matrix that results from removing the last row in M. Similarly, let $\widetilde{\alpha} \in \mathbb{R}^{n-m-1}$ denote the vector that results from removing the last entry in α. Then, we can write

$$\Delta\phi \approx H(\phi^*)\Delta\xi, \tag{3.156}$$

where

$$H(\phi^*) = \begin{bmatrix} \widetilde{M} \\ J(\phi^*) \end{bmatrix}^{-1} \begin{bmatrix} I_m \\ \widetilde{\alpha}\mathbf{1}_m^\top \\ \mathbf{0}_{(l-n+1)\times m} \end{bmatrix} \in \mathbb{R}^{l\times m}. \tag{3.157}$$

Now, define $\Delta\Xi = \Xi - m_\xi$; then, by using (3.156), we can write

$$\Delta\Phi \approx H(\phi^*)\Delta\Xi, \tag{3.158}$$

with $\Delta\Xi = [\Delta\Xi_1, \Delta\Xi_2, \dots, \Delta\Xi_m]^\top$, where the $\Delta\Xi_i$'s are independent random variables uniformly distributed over the interval $\Delta\mathcal{Z}_i = [\underline{\xi}_i, \overline{\xi}_i]$.

Next, since the rank of

$$\begin{bmatrix} \widetilde{M} \\ J(\phi^*) \end{bmatrix}^{-1} \tag{3.159}$$

is l, and the rank of

$$\begin{bmatrix} I_m \\ \widetilde{\alpha}\mathbf{1}_m^\top \\ \mathbf{0}_{(l-n+1)\times m} \end{bmatrix} \tag{3.160}$$

is m; it follows that the rank of $H(\phi^*)$ is also m, making it full rank. Thus, since in this case the dimension of the vector $\Delta\phi$ is strictly greater than the dimension of the vector $\Delta\xi$ for all $1 \leq m \leq n-1$, we must apply the techniques discussed in Case C3. To this end, we first need to find m rows in $H(\phi^*)$ that are linearly independent. As in the tree network case, we assume without loss of generality that these are the first m ones (otherwise, we reorder the rows of the matrix $H(\phi^*)$ so that this is the case and relabel the entries of $H(\phi^*)$). Then, we can write

$$\begin{bmatrix} \widetilde{\Delta\Phi}_1 \\ \widetilde{\Delta\Phi}_2 \end{bmatrix} := \begin{bmatrix} H_1(\phi^*) \\ H_2(\phi^*) \end{bmatrix} \Delta\Xi, \tag{3.161}$$

where $\widetilde{\Delta\Phi_1} \in \mathbb{R}^m$, $\widetilde{\Delta\Phi_2} \in \mathbb{R}^{l-m}$, $H_1(\phi^*) \in \mathbb{R}^{m\times m}$, and $H_2(\phi^*) \in \mathbb{R}^{(l-m)\times m}$. Next, as in the tree network case, we have that

$$\widetilde{\Delta\Phi_2} = H_2(\phi^*)H_1(\phi^*)^{-1}\widetilde{\Delta\Phi_1}, \tag{3.162}$$

and

$$f_{\widetilde{\Delta\Phi_1}}(\widetilde{\Delta\phi_1}) = \begin{cases} \dfrac{|\mathbf{det}(H_1(\phi^*)^{-1})|}{\text{vol}(\Delta\mathcal{Z})}, & \{\widetilde{\Delta\phi_1} : \widetilde{\Delta\phi_1} = H_1(\phi^*)\Delta\xi,\ \Delta\xi \in \Delta\mathcal{Z}\}, \\ 0, & \text{otherwise}, \end{cases} \tag{3.163}$$

where $\Delta\mathcal{Z} = \Delta\mathcal{Z}_1 \times \Delta\mathcal{Z}_2 \times \cdots \times \Delta\mathcal{Z}_m$, and

$$\text{vol}(\Delta\mathcal{Z}) = \text{vol}(\mathcal{Z}) = \Pi_{i=1}^m(\overline{\xi}_i - \underline{\xi}_i).$$

Then, by partitioning ϕ^* as follows:

$$\phi^* = \begin{bmatrix} \phi_1^* \\ \phi_2^* \end{bmatrix},$$

we have that $\Phi_1 = \phi_1^* + \Delta\Phi_1$; thus, the complete probabilistic description of Φ can be approximated as follows:

$$\begin{aligned} f_{\Phi_1}(\phi_1) &= f_{\Delta\Phi_1}(\phi_1 - \phi_1^*) \\ &\approx f_{\widetilde{\Delta\Phi_1}}(\phi_1 - \phi_1^*), \end{aligned} \tag{3.164}$$

together with

$$\Phi_2 \approx H_2(\phi^*)H_1(\phi^*)^{-1}\Phi_1 - H_2(\phi^*)H_1(\phi^*)^{-1}\phi_1^* + \phi_2^*. \tag{3.165}$$

Example 3.14 (Three-bus cycle network with two random injections) Consider a three-bus lossless power network, the topology of which can be described by the graph in Fig. 3.6. By using the orientation indicated in the figure, the node-to-edge incidence matrix is given by:

$$M = \begin{bmatrix} 1 & 0 & -1 \\ -1 & 1 & 0 \\ 0 & -1 & 1 \end{bmatrix}, \tag{3.166}$$

from where it follows that

$$N = \begin{bmatrix} 1 & 1 & 1 \end{bmatrix}. \tag{3.167}$$

Assume that the series reactances of the transmission lines linking bus 1 and bus 2, bus 2 and bus 3, and bus 3 and bus 1 are all equal to 1 p.u., and that the voltage magnitude at all three buses is the same and equal to 1 p.u.. Let ϕ_1, ϕ_2, and ϕ_3 denote the active power flow along the transmission line linking bus 1 and bus 2, bus 2 and bus 3, and bus 3 and bus 1, respectively, defined to be positive in the direction corresponding to the edge orientation in Fig. 3.6. Assume that

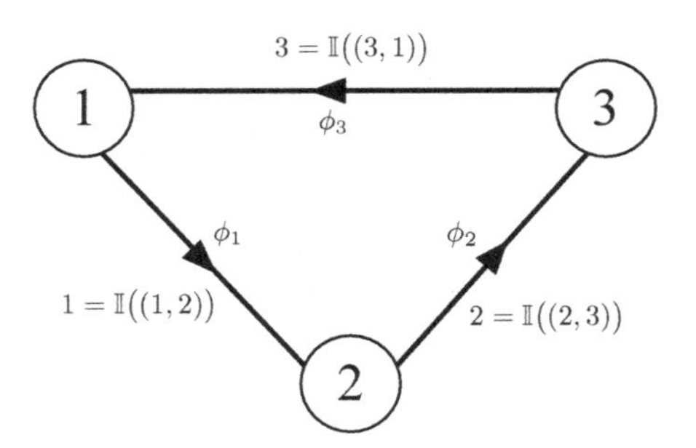

Figure 3.6 Graph describing the topology and flow assignment of a three-bus cyclic power network.

the power injections into bus 1 and bus 2, ξ_1 and ξ_2, are set extraneously, with $\xi = [\xi_1, \xi_2]^\top$ being described by a random vector, $\Xi = [\Xi_1, \Xi_2]^\top$, the entries of which are jointly uniformly distributed with joint pdf:

$$f_\Xi(\xi) = \begin{cases} 25, & 0.1 \leq \xi_1 \leq 0.3, \quad -0.6 \leq \xi_2 \leq -0.4, \\ 0, & \text{otherwise.} \end{cases} \tag{3.168}$$

Assume that the power injection into bus 3, p_3, is set as in (3.113) with $r_3 = 0$ and $\alpha_3 = -1$; thus

$$p_3 = -(\xi_1 + \xi_2), \tag{3.169}$$

which makes (3.114) reduce to

$$\begin{bmatrix} 1 & 0 \\ 0 & 1 \\ -1 & -1 \end{bmatrix} \begin{bmatrix} \xi_1 \\ \xi_2 \end{bmatrix} = M \begin{bmatrix} \phi_1 \\ \phi_2 \\ \phi_3 \end{bmatrix}, \tag{3.170}$$

with M as in (3.166). Now, since the last equation in (3.170) is linearly dependent on the other two, we can eliminate it to obtain:

$$\begin{bmatrix} \xi_1 \\ \xi_2 \end{bmatrix} = \widetilde{M} \begin{bmatrix} \phi_1 \\ \phi_2 \\ \phi_3 \end{bmatrix}, \tag{3.171}$$

where

$$\widetilde{M} = \begin{bmatrix} 1 & 0 & -1 \\ -1 & 1 & 0 \end{bmatrix}. \tag{3.172}$$

Finally, in order to complete the power flow model, we need to consider the constraint in (3.115) imposed by the presence of a cycle, namely:

$$0 = N \begin{bmatrix} \arcsin(\phi_1) \\ \arcsin(\phi_2) \\ \arcsin(\phi_3) \end{bmatrix}, \tag{3.173}$$

with N as in (3.167).

Now, by solving for ϕ in (3.171–3.173) for the value that the mean of ξ takes, i.e., for $\xi = m_\xi = [0.2, -0.5]^\top$, we obtain

$$\phi^* = [0.2337, -0.2667, 0.0337]^\top. \tag{3.174}$$

Then, by defining $\Delta\Phi = \Phi - \phi^*$ and $\Delta\Xi = \Xi - m_\xi$, and linearizing (3.171–3.173) around (m_ξ, ϕ^*), we obtain that

$$\Delta\Phi \approx H(\phi^*)\Delta\Xi, \tag{3.175}$$

where

$$\begin{aligned} H(\phi^*) &= \begin{bmatrix} 1 & 0 & -1 \\ -1 & 1 & 0 \\ \frac{1}{\sqrt{1-(\phi_1^*)^2}} & \frac{1}{\sqrt{1-(\phi_2^*)^2}} & \frac{1}{\sqrt{1-(\phi_3^*)^2}} \end{bmatrix}^{-1} \begin{bmatrix} 1 & 0 \\ 0 & 1 \\ 0 & 0 \end{bmatrix} \\ &= \begin{bmatrix} 0.3263 & -0.3383 \\ 0.3263 & 0.6617 \\ -0.6737 & -0.3383 \end{bmatrix}. \end{aligned} \tag{3.176}$$

Partition $H(\phi^*)$ as follows:

$$H(\phi^*) = \begin{bmatrix} H_1(\phi^*) \\ H_2(\phi^*) \end{bmatrix}, \tag{3.177}$$

where

$$H_1(\phi^*) = \begin{bmatrix} 0.3263 & -0.3383 \\ 0.3263 & 0.6617 \end{bmatrix}, \quad H_2(\phi^*) = \begin{bmatrix} -0.6737 & -0.3383 \end{bmatrix}. \tag{3.178}$$

Then, since

$$\mathbf{det}(H_1(\phi^*)) = \mathbf{det}\left(\begin{bmatrix} 0.3263 & -0.3383 \\ 0.3263 & 0.6617 \end{bmatrix}\right) = 0.3263, \tag{3.179}$$

we can conclude that the first two rows in (3.175) are linearly independent. Define $\widetilde{\Delta\Phi}_1 := H_1(\phi^*)\Delta\Xi$; then, by using the expression in (3.163), we obtain that

$$f_{\widetilde{\Delta\Phi}_1}(\widetilde{\Delta\phi}_1) = \begin{cases} 76.62, & \{\widetilde{\Delta\phi}_1 : \widetilde{\Delta\phi}_1 = H_1(\phi^*)\Delta\xi, \ \Delta\xi \in \Delta\mathcal{Z}\}, \\ 0, & \text{otherwise}, \end{cases} \tag{3.180}$$

where $\Delta\mathcal{Z} = [-0.1, 0.1] \times [-0.1, 0.1]$. Thus, the complete probabilistic characterization of $\Phi = [\Phi_1^\top, \Phi_2^\top]^\top$, $\Phi_1^\top \in \mathbb{R}^2$, $\Phi_2^\top \in \mathbb{R}$ is given by the pdf of Φ_1, which can be approximated by using the expression in (3.180) as follows:

$$\begin{aligned} f_{\Phi_1}(\phi_1) &\approx f_{\widetilde{\Delta\Phi}_1}(\phi_1 - \phi_1^*) \\ &= \begin{cases} 76.62, & \phi_1 \in \widetilde{\mathcal{F}}, \\ 0, & \text{otherwise}, \end{cases} \end{aligned} \tag{3.181}$$

where

$$\widetilde{\mathcal{F}} = \{\phi_1 \colon \phi_1 = H_1(\phi^*)\xi + \phi_1^* - H_1(\phi^*)m_\xi,$$
$$\xi \in [0.1, 0.3] \times [-0.6, -0.4]\}; \tag{3.182}$$

together with the relation between Φ_2 and Φ_1, which can be obtained by using (3.175) as follows:

$$\begin{aligned}\Phi_2 &\approx H_2(\phi^*)H_1(\phi^*)^{-1}\Phi_1 - H_2(\phi^*)H_1(\phi^*)^{-1}\phi_1^* + \phi_2^* \\ &= \begin{bmatrix}-1.0279 & -1.0368\end{bmatrix}\Phi_1 - 0.0026.\end{aligned} \tag{3.183}$$

3.6 Notes and References

The moment characterization of linear systems can be found in [10]. The pdf characterization of linear systems is based on standard random variable transformation results which can be found in [11, 15]. The power flow model formulation in Section 3.5.1, the derivation of which is provided in Appendix B, was inspired by [23] and has been utilized in [24]. The literature on probabilistic power flow is abundant and has its origins in the seminal work by Borkowska [25] and Sauer [26].

4 Static Systems: Probabilistic Structural Uncertainty

In this chapter, we provide tools for modeling the impact on the performance of systems described by an input-to-state mapping that can randomly change due to a finite number of structural changes in the system caused, e.g., by component failures. We assume that the random changes in the mapping are described by a Markov chain with known transition probabilities. We refer the reader to Section 2.1 for a review of notions from probability and stochastic processes used throughout this chapter.

4.1 Introduction

Consider the same DC circuit discussed in Section 3.1, which we reproduce in Fig. 4.1 (left). Assume that $w = [v_s, i_l]^\top$ takes some known value, $w^0 = [v_s^0, i_l^0]^\top$. Also, assume that there are some extraneous phenomena that may cause the value of resistance r_1 to change to a new value r_1' at some time τ, with no further changes afterwards; see Fig. 4.1 (right). Further, assume that the time τ at which the change occurs is random and can be described by some random variable, T, with known pdf, $f_T(t)$. As in Section 3.1, assume that for the circuit to perform as intended, the state, $x = [v_2, i_1]^\top$, must belong to the set $\mathcal{R} = (v_2^m, v_2^M) \times (i_1^m, i_1^M)$.

Assuming x belongs to $\mathcal{R}$ before the resistance value change occurs, a question that arises in the setting above is whether or not x will belong to $\mathcal{R}$ after the change in the resistance value. This is something that can be easily determined by plugging $w = w^0$ into

$$x = H'w, \tag{4.1}$$

where

$$H' = \begin{bmatrix} \frac{r_2}{r_1'+r_2} & -\frac{r_1' r_2}{r_1'+r_2} \\ \frac{1}{r_1'+r_2} & \frac{r_2}{r_1'+r_2} \end{bmatrix}, \tag{4.2}$$

and checking whether or not the value of x that results belongs to $\mathcal{R}$. If this is not the case, we would like to obtain some probabilistic characterization of the degree to which the system is delivering its intended function that takes into account the random change in resistance.

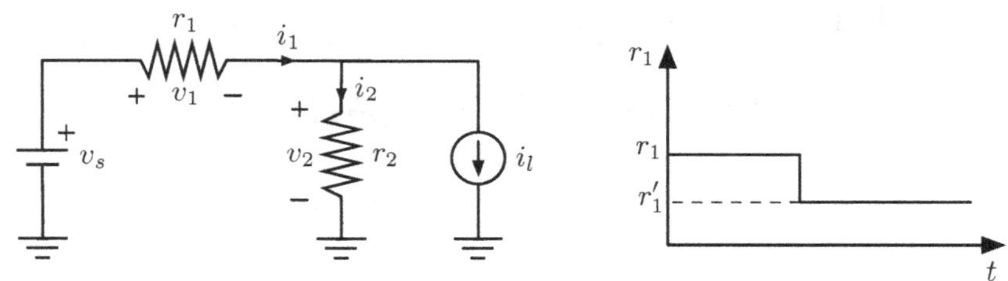

Figure 4.1 DC linear circuit and value that resistance r_1 takes.

More generally, in this chapter, we will consider systems of the form:

$$x = h_i(w), \tag{4.3}$$

where the values taken by $w \in \mathbb{R}^m$ are uncertain and described by a random vector, W, with known statistics, and $h_i \colon \mathbb{R}^m \to \mathbb{R}^n$, with the value that i takes being governed by a Markov chain with known transition probabilities. Then, the objective is to characterize the statistics of the stochastic process describing the state evolution with time.

The modeling formalism described above can be used in many applications. One such application is reliability analysis of multi-component systems, which we extensively cover later in the chapter. In this application, the $h_i(\cdot)$'s are determined by which components are operational and which components are failed, and possibly the order in which the failures occurred. Thus, transitions among the different $h_i(\cdot)$'s are determined by the failure of one or more components and/or completion of repair actions triggered in response to these failures.

4.2 System Stochastic Model

Consider a system whose input-to-state behavior is characterized by

$$x = h_i(w), \tag{4.4}$$

with $w \in \mathbb{R}^m$, $x \in \mathbb{R}^n$, and i taking values in some set $\mathcal{Q}$. Assume that the value i takes is governed by a stochastic process $Q = \{Q(t) \colon t \in \mathcal{T}\}$, with each random variable $Q(t)$ taking values in some set $\mathcal{Q}$. Since the value that i in (4.4) takes is governed by a stochastic process, the value that x takes is also governed by a stochastic process $X = \{X(t) \colon t \in \mathcal{T}\}$. The objective is then to characterize X under the following two scenarios:

S1. The values that w takes are described by a random vector, W, whose entries are jointly distributed with known first and second moments.

S2. The values that w takes are described by a random vector, W, whose entries are jointly distributed with known distribution.

Specifically, for Scenario S1, the objective is to characterize the first and second moments of $X(t)$, whereas for Scenario S2, the objective is to characterize the distribution of $X(t)$.

Characterization of First and Second Moments. Consider (4.4) and assume we know the mean of W, $\mathrm{E}[W] = m_W \in \mathbb{R}^m$, and its covariance matrix, $\mathrm{E}\big[(W - m_W)(W - m_W)^\top\big] = \Sigma_W \in \mathbb{R}^{m \times m}$. The objective is then to compute the mean of $X(t)$, $\mathrm{E}[X(t)] = m_X(t) \in \mathbb{R}^n$, and its covariance matrix $\mathrm{E}\big[(X(t) - m_X(t))(X(t) - m_X(t))^\top\big] = \Sigma_X(t) \in \mathbb{R}^{n \times n}$. By using the techniques in Chapter 3, we can compute

$$\begin{aligned} \mathrm{E}\big[X(t) \mid Q(t) = i\big] &= \mathrm{E}\big[h_i(W)\big], \quad i \in \mathcal{Q}, \\ \mathrm{E}\big[X(t)X(t)^\top \mid Q(t) = i\big] &= \mathrm{E}\Big[h_i(W)\big(h_i(W)\big)^\top\Big], \quad i \in \mathcal{Q}, \end{aligned} \tag{4.5}$$

exactly, or an approximation thereof. Then, we have that

$$\mathrm{E}\big[X(t)\big] = \sum_{i \in \mathcal{Q}} \mathrm{E}\big[X(t) \mid Q(t) = i\big] \Pr\big(Q(t) = i\big), \tag{4.6}$$

$$\mathrm{E}\big[X(t)X^\top(t)\big] = \sum_{i \in \mathcal{Q}} \mathrm{E}\big[X(t)X^\top(t) \mid Q(t) = i\big] \Pr\big(Q(t) = i\big), \tag{4.7}$$

and $\Sigma_X(t) = \mathrm{E}\big[X(t)X^\top(t)\big] - \mathrm{E}[X(t)]\mathrm{E}\big[X^\top(t)\big]$.

Distribution Characterization. Consider (4.4) and assume we know the pdf of $W(t)$, $f_{W(t)}(w)$. The objective is to compute the pdf of $X(t)$, $f_{X(t)}(x(t))$. By using the techniques in Chapter 3, we can compute

$$f_{X(t) \mid Q(t)}(x \mid i), \quad i \in \mathcal{Q}, \tag{4.8}$$

exactly, or an approximation thereof. Then, we have that

$$f_{X(t)}(x) = \sum_{i \in \mathcal{Q}} f_{X(t) \mid Q(t)}(x \mid i) \Pr\big(Q(t) = i\big). \tag{4.9}$$

From the expressions in (4.6), (4.7), and (4.9), it is clear that in order to complete the characterization of the moments or the pdf of $X(t)$, one needs to obtain the probability mass function (pmf) of $Q(t)$. In subsequent developments, we focus on solving such problem for the case when the evolution of $Q(t)$ is described by a Markov process.

4.3 Markov Process Characterization

A stochastic process $Q = \{Q(t) \colon t \in \mathcal{T}\}$, with each random variable $Q(t)$ taking a value $q(t)$ in some set $\mathcal{Q}$, is Markov if it satisfies the following property:

$$\begin{aligned} &\Pr\Big(Q(t_{k+1}) = q(t_{k+1}) \;\Big|\; Q(t_0) = q(t_0), Q(t_1) = q(t_1), \ldots, Q(t_k) = q(t_k)\Big) \\ &= \Pr\Big(Q(t_{k+1}) = q(t_{k+1}) \;\Big|\; Q(t_k) = q(t_k)\Big), \end{aligned} \tag{4.10}$$

for all $q(t_0), q(t_1), \ldots, q(t_k), q(t_{k+1}) \in \mathcal{Q}$, so that

$$\Pr\Big(Q(t_0) = q(t_0), \ldots, Q(t_k) = q(t_k)\Big) > 0,$$

and any sequence $t_0 < t_1 < t_2 < \cdots < t_k < t_{k+1}$.

This property essentially says that a Markov process is memoryless, i.e., the future evolution of the process only depends on the value that the process takes at the present time, independently of all the previous history of the process. Next, we consider the following two cases:

M1. The process $Q = \{Q(t)\colon t \in \mathcal{T}\}$, $Q(t) \in \mathcal{Q}$, is a discrete-time Markov chain (DTMC), i.e., $\mathcal{T} = \{t_0, t_1, t_2, \ldots\}$, and $\mathcal{Q} = \{1, 2, \ldots, N\}$.

M2. The process $Q = \{Q(t)\colon t \in \mathcal{T}\}$, $Q(t) \in \mathcal{Q}$, is a continuous-time Markov chain (CTMC), i.e., $\mathcal{T} = [0, \infty)$, and $\mathcal{Q} = \{1, 2, \ldots, N\}$.

4.3.1 Discrete-Time Case

Consider the system in (4.4) with the value that i takes being governed by a DTMC, $Q = \{Q(t) : t \in \mathcal{T}\}$, $\mathcal{T} = \{t_0, t_1, t_2, \ldots\}$, $Q(t) \in \mathcal{Q} = \{1, 2, \ldots, N\}$. Define $Q[k] := Q(t_k)$, then the evolution of a discrete-time Markov chain is completely determined by the one-step transition probabilities:

$$\begin{aligned} p_{i,j}(t_k) &= \Pr\left(Q[k+1] = j \mid Q[k] = i\right) \\ &=: p_{i,j}[k], \end{aligned} \tag{4.11}$$

for all $i, j \in \mathcal{Q}$. Next, we use the one-step transition probabilities together with the Markov property, as stated in (4.10), to obtain a set of linear difference equations, known as the Chapman–Kolmogorov equations, that describe the evolution of the pmf of $Q[k]$, $k = 0, 1, 2, \ldots$.

Assume that initially $Q[0] = l$ and $k+1$ time instants later, we have that $Q[k+1] = j$; then, by conditioning on $Q[k]$ and using the Markov property in (4.10), we obtain[1]

$$\begin{aligned} \Pr\Big(Q[k+1] = j \mid Q[0] = l\Big) &= \sum_{i \in \mathcal{Q}} \Pr\Big(Q[k+1] = j \mid Q[k] = i, Q[0] = l\Big) \\ &\quad \times \Pr\Big(Q[k] = i \mid Q[0] = l\Big) \\ &= \sum_{i \in \mathcal{Q}} \Pr\Big(Q[k+1] = j \mid Q[k] = i\Big) \\ &\quad \times \Pr\Big(Q[k] = i \mid Q[0] = l\Big) \\ &= \sum_{i \in \mathcal{Q}} p_{i,j}[k] \Pr\Big(Q[k] = i \mid Q[0] = l\Big). \end{aligned} \tag{4.12}$$

[1] Here we use the conditional version of the law of total probability, i.e., $\Pr(A \mid B) = \sum_{j=1}^{n} \Pr(A \mid E_j B) \Pr(E_j \mid B)$, where the collection $E_1, E_2, \ldots, E_n$ forms a partition of the sample space Ω.

Let $\pi_i[k] = \Pr\big(Q[k] = i\big)$, $i \in \mathcal{Q}$, denote the pmf of $Q[k]$; then by using (4.12) and the law of total probability, we can obtain the pmf of $Q[k+1]$ as follows:

$$\begin{aligned}
\pi_j[k+1] &= \Pr\big(Q[k+1] = j\big) \\
&= \sum_{l\in\mathcal{Q}} \Pr\big(Q[k+1] = j \mid Q[0] = l\big) \Pr\{Q[0] = l\big) \\
&= \sum_{l\in\mathcal{Q}} \sum_{i\in\mathcal{Q}} p_{i,j}[k] \Pr\big(Q[k] = i \mid Q[0] = l\big) \Pr\big(Q[0] = l\big) \\
&= \sum_{i\in\mathcal{Q}} p_{i,j}[k] \sum_{l\in\mathcal{Q}} \Pr\big(Q[k] = i \mid Q[0] = l\big) \Pr\big(Q[0] = l\big) \\
&= \sum_{i\in\mathcal{Q}} p_{i,j}[k] \Pr\big(Q[k] = i\big) \\
&= \sum_{i\in\mathcal{Q}} p_{i,j}[k]\pi_i[k]. \qquad (4.13)
\end{aligned}$$

Now, define the following row vector: $\pi[k] = \big[\pi_1[k], \pi_2[k], \ldots, \pi_n[k]\big]$; then we can rewrite (4.13) in matrix form as follows:

$$\pi[k+1] = \pi[k]P[k], \qquad (4.14)$$

where $P[k] = \big[p_{i,j}[k]\big] \in \mathbb{R}^{N\times N}$ is referred to as the transition matrix of the Markov chain. The matrix $P[k]$ is a row stochastic matrix, i.e., it satisfies

S1. the entries of $P[k]$ are probabilities, i.e., $0 \le p_{i,j}[k] \le 1$ for all $i, j \in \mathcal{Q}$; and
S2. the row-sums of $P[k]$ are equal to one, i.e., $\sum_{j\in\mathcal{Q}} p_{i,j}[k] = 1, \ \forall i \in \mathcal{Q}$.

Then, given the pmf of $Q[0]$, $\pi[0]$, it follows that:

$$\pi[k] = \pi[0]P[0]P[1]\ldots P[k-1], \quad k \ge 0. \qquad (4.15)$$

If $p_{i,j}[k] = p_{i,j}[0]$ for all i, j and k, the chain is called homogeneous; hence, in (4.12) and (4.13) we can drop the dependence on k of the one-step transition probabilities, thus (4.14) simplifies to

$$\pi[k+1] = \pi[k]P, \qquad (4.16)$$

from where it follows that

$$\pi[k] = \pi[0]P^k, \quad k \ge 0. \qquad (4.17)$$

Also, obtaining the stationary distribution of the chain, $\pi = \lim_{k\to\infty} \pi[k]$, reduces to solving the following eigenvector problem:

$$\pi = \pi P. \qquad (4.18)$$

Example 4.1 (DC circuit with parameters changing according to a DTMC) Consider the circuit in Fig. 4.1 (left), with $r_2 = 1\ \Omega$, and where r_1 can be either $1\ \Omega$ or $2\ \Omega$. Assume that the value r_1 takes is described by a discrete-time Markov process, $Q = \{Q[k],\ k \in \{0, 1, 2, \dots\}\}$, where $Q[k] = 1$ if $r_1 = 1\ \Omega$, and $Q[k] = 2$ if $r_1 = 2\ \Omega$; and one-step transition probabilities $p_{11}[k] = 0.8$, $p_{12}[k] = 0.2$, $p_{22}[k] = 1$, and $p_{21}[k] = 0$, $\forall k$. Then, if $w = [10, 5]^\top$, we have that

$$\begin{bmatrix} V_2[k] \\ I_1[k] \end{bmatrix} = H\big(Q[k]\big) \begin{bmatrix} 10 \\ 5 \end{bmatrix}, \tag{4.19}$$

where

$$H(Q[k]) = \begin{cases} \begin{bmatrix} \frac{1}{2} & -\frac{1}{2} \\ \frac{1}{2} & \frac{1}{2} \end{bmatrix}, & \text{if } Q[k] = 1, \\ \begin{bmatrix} \frac{1}{3} & -\frac{2}{3} \\ \frac{1}{3} & \frac{1}{3} \end{bmatrix}, & \text{if } Q[k] = 2. \end{cases} \tag{4.20}$$

Now by using the one-step transition probabilities, and assuming that at $k = 0$, $r_1 = 1\ \Omega$, we obtain

$$\begin{aligned} \big[\pi_1[k+1], \pi_2[k+1]\big] &= \big[\pi_1[k], \pi_2[k]\big] P, \\ \big[\pi_1[0], \pi_2[0]\big] &= [1, 0], \end{aligned} \tag{4.21}$$

where

$$P = \begin{bmatrix} 0.8 & 0.2 \\ 0 & 1 \end{bmatrix}, \tag{4.22}$$

which results in

$$\begin{aligned} [\pi_1[k], \pi_2[k]] &= [1, 0] P^k \\ &= [1, 0] \begin{bmatrix} 0.8^k & 1 - 0.8^k \\ 0 & 1 \end{bmatrix} \\ &= \big[0.8^k, 1 - 0.8^k\big]. \end{aligned} \tag{4.23}$$

Then, by using (4.19) together with (4.23), we can obtain the pmf of $X[k] = [V_2[k], I_1[k]]^\top$ as follows:

$$p_{X[k]}\big(x[k]\big) = \begin{cases} 0.8^k, & x[k] = [2.5, 7.5]^\top, \\ 1 - 0.8^k, & x[k] = [0, 5]^\top. \end{cases} \tag{4.24}$$

Example 4.2 (DC circuit with uncertain parameters and input) Consider the same setting in Example 4.1, except that here we assume $i_l = 0$, and v_s is uncertain and can be described by a random variable uniformly distributed in the interval $[9.5, 10.5]$ V. Thus, in this case, we have that $i_1 = i_2$, and by using the techniques in Section 3.3.1, we obtain that

$$f_{I_2[k] \mid Q[k]}(i_2[k] \mid 1) = \begin{cases} 2, & \text{if } \frac{9.5}{2} \leq i_2[k] \leq \frac{10.5}{2}, \\ 0, & \text{otherwise;} \end{cases} \tag{4.25}$$

and

$$f_{I_2[k] \mid Q[k]}(i_2[k] \mid 2) = \begin{cases} 3, & \text{if } \frac{9.5}{3} \leq i_2[k] \leq \frac{10.5}{3}, \\ 0, & \text{otherwise.} \end{cases} \tag{4.26}$$

Then, since $\left[\frac{9.5}{2}, \frac{10.5}{2}\right]$ and $\left[\frac{9.5}{3}, \frac{10.5}{3}\right]$ are non-overlapping intervals, we can use (4.23) to obtain the pdf of $I_2[k]$, which yields:

$$\begin{aligned} f_{I_2[k]}(i_2[k]) &= 0.8^k f_{I_2[k] \mid Q[k]}(i_2[k] \mid 1) + (1 - 0.8^k) f_{I_2[k] \mid Q[k]}(i_2[k] \mid 2) \\ &= \begin{cases} 3\,(1 - 0.8^k), & \text{if } \frac{9.5}{3} \leq i_2[k] \leq \frac{10.5}{3}, \\ 2\,(0.8^k), & \text{if } \frac{9.5}{2} \leq i_2[k] \leq \frac{10.5}{2}, \\ 0, & \text{otherwise.} \end{cases} \end{aligned} \tag{4.27}$$

4.3.2 Continuous-Time Case

Consider the system in (4.4) with the value that i takes governed by a CTMC, $\mathcal{Q} = \{Q(t) : t \in \mathcal{T}\}$, $\mathcal{T} = [0, \infty)$, $Q(t) \in \mathcal{Q} = \{1, 2, \ldots, N\}$. Suppose that at time t, $Q(t) = i$; then, within the small time interval $[t, t + \Delta t)$, the value of $Q(t)$ changes with probability proportional to the length of the interval, specifically:

$$\Pr\Big(Q(t + \Delta t) = j \mid Q(t) = i\Big) \approx r_{i,j}(t)\Delta t + \mathrm{o}(t). \tag{4.28}$$

Similarly, if $Q(t) = i$, then, within the small interval $[t, t + \Delta t)$, the value of $Q(t)$ does not change with probability also proportional to the length of the interval, i.e.,

$$\begin{aligned} \Pr\Big(Q(t + \Delta t) = i \mid Q(t) = i\Big) &\approx r_{i,i}(t)\Delta t + \mathrm{o}(t) \\ &\approx 1 - \sum_{l \neq i} r_{i,l}(t)\Delta t + \mathrm{o}(t). \end{aligned} \tag{4.29}$$

Next, we derive a set of differential equations that describes the evolution with time of the pmf of $Q(t)$. To accomplish this, we will use the Markov property, as stated in (4.10), together with (4.28) and (4.29), to derive an expression similar to that in (4.13). Then, the aforementioned set of differential equations will result from a limit operation as $\Delta t \to 0$. As in the analysis for the discrete-time case, assume that initially $Q(0) = l$ and after $t + \Delta t$ units of time have elapsed, we have that $Q(t + \Delta t) = j$; then, by conditioning on $Q(t)$ and using (4.10), we obtain

$$\begin{aligned}
\Pr\Big(Q(t+\Delta t) = j \mid Q(0) = l\Big) &= \sum_{i \in \mathcal{Q}} \Pr\Big(Q(t+\Delta t) = j \mid Q(t) = i, Q(0) = l\Big) \\
&\quad \times \Pr\Big(Q(t) = i \mid Q(0) = l\Big) \\
&= \sum_{i \in \mathcal{Q}} \Pr\Big(Q(t+\Delta t) = j \mid Q(t) = i\Big) \\
&\quad \times \Pr\Big(Q(t) = i \mid Q(0) = l\Big) \\
&\approx \Big(1 - \sum_{m \neq j} r_{j,m}(t)\Delta t\Big) \Pr\Big(Q(t) = j \mid Q(0) = l\Big) \\
&\quad + \sum_{i \in \mathcal{Q}\setminus\{j\}} r_{i,j}(t)\Delta t \Pr\Big(Q(t) = i \mid Q(0) = l\Big).
\end{aligned} \tag{4.30}$$

Let $\pi_i(t) = \Pr(Q(t) = i)$, $i \in \mathcal{Q}$, denote the pmf of $Q(t)$; then, as in the discrete-time case, by using (4.30) together with the law of total probability, we obtain that

$$\pi_j(t+\Delta t) \approx \Big(1 - \sum_{l \neq j} r_{j,l}(t)\Delta t\Big)\pi_j(t) + \sum_{i \in \mathcal{Q}\setminus\{j\}} r_{i,j}(t)\Delta t \pi_i(t), \tag{4.31}$$

and by rearranging terms and dividing throughout by Δt, it follows that

$$\frac{\pi_j(t+\Delta t) - \pi_j(t)}{\Delta t} \approx -\sum_{l \neq j} r_{j,l}(t)\pi_j(t) + \sum_{i \in \mathcal{Q}\setminus\{j\}} r_{i,j}(t)\pi_i(t). \tag{4.32}$$

Then, by taking the limit as $\Delta t \to 0$ on both sides of (4.32), we obtain that

$$\frac{d\pi_j(t)}{dt} = -\sum_{l \neq j} r_{j,l}(t)\pi_j(t) + \sum_{i \in \mathcal{Q}\setminus\{j\}} r_{i,j}(t)\pi_i(t). \tag{4.33}$$

Now, define the following row vector: $\pi(t) = \big[\pi_1(t), \pi_2(t), \ldots, \pi_n(t)\big]$; then we can rewrite (4.33) in matrix form as

$$\frac{d\pi(t)}{dt} = \pi(t)\Gamma(t), \tag{4.34}$$

with $\Gamma(t) = \big[\gamma_{i,j}(t)\big] \in \mathbb{R}^{N \times N}$, where $\gamma_{i,i}(t) = -\sum_{l \neq i} r_{i,l}(t)$, $\forall i$, and $\gamma_{i,j}(t) = r_{i,j}(t)$, $\forall j \neq i$. The set of differential equations in (4.34) describing the evolution of the pdf of $Q(t)$ is essentially the continuous-time counterpart of the set of difference equations in (4.14) describing the evolution of the pmf of $Q[k]$; thus, they are similarly referred to as the Chapman–Kolmogorov equations.

In the stochastic processes literature, the matrix $\Gamma(t)$ is called the generator matrix. Also, as in the discrete-time case, whenever the generator matrix does not depend on time, i.e., $\Gamma(t) = \Gamma$, the chain is called homogeneous; thus, (4.34) simplifies to

$$\frac{d\pi(t)}{dt} = \pi(t)\Gamma, \tag{4.35}$$

and $\pi(t) = \pi(0)e^{\Gamma t}$. Then, by making the left-hand side in (4.35) equal to zero, we can obtain the stationary distribution of the chain, $\pi = \lim_{t\to\infty} \pi(t)$, by solving

$$\pi\Gamma = 0, \tag{4.36}$$

i.e., π is the left eigenvector associated with the zero eigenvalue of the matrix Γ.

Example 4.3 (DC circuit with parameters changing according to a CTMC) Consider the same setting in Example 4.2, but assume that instead of being governed by a discrete-time Markov chain, the value that r_1 takes is governed by a CTMC $Q = \{Q(t) \colon t \geq 0\}$, where $Q(t) = 1$ if $r_1 = 1\ \Omega$, $Q(t) = 2$ if $r_1 = 2\ \Omega$, with a constant state-transition matrix

$$\Gamma = \begin{bmatrix} -\lambda_1 & \lambda_1 \\ 0 & 0 \end{bmatrix}, \tag{4.37}$$

where $\lambda_1 = 10^{-3}\ \mathrm{s}^{-1}$. Then, by solving (4.35) for $\pi(0) = [1, 0]$ and Γ as in (4.37), we obtain

$$\pi(t) = \left[\mathrm{e}^{-\lambda_1 t}, 1 - \mathrm{e}^{-\lambda_1 t}\right]. \tag{4.38}$$

Thus, following a similar development to that of Example 4.2, we have that

$$f_{I_2(t)\,|\,Q(t)}(i_2(t) \mid 1) = \begin{cases} 2, & \text{if } \frac{9.5}{2} \leq i_2(t) \leq \frac{10.5}{2}, \\ 0, & \text{otherwise}, \end{cases} \tag{4.39}$$

$$f_{I_2(t)\,|\,Q(t)}(i_2(t) \mid 2) = \begin{cases} 3, & \text{if } \frac{9.5}{3} \leq i_2(t) \leq \frac{10.5}{3}, \\ 0, & \text{otherwise}, \end{cases} \tag{4.40}$$

and

$$\begin{aligned} f_{I_2(t)}\big(i_2(t)\big) &= \mathrm{e}^{-\lambda_1 t} f_{I_2(t)\,|\,Q(t)}(i_2(t) \mid 1) + (1 - \mathrm{e}^{-\lambda_1 t})\, f_{I_2(t)\,|\,Q(t)}(i_2(t) \mid 2) \\ &= \begin{cases} 3\,(1 - \mathrm{e}^{-\lambda_1 t}), & \text{if } \frac{9.5}{3} \leq i_2(t) \leq \frac{10.5}{3}, \\ 2\,\mathrm{e}^{-\lambda_1 t}, & \text{if } \frac{9.5}{2} \leq i_2(t) \leq \frac{10.5}{2}, \\ 0, & \text{otherwise}. \end{cases} \end{aligned} \tag{4.41}$$

4.4 Performance Characterization

Consider the system in (4.4) with the value that i takes being governed by a CTMC, $\mathcal{Q} = \{Q(t);\ t \geq 0\}$, $Q(t) \in \mathcal{Q} = \{1, 2, \ldots, N\}$, and where the value that w takes is described by a random vector W with known pdf $f_W(w)$. Next, we will show how to utilize the Markov chain occupational probabilities and $f_W(w)$ to compute a probabilistic measure that characterizes the degree to which a system is performing its intended function.

Let $\mathcal{R}_i = (x_i^m, x_i^M)$, where $x_i^m, x_i^M \in \mathbb{R}$ respectively denote the minimum and maximum values that the ith entry of x can take so as to guarantee the system is performing its intended function. Let $\mathcal{R} = \mathcal{R}_1 \times \mathcal{R}_2 \times \cdots \times \mathcal{R}_n$. Then, for any $t \geq 0$, we can define a probabilistic measure of the degree to which the system is performing its intended function as follows:

$$\begin{aligned} \rho(t) &= \Pr\big(X(t) \in \mathcal{R}\big) \\ &= \sum_{i \in \mathcal{Q}} \varphi_i \pi_i(t), \end{aligned} \tag{4.42}$$

where

$$\varphi_i = \int_{x \in \mathcal{R}} f_{X(t) \mid Q(t)}(x \mid i) dx. \tag{4.43}$$

A similar approach can be utilized to define a performance measure if the value of i in (4.4) were governed by a discrete-time Markov chain.

Example 4.4 (DC circuit performance under input and parameter uncertainty) Consider the same setting as in Example 4.3. Assume that for the circuit to perform its intended function, the value of the current i_2 needs to lie in the interval $[3.25\ \text{A}, 5.75\ \text{A}]$. Then, by plugging (4.39–4.40) into (4.43), we obtain:

$$\begin{aligned} \varphi_1 &= \int_{3.25}^{5.75} f_{I_2 \mid Q(t)}(i_2 \mid 1) di_2 \\ &= 1, \\ \varphi_2 &= \int_{3.25}^{5.75} f_{I_2 \mid Q(t)}(i_2 \mid 2) di_2 \\ &= 0.75, \end{aligned} \tag{4.44}$$

from where it follows that

$$\begin{aligned} \rho(t) &= \varphi_1 \pi_1(t) + \varphi_2 \pi_2(t) \\ &= 0.75 + 0.25 \mathrm{e}^{-\lambda_1 t}. \end{aligned} \tag{4.45}$$

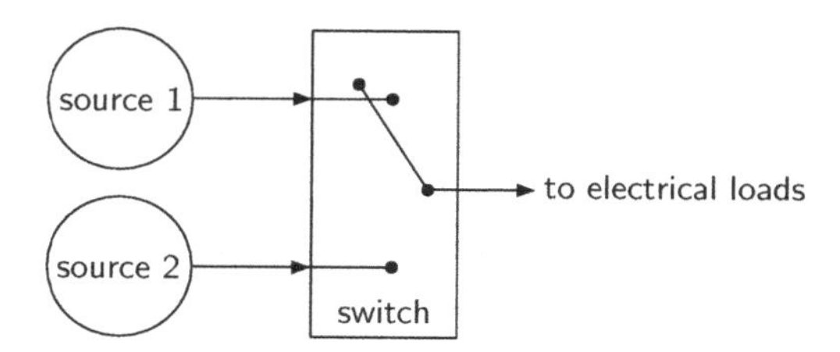

Figure 4.2 Cold standby power supply system.

4.5 Application to Reliability and Availability Analysis

In this section, we tailor the tools developed earlier to analyze the performance of multi-component systems subject to component failures and repairs. We first provide a systematic procedure to construct the system input-to-state model as this is typically the most difficult part of the analysis when handling systems with a large number of components. We then use this model together with stochastic models of the failure/repair behavior of individual components to construct Markov models describing the overall system behavior.

4.5.1 Multi-Component System Input-to-State Characterization

Consider a multi-component system comprising r components indexed by the elements in $\mathcal{C} = \{c_1, c_2, \dots, c_r\}$. Assume that each component c_i can operate in one of two modes: nominal and off-nominal (failed). Further assume that at $t = 0$, all the components are operating in their nominal mode, and as time evolves, transitions from the nominal to the failed mode may occur. Then, the system input-to-state behavior can be described by

$$x = h_i(w), \tag{4.46}$$

where $x \in \mathbb{R}^n$, $w \in \mathbb{R}^m$, and $h_i\colon \mathbb{R}^m \to \mathbb{R}^n$, $i \in \mathcal{Q} = \{1, 2, \dots, N\}$, is determined by (i) the components that are operating in their nominal mode, and (ii) the components that have transitioned from operating in their nominal to their failed mode, and the chronological order in which these transitions occurred. While the framework developed earlier allows one to capture uncertainty in the input w, in subsequent developments, we focus on the case when w is a constant vector.

Component Failure Sequences

Consider the power supply system depicted in Fig. 4.2. In this case, we have that $\mathcal{C} = \{\mathsf{source\ 1}, \mathsf{source\ 2}, \mathsf{switch}\}$. In their nominal mode, both electricity supply sources, source 1 and source 2, are able to produce any amount of electric power demanded by the electrical loads, whereas in their off-nominal mode, they are not able to produce any power. In its nominal mode, the switching element, switch, monitors the operational mode of source 1, and whenever source 1 is operating in its nominal mode, switch will connect source 1 to the electrical loads; otherwise, it will connect source 2. Thus, when all components are operating in their nominal mode, power to the electrical loads is supplied by source 1, with source 2 in cold standby, i.e., it can produce electric power if necessary.

Assume that all components are initially operating in their nominal mode, and after some time, source 1 transitions from operating in its nominal mode to its off-nominal mode. Then, according to the description above, since switch is operating in its nominal mode, it will be able to detect the mode transition and connect source 2 to the loads. Thus, since source 2 was also operating in its nominal mode before the mode transition of source 1, the loads will be supplied by source 2. Then, if after some time, switch transitions from its nominal mode of operation to its off-nominal mode, switch will not be able to monitor the mode of source 1; however, this does not have any impact on source 2, which will continue supplying power to the electrical loads.

Consider the reverse scenario to the one above, i.e., switch transitions from its nominal mode to its off-nominal mode first, and at some time later, source 1 transitions from its nominal mode to its off-nominal mode. In this case, after the first transition, switch is no longer able to monitor the mode of source 1, but the occurrence of this transition alone does not have any impact on the supply of electric power to the electrical loads since source 1 is still able to provide power. However, after the occurrence of the second transition, since switch is not able to detect a change in the operating mode of source 1, switch will not connect source 2 to the loads; therefore, the supply of power to the loads will be interrupted.

From the discussion above, it is clear the importance of capturing the chronological order in which the components of a system transition from operating in their nominal to their off-nominal mode. This can be accomplished via the notion of component failure sequence introduced next.

DEFINITION 4.1 (k-component failure sequence) Let $\mathcal{C} = \{c_1, c_2, \dots, c_r\}$ index the components of a system. A k-component failure sequence is a k-tuple

$$(c_{\ell_1}, c_{\ell_2}, c_{\ell_3}, \dots, c_{\ell_{k-1}}, c_{\ell_k}), \tag{4.47}$$

where $\ell_1, \ell_2, \ell_3, \dots, \ell_{k-1}, \ell_k$ are distinct elements of $\{1, 2, \dots, r\}$.

From the definition above, it should be clear that k-component failure sequences are formed by taking k elements in $\mathcal{C}$ and listing them in some particular order. For example, take $c_i, c_j \in \mathcal{C}$; then, the possible 2-component failure sequences that can be formed with these components are (c_i, c_j) and (c_j, c_i). In the context here, a k-component failure sequence attempts to encode the order in which components transition from operating in their nominal mode to operating in their failed mode, i.e., $(c_{\ell_1}, c_{\ell_2}, c_{\ell_3}, \dots, c_{\ell_{k-1}}, c_{\ell_k})$ is to be read as: the mode transition of component c_{ℓ_1} preceding in time the mode transition of component c_{ℓ_2}, and the mode transition of component c_{ℓ_2} preceding in time the mode transition of component c_{ℓ_3}, etc. In the remainder, we will use the 0-tuple, denoted by $()$, to capture the case when no transition has occurred, i.e., all components are operating in their nominal mode.

Example 4.5 (Failure sequences for cold standby system) Consider again the power supply system in Fig. 4.2. In order to construct all component failure sequences, we first list all nonempty subsets of $\mathcal{C} = \{\mathsf{source\ 1}, \mathsf{source\ 2}, \mathsf{switch}\}$:

$$\{\mathsf{source\ 1}\}, \quad \{\mathsf{source\ 2}\}, \quad \{\mathsf{switch}\},$$
$$\{\mathsf{source\ 1}, \mathsf{source\ 2}\}, \quad \{\mathsf{source\ 1}, \mathsf{switch}\}, \quad \{\mathsf{source\ 2}, \mathsf{switch}\},$$
$$\{\mathsf{source\ 1}, \mathsf{source\ 2}, \mathsf{switch}\}.$$

First, by considering the sets $\{\mathsf{source\ 1}\}$, $\{\mathsf{source\ 2}\}$, and $\{\mathsf{switch}\}$, it is clear that the 1-component failure sequences are:

$$(\mathsf{source\ 1}), \quad (\mathsf{source\ 2}), \quad (\mathsf{switch}).$$

Next, by ordering the elements in the sets $\{\mathsf{source\ 1}, \mathsf{source\ 2}\}$, $\{\mathsf{source\ 1}, \mathsf{switch}\}$, and $\{\mathsf{source\ 2}, \mathsf{switch}\}$, we obtain all 2-component failure sequences:

$$(\mathsf{source\ 1}, \mathsf{source\ 2}), \quad (\mathsf{source\ 2}, \mathsf{source\ 1}),$$
$$(\mathsf{source\ 1}, \mathsf{switch}), \quad (\mathsf{switch}, \mathsf{source\ 1}),$$
$$(\mathsf{source\ 2}, \mathsf{switch}), \quad (\mathsf{switch}, \mathsf{source\ 2}).$$

Finally, by ordering the elements in the set $\{\mathsf{source\ 1}, \mathsf{source\ 2}, \mathsf{switch}\}$, we obtain all 3-component failure sequences:

$(\mathsf{source\ 1}, \mathsf{source\ 2}, \mathsf{switch})$, $(\mathsf{source\ 1}, \mathsf{switch}, \mathsf{source\ 2})$, $(\mathsf{source\ 2}, \mathsf{source\ 1}, \mathsf{switch})$, $(\mathsf{source\ 2}, \mathsf{switch}, \mathsf{source\ 1})$, $(\mathsf{switch}, \mathsf{source\ 1}, \mathsf{source\ 2})$, $(\mathsf{switch}, \mathsf{source\ 2}, \mathsf{source\ 1})$.

It is standard to define a $(k+1)$-tuple $(a_1, a_2, \ldots, a_k, a_{k+1})$ iteratively as an ordered pair of the k-tuple $(a_1, a_2, \ldots, a_k)$ and a_{k+1}, i.e.,

$$(a_1, a_2, \ldots, a_{k+1}) = \big((a_1, a_2, \ldots, a_k), a_{k+1}\big). \tag{4.48}$$

This iterative definition of a tuple is very natural for recursively constructing component failure sequences as follows. Consider the following k-component failure sequence:

$$s_m^{(k)} = (c_{\ell_1}, c_{\ell_2}, c_{\ell_3}, \ldots, c_{\ell_{k-1}}, c_{\ell_k}),$$

where $\ell_1, \ell_2, \ell_3, \ldots, \ell_{k-1}, \ell_k$ are distinct elements of $\{1, 2, \ldots, r\}$. Then, we can obtain every $(k+1)$-component failure sequence in which the order of the first k component failures coincides with that of $s_m^{(k)}$ as follows:

$$s_{m_j}^{(k+1)} = \left(s_m^{(k)}, c_j\right), \quad \forall c_j \in \mathcal{C} \setminus \mathsf{set}\big(s_m^{(k)}\big), \tag{4.49}$$

where $\mathsf{set}\big(s_m^{(k)}\big)$ denotes the set that contains the elements that form $s_m^{(k)}$, i.e.,

$$\mathsf{set}\big(s_m^{(k)}\big) = \big\{c_{\ell_1}, c_{\ell_2}, c_{\ell_3}, \ldots, c_{\ell_{k-1}}, c_{\ell_k}\big\}.$$

As mentioned earlier, we can use the 0-tuple, (), to capture the case when no component mode transition has occurred. Then, we can extend the recursive relation in (4.49) to also include $s_1^{(0)} := ()$ as follows:

$$\begin{aligned} s_j^{(1)} &= \left(s_1^{(0)}, c_j\right) \\ &= (c_j), \qquad \forall c_j \in \mathcal{C} \setminus \mathsf{set}\left(s_1^{(0)}\right), \end{aligned} \tag{4.50}$$

with $\mathsf{set}\left(s_1^{(0)}\right) = \emptyset$.

Common Cause Failures. Note that in the developments thus far, we have ruled out the possibility of two components failing at the same time. However, there are certain scenarios in which the occurrence of some event causes several components to transition mode at the same time; such events are usually referred to as common cause failures. For example, a hurricane could potentially cause multiple transmission lines of a power system to fail simultaneously. With a slight abuse of notation, we can utilize the tuple representation above to also capture the occurrence of these common cause failures; the next example illustrates the approach. In the remainder, unless discussed in specific examples, we will rule out the occurrence of common cause failures so as to simplify the developments.

Example 4.6 (Common cause failure for cold standby system) Consider again the system in Fig. 4.2. Assume that the system may be subjected to a common cause failure, denoted by ccf, the occurrence of which would cause both source 1 and source 2 to fail at the same time. Then, besides the sequences involving the failure of one component at a time, we can list sequences that involve the occurrence of ccf. For example, we can encode the occurrence of ccf by the following 1-tuple:

$$(\mathsf{ccf}), \tag{4.51}$$

which is to be read as source 1 and source 2 have failed simultaneously because of the occurrence of ccf. The failure of source 1 followed by the occurrence of ccf would be encoded as

$$(\mathsf{source\ 1}, \mathsf{ccf}). \tag{4.52}$$

Note that in this case the occurrence of ccf would only cause source 2 to fail as source 1 was already in its failed mode.

System Status

Let $\mathcal{S}^{(k)}$, $k = 0, 1, \dots, r$, denote the set of all k-component failure sequences (note that $\mathcal{S}^{(0)}$ only contains the 0-tuple). Then, we can define the set of all component failure sequences as $\mathcal{S} = \cup_{k=0}^{r} \mathcal{S}^{(k)}$. Then, since the system input is assumed to be a known constant vector, the value the state takes only depends on each particular component failure sequence, i.e.,

$$x = \phi(s), \quad s \in \mathcal{S}, \tag{4.53}$$

where $\phi(\cdot)$ is some function that can be inferred from the model describing the system input-to-state behavior.

As in Section 4.4, let $\mathcal{R} = \mathcal{R}_1 \times \mathcal{R}_2 \times \cdots \times \mathcal{R}_n$, with $\mathcal{R}_i = (x_i^m, x_i^M)$, where $x_i^m, x_i^M \in \mathbb{R}$ respectively denote the minimum and maximum values that the ith entry of x can take so as to guarantee that the system is performing its intended function. Then, we can define the notion of system status as follows.

DEFINITION 4.2 (System status) Given $s \in \mathcal{S}$, a system is said to be

1. operational, if $x = \phi(s) \in \mathcal{R}$; and
2. nonoperational, if $x = \phi(s) \notin \mathcal{R}$.

Example 4.7 (Input-to-state model for cold standby system) Consider again the power supply system in Fig. 4.2. We will show how to partially construct the function $\phi(\cdot)$. Specifically, we will determine the value it takes for 0-, 1-, and 2-component failure sequences.

Take the state x to be the power delivered to the loads, and take the input w to be the sum of the power generated by both sources. Assume that when delivering power to the load, each source outputs a power of 1 p.u. (note that, even if both sources are operating in their nominal mode, a source will only deliver power if the switch position is such that the particular source is connected to the electrical loads). Thus, for the 0-component failure sequence, denoted by (), i.e., before any component mode transition has occurred, we have that $x = 1$ p.u.; thus,

$$\phi(()) = 1.$$

As discussed earlier, after the occurrence of failure sequence (source 1), switch will connect source 2 to the electrical loads. Then, since source 2 is operational, we have that $x = 1$ p.u.; therefore,

$$\phi\big((\text{source 1})\big) = 1.$$

Also, after the occurrence of a failure in source 2 or switch, if operational, we argued earlier that source 1 will still supply power to the electrical loads. Thus, for both failure sequences (source 2) and (switch), we have that $x = 1$ p.u.; therefore,

$$\phi\big((\text{source 2})\big) = 1, \quad \phi\big((\text{switch})\big) = 1.$$

It should be clear that if source 2 fails after source 1, or vice versa, it is not possible to supply any power to the loads, i.e., for failure sequences (source 1, source 2) or (source 2, source 1), we have that $x = 0$ p.u., and

$$\phi\big((\text{source 1}, \text{source 2})\big) = 0, \quad \phi\big((\text{source 2}, \text{source 1})\big) = 0.$$

Next, consider the case when source 1 fails first, followed by a failure of switch. Then, as argued earlier, since the loads are already being supplied by source 2, which in this case is operational, we have that $x = 1$ p.u.; thus,

$$\phi\big((\text{source 1}, \text{switch})\big) = 1.$$

Now, consider the case when switch fails first, followed by a failure of source 1. Then, as argued earlier, since switch stopped monitoring source 1 before the failure of the latter, switch will not connect the output of source 2 to the electrical loads. Thus, for failure sequence (switch, source 1), we have that $x = 0$ p.u., and

$$\phi\big((\text{switch}, \text{source 1})\big) = 0.$$

Finally, by following a similar analysis, one can easily see that for both failure sequences (source 2, switch) and (switch, source 2), we have that

$$\phi\big((\text{source 2}, \text{switch})\big) = 1, \quad \phi\big((\text{switch}, \text{source 2})\big) = 1.$$

Now, let $\mathcal{R} = (0.9, 1.1)$; then, according to the categorization given in Definition 4.2, it should be clear from the developments above that for (), (source 1), (source 2), (switch), (source 1, switch), (source 2, switch), and (switch, source 2), the system status is operational. On the other hand, for (source 1, source 2), (source 2, source 1), and (switch, source 1), the system status is nonoperational.

Attainable Failure Sequences

From the discussion thus far, it should be clear that each k-component failure sequence is just encoding a particular permutation of the elements of a subset of $\mathcal{C}$ formed by taking k elements. Thus, one may think that, in order to characterize the behavior of the system under analysis, it is necessary to consider all such possible permutations, which would yield a total number of $\sum_{k=0}^{r} \frac{r!}{(r-k)!}$ failure sequences; however, this is not the case in practice. This is due to the fact that not all failure sequences in $\mathcal{S}$ are *attainable*, a notion we define next.

DEFINITION 4.3 (Attainable k-component failure sequence) Consider the following k-component failure sequence:

$$(c_{\ell_1}, c_{\ell_2}, \ldots, c_{\ell_{k-1}}, c_{\ell_k}), \tag{4.54}$$

and all j-component failure sequences constructed by considering the first j elements of $(c_{\ell_1}, c_{\ell_2}, \ldots, c_{\ell_{k-1}}, c_{\ell_k})$, i.e., all failure sequences of the form

$$(c_{\ell_1}, c_{\ell_2}, \ldots, c_{\ell_{j-1}}, c_{\ell_j}), \; j = 1, 2, \ldots, k-1. \tag{4.55}$$

Then, the k-component failure sequence in (4.54) is said to be attainable if for every $s \in \big\{(c_{\ell_1}, c_{\ell_2}, \ldots, c_{\ell_{j-1}}, c_{\ell_j}), \; j = 1, 2, \ldots, k-1\big\}$, we have that

$$x = \phi(s) \in \mathcal{R}, \tag{4.56}$$

i.e., all the system configurations corresponding to the j-component failure sequences in (4.55) are operational.

Note that in Definition 4.3, we do not force the system to be operational for the k-component failure sequence $(c_{l_1}, c_{l_2}, \ldots, c_{l_{k-1}}, c_{l_k})$. However, it should be clear from the definition that if the system is nonoperational for this failure sequence, then, no $(k+l)$-component failure sequence, $l = 1, 2, \ldots, r-k$, with its k first components identical to those of $(c_{\ell_1}, c_{\ell_2}, c_{\ell_3}, \ldots, c_{l_{k-1}}, c_{\ell_k})$ is attainable. In the remainder, we assume that the system input-to-state behavior is not defined for such $(k+l)$-component failure sequences; this is a reasonable assumption because of the following argument. Consider some attainable component failure sequence after which the system is operational, and assume that an additional component failure occurs that results in the system status transitioning to being nonoperational. At this point, it is reasonable to assume that the system is shut down (de-energized). Then, once the system is de-energized, the notion of input-to-state behavior is no longer meaningful. Thus, for any subsequent component failure occurrence, since the system has already been de-energized, the notion of input-to-state behavior no longer applies.

The set of attainable component failure sequences, which we denote by $\mathcal{A}$, can be constructed recursively as follows. Let $\mathcal{A}^{(k)} \subseteq \mathcal{S}^{(k)}$, $k \geq 0$, denote the set of attainable k-component failure sequences. Also let $\mathcal{O}^{(k)}$ denote the subset of attainable k-component failure sequences for which the system is operational, and let $\mathcal{N}^{(k)}$ denote the subset of attainable k-component failure sequences for which the system is nonoperational. Then, clearly $\mathcal{O}^{(k)}$ and $\mathcal{N}^{(k)}$ form a partition of $\mathcal{A}^{(k)}$, i.e., $\mathcal{A}^{(k)} = \mathcal{O}^{(k)} \cup \mathcal{N}^{(k)}$, $\mathcal{O}^{(k)} \cap \mathcal{N}^{(k)} = \emptyset$. Assume that before the occurrence of any component failure, the system status is operational; thus, trivially, $\mathcal{S}^{(0)} = \mathcal{A}^{(0)} = \mathcal{O}^{(0)} = \{()\}$, and $\mathcal{N}^{(0)} = \emptyset$. Now, we can construct the set of attainable 1-component failure sequences, $\mathcal{A}^{(1)}$, which in this case coincides with the set of all 1-component failure sequences $\mathcal{S}^{(1)} = \{s_1^{(1)}, s_2^{(1)}, \ldots, s_r^{(1)}\}$, with $s_j^{(1)}$, $j = 1, 2, \ldots, r$, obtained using (4.50). Next, we need to partition $\mathcal{A}^{(1)} \equiv \mathcal{S}^{(1)}$ into $\mathcal{O}^{(1)}$ and $\mathcal{N}^{(1)}$ as follows:

$$\mathcal{O}^{(1)} = \left\{a_i^{(1)} \in \mathcal{A}^{(1)} : x = \phi\big(a_i^{(1)}\big) \in \mathcal{R}\right\}, \tag{4.57}$$

$$\mathcal{N}^{(1)} = \left\{a_i^{(1)} \in \mathcal{A}^{(1)} : x = \phi\big(a_i^{(1)}\big) \notin \mathcal{R}\right\}. \tag{4.58}$$

Let r_1 denote the number of attainable 1-component failure sequences for which the system is operational, i.e., $r_1 = \left|\mathcal{O}^{(1)}\right|$. Then, we can obtain elements in the set of attainable 2-component failure sequences, $\mathcal{A}^{(2)} = \left\{a_1^{(2)}, a_2^{(2)}, \ldots, a_{r_1(r-1)}^{(2)}\right\}$, recursively from the elements in the set $\mathcal{O}^{(1)} = \{o_1^{(1)}, o_2^{(1)}, \ldots, o_{r_1}^{(1)}\}$ as follows:

$$a_{m_j}^{(2)} = \big(o_m^{(1)}, c_j\big), \; \forall c_j \in \mathcal{C} \setminus \mathsf{set}\big(o_m^{(1)}\big). \tag{4.59}$$

Next, we partition $\mathcal{A}^{(2)}$ as follows:

$$\mathcal{O}^{(2)} = \{a_i^{(2)} \in \mathcal{A}^{(2)} : x = \phi\big(a_i^{(2)}\big) \in \mathcal{R}\}, \tag{4.60}$$

$$\mathcal{N}^{(2)} = \{a_i^{(2)} \in \mathcal{A}^{(2)} : x = \phi\big(a_i^{(2)}\big) \notin \mathcal{R}\}. \tag{4.61}$$

Thus, in general, given $\mathcal{A}^{(k)} = \mathcal{O}^{(k)} \cup \mathcal{N}^{(k)}$, where

$$\mathcal{O}^{(k)} = \{a_i^{(k)} \in \mathcal{A}^{(k)} : x = \phi(a_i^{(k)}) \in \mathcal{R}\}, \quad (4.62)$$

$$\mathcal{N}^{(k)} = \{a_i^{(k)} \in \mathcal{A}^{(k)} : x = \phi(a_i^{(k)}) \notin \mathcal{R}\}, \quad (4.63)$$

with $|\mathcal{O}^{(k)}| = r_k \geq 1$, we can construct the set of attainable $(k+1)$-component failure sequences, $\mathcal{A}^{(k+1)} = \{a_1^{(k+1)}, a_2^{(k+1)}, \ldots, a_{r_k(r-k)}^{(k+1)}\}$, recursively from the elements in the set $\mathcal{O}^{(k)} = \{o_1^{(k)}, o_2^{(k)}, \ldots, o_{r_k}^{(k)}\}$ as follows:

$$a_{m_j}^{(k+1)} = \left(o_m^{(k)}, c_j\right), \ \forall c_j \in \mathcal{C} \setminus \mathsf{set}\left(o_m^{(k)}\right). \quad (4.64)$$

This procedure is repeated until, for some $K \geq 1$, we have that $\mathcal{A}^{(K)} = \mathcal{N}^{(K)}$ and $\mathcal{O}^{(K)} = \emptyset$, i.e., there are no K-component failure sequences for which the system is operational. Then, the set of attainable component failure sequences, $\mathcal{A}$, is given by

$$\begin{aligned}\mathcal{A} &= \cup_{k=0}^{K} \mathcal{A}^{(k)} \\ &= \left(\cup_{k=0}^{K-1} \mathcal{O}^{(k)}\right) \cup \left(\cup_{k=1}^{K} \mathcal{N}^{(k)}\right). \end{aligned} \quad (4.65)$$

Example 4.8 (Attainable failure sequences for cold standby system) Consider again the system in Fig. 4.2. Since by assumption the system is operational for the 0-component failure sequence, (), we have $\mathcal{A}^{(0)} = \mathcal{O}^{(0)} \cup \mathcal{N}^{(0)}$, with

$$\begin{aligned}\mathcal{O}^{(0)} &= \{()\}, \\ \mathcal{N}^{(0)} &= \emptyset,\end{aligned}$$

and it trivially follows from Definition 4.3 that all 1-component failure sequences are attainable; thus,

$$\mathcal{A}^{(1)} = \{(\mathsf{source\ 1}), (\mathsf{source\ 2}), (\mathsf{switch})\}.$$

Then, since in Example 4.7 we established that the system is operational for all 1-component failure sequences, we have that $\mathcal{A}^{(1)} = \mathcal{O}^{(1)} \cup \mathcal{N}^{(1)}$, with

$$\begin{aligned}\mathcal{O}^{(1)} &= \{(\mathsf{source\ 1}), (\mathsf{source\ 2}), (\mathsf{switch})\}, \\ \mathcal{N}^{(1)} &= \emptyset;\end{aligned}$$

from where it follows that all 2-component failure sequences are attainable, i.e.,

$$\begin{aligned}\mathcal{A}^{(2)} = \{&(\mathsf{source\ 1}, \mathsf{source\ 2}), (\mathsf{source\ 2}, \mathsf{source\ 1}), (\mathsf{source\ 1}, \mathsf{switch}), \\ &(\mathsf{switch}, \mathsf{source\ 1}), (\mathsf{source\ 2}, \mathsf{switch}), (\mathsf{switch}, \mathsf{source\ 2})\}.\end{aligned}$$

Also, since for $(\mathsf{source\ 1}, \mathsf{switch})$, $(\mathsf{source\ 2}, \mathsf{switch})$, and $(\mathsf{switch}, \mathsf{source\ 2})$, the system is operational, whereas, for $(\mathsf{source\ 1}, \mathsf{source\ 2})$, $(\mathsf{source\ 2}, \mathsf{source\ 1})$, and (switch, source 1), the system is nonoperational, we have that $\mathcal{A}^{(2)} = \mathcal{O}^{(2)} \cup \mathcal{N}^{(2)}$, with

$$\begin{aligned}\mathcal{O}^{(2)} &= \{(\text{source 1}, \text{switch}), (\text{source 2}, \text{switch}), (\text{switch}, \text{source 2})\},\\ \mathcal{N}^{(2)} &= \{(\text{source 1}, \text{source 2}), (\text{source 2}, \text{source 1}), (\text{switch}, \text{source 1})\};\end{aligned}$$

thus,

(source 1, switch, source 2), (source 2, switch, source 1), (switch, source 2, source 1)

are attainable 3-component failure sequences, whereas

(source 1, source 2, switch), (source 2, source 1, switch), (switch, source 1, source 2)

are non-attainable 3-component failure sequences. Therefore,

$$\begin{aligned}\mathcal{A}^{(3)} =&\{(\text{source 1}, \text{switch}, \text{source 2}), (\text{source 2}, \text{switch}, \text{source 1}),\\ &(\text{switch}, \text{source 2}, \text{source 1})\}.\end{aligned}$$

Then, since all attainable 3-component failure sequences involve both source 1 and source 2, we have that $\mathcal{N}^{(3)} = \mathcal{A}^{(3)}$, and $\mathcal{O}^{(3)} = \emptyset$.

4.5.2 Systems with Non-Repairable Components

In this section, we develop Markov models for describing the behavior of multi-component systems comprising non-repairable components. We first consider the single-component case; this will allow the reader to become familiar with the terminology and techniques, and also to develop some intuition for the analysis of the general multi-component system case presented later.

Single-Component Case

Initially, the component is working, i.e., it is operating in its nominal mode, and as time progresses, it may fail, i.e., it may transition from operating in its nominal mode to operating in its off-nominal mode. Once this transition occurs, the component will remain in its off-nominal mode indefinitely.

Failure Rate. Assume that *the time to failure* of a non-repairable component, i.e., the time it takes the component to transition from its nominal to its off-nominal mode of operation can be described by a continuous random variable, T, with continuously differentiable cdf $F_T(t)$ and pdf $f_T(t) = dF_T(t)/dt = F'_T(t)$. The component failure rate at time t, which we denote by $\lambda(t)$, is then defined as follows:

$$\lambda(t) := \frac{f_T(t)}{1 - F_T(t)}. \tag{4.66}$$

While $F_T(t)$, $f_T(t)$, and $\lambda(t)$ convey the same information, it is more common to characterize a component by its failure rate. In the remainder we assume that $f_T(t) = 0$ for all $t < 0$; thus, $\lambda(t) = 0$ for all $t < 0$.

Reliability Function. The reliability of a non-repairable component at time t, which we denote by $R(t)$, is defined as follows:

$$\begin{aligned} R(t) &= \Pr(T > t) \\ &= 1 - F_T(t). \end{aligned} \tag{4.67}$$

It then follows from (4.66) and (4.67) that

$$\begin{aligned} \lambda(t) &= \frac{F_T'(t)}{1 - F_T(t)} \\ &= \frac{-R'(t)}{R(t)}; \end{aligned} \tag{4.68}$$

thus,

$$\frac{dR(t)}{R(t)} = -\lambda(t)dt. \tag{4.69}$$

Then, by integrating both sides of (4.69), we obtain that

$$\ln R(t) - \ln R(0) = -\int_0^t \lambda(\tau)d\tau, \tag{4.70}$$

which yields

$$R(t) = R(0)e^{-\int_0^t \lambda(\tau)d\tau}. \tag{4.71}$$

Finally, since we assumed the component is operating in its nominal mode at time $t = 0$, we have that $R(0) = \Pr(T > 0) = 1$, from where it follows that

$$R(t) = e^{-\int_0^t \lambda(\tau)d\tau}. \tag{4.72}$$

Mean Time to Failure. The mean time to failure (MTTF) of a non-repairable component is the expected value of the random variable T describing its time to failure. The MTTF is related to the reliability function, $R(\cdot)$, as follows:

$$\begin{aligned} \mathrm{E}[T] &= \int_0^\infty t f_T(t)dt \\ &= -\int_0^\infty tR'(t)dt \\ &= -tR(t)\Big|_0^\infty + \int_0^\infty R(t)dt \\ &= -\lim_{t\to\infty} tR(t) + \int_0^\infty R(t)dt \\ &= \int_0^\infty R(t)dt, \end{aligned} \tag{4.73}$$

where the last equality follows from the fact that $R(t)$ goes to zero as $t \to \infty$ exponentially fast.

Typical Failure Distributions. In reliability analysis, it is typical to assume that the failure rate is constant. This can be justified by examining the so-called

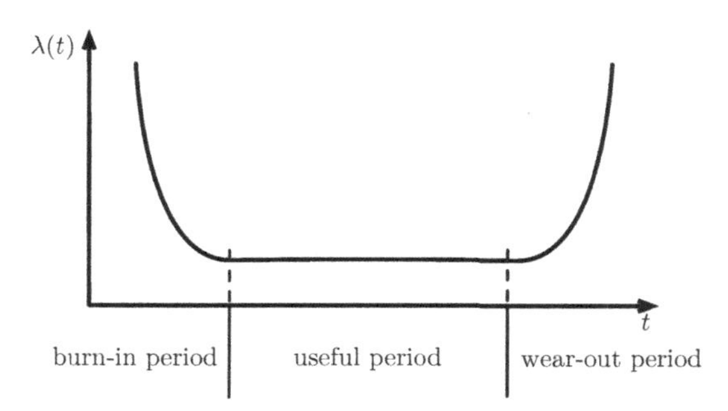

Figure 4.3 Bathtub Curve.

bathtub curve in Figure 4.3, which is just the graph of $\lambda(\cdot)$. The shape of the bathtub curve is a result of components having higher failure rates during the so-called burn-in period (shortly after their inception), and during the so-called wear-out period (near the end of their lifetime). If we assume proper testing by the manufacturer, the duration of the burn-in period for a component can be assumed to be zero from the perspective of the system in which the component is installed. In addition, components are usually designed in such a way that their useful lifetimes are much longer than the useful lifetime of the system in which they are installed, i.e., the duration of the wear-out period can also be assumed to be zero from the perspective of the system. Hence, the failure rate of a component is, in many instances, satisfactorily modeled with a constant failure rate, $\lambda(t) = \lambda$, where λ is some positive constant. In such case, (4.72) reduces to

$$R(t) = e^{-\lambda t}. \tag{4.74}$$

Thus, since $R(t) = 1 - F_T(t)$ and $f_T(t) = F'(t)$, we have that

$$f_T(t) = \lambda e^{-\lambda t}, \tag{4.75}$$

i.e., T is an exponentially-distributed random variable with parameter λ. By plugging the expression for $R(t)$ in (4.74) into (4.73), we obtain

$$\begin{aligned} \mathrm{E}[T] &= \int_0^\infty e^{-\lambda t} dt \\ &= \frac{1}{\lambda}. \end{aligned} \tag{4.76}$$

In some cases, the constant failure rate assumption may not hold as the components wear out during the projected system lifetime. In this regard, a popular model used to capture the wear-out period is the Weibull model. In this model, the failure rate is parametrized as follows:

$$\lambda(t) = \alpha \lambda^\alpha t^{\alpha-1}, \tag{4.77}$$

where $\lambda > 0$, and $\alpha \geq 1$. For $\alpha = 1$, we obtain that $\lambda(t) = \lambda$, which is the failure rate corresponding to an exponentially distributed time to failure. Now, recall from (4.66) that $\lambda(t) = \frac{f_T(t)}{1-F_T(t)}$; then

$$f_T(t) = \alpha\lambda^\alpha t^{\alpha-1} e^{-(\lambda t)^\alpha}, \; t \geq 0, \tag{4.78}$$

$$F_T(t) = 1 - e^{-(\lambda t)^\alpha}, \; t \geq 0; \tag{4.79}$$

therefore,

$$R(t) = e^{-(\lambda t)^\alpha}. \tag{4.80}$$

By plugging the expression for $R(t)$ in (4.80) into (4.73), we obtain that

$$\mathrm{E}[T] = \frac{1}{\lambda}\Gamma\Big(\frac{1}{\alpha} + 1\Big), \tag{4.81}$$

where $\Gamma\big(\frac{1}{\alpha} + 1\big) = \int_0^\infty t^{1/\alpha}\, e^{-t} dt$.

Markov Reliability Model. The behavior of a non-repairable component can also be described by a CTMC, $Q = \big\{Q(t)\colon t \geq 0\big\}$, $Q(t) \in \mathcal{Q}$, as follows. Let $\mathcal{Q} = \{1, 2\}$, with $Q(t) = 1$ if the component is operating in its nominal mode at time t, and $Q(t) = 2$ if the component is operating in its failed mode at time t. Assume that we are given the component failure rate, $\lambda(t)$, $t \geq 0$. Then, we have that

$$\begin{aligned}\Pr\Big(Q(t+\Delta t) = 2 \mid Q(t) = 1\Big) &= \Pr\Big(t \leq T \leq t + \Delta t \mid T > t\Big) \\ &\approx \frac{f_T(t)\Delta t}{1 - F_T(t)} \\ &= \lambda(t)\Delta t,\end{aligned} \tag{4.82}$$

and, since the component is not repairable, we also have that

$$\Pr\Big(Q(t+\Delta t) = 1 \mid Q(t) = 2\Big) = 0. \tag{4.83}$$

Define $\pi(t) = \big[\pi_1(t), \pi_2(t)\big]$, where $\pi_i(t) = \Pr\big(Q(t) = i\big)$, $i = 1, 2$. Then, it follows from the development in (4.28–4.34) that

$$\frac{d\pi(t)}{dt} = \pi(t)\Gamma(t), \tag{4.84}$$

where

$$\Gamma = \begin{bmatrix} -\lambda(t) & \lambda(t) \\ 0 & 0 \end{bmatrix}. \tag{4.85}$$

By solving (4.84) for $\pi_1(0) = 1$, and $\pi_2(0) = 0$ (we assume that at $t = 0$ the component is operating in its nominal mode), we obtain

$$\begin{aligned}\pi_1(t) &= e^{-\int_0^t \lambda(\tau)d\tau}, \\ \pi_2(t) &= 1 - e^{-\int_0^t \lambda(\tau)d\tau}.\end{aligned} \tag{4.86}$$

Notice that the expression for $\pi_1(t)$ in (4.86) is identical to that for $R(t)$ in (4.72). This makes sense as, for a non-repairable component, we defined its reliability at time t as $R(t) = \Pr(T > t)$, but since $\Pr(T > t) = \Pr\big(Q(t) = 1\big)$, then, clearly $R(t) = \pi_1(t)$.

Multi-Component Case

Now consider the general case of a system with r components, $c_1, c_2, \ldots, c_r$, with the value that the state $x \in \mathbb{R}^n$ takes described by

$$x = \phi(s), \tag{4.87}$$

where

$$s \in \mathcal{A} = \cup_{k=0}^{K} \mathcal{A}^{(k)}, \tag{4.88}$$

with $\mathcal{A}^{(k)} = \mathcal{O}^{(k)} \cup \mathcal{N}^{(k)}$, where $\mathcal{O}^{(k)}$ denotes the set of k-component failure sequences after which the system is operational, and $\mathcal{N}^{(k)}$ denotes the set of k-component failure sequences after which the system is nonoperational.

Assume that the time it takes for component c_i to fail, i.e., to transition from operating in its nominal mode to operating in its off-nominal (failed) mode, can be described by a random variable, T_{c_i}, with known pdf, $f_{T_{c_i}}(t)$, from which the component failure rate, $\lambda_{c_i}(t)$, can be inferred using (4.66). Also, assume that the T_{c_i}'s are pairwise independent. Then, the behavior of (4.87–4.88) is completely determined by a CTMC, $Q = \{Q(t) \colon t \geq 0\}$, with $Q(t) \in \mathcal{Q} = \{1, 2, \ldots, |\mathcal{A}|\}$, where each element in $\mathcal{Q}$ corresponds to a unique element in $\mathcal{A}$. To see this, first note that, since each element in $\mathcal{Q}$ corresponds to a unique component failure sequence, a transition from some $i \in \mathcal{Q}$ to some $j \in \mathcal{Q}$ can only occur if the attainable k- and $(k+1)$-component failure sequences corresponding to i and j, respectively denoted by $a_m^{(k)}$ and $a_n^{(k+1)}$, are such that $a_n^{(k+1)} = \big(a_m^{(k)}, c_\ell\big)$, $c_\ell \in \mathcal{C} \setminus \mathsf{set}\big(a_m^{(k)}\big)$. In words, a transition from i to j can only occur if the failure of component c_ℓ occurs after all the component failures defining the attainable failure sequence $a_m^{(k)}$ have occurred. Thus, if $Q(t) = i$, the time it takes to transition out of this state is governed by the failure of a component in $\mathcal{C} \setminus \mathsf{set}(a_m^{(k)})$; therefore, it is independent of the previous history of the process. Then the process Q satisfies the property in (4.10); thus Q is a Markov chain.

The discussion above also provides some intuition for the structure of the chain generator matrix; next, we provide a formal procedure for how to construct it. Following the notation in Section 4.5.1, let $r_k = \big|\mathcal{O}^{(k)}\big|$, $k = 0, 1, \ldots, K-1$, and denote the elements in $\mathcal{O}^{(k)}$ by $o_1^{(k)}, o_2^{(k)}, \ldots, o_{r_k}^{(k)}$. Similarly, let $q_k = \big|\mathcal{A}^{(k)}\big|$, $k = 0, 1, \ldots, K$, and denote the elements in $\mathcal{A}^{(k)}$ by $a_1^{(k)}, a_2^{(k)}, \ldots, a_{q_k}^{(k)}$. Then the generator matrix, $\Gamma(t) = \big[\gamma_{i,j}(t)\big] \in \mathbb{R}^{|\mathcal{A}|\times|\mathcal{A}|}$, of the Markov chain Q is given by

$$\begin{bmatrix} \Gamma^{(0,0)}(t) & \Gamma^{(0,1)}(t) & \cdots & \mathbf{0}_{1\times q_{K-1}} & \mathbf{0}_{1\times q_K} \\ \mathbf{0}_{q_1\times 1} & \Gamma^{(1,1)}(t) & \cdots & \mathbf{0}_{q_1\times q_{K-1}} & \mathbf{0}_{q_1\times q_K} \\ \vdots & \vdots & \cdots & \vdots & \vdots \\ \vdots & \vdots & \ddots & \vdots & \vdots \\ \mathbf{0}_{q_{K-1}\times 1} & \mathbf{0}_{q_{K-1}\times q_1} & \cdots & \Gamma^{(K-1,K-1)}(t) & \Gamma^{(K-1,K)}(t) \\ \mathbf{0}_{q_K\times 1} & \mathbf{0}_{q_K\times q_1} & \cdots & \mathbf{0}_{q_K\times q_{K-1}} & \mathbf{0}_{q_K\times q_K} \end{bmatrix}, \tag{4.89}$$

where

$$\Gamma^{(k,k)}(t) = \left[\gamma_{i,j}^{(k,k)}(t)\right] \in \mathbb{R}^{q_k \times q_k}, \ k = 0, 1, \ldots, K-1,$$

with

$$\gamma_{i,i}^{(k,k)}(t) = \begin{cases} -\displaystyle\sum_{c_\ell \in \mathcal{C}\setminus \mathsf{set}(o_i^{(k)})} \lambda_{c_\ell}(t), & i = 1, 2, \ldots, r_k, \\ 0, & i = r_k+1, r_k+2, \ldots, q_k, \end{cases}$$
$$\gamma_{i,j}^{(k,k)}(t) = 0, \quad \forall i \neq j, \tag{4.90}$$

and

$$\Gamma^{(k,k+1)}(t) = \left[\gamma_{i,j}^{(k,k+1)}(t)\right] \in \mathbb{R}^{q_k \times q_{k+1}}, \ k = 0, 1, \ldots, K-1,$$

with

$$\gamma_{i,j}^{(k,k+1)}(t) = \begin{cases} \lambda_{c_\ell}(t), & \text{if } a_j^{(k+1)} = \left(o_i^{(k)}, c_\ell\right), \ c_\ell \in \mathcal{C} \setminus \mathsf{set}\left(o_i^{(k)}\right), \\ 0, & \text{otherwise.} \end{cases} \tag{4.91}$$

Example 4.9 (Markov reliability model for cold standby system) Consider again the power supply system in Fig. 4.2. From Example 4.8, one can easily check that $|\mathcal{A}| = q_0 + q_1 + q_2 + q_3 = 13$; thus, in this case $\mathcal{Q} = \{1, 2, \ldots, 13\}$. Table 4.1 provides one possible mapping between each element $a \in \mathcal{A}$, and each element $i \in \mathcal{Q}$, along with the corresponding system status (as determined in Example 4.7).

Table 4.1 State space for cold standby system.

attainable failure sequence	i	system status
()	1	operational
(source 1)	2	operational
(source 2)	3	operational
(switch)	4	operational
(source 1,source 2)	5	nonoperational
(source 1,switch)	6	operational
(source 2,source 1)	7	nonoperational
(source 2,switch)	8	operational
(switch,source 1)	9	nonoperational
(switch,source 2)	10	operational
(source 1,switch,source 2)	11	nonoperational
(source 2,switch,source 1)	12	nonoperational
(switch,source 2,source 1)	13	nonoperational

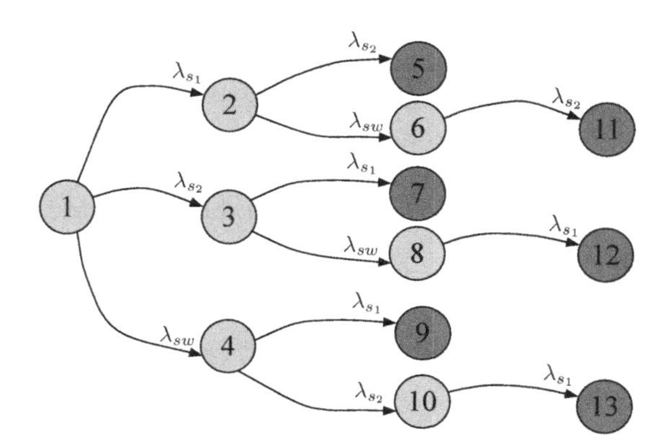

Figure 4.4 Markov reliability model for standby system.

Let λ_{s_1} and λ_{s_2} respectively denote the failure rates of source 1 and source 2; and let λ_{sw} denote the failure rate of switch. Then, by using the information in Table 4.1, and (4.90) and (4.91), we have that $\Gamma = [\gamma_{i,j}] \in \mathbb{R}^{13\times 13}$, with

- $\gamma_{1,1} = -(\lambda_{s_1} + \lambda_{s_2} + \lambda_{sw}), \quad \gamma_{1,2} = \lambda_{s_1}, \quad \gamma_{1,3} = \lambda_{s_2}, \quad \gamma_{1,4} = \lambda_{sw};$
- $\gamma_{2,2} = -(\lambda_{s_2} + \lambda_{sw}), \quad \gamma_{2,5} = \lambda_{s_2}, \quad \gamma_{2,6} = \lambda_{sw};$
- $\gamma_{3,3} = -(\lambda_{s_1} + \lambda_{sw}), \quad \gamma_{3,7} = \lambda_{s_1}, \quad \gamma_{3,8} = \lambda_{sw};$
- $\gamma_{4,4} = -(\lambda_{s_1} + \lambda_{s_2}), \quad \gamma_{4,9} = \lambda_{s_1}, \quad \gamma_{4,10} = \lambda_{s_2};$
- $\gamma_{6,6} = -\lambda_{s_2}, \quad \gamma_{6,11} = \lambda_{s_2};$
- $\gamma_{8,8} = -\lambda_{s_1}, \quad \gamma_{8,12} = \lambda_{s_1};$
- $\gamma_{10,10} = -\lambda_{s_1}, \quad \gamma_{10,13} = \lambda_{s_1};$

with all other entries being zero. Figure. 4.4 shows the state-transition diagram for this Markov reliability model.

Reliability Function. Let T denote a random variable describing the time elapsed until the system status transitions from being operational to being nonoperational. Then, at time t, the system reliability function, denoted by $R(\cdot)$, can be defined as we did earlier for a non-repairable component, i.e.,

$$R(t) = \Pr(T > t). \tag{4.92}$$

The value of $R(t)$ can be computed using the $\pi_i(t)$'s obtained by solving (4.34), with the generator matrix, $\Gamma(t)$, as defined in (4.89), as follows:

$$\begin{aligned} R(t) &= \Pr(T > t) \\ &= \sum_{i \in \mathcal{Q}_o} \pi_i(t) \\ &= 1 - \sum_{i \in \mathcal{Q}_n} \pi_i(t), \end{aligned} \tag{4.93}$$

where $\mathcal{Q}_o$ is a subset of $\mathcal{Q}$ whose elements correspond to elements in the set of attainable failure sequences for which the system status is operational, and $\mathcal{Q}_n$

is a subset of $\mathcal{Q}$ whose elements correspond to elements in the set of attainable failure sequences for which the system status is nonoperational. (Note that $\mathcal{Q}_o$ and $\mathcal{Q}_n$ form a partition of $\mathcal{Q}$.)

Failure Rate. Similar to the non-repairable component case, we can define the failure rate, $\lambda(t)$, of a non-repairable system as follows:

$$\lambda(t) = -\frac{R'(t)}{R(t)}. \tag{4.94}$$

Then, by noticing that

$$\begin{aligned} R'(t) &= \sum_{i \in \mathcal{Q}_o} \frac{d\pi_i(t)}{dt} \\ &= -\sum_{i \in \mathcal{Q}_n} \frac{d\pi_i(t)}{dt}; \end{aligned} \tag{4.95}$$

we have that

$$\begin{aligned} \lambda(t) &= -\frac{\sum_{i \in \mathcal{Q}_o} \sum_{j=1}^{|\mathcal{A}|} \gamma_{j,i} \pi_j(t)}{\sum_{i \in \mathcal{Q}_o} \pi_i(t)} \\ &= \frac{\sum_{l \in \mathcal{Q}_n} \sum_{j=1}^{|\mathcal{A}|} \gamma_{j,l} \pi_j(t)}{\sum_{i \in \mathcal{Q}_o} \pi_i(t)}. \end{aligned} \tag{4.96}$$

Mean Time to Failure. A similar development to that in (4.73) leads to the following expression for the MTTF of a non-repairable system:

$$\begin{aligned} \mathrm{E}[T] &= \int_0^\infty R(\tau) d\tau \\ &= \int_0^\infty \sum_{i \in \mathcal{Q}_o} \pi_i(\tau) d\tau. \end{aligned} \tag{4.97}$$

Example 4.10 (Reliability, failure rate, and MTTF of cold standby system) Consider again the power supply system of earlier examples. By combining the information in the last two columns of Table 4.1, it is easy to see that

$$\mathcal{Q}_o = \{1, 2, 3, 4, 6, 8, 10\}, \qquad \mathcal{Q}_n = \{5, 7, 9, 11, 12, 13\}.$$

Then, by using (4.93), we obtain

$$R(t) = \pi_1(t) + \pi_2(t) + \pi_3(t) + \pi_4(t) + \pi_6(t) + \pi_8(t) + \pi_{10}(t).$$

Also, by using (4.96), we obtain that

$$\lambda(t) = \frac{\lambda_{s_1}(\pi_3(t) + \pi_4(t) + \pi_8(t) + \pi_{10}(t)) + \lambda_{s_2}(\pi_2(t) + \pi_6(t))}{\pi_1(t) + \pi_2(t) + \pi_3(t) + \pi_4(t) + \pi_6(t) + \pi_8(t) + \pi_{10}(t)}.$$

Finally, by using (4.97), we have that

$$\mathrm{E}[T] = \int_0^\infty \big(\pi_1(\tau) + \pi_2(\tau) + \pi_3(\tau) + \pi_4(\tau) + \pi_6(\tau) + \pi_8(\tau) + \pi_{10}(\tau)\big) d\tau.$$

Failure Coverage

So far, we have assumed that a particular component failure sequence completely determines the system operational status. However, due to some phenomena not captured by the system input-to-state behavior model, there might be some uncertainty on whether or not the system remains operational for the particular failure sequence. Next, we show how to model this uncertainty and incorporate its effect into a Markov reliability model.

Let $o_m^{(k)}$ denote the mth k-component attainable failure sequence for which the system is operational. Assume that an additional component $c_\ell \in \mathcal{C} \setminus \mathsf{set}(o_m^{(k)})$ fails; thus, we can recursively define the resulting $(k+1)$-component failure sequence as $\big(o_m^{(k)}, c_\ell\big)$. Assume that, due to some unmodeled phenomena, there is some uncertainty on whether or not the system status is operational for $\big(o_m^{(k)}, c_\ell\big)$, $c_\ell \in \mathcal{C} \setminus \mathsf{set}\big(o_m^{(k)}\big)$. We can characterize this uncertainty via the notion of failure coverage.

DEFINITION 4.4 (Failure coverage) Assume the status of a system is operational for component failure sequence $o_m^{(k)}$. The failure coverage of component c_ℓ given $o_m^{(k)}$ is the probability that the system status is operational for component failure sequence $\big(o_m^{(k)}, c_\ell\big)$.

Now, since there are two possible outcomes associated with $\big(o_m^{(k)}, c_\ell\big)$, this component failure sequence cannot be mapped into a unique element of $\mathcal{Q}$ but it needs to be mapped into two different ones, j_o, j_n, with j_o corresponding to the case when the system status is operational for $\big(o_m^{(k)}, c_\ell\big)$, and j_n corresponding to the case when the system status is nonoperational for $\big(o_m^{(k)}, c_\ell\big)$. Let i_o denote the element in $\mathcal{Q}$ that corresponds to $o_m^{(k)}$, and denote by c_{i_o,j_o} the failure coverage of component c_ℓ given $o_m^{(k)}$. Then, for a small $\Delta t > 0$, we have that

$$\begin{aligned}
\Pr\Big(Q(t+\Delta t) = j_o \;\big|\; Q(t) = i_o\Big) &\approx c_{i_o,j_o}\lambda_{c_\ell}(t)\Delta t + \mathrm{o}(t), \\
\Pr\Big(Q(t+\Delta t) = j_n \;\big|\; Q(t) = i_o\Big) &\approx (1 - c_{i_o,j_o})\lambda_{c_\ell}(t)\Delta t + \mathrm{o}(t),
\end{aligned} \tag{4.98}$$

where $\lambda_{c_\ell}(t)$ denotes the failure rate of component c_ℓ. Thus, the (i_o, j_o) and (i_o, j_n) entries of the Markov reliability model generator matrix are given by:

$$\begin{aligned}
\gamma_{i_o,j_o} &= c_{i_o,j_o}\lambda_{c_\ell}(t), \\
\gamma_{i_o,j_n} &= (1 - c_{i_o,j_o})\lambda_{c_\ell}(t).
\end{aligned} \tag{4.99}$$

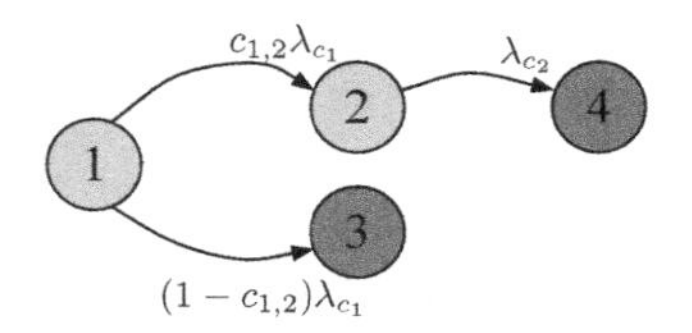

Figure 4.5 State-transition diagram for failure coverage example.

Example 4.11 (Failure coverage) Consider a system comprising two components, c_1 and c_2. For the system to be operational, either component needs to be operating in its nominal mode. Initially, only component c_1 is put into operation, whereas component c_2 is in standby and will only be put into operation in case component c_1 fails. Assume that while in operation, the failure rate of these components is λ_{c_1} and λ_{c_2}, respectively (the failure rate of component c_2 is zero when in standby). Also, let $c_{1,2}$ denote the failure coverage of component c_1 given the system is operating with no failed components.

In this case, the only 1-component failure sequence is that corresponding to the failure of component c_1 (component c_2 cannot fail before it is put into operation). However, after the failure of component c_1, there are two possible outcomes: (i) component c_2 is successfully put into operation and the system status remains operational, or (ii) component c_2 is not successfully put into operation, and the system status transitions to nonoperational. If component c_2 is successfully brought online after failure of component c_1, the system will remain operational until the failure of component c_2. Figure 4.5 shows the Markov reliability model for this system.

4.5.3 Systems with Repairable Components

In this section, we develop Markov models for analyzing multi-component systems comprising components that can be repaired. These Markov models can be obtained by adding additional transitions among states of the Markov reliability model for the same system when the components are non-repairable. As we did for non-repairable systems, we first consider the single-component case.

Single-Component Case

Repairable components are those that once they have failed, i.e., they transition from operating in their nominal mode to operating in their off-nominal mode, it is possible to restore them back to operation in their nominal mode. Such restoration process depends on the completion of a repair action. As we saw in Section 4.5.2, the failure behavior of a non-repairable component is completely characterized by its failure rate, which can be obtained from the pdf of the random variable T describing the time to failure. Similarly, assume that once in its failed mode, *the time to repair*, i.e., the time it takes to restore the component back to its operating mode, can be described by a random variable D with pdf $f_D(t)$ and cdf $F_D(t)$. Then we can define the component repair rate as follows:

$$\mu(t) = \frac{f_D(t)}{1 - F_D(t)}. \tag{4.100}$$

Availability Model. Given $\lambda(t)$ and $\mu(t)$, the failure/repair behavior of a repairable component can be described by a CTMC, $Q = \{Q(t): t \geq 0\}$, with $Q(t) \in \mathcal{Q} = \{1, 2\}$, and generator matrix

$$\Gamma = \begin{bmatrix} -\lambda(t) & \lambda(t) \\ \mu(t) & -\mu(t) \end{bmatrix}; \tag{4.101}$$

thus, $Q(t) = 1$ if the component is in its nominal mode at time t, and $Q(t) = 2$ if the component is in its failed mode at time t. This CTMC can then be used to define the notion of availability at time t, which we denote by $A(t)$, as follows:

$$\begin{aligned} A(t) &= \Pr\big(Q(t) = 1\big) \\ &= \pi_1(t); \end{aligned} \tag{4.102}$$

thus, we refer to this Markov chain as the Markov availability model. Another measure of performance that can be derived from $A(t)$ is the so-called long-term availability, $A = \lim_{t\to\infty} A(t)$; thus,

$$A = \lim_{t\to\infty} \pi_1(t). \tag{4.103}$$

Reliability Model. For a repairable component, we may also be interested in capturing the probability that at time t the component is operating in its nominal mode without having transitioned once to its off-nominal mode in the time interval $[0, t)$. This behavior can also be characterized by a CTMC, $Q = \{Q(t): t \geq 0\}$, with $Q(t) \in \mathcal{Q} = \{1, 2\}$, and generator matrix,

$$\Gamma = \begin{bmatrix} -\lambda(t) & \lambda(t) \\ 0 & 0 \end{bmatrix}. \tag{4.104}$$

This Markov chain is identical to that describing the failure behavior of a non-repairable component. This should not come as a surprise to the reader as here we are only interested in capturing the behavior of the repairable component from $t = 0$ until the first time the component transitions from operating in its nominal mode to operating in its off-nominal mode; this is also what the CTMC model describing the failure behavior of a non-repairable component captures. Thus, we can define the notion of reliability at time t, denoted by $R(t)$, as follows:

$$R(t) = \Pr\big(Q(t) = 1\big). \tag{4.105}$$

Multi-Component Case

Next, we consider the general case of a multi-component system with r components, $c_1, c_2, \ldots, c_r$, with the value that the state $x \in \mathbb{R}^n$ takes described by (4.87–4.88). As in the non-repairable case, assume that each component c_i can operate in one of two modes: nominal and off-nominal. Assume that transitions from the nominal mode to the off-nominal mode are random and occur at a rate $\lambda_{c_i}(t)$, with the individual transition times being pairwise independent.

Availability Model. First, consider the case when the system components cannot be repaired; thus, this is just the case discussed in Section 4.5.2. Then, the system failure behavior can be described by a CTMC

$$Q^{(1)} = \{Q^{(1)}(t) \colon t \geq 0\}, \; Q^{(1)}(t) \in \mathcal{Q},$$

with a generator matrix, $\Gamma^{(1)}(t) \in \mathbb{R}^{|\mathcal{A}| \times |\mathcal{A}|}$, constructed using the procedure in (4.90–4.91), and $\mathcal{Q} = \mathcal{Q}_o \cup \mathcal{Q}_n, \; \mathcal{Q}_o \cap \mathcal{Q}_n = \emptyset$, where, as before, the elements in $\mathcal{Q}_o$ correspond to elements in the set of attainable failure sequences for which the system status is operational, and where the elements in $\mathcal{Q}_n$ correspond to elements in the set of attainable failure sequences for which the system status is nonoperational.

Now, assume that all components in the system are repairable. Let $\mu_{i,j}$, $i \neq j$, denote the rate at which transitions from some state i to some other state j occur after one or more components have been repaired. These transitions include: (i) pairs of states, (i, j), for which both $i, j \in \mathcal{Q}_o$, and (ii) pairs of states, (i, j), for which $i \in \mathcal{Q}_n$, and $j \in \mathcal{Q}_o$ (we rule out transitions between pairs of states, (i, j), for which both $i, j \in \mathcal{Q}_n$). Use the $\mu_{i,j}(t)$'s to construct an $(|\mathcal{A}| \times |\mathcal{A}|)$-dimensional matrix, $\Gamma^{(2)}(t) = [\gamma^{(2)}_{i,j}(t)]$, as follows:

$$\gamma^{(2)}_{i,j}(t) = \begin{cases} \mu_{i,j}(t), & \text{if } i \neq j, \\ -\sum_{l=1}^{|\mathcal{A}|} \mu_{i,l}(t), & \text{otherwise.} \end{cases} \tag{4.106}$$

Then, the failure/repair behavior of the system can be also described by a CTMC, $Q = \{Q(t) \colon t \geq 0\}$, $Q(t) \in \mathcal{Q}$, with generator matrix, $\Gamma(t) \in \mathbb{R}^{|\mathcal{A}| \times |\mathcal{A}|}$, defined as follows:

$$\Gamma(t) = \Gamma^{(1)}(t) + \Gamma^{(2)}(t). \tag{4.107}$$

The transitions associated with component repairs depend on the repair policies and may involve more than one component. Thus, it may be possible to have more than one Markov availability model for the same system as illustrated in the example that follows.

Example 4.12 (Two repairable components in parallel) Consider a system with two components, c_1 and c_2, with failure rates λ_{c_1} and λ_{c_2}, respectively. For this system to be operational, it is necessary that at least one component is operating in its nominal mode. The attainable component failure sequences are listed in

Table 4.2 Failure sequences for system with two non-repairable components in parallel.

attainable failure sequence	i	system status
$()$	1	operational
(c_1)	2	operational
(c_2)	3	operational
(c_1, c_2)	4	nonoperational
(c_2, c_1)	5	nonoperational

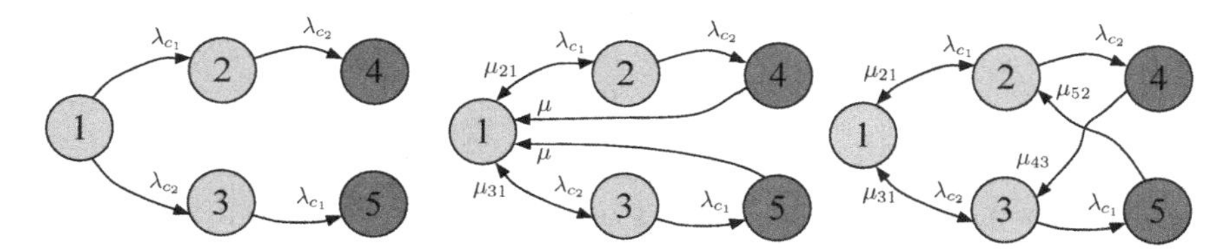

Figure 4.6 State-transition diagrams for system comprised of two components in parallel.

Table 4.2 along with the corresponding system status. The table also provides a mapping between the attainable failure sequences and each $i \in \mathcal{Q}$.

First, assume that both components are non-repairable; Fig. 4.6 (left) shows the state transition diagram for the resulting Markov reliability model, from which one can also infer that

$$\Gamma^{(1)} = \begin{bmatrix} -(\lambda_{c_1} + \lambda_{c_2}) & \lambda_{c_1} & \lambda_{c_2} & 0 & 0 \\ 0 & -\lambda_{c_2} & 0 & \lambda_{c_2} & 0 \\ 0 & 0 & -\lambda_{c_1} & 0 & \lambda_{c_1} \\ 0 & 0 & 0 & 0 & 0 \\ 0 & 0 & 0 & 0 & 0 \end{bmatrix}. \tag{4.108}$$

Next, consider the following repair strategy:

A1. If the system is operating in state 2 (state 3), i.e., component c_1 (component c_2) has failed but the system is still operational; then a repair action is triggered. When completed, this repair action will restore the system back to operating in state 1.

A2. If both components fail, i.e., the system is in either state 4 or state 5, then a repair action is triggered. When completed, this repair action restores both components back to their nominal mode and the system is therefore restored back to operating in state 1.

Assume that the time it takes to complete the repair of component c_1 if the system is operating in state 2 is exponentially distributed with parameter μ_{21}. Similarly, assume that the time it takes to complete the repair of component c_2 if the system is operating in state 3 is exponentially distributed with parameter μ_{31}. Finally, assume that if both components are failed (i.e., the system is operating in either state 4 or state 5), the time it takes to repair them both is exponentially distributed with parameter μ. Then, the system failure/repair behavior can be described by a Markov chain whose state-transition diagram is displayed in Fig. 4.6 (center), from where one can infer that

$$\Gamma = \Gamma^{(1)} + \Gamma^{(2)}, \tag{4.109}$$

with $\Gamma^{(1)}$ as in (4.108), and

$$\Gamma^{(2)} = \begin{bmatrix} 0 & 0 & 0 & 0 & 0 \\ \mu_{21} & -\mu_{21} & 0 & 0 & 0 \\ \mu_{31} & 0 & -\mu_{31} & 0 & 0 \\ \mu & 0 & 0 & -\mu & 0 \\ \mu & 0 & 0 & 0 & -\mu \end{bmatrix}. \tag{4.110}$$

Now, consider a repair strategy that exploits the fact that if both components have failed (i.e., the system status is nonoperational), it is possible to restore the system status to operational after only one component has been repaired; thus:

B1. Same as in A1 above.

B2. If both components are failed (i.e., the system is operating in either state 4 or state 5), the component that failed first is repaired first, i.e., component c_1 is repaired first if system is in state 4 and component c_2 is repaired first if system is in state 5. Then, once the particular component is repaired, the system is immediately restored back to operation. Afterwards, the other component is repaired, with a repair strategy identical to the one in B1.

Assume that the time it takes to complete the repair of component c_1 if the system is operating in state 2 is exponentially distributed with parameter μ_{21}. Similarly, assume that the time it takes to complete the repair of component c_2 if the system is operating in state 3 is exponentially distributed with parameter μ_{31}. Finally, assume that if the system is in state 4, the time it takes to repair component c_1 is exponentially distributed with parameter μ_{43}; and if the system is in state 5, the time it takes to repair component c_2 is exponentially distributed with parameter μ_{52}. Then, the system failure/repair behavior can be described by a Markov chain whose state-transition diagram is displayed in Fig. 4.6 (right).

Earlier we defined the availability function of a single repairable component, $A(t)$, as the probability that the component is operating in its nominal mode at time t. This notion can be extended to the multi-component case as follows:

$$A(t) = \sum_{i \in \mathcal{Q}_o} \pi_i(t), \tag{4.111}$$

where, following earlier notation, $\mathcal{Q}_o$ denotes the subset of $\mathcal{Q}$ the elements of which correspond to failure sequences for which the system status is operational. Also, similarly to the single-repairable component case, we can define the long-term availability of the system, A, as follows:

$$A = \lim_{t \to \infty} A(t) \tag{4.112}$$
$$= \sum_{i \in \mathcal{Q}_o} \pi_i. \tag{4.113}$$

Reliability Model. As with the single repairable component case, one may be interested in characterizing the probability that a system comprising repairable components remains in its operational status continuously. Recall that for a single repairable component, the Markov reliability model could be constructed from the Markov availability model by suppressing the transition from the failed state to the nominal (non-failed) state. The idea here is the same, i.e., the Markov reliability model can be constructed by suppressing transitions out of states for which the system status is nonoperational. Let $\Gamma^{(1)}(t)$ denote the state transition matrix of a Markov chain describing the system behavior when components cannot be repaired; this matrix can be constructed using the procedure in (4.90–4.91), and let $\Gamma^{(2)}(t)$ denote the matrix in (4.106) constructed by using the repair rates. As noted earlier, the matrix $\Gamma^{(2)}(t)$ includes repair transitions out of states for which the system status is nonoperational. Now, construct an $\big(|\mathcal{A}| \times |\mathcal{A}|\big)$-dimensional matrix $\Gamma^{(3)}(t) = \big[\gamma^{(3)}_{i,j}(t)\big]$ as follows:

$$\gamma^{(3)}_{i,j}(t) = \begin{cases} \mu_{i,j}(t), & \text{if } i \neq j, \\ -\sum_{l=1}^{|\mathcal{A}|} \mu_{i,l}(t), & \text{otherwise,} \end{cases} \tag{4.114}$$

for all $i \in \mathcal{Q}_n$, and $\gamma^{(3)}_{i,j}(t) = 0$, otherwise. Then, the Markov reliability model generator matrix, $\Gamma(t) \in \mathbb{R}^{|\mathcal{A}| \times |\mathcal{A}|}$, can be obtained as follows:

$$\Gamma(t) = \Gamma^{(1)}(t) + \Gamma^{(2)}(t) - \Gamma^{(3)}(t). \tag{4.115}$$

Example 4.13 (Reliability of two repairable components in series) Consider the system in Example 4.12 under the repair strategy in A1 – A2; see Fig. 4.6 (center) for the state-transition diagram of the resulting Markov availability model. In this case, $\mathcal{Q}_n = \{4, 5\}$; thus, $\Gamma = \Gamma^{(1)} + \Gamma^{(2)} - \Gamma^{(3)}$, with $\Gamma^{(1)}$ and $\Gamma^{(2)}$ as given in (4.108) and (4.110), respectively, and

$$\Gamma^{(3)} = \begin{bmatrix} 0 & 0 & 0 & 0 & 0 \\ 0 & 0 & 0 & 0 & 0 \\ 0 & 0 & 0 & 0 & 0 \\ \mu & 0 & 0 & -\mu & 0 \\ \mu & 0 & 0 & 0 & -\mu \end{bmatrix}. \tag{4.116}$$

4.5.4 Reduced-Order Models

In this section, we discuss two techniques to reduce the dimension of Markov reliability and availability models. The first one, referred to as truncation, is suitable for Markov models describing the failure behavior of systems with non-repairable components. The second technique, referred to as aggregation, can be applied to any class of systems, but it is best suited to the analysis of systems comprising collections of identical components.

Truncation

Let $Q = \{Q(t) \colon t \geq 0\}$, with $Q(t)$ taking values in some finite set $\mathcal{Q}$, be a Markov chain describing the failure behavior of a system with non-repairable components. Following earlier notation, we can partition the state space as follows: $\mathcal{Q} = \mathcal{Q}_o \cup \mathcal{Q}_n$, with $\mathcal{Q}_o = \cup_{k=0}^{K-1} \mathcal{Q}_o^{(k)}$, and $\mathcal{Q}_n = \cup_{k=1}^{K} \mathcal{Q}_n^{(k)}$, where each element in $\mathcal{Q}_o^{(k)}$ is uniquely associated with an element in $\mathcal{O}^{(k)}$, and where each element in $\mathcal{Q}_n^{(k)}$ is uniquely associated with an element in $\mathcal{N}^{(k)}$.

For some $K_0 \leq K - 1$, let $\widehat{\mathcal{Q}}_o^{(k)} = \mathcal{Q}_o^{(k)}$, $k = 0, 1, 2, \ldots, K_0$, and $\widehat{\mathcal{Q}}_n^{(k)} = \mathcal{Q}_n^{(k)}, k = 1, 2 \ldots K_0$; and define

$$\widehat{\mathcal{Q}} = \widehat{\mathcal{Q}}_o \cup \widehat{\mathcal{Q}}_n,$$

where $\widehat{\mathcal{Q}}_o = \cup_{k=0}^{K_0} \widehat{\mathcal{Q}}_o^{(k)}$ and $\widehat{\mathcal{Q}}_n = \cup_{k=1}^{K_0} \widehat{\mathcal{Q}}_n^{(k)}$; clearly $\widehat{\mathcal{Q}} \subseteq \mathcal{Q}$. Next, define the following Markov chain:

$$\widehat{Q} = \{\widehat{Q}(t),\ t \geq 0\},$$

where $\widehat{Q}(t)$ takes values in $\widehat{\mathcal{Q}}$, with the transitions in and out of the states in $\widehat{\mathcal{Q}} \setminus \widehat{\mathcal{Q}}_o^{(K_0)}$ being identical to those in

$$\mathcal{Q}_p = \left(\cup_{k=0}^{K_0-1} \mathcal{Q}_o^{(k)}\right) \cup \left(\cup_{k=1}^{K_0} \mathcal{Q}_n^{(k)}\right)$$

for the Markov chain Q, whereas for the states in $\widehat{\mathcal{Q}}_o^{(K_0)}$, the transitions in are the same as those for the states in $\mathcal{Q}_o^{(K_0)}$, but there are no transitions out. Let $\widehat{\pi}_i(t),\ i \in \widehat{\mathcal{Q}}$; then, from the description of the Markov chain $\widehat{Q}$, it follows that

$$\widehat{\pi}_i(t) = \pi_i(t), \quad i \in \widehat{\mathcal{Q}} \setminus \widehat{\mathcal{Q}}_o^{(K_0)} \equiv \mathcal{Q}_p, \tag{4.117}$$

and

$$\sum_{i \in \widehat{\mathcal{Q}}_o^{(K_0)}} \widehat{\pi}_i(t) = \sum_{i \in \cup_{k=K_0}^{K-1} \mathcal{Q}_o^{(k)}} \pi_i(t) + \sum_{i \in \cup_{k=K_0+1}^{K} \mathcal{Q}_n^{(k)}} \pi_i(t). \tag{4.118}$$

Now, we can utilize the Markov chain $\widehat{Q}$ to compute upper and lower bounds on $R(t) = \sum_{i \in \mathcal{Q}_o} \pi_i(t)$, for any $t \geq 0$. To this end, define

$$\widehat{R}(t) = \sum_{i \in \widehat{\mathcal{Q}}_o \setminus \widehat{\mathcal{Q}}_o^{(K_0)}} \widehat{\pi}_i(t). \tag{4.119}$$

Then,

$$R(t) = \sum_{i \in \cup_{k=0}^{K_0-1} \mathcal{Q}_o^{(k)}} \pi_i(t) + \sum_{i \in \cup_{k=K_0}^{K-1} \mathcal{Q}_o^{(k)}} \pi_i(t)$$
$$= \underbrace{\sum_{i \in \widehat{\mathcal{Q}}_o \backslash \widehat{\mathcal{Q}}_o^{(K_0)}} \widehat{\pi}_i(t)}_{= \widehat{R}(t)} + \sum_{i \in \cup_{k=K_0}^{K-1} \mathcal{Q}_o^{(k)}} \pi_i(t), \tag{4.120}$$

where the last equality follows from (4.117); thus, $R(t) \geq \widehat{R}(t)$. Also, from (4.118), we have that

$$\sum_{i \in \widehat{\mathcal{Q}}_o^{(K_0)}} \widehat{\pi}_i(t) \geq \sum_{i \in \cup_{k=K_0}^{K-1} \mathcal{Q}_o^{(k)}} \pi_i(t); \tag{4.121}$$

then it follows from (4.120) that

$$R(t) \leq \widehat{R}(t) + \sum_{i \in \widehat{\mathcal{Q}}_o^{(K_0)}} \widehat{\pi}_i(t). \tag{4.122}$$

To summarize, we can compute upper and lower bounds on $R(t)$ by using the truncated model as follows:

$$\widehat{R}(t) \leq R(t) \leq \widehat{R}(t) + \sum_{i \in \widehat{\mathcal{Q}}_o^{(K_0)}} \widehat{\pi}_i(t). \tag{4.123}$$

Example 4.14 (Truncated model for three-component system) Consider a system comprising three non-repairable components, c_1, c_2, c_3, with failure rates λ_{c_1}, λ_{c_2}, and λ_{c_3}, respectively. Assume that system structure is such that its failure behavior is described by a Markov chain, Q, with the state-transition diagram displayed in Fig. 4.7 (left).

Now consider a truncated Markov chain, $\widehat{Q}$, obtained from Q using the procedure described above for $K_0 = 2$; Fig. 4.7 (right) displays the state-transition

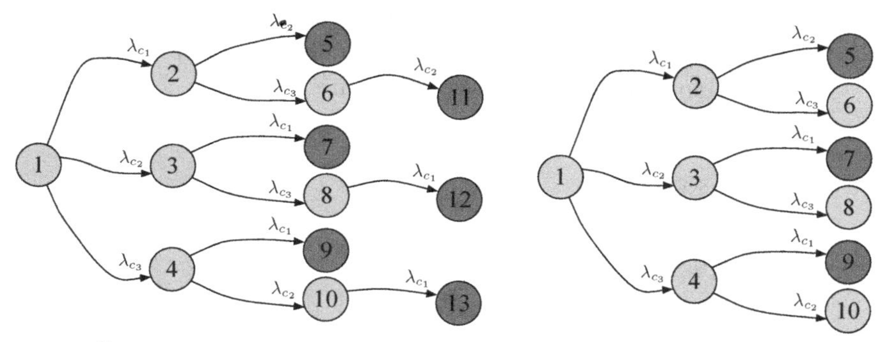

Figure 4.7 State-transition diagrams for three-component system.

diagram for this truncated model. It should be clear from the state-transition diagram in Fig. 4.7 (left) that:

$$R(t) = \pi_1(t) + \pi_2(t) + \pi_3(t) + \pi_4(t) + \pi_6(t) + \pi_8(t) + \pi_{10}(t). \tag{4.124}$$

Now, if we use (4.119), we obtain that

$$\widehat{R}(t) = \widehat{\pi}_1(t) + \widehat{\pi}_2(t) + \widehat{\pi}_3(t) + \widehat{\pi}_4(t), \tag{4.125}$$

and by using (4.123), we have that $\widehat{R}(t) \leq R(t) \leq \widehat{R}(t) + \widehat{\pi}_6(t) + \widehat{\pi}_8(t) + \widehat{\pi}_{10}(t)$.

Aggregation

Let $Q = \{Q(t) \colon t \geq 0\}$, with $Q(t)$ taking values in some finite set $\mathcal{Q}$, denote a Markov chain describing the failure/repair behavior of a multi-component system. By aggregation of states, we mean the combination of two or more states in $\mathcal{Q}$ into a single aggregated state, so as to obtain a reduced-order Markov chain, $\widetilde{Q} = \{\widetilde{Q}(t) \colon t \geq 0\}$, with $\widetilde{Q}$ taking values in some set $\widetilde{\mathcal{Q}}$. Let

$$\mathcal{Q} = \mathcal{Q}_u \cup \mathcal{Q}_{a_1} \cup \mathcal{Q}_{a_2} \cup \cdots \cup \mathcal{Q}_{a_r},$$

with $\mathcal{Q}_u \cap \mathcal{Q}_{a_i} = \emptyset$ for all i, and $\mathcal{Q}_{a_i} \cap \mathcal{Q}_{a_j} = \emptyset$, $i \neq j$. Here, $\mathcal{Q}_u$ denotes the subset of $\mathcal{Q}$ whose elements will not be aggregated, and $\mathcal{Q}_{a_i}$, $i = 1, 2, \ldots, r$, denotes a subset of $\mathcal{Q}$ whose elements will be aggregated. Let a_i, $i = 1, 2, \ldots, r$, denote the state in $\widetilde{\mathcal{Q}}$ that corresponds to the aggregation of the states in $\mathcal{Q}_{a_i}$; then, $\widetilde{\mathcal{Q}} = \mathcal{Q}_u \cup \{a_1, a_2, \ldots, a_r\}$. Let $\widetilde{\pi}_i(t) = \Pr\big(\widetilde{Q}(t) = i\big)$, $i \in \widetilde{\mathcal{Q}}$; then, the Markov chain $\widetilde{Q}$ needs to satisfy the following two properties:

P1. $\widetilde{\pi}_i(t) = \pi_i(t)$ for all $i \in \mathcal{Q}_u$, and all $t \geq 0$.
P2. $\widetilde{\pi}_{a_i}(t) = \sum_{j \in \mathcal{Q}_{a_i}} \pi_j(t)$ for all $i \in \{1, 2, \ldots, r\}$, and all $t \geq 0$.

In general, obtaining an aggregated model that satisfies Properties P1 and P2 may result in transition rates that are dependent on the probabilities of the unaggregated model as illustrated next.

Example 4.15 (Two non-repairable components in parallel) Consider a Markov reliability model with state-transition diagram as in Fig. 4.8 (a); the corresponding Chapman-Kolmogorov equations are as follows:

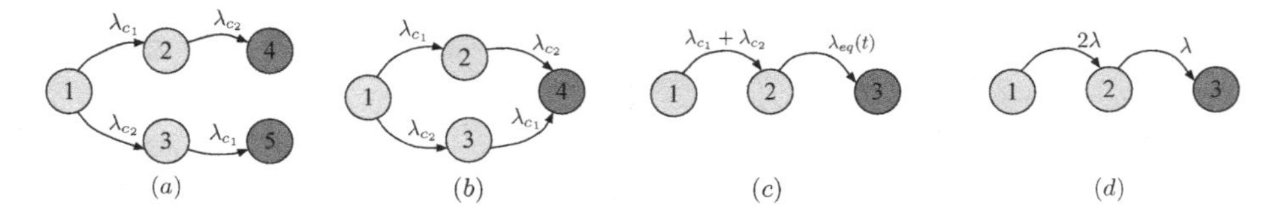

Figure 4.8 State-transition diagrams for Example 4.15.

$$
\begin{aligned}
\frac{d\pi_1(t)}{dt} &= -(\lambda_{c_1} + \lambda_{c_2})\pi_1(t), \\
\frac{d\pi_2(t)}{dt} &= \lambda_{c_1}\pi_1(t) - \lambda_{c_2}\pi_2(t), \\
\frac{d\pi_3(t)}{dt} &= \lambda_{c_2}\pi_1(t) - \lambda_{c_1}\pi_3(t), \\
\frac{d\pi_4(t)}{dt} &= \lambda_{c_2}\pi_2(t), \\
\frac{d\pi_5(t)}{dt} &= \lambda_{c_1}\pi_3(t).
\end{aligned} \tag{4.126}
$$

By adding the equations in the fourth and fifth lines of (4.126), and letting $\widetilde{\pi}_4(t) = \pi_4(t) + \pi_5(t)$, and $\widetilde{\pi}_i(t) = \pi_i(t),\ i = 1, 2, 3$, we obtain

$$
\begin{aligned}
\frac{d\widetilde{\pi}_1(t)}{dt} &= -(\lambda_{c_1} + \lambda_{c_2})\widetilde{\pi}_1(t), \\
\frac{d\widetilde{\pi}_2(t)}{dt} &= \lambda_{c_1}\widetilde{\pi}_1(t) - \lambda_{c_2}\widetilde{\pi}_2(t), \\
\frac{d\widetilde{\pi}_3(t)}{dt} &= \lambda_{c_2}\widetilde{\pi}_1(t) - \lambda_{c_1}\widetilde{\pi}_3(t), \\
\frac{d\widetilde{\pi}_4(t)}{dt} &= \lambda_{c_2}\widetilde{\pi}_2(t) + \lambda_{c_1}\widetilde{\pi}_3(t).
\end{aligned} \tag{4.127}
$$

By inspection of (4.127), one can easily see that this set of differential equations corresponds to the Chapman–Kolmogorov equations of a Markov chain, $\widetilde{\mathcal{Q}}$, with the state-transition diagram displayed in Fig. 4.8 (b). Furthermore, we can also see that this Markov chain satisfies Properties P1 and P2. In other words, state 4 in the Markov chain in Fig. 4.8 (b) results from aggregation of state 4 and state 5 in the Markov chain in Fig. 4.8 (a). Now, by adding the equations in the second and third lines of (4.127), we obtain

$$
\frac{d\big(\widetilde{\pi}_2(t) + \widetilde{\pi}_3(t)\big)}{dt} = (\lambda_{c_1} + \lambda_{c_2})\widetilde{\pi}_1(t) - \lambda_{c_2}\widetilde{\pi}_2(t) - \lambda_{c_1}\widetilde{\pi}_3(t). \tag{4.128}
$$

Next, by letting $\widehat{\pi}_2(t) = \widetilde{\pi}_2(t) + \widetilde{\pi}_3(t)$, $\widehat{\pi}_1(t) = \widetilde{\pi}_1(t)$, and $\widehat{\pi}_3(t) = \widetilde{\pi}_4(t)$, and defining

$$
\lambda_{eq}(t) = \frac{\lambda_{c_2}\widetilde{\pi}_2(t) + \lambda_{c_1}\widetilde{\pi}_3(t)}{\widetilde{\pi}_2(t) + \widetilde{\pi}_3(t)}, \tag{4.129}
$$

we obtain that

$$
\begin{aligned}
\frac{d\widehat{\pi}_1(t)}{dt} &= -(\lambda_{c_1} + \lambda_{c_2})\widehat{\pi}_1(t), \\
\frac{d\widehat{\pi}_2(t)}{dt} &= (\lambda_{c_1} + \lambda_{c_2})\widetilde{\pi}_1(t) - \lambda_{eq}(t)\widehat{\pi}_2(t), \\
\frac{d\widehat{\pi}_3(t)}{dt} &= \lambda_{eq}(t)\widehat{\pi}_2(t).
\end{aligned} \tag{4.130}
$$

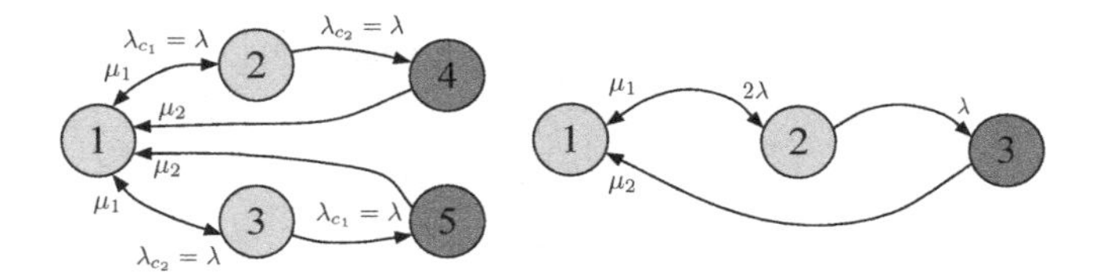

Figure 4.9 State aggregation for dual redundant system.

By inspection of (4.130), one can see that this set of differential equations corresponds to the Chapman–Kolmogorov equations of a Markov chain, $\widehat{\mathcal{Q}}$, with the state-transition diagram in Fig. 4.8 (c). This aggregated model also satisfies Properties P1 and P2; however, the transition rates now depend on the probabilities of the unaggregated model states. Thus, from a practical point of view, this model is of little use as one would need to first solve the unaggregated model and compute $\lambda_{eq}(t)$. On the other hand, if, e.g., $\lambda_{c_1} = \lambda_{c_2} = \lambda$, then, $\lambda_{eq}(t) = \lambda$; see Fig. 4.8 (d) for the state-transition diagram of the resulting Markov chain. Thus, if the transition rates out of the states we want to aggregate (in this case, state 2 and state 3) are the same, then the transition rates of the aggregated model do not depend on the probabilities of the unaggregated model.

In the example above, we saw that aggregating absorbing states, i.e., those with no transitions out, results in a reduced model with constant failure rates. We also saw in the example that aggregating states with equal transition rates out will result in a Markov reliability model with constant failure rates. One can see the former case of absorbing state aggregation as a particular example of the latter for equal transition rates out, when those are identically equal to zero. Indeed, that is the general condition that a group of states need to satisfy for them to be aggregated, i.e., the transition rate from any of these states to be aggregated to any other state (or group of states) needs to be identical.

Example 4.16 (Dual redundant system) Consider a dual-redundant system, i.e., a system comprised of two components, c_1, c_2, where it is necessary that only one component is in its nominal mode for the system to remain operational. Assume both components have identical failure rates, i.e., $\lambda_{c_1} = \lambda_{c_2} = \lambda$. Assume a similar repair strategy to that in A1 – A2 of Example 4.12, i.e., repair actions restore the system back to operating with no failed components no matter how many components have failed. The rates at which these repair actions are completed is μ_1 if either component c_1 or component c_2 has failed, and μ_2 if both component c_1 and component c_2 have failed. The state-transition diagram of the resulting Markov chain is shown in Fig. 4.9 (left). One can easily see that for state 2 and state 3 transitions out to state 4 and state 5 occur at the same rate, λ; similarly, transitions out of state 2 and state 3 to state 1, after the corresponding repair action is completed, occur at rate μ_1; thus, state 2 and state 3 can be aggregated into a single state. A similar analysis can be used to argue that state 4 and state 5 can also be aggregated. Figure 4.9 (right) shows the state-transition diagram of the Markov chain that results from this aggregation.

4.6 Notes and References

The material on general discrete- and continuous-time Markov chains is standard and can be found in [27, 28, 29, 11]. Discussions on the specific applications of Markov chains to reliability modeling can be found in standard reliability analysis textbooks, including [30, 31, 32, 33]. The notion of failure coverage is discussed in [34, 35]. The truncation technique for reducing the dimension of Markov reliability models is discussed in [36, 37]. The state aggregation technique for reducing the dimension of Markov reliability models is discussed in [31].

5 Discrete-Time Systems: Probabilistic Input Uncertainty

In this chapter, we analyze systems in which the relation between the input and the state is governed by a discrete-time state-space model. We will model the input as a stochastic process with known probability distribution (or known first few moments). The objective is then to obtain the distribution of the state (or its moments). Fundamental concepts from stochastic processes and discrete-time linear dynamical systems used throughout this chapter are reviewed in Sections 2.1 and 2.3.1, respectively.

5.1 Introduction

Jane has two bank accounts, a checking account and a savings account. Her payroll is deposited in her checking account on the first day of each month, and she uses that account to pay her recurrent monthly bills and credit card statements, and occasionally withdraw cash. On the day her payroll is deposited, she transfers to her savings account all the money that was left in her checking account before the payroll deposit. Let d_k denote the money transfer she makes on the first day of month k from her checking account to her savings account. Let s_k denote the money on her savings account on the first day of month k before she makes the monthly transfer. Assume the return of her savings account is $\alpha\%$; then the amount of money in her savings accounts evolves monthly as follows:

$$s_{k+1} = \alpha s_k + d_k. \tag{5.1}$$

Because Jane's monthly expenditures are variable and a priori unknown, the value that d_k takes is also a priori unknown but can be described by some stochastic process. She would like then to know how her savings will evolve over the next few months for which she would need to use (5.1) together with the statistical characterization of d_k.

The discussion above pertains to the class of discrete-time linear systems, which can be compactly described as follows:

$$x_{k+1} = G_k x_k + H_k w_k, \tag{5.2}$$

where $x_k \in \mathbb{R}^n$, $w_k \in \mathbb{R}^m$, and G_k and H_k are matrices of appropriate dimensions (the reader is referred to Section 2.3.1 for a review of basic notions used in later developments). In the context of the example above, $x_k = s_k$, $w_k = d_k$,

$G_k = \alpha$, and $H_k = 1$. In this chapter, we will assume the values that w_k, $k = 0, 1, 2, \ldots$, take are random and governed by a stochastic process $W = \{W_k\colon k \in \{0, 1, \ldots\}\}$; thus, the values taken by x_k, $k = 0, 1, 2, \ldots$, are also random and governed by a stochastic process $X = \{X_k\colon k \in \{0, 1, \ldots\}\}$. The objective will be to characterize X under the following two scenarios:

D1. Assuming W_k, $k \geq 0$, are uncorrelated with known mean and known covariance matrix, we will characterize the mean, covariance, and correlation functions of X.

D2. Assuming W_k, $k \geq 0$, are independent with known pdf, we will characterize the pdf of X_k, $k \geq 0$.

In addition, we will provide some techniques for approximately characterizing the first and second moments and/or distribution of X_k, $k \geq 0$, for stochastic nonlinear discrete-time systems of the form

$$X_{k+1} = h_k(X_k, W_k). \tag{5.3}$$

The analysis techniques developed in this chapter can then be utilized to compute the value of some metric characterizing the degree to which a system of the form in (5.2) or (5.3) performs its function as intended. For example, using the results developed under Scenario D1, one can use the covariance matrix to determine whether or not the system trajectories remain close to the mean (presumably describing the nominal system behavior). Furthermore, the techniques developed under Scenario D2 can be used to compute the probability that the state belongs to some set defined by performance requirements imposing constraints on the values that the state can take.

5.2 Discrete-Time Linear Systems

In this section, we study systems of the form:

$$X_{k+1} = G_k X_k + H_k W_k, \tag{5.4}$$

where $G_k \in \mathbb{R}^{n \times n}$, $H_k \in \mathbb{R}^{n \times m}$, $W = \{W_k\colon k \in \{0, 1, \ldots\}\}$ is a stochastic process, with the W_k's taking values in $\mathbb{R}^m$, and X_0 is a random vector taking values in $\mathbb{R}^n$. We first address the case when the mean and covariance functions of W are known and provide the exact characterization of the mean, covariance, and correlation functions of the stochastic process $X = \{X_k\colon k \in \{0, 1, \ldots\}\}$ describing the evolution of the system state. Then, we address the case when the distribution of W is known and characterize the distribution of the system state.

5.2.1 Characterization of First and Second Moments

Consider the system (5.4) and assume that the values taken by the mean function of the stochastic process W,

$$m_W[k] = \mathrm{E}\big[W_k\big] \in \mathbb{R}^m, \quad k = 0, 1, 2, \ldots,$$

are given, and so are the values taken by its covariance function,

$$C_W[k_1,k_2] = \mathrm{E}\Big[\big(W_{k_1} - m_W[k_1]\big)\big(W_{k_2} - m_W[k_2]\big)^\top\Big] = \begin{cases} Q_{k_1}, & \text{if } k_1 = k_2, \\ \mathbf{0}_{m\times m}, & \text{if } k_1 \neq k_2, \end{cases} \tag{5.5}$$

for all $k_1 = 0,1,2,\dots,$ and $k_2 = 0,1,2,\dots,$ with $Q_{k_1} \in \mathbb{R}^{m\times m}$. In addition, assume we are given the mean of X_0, $m_X[0] = \mathrm{E}\big[X_0\big] \in \mathbb{R}^n$, and the covariance matrix of X_0,

$$\Sigma_X[0] = \mathrm{E}\Big[\big(X_0 - \mathrm{E}[X_0]\big)\big(X_0 - \mathrm{E}[X_0]\big)^\top\Big] \in \mathbb{R}^{n\times n}.$$

Furthermore, assume that X_0 and W_k are uncorrelated for all $k = 0,1,2,\dots,$ i.e.,

$$\mathrm{E}\Big[\big(X_0 - m_X[0]\big)\big(W_k - m_W[k]\big)^\top\Big] = \mathbf{0}_{n\times m}, \quad k \geq 0. \tag{5.6}$$

Then the objective is to compute the values taken by: (i) the mean function of X,

$$m_X[k] = \mathrm{E}\big[X_k\big] \in \mathbb{R}^n, \quad k \geq 0,$$

(ii) the covariance function of X,

$$C_X[k_1,k_2] = \mathrm{E}\Big[\big(X_{k_1} - \mathrm{E}[X_{k_1}]\big)\big(X_{k_2} - \mathrm{E}[X_{k_2}]\big)^\top\Big] \in \mathbb{R}^{n\times n}, \quad k_1,k_2 \geq 0,$$

and (iii) the correlation function of X,

$$R[k_1,k_2] = \mathrm{E}\Big[X_{k_1}X_{k_2}^\top\Big] \in \mathbb{R}^{n\times n}, \quad k_1,k_2 \geq 0.$$

Mean Function. The values taken by the mean function of X, $m_X[k]$, $k \geq 0$, are determined by the following recursion

$$m_X[k+1] = G_k m_X[k] + H_k m_W[k], \tag{5.7}$$

for given $m_X[0]$.

To establish (5.7), we proceed as follows:

$$\begin{aligned} m_X[k+1] &= \mathrm{E}[X_{k+1}] \\ &= \mathrm{E}[G_k X_k + H_k W_k] \\ &= \mathrm{E}[G_k X_k] + \mathrm{E}[H_k W_k] \\ &= G_k \mathrm{E}[X_k] + H_k \mathrm{E}[W_k] \\ &= G_k m_X[k] + H_k m_W[k]. \end{aligned} \tag{5.8}$$

Covariance Function. Let $\Sigma_X[k] = \mathrm{E}\Big[\big(X_k - m_X[k]\big)\big(X_k - m_X[k]\big)^\top\Big]$, then, the values taken by the covariance function of X, $C_X[k_1,k_2]$ $k_1,k_2 \geq 0$, are determined by

$$\Sigma_X[k+1] = G_k \Sigma_X[k] G_k^\top + H_k Q_k H_k^\top, \quad k \geq 0, \tag{5.9}$$

$$C_X[k_1, k_2] = \begin{cases} \Phi_{k_1,k_2} \Sigma_X[k_2], & \text{if } k_1 \geq k_2, \\ \Sigma_X[k_1] \Phi_{k_2,k_1}^\top, & \text{if } k_1 < k_2, \end{cases} \tag{5.10}$$

for a given $\Sigma_X[0]$, where

$$\Phi_{k,\ell} = \begin{cases} G_{k-1} G_{k-2} \cdots G_\ell, & \text{if } k > \ell \geq 0, \\ I_n & \text{if } k = \ell. \end{cases} \tag{5.11}$$

In order to establish (5.9) and (5.10), we first need to establish that $X_{k'}$ and W_k are uncorrelated for any $k \geq k' \geq 0$, i.e., $\mathrm{E}\big[(X_{k'} - m_X[k'])(W_k - m_W[k])^\top\big] = 0$ for any $k \geq k' \geq 0$. To this end, recall the expression in (2.108); then we have that

$$X_k = \Phi_{k,0} X_0 + \sum_{\ell=0}^{k-1} \Phi_{k,\ell+1} H_\ell W_\ell, \quad k \geq 0; \tag{5.12}$$

thus

$$m_X[k] = \Phi_{k,0} m_X[0] + \sum_{\ell=0}^{k-1} \Phi_{k,\ell+1} H_\ell m_W[\ell], \quad k \geq 0. \tag{5.13}$$

Then, by using (5.12) and (5.13), we have that, for any $k \geq k' \geq 0$,

$$\begin{aligned}
&\mathrm{E}\Big[(X_{k'} - m_X[k'])(W_k - m_W[k])^\top\Big] \\
&= \mathrm{E}\Big[\Phi_{k',0}(X_0 - m_X[0])(W_k - m_W[k])^\top + \sum_{\ell=0}^{k'-1} \Phi_{k',\ell+1} H_\ell \\
&\quad \times (W_\ell - m_W[\ell])(W_k - m_W[k])^\top\Big] \\
&= \Phi_{k',0} \underbrace{\mathrm{E}\Big[(X_0 - m_X[0])(W_k - m_W[k])^\top\Big]}_{= \mathbf{0}_{n\times m} \text{ by (5.6)}} + \sum_{\ell=0}^{k'-1} \Phi_{k',\ell+1} H_\ell \\
&\quad \times \underbrace{\mathrm{E}\Big[(W_\ell - m_W[\ell])(W_k - m_W[k])^\top\Big]}_{= \mathbf{0}_{m\times m} \text{ by (5.5) for any } \ell \neq k} \\
&= \mathbf{0}_{n\times m}.
\end{aligned} \tag{5.14}$$

Now, to establish (5.9), we proceed as follows:

$$\begin{aligned}
\Sigma_X[k+1] &= \mathrm{E}\Big[(X_{k+1} - m_X[k+1])(X_{k+1} - m_X[k+1])^\top\Big] \\
&= \mathrm{E}\Big[\Big(G_k(X_k - m_X[k]) + H_k(W_k - m_W[k])\Big) \\
&\quad \times \Big(G_k(X_k - m_X[k]) + H_k(W_k - m_W[k])\Big)^\top\Big]
\end{aligned}$$

$$
\begin{aligned}
&= \mathrm{E}\Big[G_k\big(X_k - m_X[k]\big)\big(X_k - m_X[k]\big)^\top G_k^\top\Big] \\
&\quad + \mathrm{E}\Big[G_k\big(X_k - m_X[k]\big)\big(W_k - m_W[k]\big)^\top H_k^\top\Big] \\
&\quad + \mathrm{E}\Big[H_k\big(W_k - m_W[k]\big)\big(X_k - m_X[k]\big)^\top G_k^\top\Big] \\
&\quad + \mathrm{E}\Big[H_k\big(W_k - m_W[k]\big)\big(W_k - m_W[k]\big)^\top H_k^\top\Big] \\
&= G_k \mathrm{E}\Big[\big(X_k - m_X[k]\big)\big(X_k - m_X[k]\big)^\top\Big] G_k^\top \\
&\quad + G_k \underbrace{\mathrm{E}\Big[\big(X_k - m_X[k]\big)\big(W_k - m_W[k]\big)^\top\Big]}_{= \mathbf{0}_{n\times m} \text{ by } (5.14)} H_k^\top \\
&\quad + H_k \underbrace{\mathrm{E}\Big[(W_k - m_W[k])(X_k - m_X[k])^\top\Big]}_{= \mathbf{0}_{m\times n} \text{ by } (5.14)} G_k^\top \\
&\quad + H_k \underbrace{\mathrm{E}\Big[(W_k - m_W[k])(W_k - m_W[k])^\top\Big]}_{= Q_k \text{ by } (5.5)} H_k^\top \\
&= G_k \Sigma_X[k] G_k^\top + H_k Q_k H_k^\top.
\end{aligned} \tag{5.15}
$$

Finally, to establish (5.10), we proceed as follows. For $k_1 \geq k_2 \geq 0$, note that

$$
X_{k_1} = \Phi_{k_1,k_2} X_{k_2} + \sum_{\ell=k_2}^{k_1-1} \Phi_{k_1,\ell+1} H_\ell W_\ell, \tag{5.16}
$$

from where it follows that

$$
m_X[k_1] = \Phi_{k_1,k_2} m_X[k_2] + \sum_{\ell=k_2}^{k_1-1} \Phi_{k_1,\ell+1} H_\ell m_W[\ell]. \tag{5.17}
$$

Then, by using (5.16) and (5.17), we have that, for all $k_1 \geq k_2 \geq 0$,

$$
\begin{aligned}
C_X[k_1,k_2] &= \mathrm{E}\Big[\big(X_{k_1} - m_X[k_1]\big)\big(X_{k_2} - m_X[k_2]\big)^\top\Big] \\
&= \mathrm{E}\Big[\Phi_{k_1,k_2}\big(X_{k_2} - m_X[k_2]\big)\big(X_{k_2} - m_X[k_2]\big)^\top \\
&\quad + \sum_{\ell=k_2}^{k_1-1} \Phi_{k_1,\ell+1} H_\ell \big(W_\ell - m_W[\ell]\big)\big(X_{k_2} - m_X[k_2]\big)^\top\Big] \\
&= \Phi_{k_1,k_2} \mathrm{E}\Big[\big(X_{k_2} - m_X[k_2]\big)\big(X_{k_2} - m_X[k_2]\big)^\top\Big] \\
&\quad + \sum_{\ell=k_2}^{k_1-1} \Phi_{k_1,\ell+1} H_\ell \underbrace{\mathrm{E}\Big[\big(W_\ell - m_W[\ell]\big)\big(X_{k_2} - m_X[k_2]\big)^\top\Big]}_{= \mathbf{0}_{m\times n} \text{ by } (5.14) \text{ for any } \ell \geq k_2} \\
&= \Phi_{k_1,k_2} \Sigma_X[k_2],
\end{aligned} \tag{5.18}
$$

as claimed in the first line of (5.10). For $0 \le k_1 < k_2$, we have

$$\begin{aligned} C_X[k_1,k_2] &= \mathrm{E}\Big[\big(X_{k_1} - m_X[k_1]\big)\big(X_{k_2} - m_X[k_2]\big)^\top\Big] \\ &= \mathrm{E}\Big[\Big(\big(X_{k_2} - m_X[k_2]\big)\big(X_{k_1} - m_X[k_1]\big)^\top\Big)^\top\Big] \\ &= \Big(\underbrace{\mathrm{E}\Big[\big(X_{k_2} - m_X[k_2]\big)\big(X_{k_1} - m_X[k_1]\big)^\top\Big]}_{= \Phi_{k_2,k_1}\Sigma_X[k_1] \text{ by (5.18) because } k_2 > k_1}\Big)^\top \\ &= \Sigma_X[k_1]\Phi_{k_2,k_1}^\top, \end{aligned} \tag{5.19}$$

as claimed in the second line of (5.10).

Correlation Function. The values taken by the correlation function of X, $R_X[k_1,k_2]$, $k_1,k_2 \ge 0$, can be obtained from $C_X[k_1,k_2]$ and $m_X[k]$ as follows:

$$R_X[k_1,k_2] = C_X[k_1,k_2] + m_X[k_1]m_X^\top[k_2]; \tag{5.20}$$

this can be established by using the fact that

$$\mathrm{E}\Big[\big(X_{k_1} - m_X[k_1]\big)\big(X_{k_2} - m_X[k_2]\big)^\top\Big] = \mathrm{E}\big[X_{k_1}X_{k_2}^\top\big] - m_X[k_1]m_X^\top[k_2]. \tag{5.21}$$

Vectorization of Matrix Difference Equation

The recursion in (5.9) can be rewritten in vector form using the Kronecker product (see Appendix A). In the process, the matrices $\Sigma_X[k]$ and $H_kQ_kH_k^\top$ are replaced by vectors obtained by stacking their columns left to right. More formally, for any given matrix $A = [a_{i,j}] \in \mathbb{R}^{n\times m}$, we will associate a vector $\mathbf{vec}(A) \in \mathbb{R}^{nm}$ defined as follows:

$$\mathbf{vec}(A) = [a_{1,1}, a_{2,1}, \dots, a_{n,1}, a_{1,2}, a_{2,2}, \dots, a_{n,2}, \dots, a_{1,m}, a_{2,m}, \dots, a_{n,m}]^\top. \tag{5.22}$$

We can now write (5.9) in vector form as follows:

$$\mathbf{vec}\big(\Sigma_X[k+1]\big) = \big(G_k \otimes G_k\big)\mathbf{vec}\big(\Sigma_X[k]\big) + \mathbf{vec}\big(H_kQ_kH_k^\top\big). \tag{5.23}$$

For a time-invariant system, $H_k = H$, $G_k = G$, for all $k \ge 0$; thus

$$\mathbf{vec}\big(\Sigma_X[k+1]\big) = \big(G \otimes G\big)\mathbf{vec}\big(\Sigma_X[k]\big) + \mathbf{vec}\big(HQ_kH^\top\big). \tag{5.24}$$

Now we can determine whether or not $\mathbf{vec}\big(\Sigma_X[k]\big)$ will remain bounded (note that $\mathbf{vec}\big(HQ_kH^\top\big)$ is a bounded vector if all the entries of Q_k are finite) by computing the eigenvalues of $\big(G \otimes G\big)$; these can be easily obtained from the eigenvalues of G as follows. Let $\sigma(G)$ denote the spectrum of G, i.e., the set containing its eigenvalues; then the spectrum of $G \otimes G$ is given by:

$$\sigma\big(G \otimes G\big) = \{\lambda \colon \lambda = \lambda_i\lambda_j,\ \lambda_i \in \sigma(G), \lambda_j \in \sigma(G)\}. \tag{5.25}$$

Thus, we can guarantee that $\mathbf{vec}\big(\Sigma_X[k]\big)$ will remain bounded if the eigenvalues of G, $\lambda_1, \lambda_2, \dots, \lambda_n$, are within the unit circle, i.e., $|\lambda_i| < 1$, $i = 1, 2, \dots, n$.

Example 5.1 (One-dimensional linear time-invariant system) Consider the following system

$$X_{k+1} = \alpha X_k + \beta W_k, \tag{5.26}$$

where $\alpha, \beta \in \mathbb{R}$, $|\alpha| \leq 1$, are known, W_k, $k = 0, 1, \ldots$, are independent random variables with known mean, $\mu_W[k]$, and known standard deviation, $\sigma_W[k]$, and X_0 is a random variable with known mean, $\mu_X[0]$, and standard deviation, $\sigma_X[0]$. Assume that X_0 and W_k, $k = 0, 1, \ldots$, are independent. The objective is to characterize the mean of X_k, denoted by $\mu_X[k]$, for all $k \geq 0$, and the covariance and correlation of X_{k_1} and X_{k_2}, respectively denoted by $C_X[k_1, k_2]$ and $R_X[k_1, k_2]$, for all $k_1, k_2 \geq 0$.

First, tailoring (5.7) to the setting here yields

$$\mu_X[k+1] = \alpha \mu_X[k] + \beta \mu_W[k]; \tag{5.27}$$

then, by noting that in this case $\Phi_{k,\ell} = \alpha^{k-\ell}$, $k \geq \ell \geq 0$, it follows that

$$m_X[k] = \alpha^k m_X[0] + \sum_{\ell=0}^{k-1} \alpha^{k-1-\ell} \beta m_W[\ell], \quad k \geq 0. \tag{5.28}$$

Next, note that $C_W[k_1, k_2] = \sigma_X^2[k]$ if $k_1 = k_2$ and $C_W[k_1, k_2] = 0$ otherwise, and let $\sigma_X[k]$ denote the standard deviation of X_k. Then, by tailoring (5.9) and (5.10) to the setting here, we obtain

$$\sigma_X^2[k+1] = \alpha^2 \sigma_X^2[k] + \beta^2 \sigma_W^2[k], \tag{5.29}$$

$$\begin{aligned} C_X[k_1, k_2] &= \begin{cases} \alpha^{k_1 - k_2} \sigma_X^2[k_2], & \text{if } k_1 \geq k_2 \geq 0, \\ \sigma_X^2[k_1] \alpha^{k_2 - k_1}, & \text{if } 0 \leq k_1 < k_2, \end{cases} \\ &= \alpha^{|k_1 - k_2|} \sigma_X^2[m], \end{aligned} \tag{5.30}$$

where $m = \min\{k_1, k_2\}$, for all $k_1, k_2 \geq 0$. Note that in this case, we have a closed-form expression for $\sigma_X^2[k]$ as follows:

$$\sigma_X^2[k] = \alpha^{2k} \sigma_X^2[0] + \sum_{\ell=0}^{k-1} \alpha^{2(k-1-\ell)} \beta^2 \sigma_W^2[\ell]. \tag{5.31}$$

Example 5.2 (Two-dimensional linear time-invariant system) Consider the following system

$$X_{k+1} = G X_k + H W_k, \tag{5.32}$$

where

$$\begin{aligned} G &= \begin{bmatrix} \frac{1}{2} & 1 \\ 0 & \frac{1}{2} \end{bmatrix}, \\ H &= \begin{bmatrix} 1 \\ \frac{1}{2} \end{bmatrix}, \end{aligned} \tag{5.33}$$

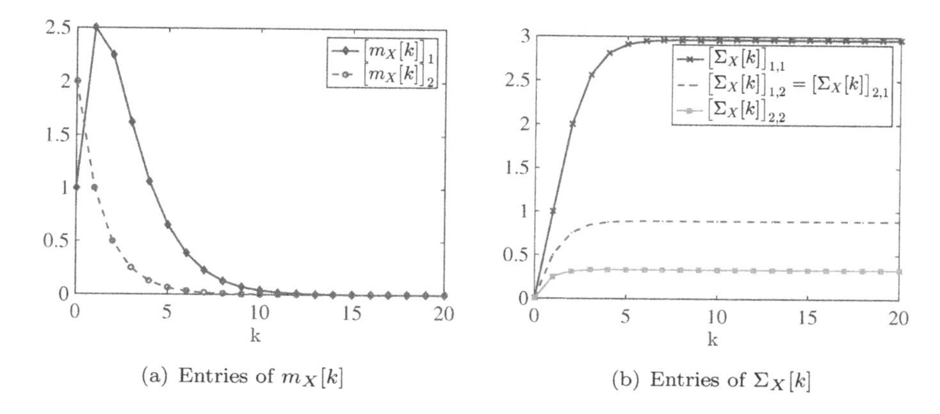

(a) Entries of $m_X[k]$

(b) Entries of $\Sigma_X[k]$

Figure 5.1 Evolution of the entries of the mean and the covariance matrix of X_k.

and $W_0, W_1, \dots$ are independent random variables with mean $\mu_W[k] = 0$ and standard deviation $\sigma_W[k] = 1$, and $X_0 = [1, 2]^\top$; thus $m_X[0] = [1, 2]^\top$ and $\Sigma_X[0]$ is the two-dimensional all-zeros matrix. Assume that X_0 is also independent of $W_0, W_1, \dots$. The objective is to obtain the mean and covariance functions of X.

Since $\mu_W[k] = 0$ for all $k \geq 0$, by using (5.7), we obtain:

$$m_X[k+1] = Gm_X[k], \tag{5.34}$$

with $m_X[0] = [1, 2]^\top$. The trajectories followed by the entries of the vector $m_X[k]$, denoted by $[m_X[k]]_i$, $i = 1, 2$, are displayed in Fig. 5.1(a). Since $\sigma_W^2[k] = 1$, in this case $Q_k = 1$; thus, by using (5.9), we obtain

$$\Sigma_X[k+1] = G\Sigma_X[k]G^\top + HH^\top, \tag{5.35}$$

with

$$\Sigma_X[0] = \begin{bmatrix} 0 & 0 \\ 0 & 0 \end{bmatrix}. \tag{5.36}$$

The trajectories followed by the entries of the matrix $\Sigma_X[k]$, which we denote by $[\Sigma_X[k]]_{i,j}$, $i = 1, 2$, $j = 1, 2$, are displayed in Fig. 5.1(b). Finally, by using (5.10), we obtain

$$C_X[k_1, k_2] = \begin{cases} G^{k_1 - k_2}\Sigma_X[k_2], & \text{if } k_1 \geq k_2, \\ \Sigma_X[k_1](G^\top)^{k_2 - k_1}, & \text{if } k_1 < k_2. \end{cases} \tag{5.37}$$

Deterministic Inputs

Note that in (5.4) we have not considered deterministic inputs; however, these can be easily incorporated as follows. Assume we are given

$$X_{k+1} = G_k X_k + H_k W_k + L_k u_k, \tag{5.38}$$

where $L_k \in \mathbb{R}^{n\times l}$ and u_k are known for all $k \geq 0$, and with G_k, H_k, X_k, and W_k as in earlier developments. Define $\Delta X_k = X_k - x_k^\circ$, with x_k° determined by

$$x_{k+1}^\circ = G_k x_k^\circ + L_k u_k, \tag{5.39}$$

with $x_0^\circ = \mathbf{0}_n$. Then we have that

$$\begin{aligned}\Delta X_{k+1} &= X_{k+1} - x_{k+1}^\circ \\ &= G_k X_k + H_k W_k + L_k u_k - \left(G_k x_k^\circ + L_k u_k\right) \\ &= G_k \Delta X_k + H_k W_k, \end{aligned} \tag{5.40}$$

which is of the form in (5.4); thus, we can use the techniques discussed earlier to compute the values taken by the mean, covariance, and correlation functions of ΔX, $m_{\Delta X}[k]$, $k \geq 0$, $C_{\Delta X}[k_1, k_2]$, $k_1, k_2 \geq 0$, and $R_{\Delta X}[k_1, k_2]$, $k_1, k_2 \geq 0$, respectively. From these, we can obtain $m_X[k]$, $C_X[k_1, k_2]$, and $R_X[k_1, k_2]$ as follows:

$$\begin{aligned} m_X[k] &= \mathrm{E}\left[x_k^\circ + \Delta X_k\right] \\ &= x_k^\circ + m_{\Delta X}[k], \quad k \geq 0, \\ C_X[k_1, k_2] &= \mathrm{E}\Big[\left(x_k^\circ + \Delta X_k - m_X[k]\right)\left(x_k^\circ + \Delta X_k - m_X[k]\right)^\top\Big] \\ &= \mathrm{E}\Big[\left(\Delta X_k - m_{\Delta X}[k]\right)\left(\Delta X_k - m_{\Delta X}[k]\right)^\top\Big] \\ &= C_{\Delta X}[k_1, k_2], \quad k_1, k_2 \geq 0, \\ R_X[k_1, k_2] &= C_X[k_1, k_2] + m_X[k] m_X^\top[k] \\ &= C_{\Delta X}[k_1, k_2] + \left(x_k^\circ + m_{\Delta X}[k]\right)\left(x_k^\circ + m_{\Delta X}[k]\right)^\top, \quad k_1, k_2 \geq 0. \end{aligned} \tag{5.41}$$

5.2.2 Probability Distribution

Consider the system in (5.4), where the W_k's are i.i.d. random vectors with known pdf, denoted by $f_{W_k}(w_k)$, and X_0 is a random vector with known pdf, denoted by $f_{X_0}(x_0)$, that is also independent of the W_k's. Assume that G_k is invertible for all $k \geq 0$. The objective here is to obtain the pdf of X_k, denoted by $f_{X_k}(x_k)$, for all $k > 0$. To this end, we will build on the techniques presented in Chapter 3.

First note that X_k and W_k are independent. This follows from (5.12), which establishes that X_k is a linear combination of X_0 and W_ℓ, $\ell = 0, \ldots, k-1$, but these are independent of W_k; thus, X_k and W_k are independent. Now define $Y_k = \left[X_k^\top, W_k^\top\right]^\top$ and $Z_k = \left[X_{k+1}^\top, W_k^\top\right]^\top$; then we have that

$$Z_k = \widetilde{H}_k Y_k, \tag{5.42}$$

where

$$\widetilde{H}_k = \begin{bmatrix} G_k & H_k \\ \mathbf{0}_{m\times n} & I_m \end{bmatrix}. \tag{5.43}$$

Since G_k is invertible, so is $\widetilde{H}_k$, and one can easily check that

$$\widetilde{H}_k^{-1} = \begin{bmatrix} G_k^{-1} & -G_k^{-1}H_k \\ \mathbf{0}_{m\times n} & I_m \end{bmatrix} \tag{5.44}$$

and $\mathbf{det}(\widetilde{H}_k^{-1}) = \mathbf{det}(G_k^{-1})$. Then, as shown in Chapter 3, the pdf of Z_k, denoted by $f_{Z_k}(z_k)$, is given by

$$\begin{aligned} f_{Z_k}(z_k) &= \begin{cases} \left|\mathbf{det}(\widetilde{H}_k^{-1})\right| f_{Y_k}(\widetilde{H}_k^{-1}z_k), & z_k \in \left\{z_k\colon f_{Y_k}(\widetilde{H}_k^{-1}z_k) \neq 0\right\}, \\ 0, & \text{otherwise,} \end{cases} \\ &= \begin{cases} \left|\mathbf{det}(G_k^{-1})\right| f_{Y_k}(\widetilde{H}_k^{-1}z_k), & z_k \in \left\{z_k\colon f_{Y_k}(\widetilde{H}_k^{-1}z_k) \neq 0\right\}, \\ 0, & \text{otherwise.} \end{cases} \end{aligned} \tag{5.45}$$

Then since X_k and W_k are independent, we have that $f_{Y_k}(y_k) = f_{X_k}(x_k)f_{W_k}(w_k)$, and we can write (5.45) as

$$f_{Z_k}(z_k) = \begin{cases} \left|\mathbf{det}(G_k^{-1})\right| f_{X_k}\left(G_k^{-1}(x_{k+1} - H_k w_k)\right) f_{W_k}(w_k), & \begin{bmatrix} x_{k+1} \\ w_k \end{bmatrix} \in \mathcal{Z}_k, \\ 0, & \text{otherwise,} \end{cases} \tag{5.46}$$

where

$$\mathcal{Z}_k = \left\{ \begin{bmatrix} x_{k+1} \\ w_k \end{bmatrix} : f_{X_k}\left(G_k^{-1}(x_{k+1} - H_k w_k)\right) f_{W_k}(w_k) \neq 0 \right\}. \tag{5.47}$$

Now, as shown in Chapter 3, we just need to integrate $f_{Z_k}\left(\left[x_{k+1}^\top, w_k^\top\right]^\top\right)$ over the entries of w_k in order to obtain $f_{X_{k+1}}(x_{k+1})$:

$$f_{X_{k+1}}(x_{k+1}) = \begin{cases} \left|\mathbf{det}(G_k^{-1})\right| \zeta(x_{k+1}), & x_{k+1} \in \mathcal{X}_{k+1}, \\ 0, & \text{otherwise,} \end{cases} \tag{5.48}$$

with

$$\mathcal{X}_{k+1} = \left\{x_{k+1}\colon f_{X_k}\left(G_k^{-1}(x_{k+1} - H_k w_k)\right) \neq 0\right\}, \tag{5.49}$$

and

$$\zeta(x_{k+1}) = \int_{-\infty}^{\infty} \cdots \int_{-\infty}^{\infty} f_{X_k}\left(G_k^{-1}(x_{k+1} - H_k w_k)\right) f_{W_k}(w_k) dw_{k,1} \cdots dw_{k,m}, \tag{5.50}$$

where $dw_{k,i}$ denotes the ith entry of dw_k.

Example 5.3 (One-dimensional linear time-invariant Gaussian system) Here we consider the same one-dimensional system as in Example 5.1 with the additional assumption that X_0 and W_k, $k = 0, 1, \ldots$, are Gaussian. The objective is to obtain the pdf of X_k for all k.

First, note that, by using (5.12), we have

$$X_k = \alpha^k X_0 + \sum_{\ell=0}^{k-1} \alpha^{k-1-\ell} \beta W_\ell, \tag{5.51}$$

i.e., X_k is a linear combination of X_0 and $W_0, W_1, \ldots, W_{k-1}$, and since W_k is independent of X_0 and W_ℓ, $\ell = 0, 1, \ldots, k-1$, we have that X_{k+1} is the sum of two independent random variables, αX_k and βW_k. Next, we show by induction that since X_0 and W_k, $k \geq 0$, are Gaussian, X_{k+1} is also Gaussian.

We proceed directly to the inductive step since the procedure is identical to that one would follow for $k = 0$, in which case X_0 is known to be Gaussian. Assume that X_k is Gaussian with mean $\mu_X[k]$ and standard deviation $\sigma_X[k]$. Define $\widetilde{X}_{k+1} = \widetilde{Y}_k + \widetilde{Z}_k$, where $\widetilde{Y}_k = \alpha\big(X_k - \mu_X[k]\big)$ and $\widetilde{Z}_k = \beta\big(W_k - \mu_W[k]\big)$. Then, we have that

$$\begin{aligned} f_{\widetilde{Y}_k}(\widetilde{y}_k) &= \frac{1}{\alpha} f_{X_k}\left(\frac{\widetilde{y}_k}{\alpha} + \mu_X[k]\right) \\ &= \frac{1}{\sqrt{2\pi\alpha^2\sigma_X^2[k]}} e^{-\frac{\widetilde{y}_k^2}{2\alpha^2\sigma_X^2[k]}}, \end{aligned} \tag{5.52}$$

which corresponds to the pdf of a Gaussian random variable with zero mean and standard deviation $a := \alpha\sigma_X[k]$. Similarly, we have that

$$\begin{aligned} f_{\widetilde{Z}_k}(z_k) &= \frac{1}{\beta} f_{W_k}\left(\frac{\widetilde{z}_k}{\beta} + \mu_W[k]\right) \\ &= \frac{1}{\sqrt{2\pi\beta^2\sigma_W^2[k]}} e^{-\frac{\widetilde{z}_k^2}{2\beta^2\sigma_W^2[k]}}, \end{aligned} \tag{5.53}$$

which corresponds to the pdf of a Gaussian random variable with zero mean and standard deviation $b := \beta\sigma_W[k]$. Now, we can utilize the expression in (5.48) to obtain the pdf of $\widetilde{X}_{k+1}$ as follows:

$$f_{\widetilde{X}_{k+1}}(\widetilde{x}_{k+1}) = \frac{1}{\sqrt{(2\pi)^2 a^2 b^2}} \int_{-\infty}^{\infty} e^{-\frac{(\widetilde{x}_{k+1} - \widetilde{z}_k)^2}{2a^2}} e^{-\frac{\widetilde{z}_k^2}{2b^2}} \, d\widetilde{z}_k. \tag{5.54}$$

The exponent in the integrand of (5.54) is of the form $-(c_1\widetilde{z}_k^2 + 2c_2\widetilde{z}_k + c_3)/2$, where

$$c_1 = \frac{1}{a^2} + \frac{1}{b^2},$$

$$c_2 = \frac{-\widetilde{x}_{k+1}}{a^2},$$

$$c_3 = \frac{\widetilde{x}_{k+1}^2}{a^2}. \tag{5.55}$$

Then, by completing the square, we obtain

$$e^{-\frac{c_1\widetilde{z}_k^2 + 2c_2\widetilde{z}_k + c_3}{2}} = e^{\frac{c_2^2}{2c_1} - \frac{c_3}{2}} e^{-\frac{(\widetilde{z}_k + c_2/c_1)^2}{2/c_1}}, \tag{5.56}$$

thus

$$\begin{aligned}
\int_{-\infty}^{\infty} e^{-\frac{c_1\widetilde{z}_k^2 + 2c_2\widetilde{z}_k + c_3}{2}} d\widetilde{z}_k &= e^{\frac{c_2^2}{2c_1} - \frac{c_3}{2}} \int_{-\infty}^{\infty} e^{-\frac{(\widetilde{z}_k + c_2/c_1)^2}{2/c_1}} d\widetilde{z}_k \\
&= \sqrt{\frac{2\pi}{c_1}} e^{\frac{c_2^2}{2c_1} - \frac{c_3}{2}} \int_{-\infty}^{\infty} \frac{1}{\sqrt{2\pi/c_1}} e^{-\frac{(\widetilde{z}_k + c_2/c_1)^2}{2/c_1}} d\widetilde{z}_k \\
&= \sqrt{\frac{2\pi}{c_1}} e^{\frac{c_2^2}{2c_1} - \frac{c_3}{2}},
\end{aligned} \tag{5.57}$$

where the last equality follows from the fact that the integrand in the second equality can be thought of as the pdf of a Gaussian random variable with mean $-c_2/c_1$ and standard deviation $1/\sqrt{c_1}$; thus, the integral evaluates to one. Now, by plugging the expressions for c_1, c_2, and c_3 into (5.57) and manipulating, we obtain

$$\begin{aligned}
f_{\widetilde{X}_{k+1}}(\widetilde{x}_{k+1}) &= \frac{1}{\sqrt{2\pi(a^2+b^2)}} e^{-\frac{\widetilde{x}_{k+1}^2}{2(a^2+b^2)}} \\
&= \frac{1}{\sqrt{2\pi\left(\alpha^2\sigma_X^2[k] + \beta^2\sigma_W^2[k]\right)}} e^{-\frac{\widetilde{x}_{k+1}^2}{2\left(\alpha^2\sigma_X^2[k] + \beta^2\sigma_W^2[k]\right)}},
\end{aligned} \tag{5.58}$$

which corresponds to the pdf of a Gaussian random variable with zero mean and standard deviation $\sqrt{\alpha^2\sigma_X^2[k] + \beta^2\sigma_W^2[k]}$. Now since $\widetilde{X}_{k+1} = X_{k+1} - \alpha\mu_X[k] - \beta\mu_W[k]$, we conclude that X_{k+1} is a Gaussian random variable with mean $\mu_X[k+1]$ and standard deviation $\sigma_X[k+1]$ defined recursively as follows:

$$\begin{aligned}
\mu_X[k+1] &= \alpha\mu_X[k] + \beta\mu_W[k], \\
\sigma_X^2[k+1] &= \alpha^2\sigma_X^2[k] + \beta^2\sigma_W^2[k],
\end{aligned} \tag{5.59}$$

expressions that are consistent with those in (5.27) and (5.29), respectively.

The example above highlights the calculations involved in obtaining $f_{X_k}(x_k)$ by using (5.48). An alternative method is to use characteristic functions, which greatly simplifies the calculations if, for example, the W_k's are Gaussian i.i.d. random vectors and X_0 is also Gaussian and independent of the W_k's, as in the example above; we discuss this case next.

Gaussian Systems

Consider the system in (5.4) and assume the W_k's are Gaussian i.i.d. random vectors with known mean $m_W[k] = \mathrm{E}\big[W_k\big] \in \mathbb{R}^m$ and known covariance matrix $\Sigma_W[k] = \mathrm{E}\big[\big(W_k - m_W[k]\big)\big(W_k - m_W[k]\big)^\top\big] \in \mathbb{R}^{m\times m}$. Further, assume X_0 is also a Gaussian random vector with known mean $m_X[0] = \mathrm{E}\big[X_0\big] \in \mathbb{R}^n$, and known covariance matrix $\Sigma_X[0] = \mathrm{E}\big[\big(X_0 - m_X[0]\big)\big(X_0 - m_X[0]\big)^\top\big] \in \mathbb{R}^{n\times n}$. Here, instead of using the expression in (5.48) to obtain $f_{X_k}(x_k)$ for all $k > 0$, we will provide an alternative method based on the use of characteristic functions.

The characteristic function of a random vector X taking values in $\mathbb{R}^n$ with pdf $f_X(x)$, denoted by $\phi_X(\eta)$, $\eta \in \mathbb{R}^n$, is defined as follows:

$$\begin{aligned}\phi_X(\eta) &= \mathrm{E}\big[e^{\mathrm{i}\eta^\top X}\big] \\ &= \int_{-\infty}^{\infty} e^{\mathrm{i}\eta^\top x} f_X(x)dx, \end{aligned} \tag{5.60}$$

where $\mathrm{i} = \sqrt{-1}$. For the case when X is a Gaussian random vector with mean $m_X \in \mathbb{R}^n$ and covariance matrix $\Sigma_X \in \mathbb{R}^{n\times n}$, we have that

$$\begin{aligned}\phi_X(\eta) &= \mathrm{E}\big[e^{\mathrm{i}\eta^\top X}\big] \\ &= \int_{-\infty}^{\infty} \cdots \int_{-\infty}^{\infty} e^{\mathrm{i}\eta^\top x} \frac{1}{\sqrt{(2\pi)^n \mathbf{det}(\Sigma_X)}} e^{-\frac{(x-m_X)^\top \Sigma_X^{-1}(x-m_X)}{2}} dx_1 \ldots dx_n \\ &= e^{\mathrm{i}\eta^\top m_X - \frac{1}{2}\eta^\top \Sigma_X \eta}, \quad \eta \in \mathbb{R}^n. \end{aligned} \tag{5.61}$$

The following two properties of characteristic functions will be used in subsequent developments:

P1. Let $Y = KX$, where X is a random vector taking values in $\mathbb{R}^n$ with pdf $f_X(x)$ and $K \in \mathbb{R}^{q\times n}$; then, for $\eta \in \mathbb{R}^q$, we have that

$$\begin{aligned}\phi_Y(\eta) &= \mathrm{E}\big[e^{\mathrm{i}\eta^\top Y}\big] \\ &= \int_{-\infty}^{\infty} e^{\mathrm{i}\eta^\top (Kx)} f_X(x)dx \\ &= \int_{-\infty}^{\infty} e^{\mathrm{i}(K^\top \eta)^\top x} f_X(x)dx \\ &= \phi_X(K^\top \eta). \end{aligned} \tag{5.62}$$

P2 Let $Z = X + Y$, where X and Y are independent random vectors taking values in $\mathbb{R}^n$ with pdf $f_X(x)$ and $f_Y(y)$, respectively; then, for $\eta \in \mathbb{R}^n$, we have that

$$\begin{aligned}
\phi_Z(\eta) &= \mathrm{E}\big[e^{\mathrm{i}\eta^\top Z}\big] \\
&= \mathrm{E}\big[e^{\mathrm{i}\eta^\top (X+Y)}\big] \\
&= \int_{-\infty}^{\infty}\int_{-\infty}^{\infty} e^{\mathrm{i}\eta^\top (x+y)} f_{X,Y}(x,y)dxdy \\
&= \int_{-\infty}^{\infty}\int_{-\infty}^{\infty} e^{\mathrm{i}\eta^\top (x+y)} f_X(x)f_Y(y)dxdy \\
&= \int_{-\infty}^{\infty} e^{\mathrm{i}\eta^\top x} f_X(x)dx \int_{-\infty}^{\infty} e^{\mathrm{i}\eta^\top y} f_Y(y)dy \\
&= \phi_X(\eta)\phi_Y(\eta).
\end{aligned} \tag{5.63}$$

Now, we will use induction to show that X_k, $k \geq 0$, are Gaussian random vectors with mean $m_X[k]$ and covariance matrix $\Sigma_X[k]$ that can be obtained with the following recursions:

$$m_X[k+1] = G_k m_X[k] + H_k m_W[k], \tag{5.64}$$

$$\Sigma_X[k+1] = G_k \Sigma_X[k] G_k^\top + H_k \Sigma_W[k] H_k^\top. \tag{5.65}$$

Define $Y_0 = G_0 X_0$ and $Z_0 = H_0 W_0$, where X_0 and W_0 are Gaussian random vectors with mean $m_X[0] \in \mathbb{R}^n$ and $m_W[0] \in \mathbb{R}^m$, respectively, and covariance matrix $\Sigma_X[0] \in \mathbb{R}^{n\times n}$ and $\Sigma_W[0] \in \mathbb{R}^{m\times m}$, respectively. By using (5.61) and (5.62), we have that

$$\begin{aligned}
\phi_{Y_0}(\eta) &= \phi_{X_0}\big(G_0^\top \eta\big) \\
&= e^{\mathrm{i}\eta^\top G_0 m_X[0] - \frac{1}{2}\eta^\top G_0 \Sigma_X[0] G_0^\top \eta}, \quad \eta \in \mathbb{R}^n,
\end{aligned} \tag{5.66}$$

$$\begin{aligned}
\phi_{Z_0}(\eta) &= \phi_{W_0}\big(H_0^\top \eta\big) \\
&= e^{\mathrm{i}\eta^\top H_0 m_W[0] - \frac{1}{2}\eta^\top H_0 \Sigma_W[0] H_0^\top \eta}, \quad \eta \in \mathbb{R}^n,
\end{aligned} \tag{5.67}$$

thus Y_0 and Z_0 are Gaussian random vectors with mean $m_Y[0] = G_0 m_X[0]$ and $m_Z[0] = H_0 m_W[0]$, respectively, and covariance matrix $\Sigma_Y[0] = G_0 \Sigma_X[0] G_0^\top$ and $\Sigma_Z[0] = H_0 \Sigma_W[0] H_0^\top$, respectively. Now, since X_0 and W_0 are independent by assumption, so are Y_0 and Z_0, thus by using (5.66), (5.67), and (5.63), we can obtain the characteristic function of $X_1 = Y_0 + Z_0$ as follows:

$$\begin{aligned}
\phi_{X_1}(\eta) &= \phi_{Y_0}(\eta)\phi_{Z_0}(\eta) \\
&= \Big(e^{\mathrm{i}\eta^\top G_0 m_X[0] - \frac{1}{2}\eta^\top G_0 \Sigma_X[0] G_0^\top \eta}\Big)\Big(e^{\mathrm{i}\eta^\top H_0 m_W[0] - \frac{1}{2}\eta^\top H_0 \Sigma_W[0] H_0^\top \eta}\Big) \\
&= e^{\mathrm{i}\eta^\top \big(G_0 m_X[0] + H_0 m_W[0]\big) - \frac{1}{2}\eta^\top \big(G_0 \Sigma_X[0] G_0^\top + H_0 \Sigma_W[0] H_0^\top\big)\eta}, \quad \eta \in \mathbb{R}^n,
\end{aligned} \tag{5.68}$$

therefore, X_1 is also a Gaussian random vector with mean $m_X[1]$, and covariance matrix $\Sigma_X[1]$ given as follows:

$$m_X[1] = G_0 m_X[0] + H_0 m_W[0], \tag{5.69}$$

$$\Sigma_X[1] = G_0 \Sigma_X[0] G_0^\top + H_0 \Sigma_W[0] H_0^\top, \tag{5.70}$$

expressions that match those in (5.64–5.65) for $k = 0$.

Assume X_k is a Gaussian random vector with mean $m_X[k]$ and covariance matrix $\Sigma_X[k]$ and define $Y_k = G_k X_k$ and $Z_k = H_k W_k$, where W_k is a Gaussian random vector with mean $m_W[k]$ and covariance matrix $\Sigma_W[k]$. Then, by following the same steps as above for $k = 0$, we can establish that Y_k and Z_k are Gaussian random vectors with mean $m_Y[k] = G_k m_X[k]$ and $m_Z[k] = H_k m_W[k]$, respectively, and covariance matrix $\Sigma_Y[k] = G_k \Sigma_X[k] G_k^\top$ and $\Sigma_Z[k] = H_k \Sigma_W[k] H_k^\top$, respectively. Furthermore, since X_0 and W_k, $k \geq 0$, are independent, we established earlier that X_k and W_k are also independent and thus so are Y_k and Z_k. Therefore, similar to the development above for $k = 0$, we obtain that $X_{k+1} = Y_k + Z_k$ is also a Gaussian random vector with mean $m_X[k+1]$ and covariance matrix $\Sigma_X[k+1]$ given as follows:

$$m_X[k+1] = G_k m_X[k] + H_k m_W[k], \tag{5.71}$$

$$\Sigma_X[k+1] = G_k \Sigma_X[k] G_k^\top + H_k \Sigma_W[k] H_k^\top, \tag{5.72}$$

expressions that match those in (5.64–5.65).

5.3 Discrete-Time Nonlinear Systems

In this section, we study systems of the form

$$X_{k+1} = h_k(X_k, W_k), \tag{5.73}$$

where $h_k : \mathbb{R}^n \times \mathbb{R}^m \to \mathbb{R}^n$, $W = \{W_k \colon k \geq 0\}$ is a stochastic process with the W_k's taking values in $\mathbb{R}^m$, and X_0 being a random vector in $\mathbb{R}^n$. We first address the case when the mean and covariance functions of the process W are known and use them to characterize the mean, covariance, and correlation functions of the process $X = \{X_k \colon k \geq 0\}$ describing the evolution of the system state. Then, we address the case when the distribution of W is known and characterize the distribution of the system state. However, unlike the linear case, we cannot obtain simple recursive relations because of the nonlinear dependence of X_{k+1} on X_k and W_k. Thus, we will settle for approximate characterizations instead obtained by the linearization technique discussed next.

Let x_k^*, $k \geq 0$, denote the trajectory followed by the system

$$x_{k+1} = h_k(x_k, w_k) \tag{5.74}$$

for $w_k = w_k^*$, $k \geq 0$, and $x_0 = x_0^*$. Let Δx_k denote a small variation in x_k around x_k^* that results from a small variation in w_k around w_k^*, which we denote by Δw_k. Then we have

$$x_{k+1}^* + \Delta x_{k+1} = h_k(x_k^* + \Delta x_k, w_k^* + \Delta w_k). \tag{5.75}$$

Now, by linearizing $h_k(\cdot,\cdot)$ around x_k^* and w_k^*, we have that Δx_k can be approximated by some $\widetilde{\Delta x}_k$ the evolution of which is given by

$$\widetilde{\Delta x}_{k+1} = G_k \widetilde{\Delta x}_k + H_k \Delta w_k, \tag{5.76}$$

where

$$G_k = \left. \frac{\partial h_k(x,w)}{\partial x} \right|_{x=x_k^*, w=w_k^*} \in \mathbb{R}^{n\times n},$$
$$H_k = \left. \frac{\partial h_k(x,w)}{\partial w} \right|_{x=x_k^*, w=w_k^*} \in \mathbb{R}^{n\times m}.$$

For the system in (5.76) to provide a good approximation, Δw_k needs to be sufficiently small. In the context of the system in (5.73), this translates into the support of f_{W_k}, i.e., the set

$$\mathcal{W}_k = \{w_k \colon f_{W_k}(w_k) > 0\} \tag{5.77}$$

being small enough. In addition, since in (5.73), X_0 is a random vector, we will also require the support of $f_{X_0}(x_0)$, i.e., the set

$$\mathcal{X}_0 = \{x_0 \colon f_{X_0}(x_0) > 0\} \tag{5.78}$$

to be small enough. Thus, the techniques to be presented are not suited to handle, e.g., Gaussian systems, because the support of $f_{W_k}(w_k)$ and $f_{X_0}(x_0)$ is the whole real line. Therefore, unless stated otherwise, in the remainder of this section we assume the following assumption holds.

ASSUMPTION 5.1 *Let $\overline{x}_k$, $k = 0, 1, \ldots,$ denote the trajectory followed by* (5.74) *for $w_k = m_W[k]$, $k = 0, 1, \ldots,$ and $x_0 = m_X[0]$. Define*

$$\Delta\mathcal{W}_k = \big\{\Delta w_k \colon f_{W_k}\big(m_W[k] + \Delta w_k\big) > 0\big\}, \quad k \geq 0, \tag{5.79}$$
$$\Delta\mathcal{X}_0 = \big\{\Delta x_0 \colon f_{X_0}\big(m_X[0] + \Delta x_0\big) > 0\big\}. \tag{5.80}$$

For any sequence $\big\{\Delta w_k \in \Delta\mathcal{W}_k,\ k \geq 0\big\}$ and any $\Delta x_0 \in \Delta\mathcal{X}_0$, the linearized model in (5.76) *with $x_k^* = \overline{x}_k$ and $w_k^* = m_W[k]$ provides a good approximation to the nonlinear system in* (5.75)*, i.e., $x_k = \overline{x}_k + \Delta x_k \approx \overline{x}_k + \widetilde{\Delta x}_k$.*

5.3.1 Characterization of First and Second Moments

Consider the system in (5.73) and assume that the values taken by the mean and covariance functions of W, $m_W[k]$, $k \geq 0$, and $C_W[k_1,k_2]$, $k_1, k_2 \geq 0$, respectively, are given, with $C_W[k_1,k_2]$ defined as follows:

$$C_W[k_1,k_2] = \begin{cases} Q_{k_1}, & \text{if } k_1 = k_2, \\ \mathbf{0}_{m\times m}, & \text{if } k_1 \neq k_2, \end{cases} \tag{5.81}$$

with $Q_{k_1} \in \mathbb{R}^{m \times m}$, and where the mean of X_0, $m_X[0]$, and the covariance matrix of X_0, $\Sigma_X[0]$, are also given. Assume that X_0 and W_k are uncorrelated for all $k \geq 0$, i.e.,

$$\mathrm{E}\Big[\big(X_0 - m_X[0]\big)\big(W_k - m_W[k]\big)^\top\Big] = \mathbf{0}_{n \times m}, \quad k \geq 0. \tag{5.82}$$

Then, as in the linear case discussed earlier, the objective is to compute the values taken by (i) the mean function of X, $m_X[k]$, $k > 0$, (ii) the covariance function of X, $C_X[k_1, k_2]$, $k_1, k_2 \geq 0$, and (iii) the correlation function of X, $R[k_1, k_2]$, $k_1, k_2 \geq 0$. Since obtaining an exact characterization is not possible in general, we will resort to the linearization technique presented earlier to obtain an approximate characterization.

Let $\overline{x}_k$, $k \geq 0$, denote the trajectory followed by (5.74) for $w_k = m_W[k]$, $k \geq 0$, and $x_0 = m_X[0]$. Define $\Delta W_k = W_k - m_W[k]$ and $\Delta X_k = X_k - \overline{x}_k$. Assume that the conditions in Assumption 5.1 are satisfied; then, by using (5.76), we have that ΔX_k can be approximated by some $\widetilde{\Delta X}_k$ whose evolution is given by

$$\widetilde{\Delta X}_{k+1} = G_k \widetilde{\Delta X}_k + H_k \Delta W_k, \tag{5.83}$$

with

$$G_k = \left.\frac{\partial h_k(x, w)}{\partial x}\right|_{x = \overline{x}_k, w = m_W[k]} \in \mathbb{R}^{n \times n},$$

$$H_k = \left.\frac{\partial h_k(x, w)}{\partial w}\right|_{x = \overline{x}_k, w = m_W[k]} \in \mathbb{R}^{n \times m},$$

and where

$$\begin{aligned} m_{\Delta W}[k] &= \mathrm{E}\big[\Delta W_k\big] \\ &= \mathrm{E}\big[W_k - m_W[k]\big] \\ &= \mathbf{0}_m, \end{aligned} \tag{5.84}$$

$$\begin{aligned} C_{\Delta W}[k_1, k_2] &= \mathrm{E}\Big[\big(\Delta W_{k_1} - m_{\Delta W}[k_1]\big)\big(\Delta W_{k_2} - m_{\Delta W}[k_2]\big)^\top\Big] \\ &= \mathrm{E}\Big[\big(W_{k_1} - m_W[k_1]\big)\big(W_{k_2} - m_W[k_2]\big)^\top\Big] \\ &= C_W[k_1, k_2], \end{aligned} \tag{5.85}$$

and

$$\begin{aligned} m_{\widetilde{\Delta X}}[0] &= \mathrm{E}\big[\widetilde{\Delta X}_0\big] \\ &= \mathrm{E}\big[X_0 - m_X[0]\big] \\ &= \mathbf{0}_n, \end{aligned} \tag{5.86}$$

$$\begin{aligned} \Sigma_{\widetilde{\Delta X}}[0] &= \mathrm{E}\big[\big(\widetilde{\Delta X}_0 - m_{\widetilde{\Delta X}}[0]\big)\big(\widetilde{\Delta X}_0 - m_{\widetilde{\Delta X}}[0]\big)^\top\big] \\ &= \mathrm{E}\big[\big(X_0 - m_X[0]\big)\big(X_0 - m_X[0]\big)^\top\big] \\ &= \Sigma_X[0]. \end{aligned} \tag{5.87}$$

Now, we can apply the results for the linear case to obtain expressions for $m_{\widetilde{\Delta X}}[k]$, $R_{\widetilde{\Delta X}}[k_1, k_2]$, and $C_{\widetilde{\Delta X}}[k_1, k_2]$. Specifically, by using (5.13) together with (5.84) and (5.86), we obtain

$$m_{\widetilde{\Delta X}}[k] = \mathbf{0}_n. \tag{5.88}$$

Also, by using (5.9) and (5.10), together with (5.85) and (5.87), we obtain

$$\begin{aligned} \Sigma_{\widetilde{\Delta X}}[k+1] &= G_k \Sigma_{\widetilde{\Delta X}}[k] G_k^\top + H_k Q_k H_k^\top, \\ C_{\widetilde{\Delta X}}[k_1, k_2] &= \begin{cases} \Phi_{k_1,k_2} \Sigma_{\widetilde{\Delta X}}[k_2], & \text{if } k_1 \geq k_2, \\ \Sigma_{\widetilde{\Delta X}}[k_1] \Phi_{k_2,k_1}^\top, & \text{if } k_1 < k_2, \end{cases} \end{aligned} \tag{5.89}$$

with $\Sigma_{\widetilde{\Delta X}}[0] = \Sigma_X[0]$. Finally, by using (5.20) together with (5.88), we obtain

$$R_{\widetilde{\Delta X}}[k_1, k_2] = C_{\widetilde{\Delta X}}[k_1, k_2]. \tag{5.90}$$

Now, we can approximate $m_X[k]$, $k \geq 0$, as follows:

$$\begin{aligned} m_X[k] &= \mathrm{E}\big[X_k\big] \\ &= \mathrm{E}\big[\overline{x}_k\big] + \mathrm{E}\big[\Delta X_k\big] \\ &\approx \mathrm{E}\big[\overline{x}_k\big] + m_{\widetilde{\Delta X}}[k] \\ &= \overline{x}_k. \end{aligned} \tag{5.91}$$

In addition, we can obtain approximate expressions for $C_X[k_1, k_2]$, $k_1, k_2 \geq 0$, as follows:

$$\begin{aligned} C_X[k_1, k_2] &= \mathrm{E}\Big[\big(X_{k_1} - m_X[k_1]\big)\big(X_{k_2} - m_X[k_2]\big)^\top\Big] \\ &= \mathrm{E}\Big[\big(\overline{x}_{k_1} + \Delta X_{k_1} - m_X[k_1]\big)\big(\overline{x}_{k_2} + \Delta X_{k_2} - m_X[k_2]\big)^\top\Big] \\ &\approx \mathrm{E}\Big[\Delta X_{k_1} \Delta X_{k_2}^\top\Big] \\ &\approx \mathrm{E}\Big[\widetilde{\Delta X}_{k_1} \widetilde{\Delta X}_{k_2}^\top\Big] \\ &= R_{\widetilde{\Delta X}}[k_1, k_2] \\ &= C_{\widetilde{\Delta X}}[k_1, k_2], \end{aligned} \tag{5.92}$$

and for $R_X[k_1, k_2]$, $k_1, k_2 \geq 0$, as follows:

$$\begin{aligned} R_X[k_1, k_2] &= C_X[k_1, k_2] + m_X[k_1] m_X[k_2]^\top \\ &\approx C_{\widetilde{\Delta X}}[k_1, k_2] + \overline{x}_k \overline{x}_k^\top. \end{aligned} \tag{5.93}$$

Example 5.4 Consider a system of the form:

$$\Theta_{k+1} = \Theta_k - \sin\left(\Theta_k\right) - P_k \tag{5.94}$$

where $P = \{P_0, P_1, \ldots\}$ is a stochastic process with mean function defined as follows:

$$m_P[k] = \frac{\sqrt{3}}{2}, \quad k \geq 0, \tag{5.95}$$

and covariance function defined as follows:

$$C_P[k_1, k_2] = \begin{cases} 0.01, & \text{if } k_1 = k_2, \\ 0, & \text{if } k_1 \neq k_2, \end{cases} \tag{5.96}$$

$k_1, k_2 \geq 0$, and where Θ_0 is a random variable with mean

$$m_\Theta[0] = -\frac{\pi}{3} \tag{5.97}$$

and variance

$$\sigma_\Theta^2[0] = 0.05. \tag{5.98}$$

The objective is to approximately characterize the mean and covariance functions of $\Theta = \{\Theta_0, \Theta_1, \ldots\}$.

Let $\overline{\theta}_k,\ k \geq 0$, denote the trajectory followed by

$$\begin{aligned} \theta_{k+1} &= \theta_k - \sin\left(\theta_k\right) - m_P[k] \\ &= \theta_k - \sin\left(\theta_k\right) - \frac{\sqrt{3}}{2}, \end{aligned} \tag{5.99}$$

for $\theta_0 = -\frac{\pi}{3}$. Then, since $\sin\left(\theta_0\right) + \frac{\sqrt{3}}{2} = 0$, it follows that $\overline{\theta}_k = -\frac{\pi}{3}$ for all k. Define $\Delta P_k = P_k - m_P[k]$ and $\Delta\Theta_k = \Theta_k - \overline{\theta}_k$. Then, by tailoring (5.83) to the setting here, $\Delta\Theta_k$ can be approximated by some $\widetilde{\Delta\Theta}_k$ the evolution of which is given by

$$\begin{aligned} \widetilde{\Delta\Theta}_{k+1} &= \left(1 - \cos(\overline{\theta}_k)\right)\widetilde{\Delta\Theta}_k - \Delta P_k \\ &= \left(1 - \cos(-\pi/3)\right)\widetilde{\Delta\Theta}_k - \Delta P_k \\ &= 0.5\widetilde{\Delta\Theta}_k - \Delta P_k, \end{aligned} \tag{5.100}$$

with

$$m_{\Delta P}[k] = 0, \tag{5.101}$$

$$\begin{aligned} C_{\Delta P}[k_1, k_2] &= C_P[k_1, k_2] \\ &= \begin{cases} 0.01, & \text{if } k_1 = k_2, \\ 0, & \text{if } k_1 \neq k_2, \end{cases} \end{aligned} \tag{5.102}$$

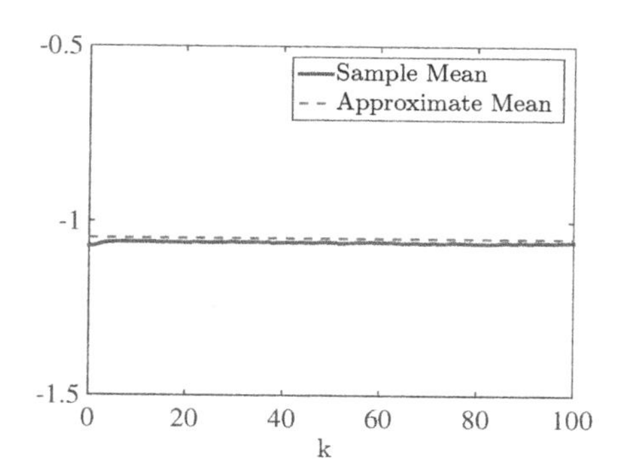

(a) Sample and approximate mean of Θ_k

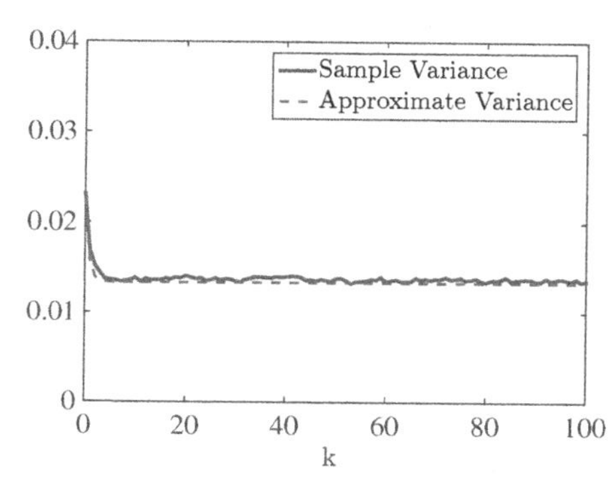

(b) Sample and approximate variance of Θ_k

Figure 5.2 Evolution of the sample and approximate mean and variance of Θ_k.

and $\widetilde{\Delta\Theta}_0 = 0$. Because $m_{\Delta P}[k] = 0$ for all $k \geq 0$ and $m_{\widetilde{\Delta\Theta}}[0] = 0$, we have that

$$m_{\widetilde{\Delta\Theta}}[k] = 0, \quad k \geq 0. \tag{5.103}$$

Let $\sigma^2_{\widetilde{\Delta\Theta}}[k] = \mathrm{E}\Big[\big(\widetilde{\Delta\Theta}_k - m_{\widetilde{\Delta\Theta}}[k]\big)^2\Big]$. Then, by using (5.89), we obtain

$$\begin{aligned} \sigma^2_{\widetilde{\Delta\Theta}}[k+1] &= 0.25\sigma^2_{\widetilde{\Delta\Theta}}[k] + 0.01, \\ C_{\widetilde{\Delta\Theta}}[k_1, k_2] &= \begin{cases} 0.5^{k_1-k_2}\sigma^2_{\widetilde{\Delta\Theta}}[k_2], & \text{if } k_1 \geq k_2, \\ 0.5^{k_2-k_1}\sigma^2_{\widetilde{\Delta\Theta}}[k_1], & \text{if } k_1 \leq k_2. \end{cases} \end{aligned} \tag{5.104}$$

Finally, by using (5.91) and (5.92), we obtain

$$\begin{aligned} m_\Theta[k] &\approx \overline{\theta}_k \\ &= -\frac{\pi}{3}, \end{aligned} \tag{5.105}$$

$$\begin{aligned} C_\Theta[k] &\approx C_{\widetilde{\Delta\Theta}}[k] \\ &= \begin{cases} 0.5^{k_1-k_2}\sigma^2_{\widetilde{\Delta\Theta}}[k_2], & \text{if } k_1 \geq k_2, \\ 0.5^{k_2-k_1}\sigma^2_{\widetilde{\Delta\Theta}}[k_1], & \text{if } k_1 \leq k_2. \end{cases} \end{aligned} \tag{5.106}$$

Figure 5.2 displays the approximate mean and variance of Θ_k, $k = 0, 1, \ldots,$ computed using (5.105–5.106), respectively, and the sample mean and variance obtained from 10,000 simulations of (5.94) assuming P_k, $k \geq 0$, are uniformly distributed with the mean and covariance functions of $P = \{P_0, P_1, \ldots\}$ as in (5.95) and (5.96), respectively, and Θ_0 is also uniformly distributed with mean and variance as in (5.97) and (5.98), respectively.

5.3.2 Probability Distribution

Consider the system in (5.73), where the W_k's are i.i.d. random vectors with known pdf, $f_{W_k}(w_k)$, and X_0 is a random vector with known pdf, $f_{X_0}(x_0)$, that is also independent of the W_k's. The objective here is to approximately characterize the pdf of X_k, $f_{x_k}(x_k)$, $k \geq 0$, by using the same linearization technique used for approximately characterizing the first and second moments of X_k, $k \geq 0$.

Let $\overline{x}_k$, $k \geq 0$, denote the trajectory followed by (5.74) for $w_k = m_W[k]$, $k \geq 0$, and $x_0 = m_X[0]$. Define $\Delta W_k = W_k - m_X[k]$ and $\Delta X_k = X_k - \overline{x}_k$. Assume that the conditions in Assumption 5.1 are satisfied; then, by using (5.76), we have that ΔX_k can be approximated by some $\widetilde{\Delta X}_k$, the evolution of which is given by

$$\widetilde{\Delta X}_{k+1} = G_k \widetilde{\Delta X}_k + H_k \Delta W_k, \tag{5.107}$$

with

$$G_k = \frac{\partial h_k(x,w)}{\partial x}\bigg|_{x=\overline{x}_k, w=m_W[k]} \in \mathbb{R}^{n\times n},$$

$$H_k = \frac{\partial h_k(x,w)}{\partial w}\bigg|_{x=\overline{x}_k, w=m_W[k]} \in \mathbb{R}^{n\times m},$$

and where

$$f_{\Delta W_k}(\Delta w_k) = f_{W_k}\big(\Delta w_k + m_W[k]\big) \tag{5.108}$$

and

$$f_{\widetilde{\Delta X}_0}(\widetilde{\Delta x}_0) = f_{X_0}\big(\widetilde{\Delta x}_0 + m_X[0]\big). \tag{5.109}$$

Now we can apply the results for the discrete-time linear time-varying system in Section 5.2.2 to obtain $f_{\widetilde{\Delta X}_k}(\widetilde{\Delta x}_k)$ for all $k \geq 0$. Specifically, by tailoring (5.48) to the setting here, we obtain the following recursion:

$$f_{\widetilde{\Delta X}_{k+1}}(\widetilde{\Delta x}_{k+1}) = \begin{cases} |\mathbf{det}(G_k^{-1})|\zeta(\widetilde{\Delta x}_{k+1}), & \widetilde{\Delta x}_{k+1} \in \widetilde{\Delta \mathcal{X}}_{k+1}, \\ 0, & \text{otherwise}, \end{cases} \tag{5.110}$$

with

$$\widetilde{\Delta \mathcal{X}}_{k+1} = \left\{\widetilde{\Delta x}_{k+1} \colon f_{\widetilde{\Delta X}_k}\big(G_k^{-1}(\widetilde{\Delta x}_{k+1} - H_k \Delta w_k)\big) \neq 0\right\}, \tag{5.111}$$

and

$$\begin{aligned}\zeta(\widetilde{\Delta x}_{k+1}) = \int_{-\infty}^{\infty} \cdots \int_{-\infty}^{\infty} & f_{\widetilde{\Delta X}_k}\big(G_k^{-1}(\widetilde{\Delta x}_{k+1} - H_k \Delta w_k)\big) \\ & \times f_{\Delta W_k}(\Delta w_k) d\Delta w_{k,1} \ldots d\Delta w_{k,m},\end{aligned} \tag{5.112}$$

where $d\Delta w_{k,i}$ denotes the ith entry of $d\Delta w_k$. Finally, by noting that

$$\begin{aligned} X_k &= \overline{x}_k + \Delta X_k \\ &\approx \overline{x}_k + \widetilde{\Delta X}_k, \end{aligned}$$

we can use $f_{\widetilde{\Delta X}_k}(\widetilde{\Delta x}_k)$ to approximate the pdf of X_k, $f_{X_k}(x_k)$, as follows:

$$\begin{aligned} f_{X_k}(x_k) &= f_{\Delta X_k}(x_k - \overline{x}_k) \\ &\approx f_{\widetilde{\Delta X}_k}(x_k - \overline{x}_k). \end{aligned} \tag{5.113}$$

5.4 Analysis of Microgrids under Power Injection Uncertainty

In this section, we apply the techniques developed earlier in this chapter to characterize the performance of inertia-less AC microgrids, i.e., small power systems whose generators and loads are interfaced with the network via power electronics. The relevance of this class of microgrids is rapidly growing, as there is a move towards relying more on small-scale renewable-based generation and electrical storage in small power systems for providing electricity to, e.g., rural communities in developing countries, disaster areas, and forward operating bases.

5.4.1 System Model

Consider a three-phase microgrid comprising n buses ($n > 1$) and l transmission lines $\big(n - 1 \leq l \leq n(n-1)/2\big)$. Assume that there is at most one transmission line connecting each pair of buses. Assign an arbitrary direction for the positive flow of power between any two buses connected by a transmission line. Then, the microgrid network topology together with the chosen orientation can be described by a directed graph, $\mathcal{G} = \{\mathcal{V}, \mathcal{E}\}$, where each element in the set $\mathcal{V} = \{1, 2, \ldots, n\}$ corresponds to a bus in the system, and $\mathcal{E} \subset \mathcal{V} \times \mathcal{V} \setminus \{(i,i) \colon i \in \mathcal{V}\}$ so that $(i,j) \in \mathcal{E}$ if there is a transmission line connecting bus i and bus j with the flow of power from bus i to bus j assigned to be positive. Let $\mathbb{I}$ denote a one-to-one mapping that assigns each element $(i,j) \in \mathcal{E}$ to an element $e \in \mathcal{L} = \{1, 2, \ldots, l\}$. Then, we can define the node-to-edge incidence matrix of $\mathcal{G}$, denoted by $M = [m_{i,e}] \in \mathbb{R}^{n \times l}$, as follows:

$$m_{i,e} = \begin{cases} 1, & \text{if } e = \mathbb{I}\big((i,j)\big),\ (i,j) \in \mathcal{E}, \\ -1, & \text{if } e = \mathbb{I}\big((j,i)\big),\ (j,i) \in \mathcal{E}, \\ 0, & \text{otherwise}. \end{cases} \tag{5.114}$$

Assume the microgrid is balanced and operating in sinusoidal regime. Further, assume that all transmission lines in the network are short and lossless. Thus, the lumped-parameter circuit describing the terminal behavior of the line connecting bus i and bus j reduces to the series element of the Π-equivalent circuit model (see Appendix B.1), which in this case is a purely reactive impedance with a reactance value $X_{ij} > 0$. Let $p_i(t)$ denote the active power injected into the microgrid via bus i at time instant t, and let $V_i(t)$ and $\theta_i(t)$ respectively denote the magnitude and phase angle of the phasor associated with bus i's voltage at time instant t (measured relatively to a reference frame rotating at ω_0 rad/s corresponding to the microgrid nominal frequency). Define $p(t) = \left[p_1(t), p_2(t), \dots, p_n(t)\right]^\top$ and $\theta(t) = \left[\theta_1(t), \theta_2(t), \dots, \theta_n(t)\right]^\top$. Then, we have that

$$p(t) = M\Gamma(t)\,\mathbf{sin}\big(M^\top \theta(t)\big), \tag{5.115}$$

with $\Gamma(t) = \mathbf{diag}\big(\gamma_1(t), \gamma_2(t), \dots, \gamma_l(t)\big)$, where

$$\gamma_e(t) = \frac{V_i(t)V_j(t)}{X_{ij}}, \quad e = \mathbb{I}\big((i,j)\big),\ (i,j) \in \mathcal{E},$$

and for $x = M^\top \theta(t) \in \mathbb{R}^l$, $\mathbf{sin}(x)$ is defined as follows:

$$\mathbf{sin}(x) = \left[\sin(x_1), \sin(x_2), \dots, \sin(x_l)\right]^\top.$$

Open-Loop Dynamics

Assume that connected to each bus i, $i = 1, 2, \dots, n$, there is either a generating- or a load-type resource interfaced via a three-phase voltage source inverter whose controls will attempt to synthesize a sinusoidal voltage waveform at its output terminals. As a result, the inverter connected to bus i can be described by a per-phase equivalent circuit model comprising a controlled voltage source with terminal voltage

$$e_i(t) = \sqrt{2}E_i(t)\sin\left(\omega_0 t + \delta_i(t)\right), \tag{5.116}$$

in series with a reactance resulting from the inverter output filter. The value of such reactance is typically small when compared to the reactance values of the microgrid network transmission lines and therefore can be neglected; thus, the magnitude and phase angle of bus i's voltage phasor at time t, $V_i(t)$ and $\theta_i(t)$, respectively, can be approximated by the magnitude and phase angle of $e_i(t)$, i.e., $V_i(t) \approx E_i(t)$ and $\theta_i(t) \approx \delta_i(t)$, $i = 1, 2, \dots, n$.

Assume that the phase angle of the inverter connected to bus i is regulated via a discrete-time frequency-droop control scheme that gets updated every T_s [s]. Then, since $\theta_i(t) \approx \delta_i(t)$, the evolution of $\theta_i(t)$ is governed by the same update rule that governs the update of $\delta_i(t)$ in the frequency-droop control scheme; thus,

$$\theta_{i,k+1} = \theta_{i,k} + \frac{T_s}{D_i}\big(u_{i,k} - p_{i,k}\big), \quad i = 1, 2, \dots, n, \tag{5.117}$$

where D_i [s/rad] is the so-called droop coefficient, $p_{i,k} := p_i(kT_s)$, $\theta_{i,k} := \theta_i(kT_s)$, $k = 0, 1, 2, \dots$, and where $u_{i,k}$ can be either extraneously set (thus, from the system point of view acts as a disturbance) or set by a control system. For the case when the resource connected to bus i is of the generating type, we have that $u_{i,k} \geq 0$, whereas for the case when the resource is of the load type, we have that $u_{i,k} \leq 0$. Without loss of generality, assume that connected to each bus i, $i = 1, 2, \dots, m$, $1 \leq m \leq n-1$, there is a resource whose $u_{i,k}$ is set extraneously to some value $\xi_{i,k}$, whereas connected to each bus $i = m+1, m+2, \dots, n$, there is a resource whose $u_{i,k}$ is set to some value $r_{i,k}$ by a control system, i.e.,

$$u_{i,k} = \begin{cases} \xi_{i,k}, & i = 1, 2, \ldots, m, \\ r_{i,k}, & i = m+1, m+2, \ldots, n, \end{cases} \tag{5.118}$$

with $1 \leq m \leq n-1$.

Assume that, as a result of the action of the inverter outer voltage and inner current control loops, we have that $E_i(t) = E_i$, $i = 1, 2, \ldots, n$, where E_i is a positive constant. (Such control loops are much faster than those of the frequency-droop control scheme; thus, they can be abstracted out from the model.) Then, it follows that $V_i(t) \approx E_i =: V_i$ $i = 1, 2, \ldots, n$; thus,

$$V_{i,k} \approx V_i, \quad i = 1, 2, \ldots, n, \tag{5.119}$$

where $V_{i,k} := V_i(kT_s)$, $k \geq 0$. Therefore,

$$\begin{aligned} \Gamma(kT_s) &= \mathbf{diag}(\gamma_1, \gamma_2, \ldots, \gamma_l) \\ &=: \Gamma, \quad k \geq 0, \end{aligned} \tag{5.120}$$

where

$$\gamma_e = \frac{V_i V_j}{X_{ij}}, \quad e = \mathbb{I}\big((i,j)\big), \ (i,j) \in \mathcal{E}. \tag{5.121}$$

Define the following vectors: $\theta_k = \big[\theta_{1,k}, \theta_{2,k}, \ldots, \theta_{n,k}\big]^\top$, $\xi_k = \big[\xi_{1,k}, \xi_{2,k}, \ldots, \xi_{m,k}\big]^\top$, and $r_k = \big[r_{m+1,k}, r_{m+2,k}, \ldots, r_{n,k}\big]^\top$. Then, combining (5.115–5.121) yields

$$\theta_{k+1} = \theta_k + T_s D^{-1}\Big(K_\xi \xi_k + K_r r_k - M\Gamma \mathbf{sin}\big(M^\top \theta_k\big)\Big), \tag{5.122}$$

where $D = \mathbf{diag}\big(D_1, D_2, \ldots, D_n\big)$, and

$$K_\xi = \begin{bmatrix} I_m \\ \mathbf{0}_{(n-m)\times m} \end{bmatrix}, \quad K_r = \begin{bmatrix} \mathbf{0}_{m\times(n-m)} \\ I_{n-m} \end{bmatrix}. \tag{5.123}$$

Next we provide some intuition for what the frequency-droop control is attempting to achieve. Assume that (5.122) is operating in some steady state with $\sum_{i=1}^m \xi_{i,k} + \sum_{i=m+1}^n r_{i,k} = \mu$, $k \geq 0$, where μ is some constant; then,

$$\theta_{i,k+1} - \theta_{i,k} = \frac{T_s}{\sum_{i=1}^n D_i}\mu, \quad i = 1, 2, \ldots, n, \tag{5.124}$$

i.e., the incremental change of all the bus angles is the same for all buses, and its value is proportional to the sum of the $\xi_{i,k}$'s and the $r_{i,k}$'s. One can see this by (i) multiplying (5.122) throughout by D, (ii) adding all the equations, and (iii) noting that in steady state, $\theta_{i,k+1} - \theta_{i,k}$, must be equal to some constant for all $i = 1, 2, \ldots, n$. Then, by using (5.117), one can see that the steady-state power injected into the network via bus i, denoted by p_i, is as follows:

$$p_i = \begin{cases} \xi_{i,k} - \frac{D_i}{\sum_{j=1}^n D_j}\mu, & i = 1, 2, \ldots, m, \\ r_{i,k} - \frac{D_i}{\sum_{j=1}^n D_j}\mu, & i = m+1, m+2, \ldots, n, \end{cases} \tag{5.125}$$

i.e., the frequency-droop control is attempting to apportion μ according to the magnitude of the inverter droop coefficients. Note that the expression in (5.125) is consistent with the power injection model adopted in Section 3.5 when analyzing the steady-state behavior of a power system. Also, by noting (5.116) and (5.124), we can conclude that in steady state, the voltage synthesized by each voltage source inverter will have a frequency $\omega_0 + \Delta\omega$ rad/s, where

$$\Delta\omega = \frac{\mu}{\sum_{i=1}^{n} D_i}. \tag{5.126}$$

Thus, the frequency-droop control scheme attempts to compensate for any deviation from zero of the sum of the $\xi_{i,k}$'s and the $r_{i,k}$'s by increasing or decreasing the frequency of the sinusoidal voltage synthesized by the voltage source inverter.

Closed-Loop Dynamics

In light of (5.126), it is clear that if we want the bus voltage frequencies to be close to ω_0, we need to regulate the $r_{i,k}$'s so that their sum closely matches the sum of the $\xi_{i,k}$'s. Next, we describe a closed-loop control scheme that can be used to accomplish this task.

Define the *average frequency error* as follows:

$$\overline{\Delta\omega}_k := \frac{\sum_{i=1}^{n} D_i(\theta_{i,k+1} - \theta_{i,k})}{T_s\left(\sum_{i=1}^{n} D_i\right)}. \tag{5.127}$$

Since the microgrid network is lossless, we have that $\sum_{i=1}^{n} p_{i,k} = 0$ for all $k \geq 0$; then it follows from (5.117–5.118) that

$$\begin{aligned}\overline{\Delta\omega}_k &= \frac{\sum_{i=1}^{m} \xi_{i,k} + \sum_{i=m+1}^{n} r_{i,k}}{\sum_{i=1}^{n} D_i}\\ &= \frac{1}{n\overline{D}}\left(\mathbf{1}_m^\top \xi_k + \mathbf{1}_{n-m}^\top r_k\right),\end{aligned} \tag{5.128}$$

where $\overline{D} := \frac{\sum_{i=1}^{n} D_i}{n}$. Assume $r_{i,k}$, $i = m+1, m+2, \ldots, n$, are regulated using the following discrete-time integral control scheme:

$$\begin{aligned}z_{i,k+1} &= z_{i,k} + \alpha_i \overline{\Delta\omega}_k,\\ r_{i,k} &= r_i^* + \beta_i z_{i,k},\end{aligned} \tag{5.129}$$

where r_i^* [p.u.] is constant, and so are the controller gains α_i [s/rad] and β_i [p.u.], which will be appropriately chosen later. Define $z_k = \left[z_{m+1,k}, z_{m+2,k}, \ldots, z_{n,k}\right]^\top$ and $r^* = \left[r_{m+1}^*, r_{m+2}^*, \ldots, r_n^*\right]^\top$; then, we can rewrite (5.129) in matrix form as follows:

$$z_{k+1} = z_k + \alpha\overline{\Delta\omega}_k, \tag{5.130}$$

$$r_k = r^* + Bz_k, \tag{5.131}$$

where $\alpha = [\alpha_{m+1}, \alpha_{m+2}, \ldots, \alpha_n]^\top$, $B = \mathbf{diag}(\beta_{m+1}, \beta_{m+2}, \ldots, \beta_n)$, which after some manipulation yields:

$$z_{k+1} = \left(I_{n-m} + \frac{1}{n\overline{D}}\alpha\beta^\top\right)z_k + \frac{1}{n\overline{D}}\alpha\left(\mathbf{1}_m^\top \xi_k + \mathbf{1}_{n-m}^\top r^*\right), \tag{5.132}$$

where $\beta = [\beta_{m+1}, \beta_{m+2}, \ldots, \beta_n]^\top$. Then, combining (5.122), (5.131), and (5.132) yields

$$\theta_{k+1} = \theta_k + T_s D^{-1}\Big(K_\xi \xi_k + K_r(r^* + Bz_k) - M\Gamma\mathbf{sin}(M^\top \theta_k)\Big),$$
$$z_{k+1} = \Big(I_{n-m} + \frac{1}{n\overline{D}}\alpha\beta^\top\Big)z_k + \frac{1}{n\overline{D}}\alpha\Big(\mathbf{1}_m^\top \xi_k + \mathbf{1}_{n-m}^\top r^*\Big). \tag{5.133}$$

In the remainder we will assume that

$$z_0 = \mathbf{0}_{n-m}, \tag{5.134}$$

while θ_0 satisfies

$$K_\xi \xi^0 + K_r r^* = M\Gamma\mathbf{sin}(M^\top \theta_0), \tag{5.135}$$

for some constant $\xi^0 = [\xi_1^0, \xi_2^0, \ldots, \xi_m^0]^\top$. This choice of initial conditions is by no means restrictive but it attempts to capture the scenario in which the system is operating in some steady state corresponding to extraneous disturbance, ξ_k, being equal to ξ^0 for all $k \geq 0$. By adding all the equations in (5.135), one can check that the following condition is then satisfied:

$$\sum_{i=1}^{m} \xi_i^0 + \sum_{i=m+1}^{n} r_i^* = 0. \tag{5.136}$$

Note that only $n-1$ equations in (5.135) are linearly independent; thus, when solving for θ_0 for some given ξ^0 and r^*, it is necessary to fix one of the angles; thus, in subsequent developments, and without loss of generality, we assume that $\theta_{n,0} = 0$.

Controller Gain Selection. Consider the system in (5.132) with the entries of $\alpha = [\alpha_{m+1}, \alpha_{m+2}, \ldots, \alpha_n]^\top$ and $\beta = [\beta_{m+1}, \beta_{m+2}, \ldots, \beta_n]^\top$ chosen so as to satisfy:

$$\begin{aligned} \alpha_i &< 0, \quad i = m+1, m+2, \ldots, n, \\ \beta_i &> 0, \quad i = m+1, m+2, \ldots, n, \\ \alpha^\top \beta &> -2n\overline{D}. \end{aligned} \tag{5.137}$$

Then the trajectories followed by (5.132) remain bounded as long as the disturbance ξ_k remains bounded for all $k \geq 0$. To see this, define $\zeta_k = \beta^\top z_k$; then, multiplying both sides of (5.132) by $\beta^\top$ yields:

$$\zeta_{k+1} = \Big(1 + \frac{1}{n\overline{D}}\beta^\top \alpha\Big)\zeta_k + \frac{1}{n\overline{D}}\beta^\top \alpha\Big(\mathbf{1}_m^\top \xi_k + \mathbf{1}_{n-m}^\top r^*\Big). \tag{5.138}$$

Now, it follows from (5.137) that

$$\left|1 + \frac{1}{n\overline{D}}\beta^\top \alpha\right| < 1;$$

thus the homogeneous part of (5.138) is asymptotically stable, and therefore, ζ_k will remain bounded for all $k \geq 0$ if ξ_k is bounded for all $k \geq 0$ (see Section 2.3.1). Finally, since $\zeta_k = \beta^\top z_k$ and the entries of β are all positive, the entries of z_k must remain bounded for all $k \geq 0$.

Extraneous Disturbance Model

In the remainder, we will assume that the value taken by ξ_k in (5.133) is random and governed by a stochastic process, $\Xi = \{\Xi_k \colon k \geq 0\}$, with known mean, $m_\Xi[k] \in \mathbb{R}^m$, $k \geq 0$, and known covariance function defined as follows:

$$C_\Xi[k_1, k_2] = \begin{cases} Q_{k_1}, & \text{if } k_1 = k_2, \\ \mathbf{0}_{m\times m}, & \text{if } k_1 \neq k_2, \end{cases} \tag{5.139}$$

for all $k_1, k_2 \geq 0$, and where Q_{k_1} is an $(m \times m)$-dimensional symmetric positive definite matrix.

Alternatively, we can characterize the extraneous disturbance as a function of another stochastic process obtained using the same ξ^0 in (5.135) determining the initial phase angle vector, θ_0. To this end, write ξ_k as follows:

$$\xi_k = \xi^0 + w_k, \quad k = 0, 1, 2, \ldots, \tag{5.140}$$

where the values taken by w_k, $k \geq 0$, are governed by a stochastic process $W = \{W_k \colon k \geq 0\}$ with mean and covariance function given by

$$m_W[k] = m_\Xi[k] - \xi^0, \quad k \geq 0, \tag{5.141}$$

$$\begin{aligned} C_W[k_1, k_2] &= C_\Xi[k_1, k_2] \\ &= \begin{cases} Q_{k_1}, & \text{if } k_1 = k_2, \\ \mathbf{0}_{m\times m}, & \text{if } k_1 \neq k_2, \end{cases} \end{aligned} \tag{5.142}$$

for all $k_1, k_2 \geq 0$.

5.4.2 Average Frequency Error Statistical Characterization

By inspection of (5.138), one can see that if ξ_k were to remain constant throughout time, $\zeta_k = \beta^\top z_k$ would converge asymptotically to some constant value, and as a result, z_k would also converge to some constant vector, which in light of (5.130) would result in $\overline{\Delta\omega}_k$ converging to zero. In reality, since the values taken by the ξ_k's are governed by a stochastic process $\Xi = \{\Xi_k \colon k \geq 0\}$ with known mean and covariance functions, they will not necessarily stay constant throughout time. Thus, in general, z_k will not converge to some constant vector and its value will be random and determined by some stochastic process, $Z = \{Z_k,\ k \geq 0\}$. As a result, $\overline{\Delta\omega}_k$ will not converge to zero either, and its value will also be random and determined by some stochastic process, $\overline{\Delta\Omega} = \{\overline{\Delta\Omega}_k,\ k \geq 0\}$. The objective then is to characterize the statistics of such processes.

Manipulating (5.128), (5.131) and (5.132), together with (5.140) and the condition in (5.136), yields

$$z_{k+1} = Gz_k + Hw_k, \tag{5.143}$$

$$\overline{\Delta\omega}_k = \frac{1}{n\overline{D}}\left(\mathbf{1}_m^\top w_k + \beta^\top z_k\right), \tag{5.144}$$

with

$$G = I_{n-m} + \frac{1}{n\overline{D}}\alpha\beta^\top, \quad H = \frac{1}{n\overline{D}}\alpha\mathbf{1}_m^\top, \tag{5.145}$$

and where values taken by w_k, $k \geq 0$, are determined by a stochastic process $W = \{W_k \colon k \geq 0\}$ with mean and covariance functions as given in (5.141–5.142), and $z_0 = \mathbf{0}_{n-m}$. Note that Z_0 and W_k are uncorrelated for any $k \geq 0$ because $Z_0 = \mathbf{0}_{n-m}$. Then it follows that $Z_{k'}$ and W_k are also uncorrelated for any $k \geq k' \geq 0$, i.e.,

$$\mathrm{E}\Big[\big(Z_{k'} - m_Z[k']\big)\big(W_k - m_W[k]\big)^\top\Big] = 0, \tag{5.146}$$

for any $k \geq k' \geq 0$. Thus, we can use the techniques developed earlier in the chapter to characterize the mean and covariance functions of the stochastic process Z, which can then be used to characterize the mean and covariance functions of the stochastic process $\overline{\Delta\Omega}$.

Mean Function. The values taken by the mean function of Z, $m_Z[k]$, $k \geq 0$, and the mean function of $\overline{\Delta\Omega}$, $m_{\overline{\Delta\Omega}}[k]$, $k \geq 0$, are given by

$$m_Z[k+1] = Gm_Z[k] + Hm_W[k], \quad k \geq 0, \tag{5.147}$$

$$m_{\overline{\Delta\Omega}}[k] = \frac{1}{n\overline{D}}\left(\mathbf{1}_m^\top m_W[k] + \beta^\top m_Z[k]\right), \quad k \geq 0, \tag{5.148}$$

with $m_Z[0] = \mathbf{0}_{n-m}$ (because $Z_0 = \mathbf{0}_{n-m}$). The recursion in (5.147) follows from tailoring (5.7) to the model in (5.143), whereas the expression in (5.148) can be obtained by taking expectations on both sides of (5.144).

Covariance Function. The values taken by the covariance function of Z, $C_Z[k_1, k_2]$, $k_1, k_2 \geq 0$, and the covariance function of $\overline{\Delta\Omega}$, $C_{\overline{\Delta\Omega}}[k_1, k_2]$, $k_1, k_2 \geq 0$, are given by

$$\Sigma_Z[k+1] = G\Sigma_Z[k]G^\top + HQ_kH^\top, \tag{5.149}$$

$$C_Z[k_1, k_2] = \begin{cases} G^{k_1-k_2}\Sigma_Z[k_2], & \text{if } k_1 \geq k_2, \\ \Sigma_Z[k_1]\left(G^\top\right)^{k_2-k_1}, & \text{if } k_1 < k_2, \end{cases} \tag{5.150}$$

with $\Sigma_Z[0] = \mathbf{0}_{(n-m)\times(n-m)}$ (because $Z_0 = \mathbf{0}_{n-m}$), and

$$\begin{aligned} C_{\overline{\Delta\Omega}}[k_1,k_2] &= \frac{1}{n^2\overline{D}^2}\mathbf{1}_m^\top C_W[k_1,k_2]\mathbf{1}_m \\ &+ \frac{1}{n^2\overline{D}^2}\mathbf{1}_m^\top \underbrace{\mathrm{E}\Big[\big(W_{k_1}-m_W[k_1]\big)\big(Z_{k_2}-m_Z[k_2]\big)^\top\Big]}_{=C_{W_{k_1},Z_{k_2}}}\beta \\ &+ \frac{1}{n^2\overline{D}^2}\beta^\top \underbrace{\mathrm{E}\Big[\big(Z_{k_1}-m_Z[k_1]\big)\big(W_{k_2}-m_W[k_2]\big)^\top\Big]}_{=C_{Z_{k_1},W_{k_2}}}\mathbf{1}_m \\ &+ \frac{1}{n^2\overline{D}^2}\beta^\top C_Z[k_1,k_2]\beta, \quad k_1,k_2\geq 0, \end{aligned} \tag{5.151}$$

with $C_W[k_1,k_2]$ as in (5.142), $C_Z[k_1,k_2]$ as in (5.150), and

$$C_{W_{k_1},Z_{k_2}} = \begin{cases} \mathbf{0}_{m\times(n-m)}, & \text{if } k_1 \geq k_2, \\ Q_{k_1}H^\top\left(G^{k_2-k_1-1}\right)^\top, & \text{if } k_1 < k_2, \end{cases} \tag{5.152}$$

$$C_{Z_{k_1},W_{k_2}} = \begin{cases} G^{k_1-k_2-1}HQ_{k_2}, & \text{if } k_1 > k_2, \\ \mathbf{0}_{(n-m)\times m}, & \text{if } k_1 \leq k_2. \end{cases} \tag{5.153}$$

The expressions in (5.149–5.150) follow from tailoring (5.9–5.11) to the model in (5.143). The expression in (5.151) can be obtained by rewriting (5.144) as follows:

$$\overline{\Delta\omega}_k = \frac{1}{n\overline{D}}\begin{bmatrix}\mathbf{1}_m^\top & \beta^\top\end{bmatrix} y_k, \tag{5.154}$$

where $y_k = [w_k^\top, z_k^\top]^\top$, and noting that

$$C_Y[k_1,k_2] = \begin{bmatrix} C_W[k_1,k_2] & C_{W_{k_1},Z_{k_2}} \\ C_{Z_{k_1},W_{k_2}} & C_Z[k_1,k_2] \end{bmatrix}, \tag{5.155}$$

where

$$\begin{aligned} C_{W_{k_1},Z_{k_2}} &= \mathrm{E}\Big[\big(W_{k_1}-m_W[k_1]\big)\big(Z_{k_2}-m_Z[k_2]\big)^\top\Big], \\ C_{Z_{k_1},W_{k_2}} &= \mathrm{E}\Big[\big(Z_{k_1}-m_Z[k_1]\big)\big(W_{k_2}-m_W[k_2]\big)^\top\Big]; \end{aligned}$$

thus,

$$\begin{aligned} C_{\overline{\Delta\Omega}}[k_1,k_2] &= \frac{1}{n^2\overline{D}^2}\begin{bmatrix}\mathbf{1}_m^\top & \beta^\top\end{bmatrix} C_Y[k_1,k_2]\begin{bmatrix}\mathbf{1}_m^\top & \beta^\top\end{bmatrix}^\top \\ &= \frac{1}{n^2\overline{D}^2}\mathbf{1}_m^\top C_W[k_1,k_2]\mathbf{1}_m + \frac{1}{n^2\overline{D}^2}\mathbf{1}_m^\top C_{W_{k_1},Z_{k_2}}\beta \\ &+ \frac{1}{n^2\overline{D}^2}\beta^\top C_{Z_{k_1},W_{k_2}}\mathbf{1}_m + \frac{1}{n^2\overline{D}^2}\beta^\top C_Z[k_1,k_2]\beta. \end{aligned} \tag{5.156}$$

Now, since $Z_{k'}$ and W_k are uncorrelated for any $k \geq k' \geq 0$, i.e.,

$$\mathrm{E}\Big[\big(Z_{k'}-m_Z[k']\big)\big(W_k-m_W[k]\big)^\top\Big] = 0$$

for any $k \geq k' \geq 0$, it follows that

$$\begin{aligned} C_{W_{k_1},Z_{k_2}} &= \mathrm{E}\Big[\big(W_{k_1} - m_W[k_1]\big)\big(Z_{k_2} - m_Z[k_2]\big)^\top\Big] \\ &= \mathbf{0}_{m\times(n-m)}, \quad k_1 \geq k_2, \\ C_{Z_{k_1},W_{k_2}} &= \mathrm{E}\Big[\big(Z_{k_1} - m_Z[k_1]\big)\big(W_{k_2} - m_W[k_2]\big)^\top\Big] \\ &= \mathbf{0}_{(n-m)\times m}, \quad k_2 \geq k_1, \end{aligned}$$

as claimed in the first line of (5.152) and second line of (5.153), respectively. Additionally, since

$$Z_k - m_Z[k] = G^{k-k'}\big(Z_{k'} - m_Z[k']\big) + \sum_{\ell=k'}^{k-1} G^{k-\ell-1}H\big(W_\ell - m_W[\ell]\big), \quad k \geq k' \geq 0,$$

it follows that, for $k_1 < k_2$,

$$\begin{aligned} C_{W_{k_1},Z_{k_2}} &= \mathrm{E}\Big[\big(W_{k_1} - m_W[k_1]\big)\big(Z_{k_2} - m_Z[k_2]\big)^\top\Big] \\ &= \underbrace{\mathrm{E}\Big[\big(W_{k_1} - m_W[k_1]\big)\big(Z_{k_1} - m_Z[k_1]\big)^\top\Big]}_{=0 \text{ because } W_{k_1} \text{ and } Z_{k_1} \text{ are uncorrelated}}\big(G^{k_2-k_1}\big)^\top \\ &\quad + \underbrace{\mathrm{E}\Big[\big(W_{k_1} - m_W[k_1]\big)\big(W_{k_1} - m_W[k_1]\big)^\top\Big]}_{=C_W[k_1,k_1]=Q_{k_1} \text{ by (5.142)}} H^\top\big(G^{k_2-k_1-1}\big)^\top \\ &\quad + \sum_{\ell=k_1+1}^{k_2-1} \underbrace{\mathrm{E}\Big[\big(W_{k_1} - m_W[k_1]\big)\big(W_\ell - m_W[\ell]\big)^\top\Big]}_{=C_W[k_1,\ell]=0 \text{ by (5.142) for } \ell\neq k_1} H^\top\big(G^{k_2-\ell-1}\big)^\top \\ &= Q_{k_1}H^\top\big(G^{k_2-k_1-1}\big)^\top, \end{aligned}$$

as claimed in the second line of (5.152). A similar development would yield

$$C_{Z_{k_1},W_{k_2}} = G^{k_1-k_2-1}HQ_{k_2}, \quad k_2 < k_1,$$

(as claimed in the first line of (5.153)), and therefore we omit it.

Example 5.5 (Three-bus microgrid average frequency error statistics) Consider a three-bus microgrid whose network topology and line flow direction assignment is captured by the graph in Fig. 5.3. Connected to bus 1 there is a load interfaced via a droop-controlled voltage source inverter with droop coefficient $D_1 = 0.1$ s/rad and update rate $T_s = 0.1$ s. Assume that power consumed by the load cannot be controlled; thus, in this case,

$$\theta_{1,k+1} = \theta_{1,k} + (\xi_{1,k} - p_{1,k}), \tag{5.157}$$

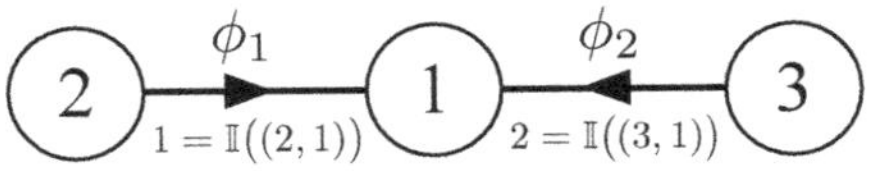

$$\text{Graph: } (2) \xrightarrow{\phi_1,\ 1=\mathbb{I}((2,1))} (1) \xleftarrow{\phi_2,\ 2=\mathbb{I}((3,1))} (3)$$

Figure 5.3 Graph describing the topology and line flow direction assignment for the microgrid in Example 5.5.

where the value that $\xi_{1,k}$ [p.u.] takes is determined by a stochastic process, $\Xi_1 = \{\Xi_{1,k} \colon k \geq 0\}$, with mean and covariance functions given as follows:

$$\begin{aligned} m_{\Xi_1}[k] &= -1, \quad k \geq 0, \\ C_{\Xi_1}[k_1, k_2] &= \begin{cases} 0.01, & \text{if } k_1 = k_2, \\ 0, & \text{if } k_1 \neq k_2, \end{cases} \end{aligned} \tag{5.158}$$

for all $k_1, k_2 \geq 0$.

Connected to bus 2 and bus 3, there are two generators also interfaced via droop-controlled voltage source inverters with droop coefficients $D_2 = 0.25$ s/rad and $D_3 = 1$ s/rad, respectively. Assume that the power generated by both generators can be regulated using the control scheme in (5.128–5.129) with $\alpha_2 = -0.5$ s/rad, $\beta_2 = 1$ p.u., and $r_2^* = 0.75$ p.u. for the generator connected to bus 2, and $\alpha_3 = -1$ s/rad, $\beta_3 = 0.5$ p.u., and $r_3^* = 0.25$ p.u. for the generator connected to bus 3. Thus, in this case, we have that

$$\begin{aligned} \theta_{2,k+1} &= \theta_{2,k} + 0.4(r_{2,k} - p_{2,k}), \\ z_{2,k+1} &= z_{2,k} - 0.3704(\xi_{1,k} + r_{2,k} + r_{3,k}), \\ r_{2,k} &= 0.75 + z_{2,k}, \end{aligned} \tag{5.159}$$

and

$$\begin{aligned} \theta_{3,k+1} &= \theta_{3,k} + 0.1(r_{3,k} - p_{3,k}), \\ z_{3,k+1} &= z_{3,k} - 0.7407(\xi_{1,k} + r_{2,k} + r_{3,k}), \\ r_{3,k} &= 0.25 + 0.5 z_{3,k}. \end{aligned} \tag{5.160}$$

By using (5.136), it follows that $\xi_1^0 = -(r_2^* + r_3^*) = -1$ p.u.; then, in light of (5.141–5.142), we can define a stochastic process, $W = \{W_k \colon k \geq 0\}$, with mean and covariance functions defined as follows:

$$\begin{aligned} m_W[k] &= m_{\Xi_1}[k] - \xi_1^0 \\ &= 0, \end{aligned} \tag{5.161}$$

for all $k \geq 0$, and

$$\begin{aligned} C_W[k_1, k_2] &= C_{\Xi_1}[k_1, k_2] \\ &= \begin{cases} 0.01, & \text{if } k_1 = k_2, \\ 0, & \text{if } k_1 \neq k_2, \end{cases} \end{aligned} \tag{5.162}$$

for all $k_1, k_2 \geq 0$, respectively.

Now, by tailoring (5.143–5.144) to the setting here, we obtain

$$\underbrace{\begin{bmatrix} z_{2,k+1} \\ z_{3,k+1} \end{bmatrix}}_{=z_{k+1}} = \underbrace{\begin{bmatrix} 0.6296 & -0.1852 \\ -0.7407 & 0.6296 \end{bmatrix}}_{=G} \underbrace{\begin{bmatrix} z_{2,k} \\ z_{3,k} \end{bmatrix}}_{=z_k} + \underbrace{\begin{bmatrix} -0.3704 \\ -0.7407 \end{bmatrix}}_{=H} w_k,$$

$$\overline{\Delta\omega_k} = 0.7407 w_k + [0.7407, 0.3704] z_k, \tag{5.163}$$

where $z_0 = [0, 0]^\top$, and with the values taken by w_k, $k \geq 0$, being determined by the stochastic process W with mean and covariance functions as given in (5.161) and (5.162), respectively. In this case, since $m_W[k] = 0$, $k \geq 0$, it follows from (5.147–5.148) that

$$m_Z[k] = 0, \quad k \geq 0, \tag{5.164}$$

$$m_{\overline{\Delta\Omega}}[0] = 0, \quad k \geq 0. \tag{5.165}$$

In addition, the values taken by the covariance matrix of Z_k, $\Sigma_Z[k] \in \mathbb{R}^2$, can be computed via (5.149) with G and H as in (5.163), and $Q_k = 0.01$ for all $k \geq 0$, which together with the expression in (5.150) completes the characterization of the covariance function of Z. Finally, by tailoring (5.151) to the setting here, we can characterize the covariance function of $\Delta\Omega$ as follows:

$$C_{\Delta\Omega}[k_1, k_2]$$

$$= \begin{cases} \frac{1}{n^2\overline{D}^2}\Big(0.01\beta^\top G^{k_1-k_2-1} H + \beta^\top G^{k_1-k_2}\Sigma_Z[k_2]\beta\Big), & \text{if } k_1 > k_2, \\ \frac{1}{n^2\overline{D}^2}\Big(0.01 + \beta^\top \Sigma_Z[k_1]\beta\Big), & \text{if } k_1 = k_2, \\ \frac{1}{n^2\overline{D}^2}\Big(0.01 H^\top \big(G^{k_2-k_1-1}\big)^\top \beta + \beta^\top \Sigma_Z[k_1]\big(G^\top\big)^{k_2-k_1}\beta\Big), & \text{if } k_1 < k_2, \end{cases} \tag{5.166}$$

where $n = 3$, $\overline{D} = 0.45$, and $\beta = [1, 0.5]^\top$, and G and H as in (5.163). Figure 5.4 shows the evolution of the entries of $\Sigma_Z[k]$ together with $C_{\Delta\Omega}[k, k] = \Sigma_{\Delta\Omega}[k]$.

5.4.3 Phase Angle Statistical Characterization

Consider (5.133) with the extraneous disturbance model described earlier. Here we will use the techniques developed in Section 5.3.1 to approximately characterize the mean and covariance functions of the stochastic process describing the evolution of θ_k and z_k.

Let $\overline{x}_k = \big[\overline{\theta}_k^\top, \overline{z}_k^\top\big]^\top$, $k \geq 0$, denote the trajectory followed by (5.133) for $\xi_k = m_\Xi[k]$ and $x_0 = \big[\theta_0^\top, \mathbf{0}_{n-m}^\top\big]^\top$. Define $\Delta\Xi_k = \Xi_k - m_\Xi[k]$ and

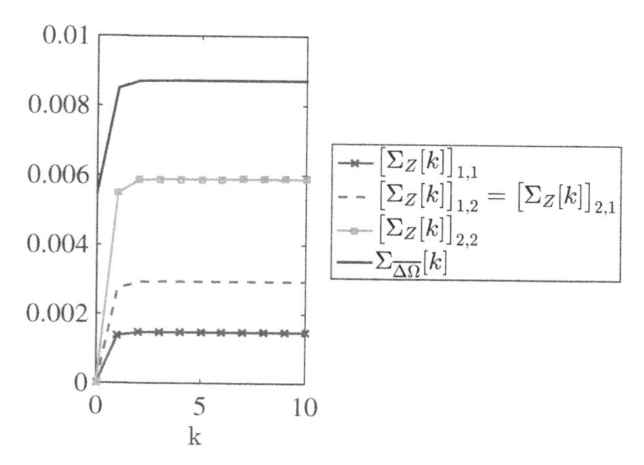

Figure 5.4 Evolution of the entries of $\Sigma_Z[k]$ and $\Sigma_{\overline{\Delta\Omega}}[k]$ for the microgrid in Example 5.5.

$$\begin{aligned}\Delta X_k &= X_k - \overline{x}_k \\ &= \left[\Delta\Theta_k^\top, \Delta Z_k^\top\right]^\top \\ &= \left[\Theta_k^\top, Z_k^\top\right]^\top - \left[\overline{\theta}_k^\top, \overline{z}_k^\top\right]^\top,\end{aligned}$$

where Θ_k is a random variable describing the values taken by θ_k, and Z_k is a random vector describing the values taken by z_k. Then, by linearizing the right-hand side of (5.133) around $(\overline{\theta}_k, \overline{z}_k, m_\Xi[k])$, the random vector $\Delta X_k = \left[\Delta\Theta_k^\top, \Delta Z_k^\top\right]^\top$ can be approximated by another random vector,

$$\widetilde{\Delta X}_k = [\widetilde{\Delta\Theta_k}^\top, \widetilde{\Delta Z}_k^\top]^\top,$$

whose evolution is determined by

$$\begin{bmatrix}\widetilde{\Delta\Theta}_{k+1} \\ \widetilde{\Delta Z}_{k+1}\end{bmatrix} = \underbrace{\begin{bmatrix} I_n - T_s D^{-1} J(\overline{\theta}_k) & T_s D^{-1} K_r B \\ \mathbf{0}_{(n-m)\times n} & I_{n-m} + \frac{1}{nD}\alpha\beta^\top \end{bmatrix}}_{=:\, G_k} \begin{bmatrix}\widetilde{\Delta\Theta}_k \\ \widetilde{\Delta Z}_k\end{bmatrix} + \underbrace{\begin{bmatrix} T_s D^{-1} K_\xi \\ \frac{1}{nD}\alpha \mathbf{1}_m^\top \end{bmatrix}}_{=:\, H} \Delta\Xi_k, \tag{5.167}$$

with

$$J(\overline{\theta}_k) = M\Gamma\mathbf{diag}(\phi_{1,k}, \phi_{2,k}, \ldots, \phi_{l,k})M^\top, \tag{5.168}$$

where

$$\phi_{e,k} = \cos\left(\overline{\theta}_{i,k} - \overline{\theta}_{j,k}\right), \quad e = \mathbb{I}((i,j)),\ (i,j) \in \mathcal{E},$$

and

$$m_{\Delta\Xi}[k] = 0, \quad k \geq 0,$$
$$C_{\Delta\Xi}[k_1, k_2] = C_{\Xi}[k_1, k_2], \quad k_1, k_2 \geq 0, \tag{5.169}$$

$m_{\widetilde{\Delta X}}[0] = \mathbf{0}_{2n-m}$, and $\Sigma_{\widetilde{\Delta X}}[0] = \mathbf{0}_{(2n-m)\times(2n-m)}$ (recall that $x_0 = \left[\theta_0^\top, \mathbf{0}_{n-m}^\top\right]^\top$). Note that in (5.167), the transition matrix, G_k, is indeed time-varying because of its dependence on $J(\overline{\theta}_k)$; however, the input matrix, H, is constant. Then, by tailoring (5.88–5.89) to the setting here, we obtain

$$m_{\widetilde{\Delta X}}[k] = 0, \tag{5.170}$$

and

$$\Sigma_{\widetilde{\Delta X}}[k+1] = G_k \Sigma_{\widetilde{\Delta X}}[k] G_k^\top + H Q_k H^\top,$$
$$C_{\widetilde{\Delta X}}[k_1, k_2] = \begin{cases} \Phi_{k_1,k_2} \Sigma_{\widetilde{\Delta X}}[k_2], & \text{if } k_1 \geq k_2, \\ \Sigma_{\widetilde{\Delta X}}[k_1] \Phi_{k_2,k_1}^\top, & \text{if } k_1 < k_2, \end{cases} \tag{5.171}$$

with G_k and H as defined in (5.167). Finally, by using (5.91) and (5.92), we obtain that

$$m_X[k] \approx \overline{x}_k,$$
$$C_X[k_1, k_2] \approx C_{\widetilde{\Delta X}}[k_1, k_2]. \tag{5.172}$$

Example 5.6 (Three-bus power system phase angle statistics) Consider again the three-bus microgrid featured in Example 5.5. By inspecting Fig. 5.3, one can easily check that

$$M = \begin{bmatrix} -1 & -1 \\ 1 & 0 \\ 0 & 1 \end{bmatrix}. \tag{5.173}$$

Assume that $X_{12} = 1$ p.u., $X_{13} = 2$ p.u., and $V_1 = V_2 = V_3 = 1$ p.u.; then, we have that

$$\Gamma = \begin{bmatrix} 1 & 0 \\ 0 & 1/2 \end{bmatrix}. \tag{5.174}$$

Let $\theta_k = [\theta_{1,k}, \theta_{2,k}, \theta_{3,k}]^\top$ and $z_k = [z_{2,k}, z_{3,k}]^\top$, $r^* = [0.75, 0.25]^\top$, and $\xi_k = \xi_{1,k}$. Then, the evolution of θ_k and z_k is described by (5.133) with M and Γ as given in (5.173–5.174), and B, D, K_ξ, and K_r as follows:

$$B = \begin{bmatrix} 1 & 0 \\ 0 & 1/2 \end{bmatrix}, \quad D = \begin{bmatrix} 0.1 & 0 & 0 \\ 0 & 0.25 & 0 \\ 0 & 0 & 1 \end{bmatrix}, \quad K_\xi = \begin{bmatrix} 1 \\ 0 \\ 0 \end{bmatrix}, \quad K_r = \begin{bmatrix} 0 & 0 \\ 1 & 0 \\ 0 & 1 \end{bmatrix}. \tag{5.175}$$

Recall that $\xi^0 = -1$; then, by solving for θ_0 in (5.135), with $\theta_{3,0} = 0$, we obtain $\theta_{1,0} = -\pi/6$ rad and $\theta_{2,0} = 0.3245$ rad.

Note that $m_\Xi[k] = \xi^0 = -1$ p.u.; thus, for $\xi_k = m_\Xi[k]$ and $x_0 = \left[\theta_0^\top, \mathbf{0}_2^\top\right]^\top$, with $\theta_0 = [-0.5236, 0.3245, 0]^\top$, the trajectory of (5.133), $\overline{x}_k = [\overline{\theta}_k^\top, \overline{z}_k^\top]^\top$, $k \geq 0$, is given by

$$\begin{aligned}\overline{\theta}_k &= \theta_0 \\ &= [-0.5236, 0.3245, 0]^\top, \quad k \geq 0, \end{aligned} \tag{5.176}$$

$$\begin{aligned}\overline{z}_k &= \overline{z}_0 \\ &= [0, 0]^\top, \quad k \geq 0, \end{aligned} \tag{5.177}$$

i.e., the system remains in the steady state operating point corresponding to the initial conditions. Now, given M as in (5.173), Γ as in (5.174), and $\overline{\theta}_k$, $k \geq 0$, as in (5.176), we can evaluate $J(\overline{\theta}_k)$, $k \geq 0$, as defined in (5.168), which in this case yields

$$\begin{aligned}J(\overline{\theta}_k) &= J(\theta_0) \\ &= \begin{bmatrix} 1.09 & -0.66 & -0.43 \\ -0.66 & 0.66 & 0 \\ -0.43 & 0 & 0.43 \end{bmatrix}. \end{aligned} \tag{5.178}$$

Then, by plugging B, D, K_r, and K_ξ as given in (5.175), $J(\overline{\theta}_k)$ as given in (5.178), $\alpha = [-1/2, -1]^\top$ and $\overline{D} = 0.45$ rad^{-1} into (5.167), we obtain

$$G_k = \begin{bmatrix} -0.0945 & 0.6614 & 0.4330 & 0 & 0 \\ 0.2646 & 0.7354 & 0 & 0.4000 & 0 \\ 0.0433 & 0 & 0.9567 & 0 & 0.0500 \\ 0 & 0 & 0 & 0.6296 & -0.1852 \\ 0 & 0 & 0 & -0.7407 & 0.6296 \end{bmatrix},$$

$$H_k = \begin{bmatrix} 1 \\ 0 \\ 0 \\ -0.3704 \\ -0.7407 \end{bmatrix}, \tag{5.179}$$

for all $k \geq 0$.

We can now use (5.170–5.172) together with G_k and H_k as in (5.179), $\Sigma_X[0] = \mathbf{0}_{5\times 5}$ (recall that θ_0 and z_0 are not random), and $Q_k = 0.01$, $k \geq 0$, to compute the mean and covariance functions of the stochastic process $X = [\Theta^\top, Z^\top]^\top$. In this case, since $\overline{x}_k = \left[\theta_0^\top, \mathbf{0}_2^\top\right]^\top$, $k \geq 0$, we have that

$$m_X[k] \approx \left[\theta_0^\top, \mathbf{0}_2^\top\right]^\top, \; k \geq 0. \tag{5.180}$$

The approximate value of $\Sigma_X[k]$ as $k \to \infty$, denoted by Σ_X^*, is given by

$$\Sigma_k^* = \begin{bmatrix} 1.02 \cdot 10^{-2} & -8.03 \cdot 10^{-5} & -1.57 \cdot 10^{-5} & -3.63 \cdot 10^{-3} & -7.27 \cdot 10^{-3} \\ -8.03 \cdot 10^{-5} & 1.64 \cdot 10^{-4} & -6.81 \cdot 10^{-7} & -1.20 \cdot 10^{-4} & -2.39 \cdot 10^{-4} \\ -1.57 \cdot 10^{-5} & -6.81 \cdot 10^{-7} & 2.70 \cdot 10^{-6} & -3.57 \cdot 10^{-6} & -7.15 \cdot 10^{-6} \\ -3.63 \cdot 10^{-3} & -1.20 \cdot 10^{-4} & -3.57 \cdot 10^{-6} & 1.47 \cdot 10^{-3} & 2.94 \cdot 10^{-3} \\ -7.27 \cdot 10^{-3} & -2.39 \cdot 10^{-4} & -7.15 \cdot 10^{-6} & 2.94 \cdot 10^{-3} & 5.88 \cdot 10^{-3} \end{bmatrix}.$$

5.5 Notes and References

The recursive characterization of the mean, covariance, and correlation functions of a linear system is standard and can be found in many references, including [10] and [38]. The recursive characterization of the pdf of a linear system for general random inputs is based on standard random variable transformation results which can be found in [11, 15]. The material on characteristic functions used for analyzing Gaussian systems can be found in [11]. The microgrid model in Section 5.4 has been adopted from [39].

6 Continuous-Time Systems: Probabilistic Input Uncertainty

In this chapter, we analyze systems in which the relation between the input and the state is governed by a continuous-time state-space model. We will model the input as a stochastic process with known mean and covariance functions, or known distribution. The objective is then to obtain the mean and covariance functions of the stochastic process governing the values taken by the state, or its distribution. The reader is referred to Sections 2.1 and 2.3.2 for important concepts from stochastic processes and continuous-time linear dynamical systems used throughout this chapter.

6.1 Introduction

Consider the circuit in Fig. 6.1 where the function of the voltage source on the left is to provide power to the current source on the right. Let $v_s(t)$ denote the value of the voltage across the terminals of the voltage source at time t. Let $i_l(t)$ denote the value of the current delivered by the current source at time t. Assume that the values that $v_s(t)$ and $i_l(t)$ take are not a priori known but governed by some stochastic processes with known mean and covariance functions, or known distribution. For the circuit to perform its intended function, the value at time t of the voltage across the terminals of the current source, denoted by $v_2(t)$, must remain within some interval (v_2^m, v_2^M) with some probability. In order to determine if this is the case, one needs to characterize the stochastic process describing the values that $v_2(t)$ can take. To this end, we need to use the joint statistical characterization of $v_s(t)$ and $i_l(t)$ together with the following differential equations obtained from Kirchhoff's laws:

$$\frac{d}{dt}\begin{bmatrix} v_2(t) \\ i_1(t) \end{bmatrix} = \begin{bmatrix} 0 & \frac{1}{c} \\ -\frac{1}{\ell} & -\frac{r}{\ell} \end{bmatrix} \begin{bmatrix} v_2(t) \\ i_1(t) \end{bmatrix} + \begin{bmatrix} 0 & -\frac{1}{c} \\ \frac{1}{\ell} & 0 \end{bmatrix} \begin{bmatrix} v_s(t) \\ i_l(t) \end{bmatrix}. \tag{6.1}$$

The discussion above pertains to the class of linear dynamical systems, which can be compactly described in state-space form as follows:

$$\frac{d}{dt}x(t) = A(t)x(t) + B(t)\dot{v}(t), \tag{6.2}$$

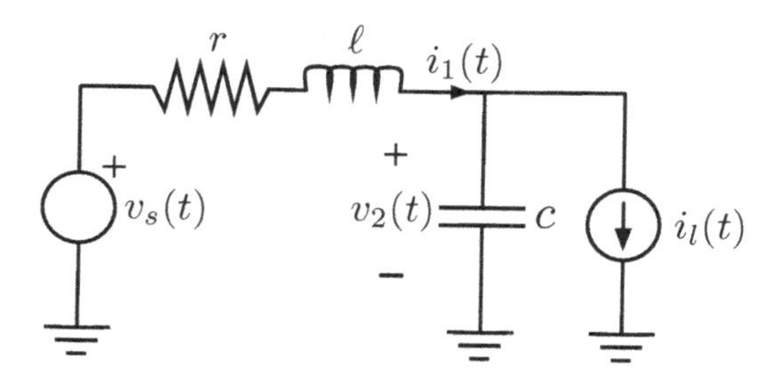

Figure 6.1 AC linear circuit.

where $x(t) \in \mathbb{R}^n$, $\dot{v}(t) \in \mathbb{R}^m$, $A(t) \in \mathbb{R}^{n\times n}$, and $B(t) \in \mathbb{R}^{n\times m}$. In the example above $x(t) = \left[v_2(t), i_1(t)\right]^\top$, $\dot{v}(t) = \left[v_s(t), i_l(t)\right]^\top$, and

$$A(t) = \begin{bmatrix} 0 & \frac{1}{c} \\ -\frac{1}{\ell} & -\frac{r}{\ell} \end{bmatrix}, \quad t \geq 0,$$

$$B(t) = \begin{bmatrix} 0 & -\frac{1}{c} \\ \frac{1}{\ell} & 0 \end{bmatrix}, \quad t \geq 0. \tag{6.3}$$

In this chapter, we will characterize the behavior of systems of the form in (6.2) for the case when the values that $\dot{v}(t)$, $t \geq 0$, takes are governed by a "white noise" vector process, i.e., a stochastic vector process $\dot{V} = \{\dot{V}(t) : t \in \mathcal{T}\}$, $\mathcal{T} = [0, \infty)$, with the following properties:

N1. $\mathrm{E}\left[\dot{V}(t)\right] = 0$ for all $t \geq 0$.

N2. $\mathrm{E}\left[\dot{V}(t)\dot{V}^\top(s)\right] = \delta(t-s)\,Q(t)$, $t, s \geq 0$, where $\delta(\cdot)$ denotes the Dirac delta function, i.e., $\delta(0) = \infty$, and $\delta(x) = 0$, $x \neq 0$, and $Q(t) \in \mathbb{R}^{m\times m}$ is symmetric and positive definite.

Because the values taken by $\dot{v}(t)$ are random and governed by the vector-valued stochastic process $\dot{V}$, the values taken by $x(t)$ are also random and governed by some vector-valued stochastic process $X = \{X(t) : t \in \mathcal{T}\}$, $\mathcal{T} = [0, \infty)$; the objective is then to characterize such a process. We will consider the following two scenarios:

C1. Given the mean and covariance functions of the process $\dot{V}$, and without making any assumptions on the distribution of $\dot{V}(t)$, $t \geq 0$, we will characterize the mean and covariance functions of X.

C2. Assuming that $\dot{V}(t)$, $t \geq 0$, are normally distributed ($\dot{V}$ is then referred to as a white Gaussian noise vector process), we will characterize the pdf of $X(t)$ for all $t \geq 0$.

We will also extend the techniques for Scenario C2 to analyze systems of the form

$$\frac{dX(t)}{dt} = \alpha\big(t, X(t)\big) + \beta\big(t, X(t)\big)\dot{V}(t). \tag{6.4}$$

6.2 Continuous-Time Linear Systems

In this section, we study systems of the form

$$\frac{d}{dt}x(t) = A(t)x(t) + B(t)\dot{v}(t), \tag{6.5}$$

with $A(t) \in \mathbb{R}^{n\times n}$, $B(t) \in \mathbb{R}^{n\times m}$, $x(t) \in \mathbb{R}^n$, and $\dot{v}(t) \in \mathbb{R}^m$. Assume the values taken by $\dot{v}(t)$, $t \geq 0$, are random and governed by the following vector-valued stochastic process

$$\dot{V} = \{\dot{V}(t) : t \geq 0\},$$

with $\dot{V}(t) = \sqrt{Q(t)}\dot{W}(t)$, where $Q(t)$ is a given $(m \times m)$-dimensional symmetric positive definite matrix, and $\dot{W} = \{\dot{W}(t) : t \geq 0\}$ is a white noise vector process, i.e., the components of $\dot{W}$ are independent white noise processes; thus the stochastic process $\dot{W}$ satisfies the following two properties:

N1. $m_{\dot{W}}(t) = \mathrm{E}\big[\dot{W}(t)\big] = \mathbf{0}_m$, i.e., $\dot{W}$ is a zero mean process.
N2. $C_{\dot{W}}(t,s) = \mathrm{E}\big[\dot{W}(t)\dot{W}^\top(s)\big] = \delta(t-s)\, I_m, \quad t,s \geq 0.$

Then, the values taken by $x(t)$ are also random and governed by some stochastic process $X = \big\{X(t) \colon t \geq 0\big\}$; the objective here is to characterize X.

White Noise as the Time Derivative of Another Process

The process $\dot{W}$ is extremely ill-behaved in the sense that its paths are discontinuous everywhere, which makes it difficult to work with when manipulating (6.5). Instead, one can interpret $\dot{W}$ as the time derivative of some vector-valued process $W = [W_1, W_2, \ldots, W_m]^\top$, i.e.,

$$\dot{W}(t) = \lim_{dt\to 0} \frac{1}{dt} dW(t) = \lim_{dt\to 0} \frac{1}{dt}\Big(W(t+dt) - W(t)\Big),$$

satisfying the following properties:

W1. $W_1, W_2, \ldots, W_m$ are independent.
W2. $W_i(0) = 0,\ i = 1, 2, \ldots, m.$
W3. $W_i,\ i = 1, 2, \ldots, m$, have independent increments, i.e., the distribution of $W_i(t) - W_i(s)$ depends on $t - s$ alone, and $W_i(t_j) - W_i(s_j),\ j = 1, 2, \ldots, n$, are independent whenever the intervals $(s_j, t_j]$ are disjoint.
W4. $W_i(s+t) - W_i(s),\ i = 1, 2, \ldots, m$, have zero mean and variance t for all $s, t \geq 0$.

A vector-valued stochastic process W satisfying Properties W1 – W4 is a zero mean process, i.e.,

$$m_W(t) = \mathbf{0}_m, \tag{6.6}$$

for all $t \geq 0$; this can be easily established as follows. First note that $\mathrm{E}\big[W(0)\big] = \mathbf{0}_m$ by Property W2; then, we have that

$$\begin{aligned} m_W(t) &= \mathrm{E}\big[W(t)\big] \\ &= \mathrm{E}\big[W(t) - W(0)\big] \\ &= \mathbf{0}_m, \end{aligned} \tag{6.7}$$

where the last equality follows from Property W4. In addition, the values taken by the covariance and correlation functions of W,

$$C_W(t,s) = \mathrm{E}\Big[\big(W(t) - m_W(t)\big)\big(W(s) - m_W(s)\big)^\top\Big], \quad s,t \geq 0,$$

and

$$R_W(t,s) = \mathrm{E}\Big[W(t)W^\top(s)\Big], \quad s,t \geq 0,$$

respectively, are equal and given by

$$\begin{aligned} C_W(t,s) &= R_W(t,s) \\ &= \min\{t,s\}I_m, \quad s,t \geq 0; \end{aligned} \tag{6.8}$$

this formula can be established as follows. Clearly $C_W(t,s) = R_W(t,s)$ since $m_W(t) = \mathbf{0}_m$ for all $t \geq 0$. By Property W1, we have that

$$\begin{aligned} \mathrm{E}\Big[\big(W_i(t) - m_{W_i}(t)\big)\big(W_j(s) - m_{W_j}(s)\big)\Big] &= \mathrm{E}\big[W_i(t)W_j(s)\big] \\ &= \mathrm{E}\big[W_i(t)\big]\mathrm{E}\big[W_j(s)\big] \\ &= 0, \end{aligned}$$

for any $i \neq j$; thus, the off-diagonal entries of $C_W(t,s)$ are all identically equal to zero. Now, assume $t \geq s$; then since $W_i(0) = 0$, $i = 1, 2, \dots, n$, by Property W2, we have that

$$\begin{aligned} \mathrm{E}\big[W_i(t)W_i(s)\big] &= \mathrm{E}\Big[\big(W_i(t) - W_i(0)\big)\big(W_i(s) - W_i(0)\big)\Big] \\ &= \mathrm{E}\Big[\big(W_i(s) - W_i(0)\big)^2 + \big(W_i(t) - W_i(s)\big)\big(W_i(s) - W_i(0)\big)\Big] \\ &= \mathrm{E}\Big[\big(W_i(s) - W_i(0)\big)^2\Big] + \underbrace{\mathrm{E}\Big[\big(W_i(t) - W_i(s)\big)\big(W_i(s) - W_i(0)\big)\Big]}_{=0 \text{ by Properties W3 and W4}} \\ &= \mathrm{E}\Big[\big(W_i(s) - W_i(0)\big)^2\Big] \\ &= s, \end{aligned} \tag{6.9}$$

where the last equality follows from Property W4. A similar derivation for the case when $t \leq s$ results in $\mathrm{E}\big[W_i(t)W_i(s)\big] = t$; thus, all the diagonal entries of $C_W(t,s)$ and $R_W(t,s)$ are identically equal to $\min\{t,s\}$.

Next we provide some heuristic arguments to justify the notion that one can interpret a white noise vector process $\dot{W}$ as the time derivative of another process W satisfying Properties W1 – W4. First, for a small $dt > 0$ and any $t \geq 0$, we have that

$$\begin{aligned}
\mathrm{E}\left[\frac{1}{dt}dW(t)\right] &= \mathrm{E}\left[\frac{1}{dt}\Big(W(t+dt) - W(t)\Big)\right] \\
&= \frac{1}{dt}\Big(\mathrm{E}\big[W(t+dt)\big] - \mathrm{E}\big[W(t)\big]\Big) \\
&= \frac{1}{dt}\Big(m_W(t+dt) - m_W(t)\Big) \\
&= \mathbf{0}_m
\end{aligned} \tag{6.10}$$

by (6.6). Thus, we have that

$$\begin{aligned}
\mathrm{E}\big[\dot{W}(t)\big] &= \lim_{dt \to 0} \mathrm{E}\left[\frac{1}{dt}dW(t)\right] \\
&= \mathbf{0}_m,
\end{aligned} \tag{6.11}$$

which is consistent with Property N1. Also, for a small $dt > 0$, a fixed $t \geq 0$, and any $s \geq 0$, by using (6.8), we can write

$$\begin{aligned}
\mathrm{E}\left[\frac{1}{dt}dW(t)\frac{1}{dt}dW^\top(s)\right] &= \mathrm{E}\left[\frac{1}{dt}\Big(W(t+dt) - W(t)\Big)\frac{1}{dt}\Big(W(s+dt) - W(s)\Big)^\top\right] \\
&= \frac{1}{dt^2}\mathrm{E}\Big[W(t+dt)W^\top(s+dt) - W(t+dt)W^\top(s) \\
&\qquad W(t)W^\top(s+dt) + W(t)W^\top(s)\Big] \\
&= \frac{1}{dt^2}\Big(\min\{t+dt, s+dt\} - \min\{t+dt, s\} \\
&\qquad - \min\{t, s+dt\} + \min\{t, s\}\Big) I_m \\
&= \phi_{dt}(s) I_m,
\end{aligned} \tag{6.12}$$

where

$$\phi_{dt}(s) = \begin{cases} 0, & s \leq t - dt, \\ \frac{s-(t-dt)}{dt^2}, & t - dt < s \leq t, \\ \frac{-s+(t+dt)}{dt^2}, & t < s \leq t + dt, \\ 0, & s > t + dt; \end{cases} \tag{6.13}$$

see Fig. 6.2 (left) for a depiction of the graph of $\phi_{dt}(\cdot)$. Note that for each $s \neq t$, $\lim_{dt \to 0} \phi_{dt}(s) = 0$. Also note that $\phi_{dt}(s) \geq 0$ for any s and $\int_{-\infty}^{\infty} \phi_{dt}(s) ds = 1$; therefore, in the limit as dt goes to zero, it is reasonable to assume that $\phi_{dt}(s)$ will converge to $\delta(t - s)$ in some sense (see Fig. 6.2 for a depiction). Thus, we can write

$$\begin{aligned}
\mathrm{E}\Big[\dot{W}(t)\dot{W}^\top(s)\Big] &= \lim_{dt \to 0} \mathrm{E}\left[\frac{1}{dt}dW(t)\frac{1}{dt}dW^\top(s)\right] \\
&= \delta(t - s) I_m,
\end{aligned} \tag{6.14}$$

which is consistent with Property N2.

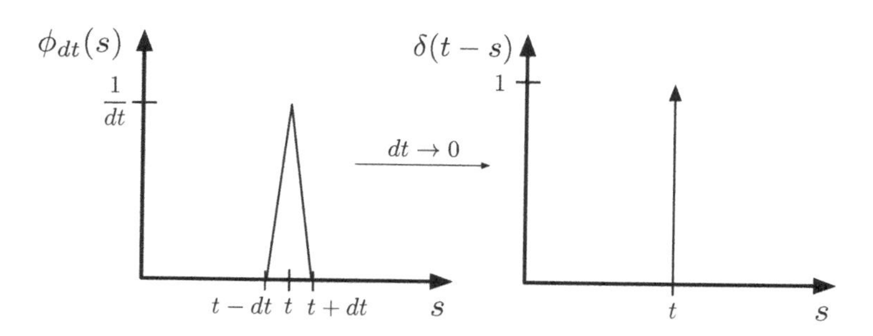

Figure 6.2 Depiction of triangle function $\phi_{dt}(\cdot)$ approximating the Dirac delta function $\delta(\cdot)$ as dt goes to zero.

Differential Representation of a Linear Stochastic System

Consider the system in (6.5) and define $dX(t) := X(t+dt) - X(t)$, where dt is positive and small. Let $V = \{V(t) : t \geq 0\}$, $V(t) \in \mathbb{R}^m$, denote a vector-valued stochastic process, so that $V(t+dt) - V(t) = \sqrt{Q(t)}\big(W(t+dt) - W(t)\big)$, where $Q(t) \in \mathbb{R}^{m\times m}$ is symmetric and positive definite, and $W = \{W(t) : t \geq 0\}$, $W(t) \in \mathbb{R}^m$, is a vector-valued process satisfying Properties W1 – W4. Then, we can rewrite (6.5) in differential form as follows:

$$dX(t) = A(t)X(t)dt + B(t)dV(t), \tag{6.15}$$

with $dV(t) := V(t+dt) - V(t)$. In subsequent developments, we will use the representation in (6.15) as it is much easier to work with the stochastic process W than to work with the stochastic process $\dot{W}$ when attempting to characterize the stochastic process $X = \{X(t) \colon t \geq 0\}$.

6.2.1 Characterization of First and Second Moments

Consider the system in (6.15) and assume we are given the mean of $X(0)$,

$$\mathrm{E}\big[X(0)\big] = m_X(0),$$

and its covariance matrix,

$$\mathrm{E}\Big[\big(X(0) - m_X(0)\big)\big(X(0) - m_X(0)\big)^\top\Big] = \Sigma_X(0).$$

In the remainder, we will assume that $X(0)$ and $W(t)$ are uncorrelated for any $t \geq 0$, i.e.,

$$\mathrm{E}\Big[\big(X(0) - m_X(0)\big)\big(W(t) - \underbrace{m_W(t)}_{=\mathbf{0}_m}\big)^\top\Big] = \mathbf{0}_{n\times m}, \quad t \geq 0. \tag{6.16}$$

Then, since $dV(t) = \sqrt{Q(t)}dW(t)$, where $dW(t) := W(t+dt) - W(t)$, we have that $m_{dV}(t) = \sqrt{Q(t)}\big(m_W(t+dt) - m_W(t)\big) = 0$, and

$$
\begin{aligned}
\mathrm{E}\Big[\big(X(0)-m_X(0)\big)\big(dV(t)-\underbrace{m_{dV}(t)}_{=\mathbf{0}_m}\big)^{\top}\Big] &= \underbrace{\mathrm{E}\Big[\big(X(0)-m_X(0)\big)W^{\top}(t+dt)\Big]}_{=\mathbf{0}_{n\times m}\text{ by (6.16)}}\sqrt{Q(t)} \\
&\quad - \underbrace{\mathrm{E}\Big[\big(X(0)-m_X(0)\big)W^{\top}(t)\Big]}_{=\mathbf{0}_{n\times m}\text{ by (6.16)}}\sqrt{Q(t)} \\
&= \mathbf{0}_{n\times m}, \quad t \geq 0, \qquad (6.17)
\end{aligned}
$$

i.e., $X(0)$ and $dV(t)$, $t \geq 0$, are uncorrelated. The objective here is to characterize the mean, covariance, and correlation functions of the vector-valued stochastic process $X = \{X(t) : t \geq 0\}$ governing the values taken by the state.

Mean Function. The mean of $X(t)$, $m_X(t)$, evolves according to the following differential equation:

$$
\frac{d}{dt}m_X(t) = A(t)m_X(t), \qquad (6.18)
$$

for a given $m_X(0)$.

To establish (6.18) we use the definition of the incremental process in (6.15) to relate the mean of $X(t+dt)$, $m_X(t+dt)$, to the mean of $X(t)$, $m_X(t)$, as follows:

$$
\begin{aligned}
m_X(t+dt) &= \mathrm{E}\big[X(t+dt)\big] \\
&= \mathrm{E}\big[X(t) + dX(t)\big] \\
&= \mathrm{E}\big[X(t) + A(t)X(t)dt + B(t)dV(t)\big] \\
&= \mathrm{E}\big[X(t) + A(t)X(t)dt + B(t)\sqrt{Q(t)}dW(t)\big] \\
&= \mathrm{E}\big[X(t)\big] + dt\,A(t)\mathrm{E}\big[X(t)\big] + B(t)\sqrt{Q(t)}\;\underbrace{\mathrm{E}\big[dW(t)\big]}_{=\mathbf{0}_m\text{ by (6.6)}} \\
&= m_X(t) + dt\,A(t)m_X(t). \qquad (6.19)
\end{aligned}
$$

Now, by rearranging terms in (6.19) and dividing by dt on both sides, we obtain

$$
\frac{1}{dt}\big(m_X(t+dt) - m_X(t)\big) = A(t)m_X(t); \qquad (6.20)
$$

then, by taking the limit as dt goes to zero on both sides, it follows that

$$
\frac{d}{dt}m_X(t) = A(t)m_X(t), \qquad (6.21)
$$

as claimed in (6.18).

Covariance Function. Let $\Sigma_X(t) = \mathrm{E}\Big[\big(X(t) - m_X(t)\big)\big(X(t) - m_X(t)]\big)^{\top}\Big]$; then, for a given $\Sigma_X(0)$, the values taken by the covariance function of X, $C_X(t,s)$, $t, s \geq 0$, are given by:

$$\frac{d}{dt}\Sigma_X(t) = A(t)\Sigma_X(t) + \Sigma_X(t)A^\top(t) + B(t)Q(t)B^\top(t), \tag{6.22}$$

$$C_X(t,s) = \begin{cases} \Phi(t,s)\Sigma_X(s), & \text{if } t \geq s, \\ \Sigma_X(t)\Big(\Phi^{-1}(t,s)\Big)^\top, & \text{if } t < s, \end{cases} \tag{6.23}$$

where $\Phi(t, s)$ satisfies the following linear matrix partial differential equation:

$$\frac{\partial}{\partial t}\Phi(t,s) = A(t)\Phi(t,s), \quad t, s \geq 0, \tag{6.24}$$

with $\Phi(s, s) = I_n$ (see (2.117) and (2.120)).

In order to establish (6.22) and (6.23), we need first to show that $X(t_{k'})$ and $dV^\top(t_k)$ are uncorrelated for any $t_k \geq t_{k'} \geq 0$. Let $t_k = k\,dt$, $k = 0, 1, 2, \ldots$, and $dt \geq 0$ sufficiently small; then, we have that $dX(t_k) = X(t_{k+1}) - X(t_k)$ and we can use (6.15) to write

$$X(t_{k+1}) = \big(I_n + dt\, A(t_k)\big)X(t_k) + B(t_k)dV(t_k). \tag{6.25}$$

Note that (6.25) is essentially a discrete-time version of (6.15); then, by tailoring the expression in (2.108) to the setting here, we obtain that

$$X(t_k) = \widetilde{\Phi}(t_k, 0)X(0) + \sum_{\ell=0}^{k-1} \widetilde{\Phi}(t_k, t_{\ell+1})B(t_\ell)dV(t_\ell), \tag{6.26}$$

where

$$\widetilde{\Phi}(t_k, t_\ell) = \begin{cases} \displaystyle\prod_{m=\ell}^{k-1} \Big(I_n + dt\, A(t_{k-1+\ell-m})\Big), & \text{if } k > \ell \geq 0, \\ I_n, & \text{if } k = \ell. \end{cases} \tag{6.27}$$

Now, by taking expectations on both sides of (6.26) and taking into account that $\mathrm{E}\big[dV(t_k)\big] = \sqrt{Q(t_k)}\mathrm{E}\big[dW(t_k)\big] = \mathbf{0}_m,\ k \geq 0$, it follows that

$$m_X(t_k) = \widetilde{\Phi}(t_k, 0)m_X(0). \tag{6.28}$$

Then, by using (6.26) and (6.28), we have that, for any $t_k, t_{k'} \geq 0$ such that $t_k \geq t_{k'}$,

$$\begin{aligned} \mathrm{E}\Big[\big(X(t_{k'}) - m_X(t_{k'})\big)dV^\top(t_k)\Big] &= \mathrm{E}\Big[\widetilde{\Phi}(t_{k'}, 0)\big(X(0) - m_X(0)\big)dV^\top(t_k) \\ &\quad + \sum_{\ell=0}^{k'-1} \widetilde{\Phi}(t_{k'}, t_{\ell+1})B(t_\ell)dV(t_\ell)dV^\top(t_k)\Big] \\ &= \widetilde{\Phi}(t_{k'}, 0)\mathrm{E}\Big[\big(X(0) - m_X(0)\big)dV^\top(t_k)\Big] \\ &\quad + \sum_{\ell=0}^{k'-1} \widetilde{\Phi}(t_{k'}, t_{\ell+1})B(t_\ell)\mathrm{E}\Big[dV(t_\ell)dV^\top(t_k)\Big] \\ &= \mathbf{0}_{n\times m}, \end{aligned} \tag{6.29}$$

where the last equality follows from (6.17) and the fact that, for any $t_\ell \neq t_k$, we have that

$$\begin{aligned} \mathrm{E}\big[dV(t_\ell)dV^\top(t_k)\big] &= \sqrt{Q(t_\ell)}\mathrm{E}\big[dW(t_\ell)dW^\top(t_k)\big]\sqrt{Q(t_k)} \\ &= \sqrt{Q(t_\ell)}\mathrm{E}\big[dW(t_\ell)\big]\mathrm{E}\big[dW^\top(t_k)\big]\sqrt{Q(t_k)} \\ &= \mathbf{0}_{m\times m} \end{aligned} \tag{6.30}$$

because $dW(t_\ell)$ and $dW(t_k)$ are independent by Property W3, and have zero mean by Property W4. Therefore, $X(t_{k'})$ and $dV^\top(t_k)$ are uncorrelated for any $t_k \geq t_{k'} \geq 0$.

Now, to establish (6.22), we proceed as follows:

$$\begin{aligned} \Sigma_X(t+dt) &= \mathrm{E}\Big[\big(X(t+dt) - m_X(t+dt)\big)\big(X(t+dt) - m_X(t+dt)\big)^\top\Big] \\ &= \mathrm{E}\Big[\big(X(t) - m_X(t) + dX(t) - dm_X(t)\big) \\ &\quad \times \big(X(t) - m_X(t) + dX(t) - dm_X(t)\big)^\top\Big] \\ &= \mathrm{E}\Big[\big(X(t) - m_X(t) + A(t)\big(X(t) - m_X(t)\big)dt + B(t)dV(t)\big) \\ &\quad \times \big(X(t) - m_X(t) + A(t)\big(X(t) - m_X(t)\big)dt + B(t)dV(t)\big)^\top\Big] \\ &= \mathrm{E}\Big[\big(X(t) - m_X(t)\big)\big(X(t) - m_X(t)\big)^\top\Big] \\ &\quad + dt\,\mathrm{E}\Big[\big(X(t) - m_X(t)\big)\big(X(t) - m_X(t)\big)^\top\Big]A^\top(t) \\ &\quad + \underbrace{\mathrm{E}\Big[\big(X(t) - m_X(t)\big)dV^\top(t)\Big]}_{=\mathbf{0}_{n\times m}\text{ by (6.29)}}B^\top(t) \\ &\quad + dt\,A(t)\mathrm{E}\Big[\big(X(t) - m_X(t)\big)\big(X(t) - m_X(t)\big)^\top\Big] \\ &\quad + (dt)^2\,A(t)\mathrm{E}\Big[\big(X(t) - m_X(t)\big)\big(X(t) - m_X(t)\big)^\top\Big]A^\top(t) \\ &\quad + dt\,A(t)\underbrace{\mathrm{E}\Big[\big(X(t) - m_X(t)\big)dV^\top(t)\Big]}_{=\mathbf{0}_{n\times m}\text{ by (6.29)}}B^\top(t) \\ &\quad + B(t)\underbrace{\mathrm{E}\Big[dV(t)\big(X(t) - m_X(t)\big)^\top\Big]}_{=\mathbf{0}_{m\times n}\text{ by (6.29)}} \\ &\quad + dt\,B(t)\underbrace{\mathrm{E}\Big[dV(t)\big(X(t) - m_X(t)\big)^\top\Big]}_{=\mathbf{0}_{m\times n}\text{ by (6.29)}}A^\top(t) \\ &\quad + B(t)\mathrm{E}\Big[dV(t)dV^\top(t)\Big]B^\top(t) \\ &= \Sigma_X(t) + dt\,\Sigma_X(t)A^\top(t) + dt\,A(t)\Sigma_X(t) \\ &\quad + (dt)^2\,A(t)\Sigma_X(t)A^\top(t) \\ &\quad + dt\,B(t)Q(t)B^\top(t), \end{aligned} \tag{6.31}$$

where the last equality follows from the fact that

$$\begin{aligned}
\mathrm{E}\big[dV(t)dV^\top(t)\big] &= \sqrt{Q(t)}\mathrm{E}\big[\big(W(t+dt)-W(t)\big)\big(W(t+dt)-W(t)\big)^\top\big]\sqrt{Q(t)} \\
&= dt\,Q(t) \qquad (6.32)
\end{aligned}$$

by Properties W1 and W4. Then, by rearranging terms in (6.31) and dividing by dt throughout, we obtain that

$$\begin{aligned}
\frac{1}{dt}\big(\Sigma_X(t+dt)-\Sigma_X(t)\big) &= A(t)\Sigma_X(t)+\Sigma_X(t)A^\top(t)+B(t)Q(t)B^\top(t) \\
&\quad + dt\,A\Sigma_X(t)A^\top; \qquad (6.33)
\end{aligned}$$

then, by taking the limit on both sides as dt goes to zero, it follows that:

$$\frac{d}{dt}\Sigma_X(t) = A(t)\Sigma_X(t)+\Sigma_X(t)A^\top(t)+B(t)Q(t)B^\top(t), \qquad (6.34)$$

as claimed in the first line of (6.22).

Finally, to establish (6.23), we proceed as follows. For $t \geq s$ and $dt \geq 0$, we have that

$$\begin{aligned}
C_X(t+dt,s) &= \mathrm{E}\Big[\big(X(t+dt)-m_X(t+dt)\big)\big(X(s)-m_X(s)\big)^\top\Big] \\
&= \mathrm{E}\Big[\big(X(t)-m_X(t)\big)\big(X(s)-m_X(s)\big)^\top\Big] \\
&\quad + \mathrm{E}\Big[\Big(dtA(t)\big(X(t)-m_X(t)\big)+B(t)dV(t)\Big)\Big(X(s)-m_X(s)\Big)^\top\Big] \\
&= \mathrm{E}\Big[\big(X(t)-m_X(t)\big)\big(X(s)-m_X(s)\big)^\top\Big] \\
&\quad + dtA(t)\mathrm{E}\Big[\big(X(t)-m_X(t)\big)\big(X(s)-m_X(s)\big)^\top\Big] \\
&\quad + B(t)\underbrace{\mathrm{E}\Big[dV(t)\big(X(s)-m_X(s)\big)^\top\Big]}_{\mathbf{0}_{m\times n}\text{ by (6.29)}} \\
&= C_X(t,s)+dt\,A(t)C_X(t,s), \qquad (6.35)
\end{aligned}$$

which, by rearranging, yields

$$\frac{1}{dt}\Big(C_X(t+dt,s)-C_X(t,s)\Big) = A(t)C_X(t,s); \qquad (6.36)$$

thus, by taking limits on both sides of the equation as dt goes to zero, it follows that $C_X(t,s)$, $t \geq s \geq 0$, must satisfy the following linear matrix partial differential equation:

$$\frac{\partial}{\partial t} C_X(t,s) = A(t) C_X(t,s), \tag{6.37}$$

with $C_X(s,s) = \Sigma_X(s)$. Now, we can use (6.24) to check that

$$C_X(t,s) = \Phi(t,s)\Sigma_X(s)$$

satisfies (6.37):

$$\begin{aligned} \frac{\partial}{\partial t} C_X(t,s) &= \frac{\partial}{\partial t}\Phi(t,s)\Sigma_X(s) \\ &= A(t)\Phi(t,s)\Sigma_X(s) \\ &= A(t) C_X(t,s). \end{aligned} \tag{6.38}$$

For $t < s$, we have that

$$\begin{aligned} C_X(t,s) &= \Big(C_X(s,t)\Big)^\top \\ &= \Big(\Phi(s,t)\Sigma_X(t)\Big)^\top \\ &= \Sigma_X(t)\Phi^\top(s,t) \\ &= \Sigma_X(t)\Big(\Phi^{-1}(t,s)\Big)^\top, \end{aligned} \tag{6.39}$$

where the last equality follows from (2.118), as claimed in the second line of (6.23).

Correlation Function. The values taken by the correlation function of X, $R_X(t,s), t, s \geq 0$, can be obtained from $C_X(t,s)$, $m_X(t)$, and $m_X(s)$ as follows:

$$R_X(t,s) = C_X(t,s) + m_X(t) m_X^\top(s); \tag{6.40}$$

this follows from the fact that

$$\mathrm{E}\Big[\big(X(t) - m_X(t)\big)\big(X(s) - m_X(s)\big)^\top\Big] = \mathrm{E}\Big[X(t)X^\top(s)\Big] - m_X(t)m_X^\top(s). \tag{6.41}$$

Vectorization of Matrix Differential Equation

In terms of numerically solving the matrix differential equation (6.22), it is convenient to rewrite it in vector form by using the Kronecker product (see Appendix A). When doing so, the matrices $\Sigma_X(t)$ and $B(t)Q(t)B^\top(t)$ are replaced by vectors obtained by stacking their columns left to right as follows.

Given a matrix $A = [a_{i,j}] \in \mathbb{R}^{n\times m}$, we will associate an nm-dimensional vector to it, denoted by $\mathbf{vec}(A)$, defined as follows:

$$\mathbf{vec}(A) = [a_{1,1}, a_{2,1}, \ldots, a_{n,1}, a_{1,2}, a_{2,2}, \ldots, a_{n,2}, \ldots, a_{1,m}, a_{2,m}, \ldots, a_{n,m}]^\top. \tag{6.42}$$

Then, (6.22) can be rewritten as follows:

$$\frac{d}{dt}\mathbf{vec}(\Sigma_X(t)) = \Big((I_n \otimes A(t)) + (A(t) \otimes I_n)\Big)\mathbf{vec}(\Sigma_X(t)) + \mathbf{vec}\big(B(t)Q(t)B^\top(t)\big), \tag{6.43}$$

where $\mathbf{vec}\big(\Sigma_X(t)\big) \in \mathbb{R}^{n^2}$, $\mathbf{vec}\big(B(t)Q(t)B^\top(t)\big) \in \mathbb{R}^{n^2}$, and

$$\Big((I_n \otimes A(t)) + \big(A(t) \otimes I_n\big)\Big) \in \mathbb{R}^{n^2\times n^2}.$$

For a linear time-invariant system, we have that $A(t) = A$ and $B(t) = B$ for all $t \geq 0$, where A and B are constant matrices; then (6.43) reduces to

$$\frac{d}{dt}\mathbf{vec}(\Sigma_X(t)) = \Big((I_n \otimes A) + (A \otimes I_n)\Big)\mathbf{vec}(\Sigma_X(t)) + \mathbf{vec}\big(BQ(t)B^\top\big). \tag{6.44}$$

Now, since the entries of $\mathbf{vec}\big(BQ(t)B^\top\big)$ are bounded, we can determine whether or not $\mathbf{vec}\big(\Sigma_X(t)\big)$ will remain bounded by computing the eigenvalues of the matrix $\Big((I_n \otimes A) + (A \otimes I_n)\Big)$; these can be easily obtained from the eigenvalues of A as follows. Let $\sigma(A)$ denote the spectrum of A, i.e., the set of the eigenvalues of A; then the spectrum of $(I_n \otimes A) + (A \otimes I_n)$ is as follows:

$$\sigma\big((I_n \otimes A) + (A \otimes I_n)\big) = \big\{\lambda\colon \lambda = \lambda_i + \lambda_j, \quad \lambda_i \in \sigma(A),\ \lambda_j \in \sigma(A)\big\}. \tag{6.45}$$

Thus, we can guarantee the entries of $\mathbf{vec}\big(\Sigma_X(t)\big)$ will remain bounded if the real part of every eigenvalue of A is strictly negative.

Example 6.1 (Two-dimensional continuous-time linear system) Consider the following system:

$$\underbrace{\begin{bmatrix} dX_1(t) \\ dX_2(t) \end{bmatrix}}_{dX(t)} = \underbrace{\begin{bmatrix} -\frac{1}{2} & \frac{1}{2} \\ 0 & -\frac{1}{2} \end{bmatrix}}_{A} \underbrace{\begin{bmatrix} X_1(t) \\ X_2(t) \end{bmatrix}}_{X(t)} dt + \underbrace{\begin{bmatrix} 1 & 0 \\ 0 & \frac{1}{2} \end{bmatrix}}_{B} \underbrace{\begin{bmatrix} dV_1(t) \\ dV_2(t) \end{bmatrix}}_{dV(t)}, \tag{6.46}$$

with $V(t) = \big[V_1(t), V_2(t)\big]^\top = \sqrt{Q}W(t)$, where $W = \big\{W(t) : t \geq 0\big\}$ is a two-dimensional vector-valued stochastic process satisfying Properties W1 – W4, and

$$Q = \begin{bmatrix} 1 & 0 \\ 0 & 2 \end{bmatrix}. \tag{6.47}$$

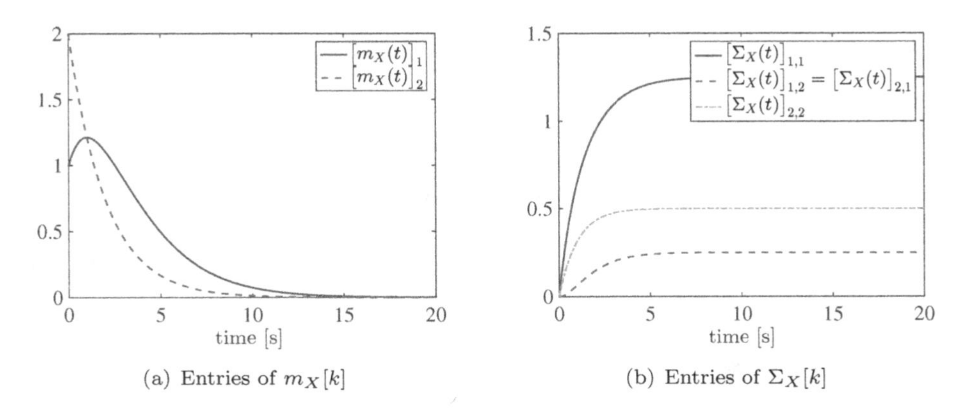

(a) Entries of $m_X[k]$

(b) Entries of $\Sigma_X[k]$

Figure 6.3 Evolution of the entries of the mean and the covariance matrix of $X(t)$.

Assume $X(0) = [1, 2]^\top$ and $V(t)$, $t \geq 0$, are uncorrelated. The objective is to obtain $m_X(t)$ and $\Sigma_X(t)$ for all $t \geq 0$.

By tailoring (6.18) to the setting here, we obtain:

$$\frac{d}{dt} m_X(t) = \underbrace{\begin{bmatrix} -\frac{1}{2} & \frac{1}{2} \\ 0 & -\frac{1}{2} \end{bmatrix}}_{= A} m_X(t), \tag{6.48}$$

with $m_X(0) = [1, 2]^\top$. The trajectories followed by the entries of $m_X(t)$, denoted by $[m_X(t)]_i$, $i = 1, 2$, are displayed in Fig. 6.3(a). Similarly, by tailoring (6.22) to the setting here, we have

$$\frac{d}{dt}\Sigma_X(t) = \underbrace{\begin{bmatrix} -\frac{1}{2} & \frac{1}{2} \\ 0 & -\frac{1}{2} \end{bmatrix}}_{= A} \Sigma_X(t) + \Sigma_X(t) \underbrace{\begin{bmatrix} -\frac{1}{2} & 0 \\ \frac{1}{2} & -\frac{1}{2} \end{bmatrix}}_{= A^\top} + \underbrace{\begin{bmatrix} 1 & 0 \\ 0 & \frac{1}{2} \end{bmatrix}}_{= BQB^\top},$$

with

$$\Sigma_X(0) = \begin{bmatrix} 0 & 0 \\ 0 & 0 \end{bmatrix}, \tag{6.49}$$

or by using (6.43), we can alternatively write

$$\frac{d}{dt}\mathbf{vec}(\Sigma_X(t)) = \Big((I_n \otimes A) + (A \otimes I_n)\Big)\mathbf{vec}(\Sigma_X(t)) + \mathbf{vec}(BQB^\top), \tag{6.50}$$

where

$$\Big((I_n \otimes A) + (A \otimes I_n)\Big) = \begin{bmatrix} -1 & \frac{1}{2} & \frac{1}{2} & 0 \\ 0 & -1 & 0 & \frac{1}{2} \\ 0 & 0 & -1 & \frac{1}{2} \\ 0 & 0 & 0 & -1 \end{bmatrix}, \quad \mathbf{vec}(BQB^\top) = \begin{bmatrix} 1 \\ 0 \\ 0 \\ \frac{1}{2} \end{bmatrix}. \tag{6.51}$$

Because of the structure of A and $\Big((I_n \otimes A) + (A \otimes I_n)\Big)$, one can easily see that $\sigma(A) = \{-1/2, -1/2\}$ and $\sigma\big((I_n \otimes A) + (A \otimes I_n)\big) = \{-1, -1, -1, -1\}$, which is consistent with (6.45). The trajectories followed by the entries of $\Sigma_X(t)$, denoted by $\big[\Sigma_X(t)\big]_{i,j}$, $i = 1, 2$, $j = 1, 2$, are displayed in Fig. 6.3(b).

Deterministic Inputs

In the developments so far, we have not considered deterministic driving terms; we will show next how to handle these. Consider the system in (6.5) and add the following term to the right-hand side:

$$C(t)u(t),$$

where $C(t) \in \mathbb{R}^{n \times l}$ and $u(t) \in \mathbb{R}^l$ are known for all $t \geq 0$. Following a similar procedure to that used when deriving (6.15), we can rewrite the modified system as follows:

$$dX(t) = \big(A(t)X(t) + C(t)u(t)\big)\, dt + BdV(t). \tag{6.52}$$

Define the following stochastic process: $\Delta X(t) = X(t) - x^\circ(t)$, where $x^\circ(t)$ is governed by

$$\frac{dx^\circ(t)}{dt} = A(t)x^\circ(t) + C(t)u(t), \tag{6.53}$$

with $x^\circ(0) = \mathbf{0}_n$, which can also be written as follows:

$$dx^\circ(t) = \big(A(t)x^\circ(t) + C(t)u(t)\big)\, dt. \tag{6.54}$$

Then, we have that

$$\begin{aligned} d\Delta X(t) &= dX(t) - dx^\circ(t) \\ &= \big(A(t)X(t) + C(t)u(t)\big)\, dt + BdV(t) - \big(A(t)x^\circ(t) + C(t)u(t)\big)\, dt \\ &= \Big(A(t)\big(X(t) - x^\circ(t)\big)\Big)\, dt + BdV(t) \\ &= A(t)\Delta X(t)\, dt + BdV(t), \end{aligned} \tag{6.55}$$

with $m_{\Delta X}(0) = m_X(0)$ and $\Sigma_{\Delta X}(0) = \Sigma_X(0)$ (because $x^\circ(0) = \mathbf{0}_n$), which is of the form in (6.15); thus, we can use the techniques discussed earlier to compute the values taken by the mean, covariance, and correlation functions of ΔX, $m_{\Delta X}(t)$, $t \geq 0$, $C_{\Delta X}(t, s)$, $t, s \geq 0$, and $R_{\Delta X}(t, s)$, $t, s \geq 0$, respectively. Then, we can use these to obtain the mean, covariance, and correlation functions of X as follows:

$$m_X(t) = x^\circ(t) + m_{\Delta X}(t), \tag{6.56}$$

$$C_X(t, s) = C_{\Delta X}(t, s), \tag{6.57}$$

$$R_X(t, s) = C_{\Delta X}(t, s) + \Big(x^\circ(t) + m_{\Delta X}(t)\Big)\Big(x^\circ(s) + m_{\Delta X}(s)\Big)^\top. \tag{6.58}$$

Example 6.2 Consider the circuit in Fig. 6.1 with $v_s(t) = v_s^*$, where $v_s^* > 0$ is known. Assume the system is initially at rest, i.e., $v_2(0) = 0$ and $i_1(0) = 0$. Further, assume that $i_l(t) = i_l^* + I_l(t)$, where $i_l^* > 0$ is known but the values that $I_l(t)$ takes are random and governed by a stochastic process $I_l = \{I_l(t) : t \geq 0\}$ satisfying the following differential equation:

$$dI_l(t) = aI_l(t)\,dt + bdW(t), \tag{6.59}$$

with $a < 0$, $b > 0$, and where $W = \{W(t),\ t \geq 0\}$ is a real-valued stochastic process satisfying Properties W1 – W4. Now, by taking into account (6.1) and (6.59), and slightly rearranging, we obtain

$$\underbrace{\begin{bmatrix} dV_2(t) \\ dI_1(t) \\ dI_l(t) \end{bmatrix}}_{=dX(t)} = \left(\underbrace{\begin{bmatrix} 0 & \frac{1}{c} & -\frac{1}{c} \\ -\frac{1}{\ell} & -\frac{r}{\ell} & 0 \\ 0 & 0 & a \end{bmatrix}}_{=A} \underbrace{\begin{bmatrix} V_2(t) \\ I_1(t) \\ I_l(t) \end{bmatrix}}_{=X(t)} + \underbrace{\begin{bmatrix} 0 & -\frac{1}{c} \\ \frac{1}{\ell} & 0 \\ 0 & 0 \end{bmatrix}}_{=C} \underbrace{\begin{bmatrix} v_s^* \\ i_l^* \end{bmatrix}}_{=u(t)} \right) dt + \underbrace{\begin{bmatrix} 0 \\ 0 \\ b \end{bmatrix}}_{=B} dW(t), \tag{6.60}$$

which has the same form as (6.52) for the case when $Q(t) = 1$, for all $t \geq 0$, $m_X(0) = \mathbf{0}_3$, and $\Sigma_X(0) = \mathbf{0}_{3\times 3}$ (because the system is initially at rest). Then, we can use the expressions in (6.56–6.58) to characterize the first and second moments of the stochastic process $X = [V_2, I_1, I_l]^\top$. To this end, we first need to use (6.53) to characterize $x^\circ(t) = [v_2^\circ(t), i_1^\circ(t), i_l^\circ(t)]^\top$. In this case, because of the structure of the matrices A and C and the fact that $i_l^\circ(0) = 0$, we have that $i_l^\circ(t) = 0$ for all $t \geq 0$, and $v_2^\circ(t)$ and $i_1^\circ(t)$ are governed by

$$\frac{d}{dt}\begin{bmatrix} v_2^\circ(t) \\ i_1^\circ(t) \end{bmatrix} = \begin{bmatrix} 0 & \frac{1}{c} \\ -\frac{1}{\ell} & -\frac{r}{\ell} \end{bmatrix} \begin{bmatrix} v_2^\circ(t) \\ i_1^\circ(t) \end{bmatrix} + \begin{bmatrix} 0 & -\frac{1}{c} \\ \frac{1}{\ell} & 0 \end{bmatrix} \begin{bmatrix} v_s^* \\ i_l^* \end{bmatrix}, \tag{6.61}$$

with $v_2^\circ(0) = 0$ and $i_1^\circ(0) = 0$. Now, let

$$\begin{aligned} \Delta X(t) &= X(t) - x^\circ(t) \\ &= \left[V_2(t), I_1(t), I_l(t)\right]^\top - \left[v_2^\circ(t), i_1^\circ(t), i_l^\circ(t)\right]^\top; \end{aligned}$$

then, $m_{\Delta X}(t)$ and $\Sigma_{\Delta X}(t)$ are governed by

$$\begin{aligned} \frac{d}{dt} m_{\Delta X}(t) &= A m_{\Delta X}(t), \\ \frac{d}{dt} \Sigma_{\Delta X}(t) &= A\Sigma_{\Delta X}(t) + \Sigma_{\Delta X}(t)A^\top + BB^\top, \end{aligned} \tag{6.62}$$

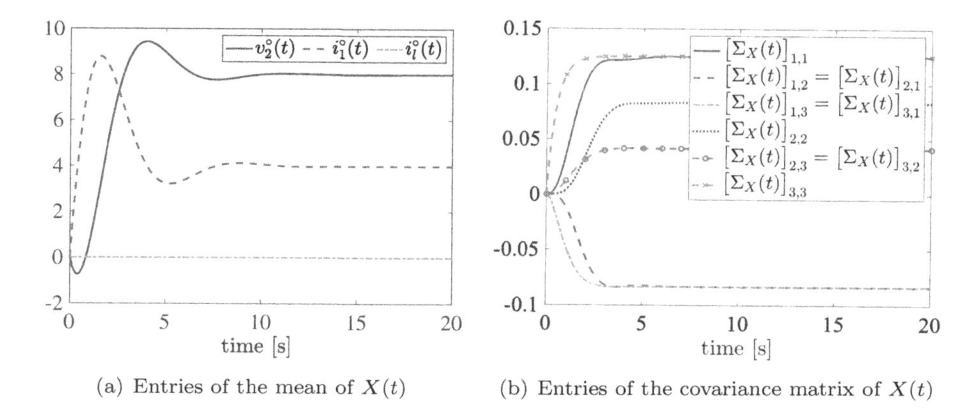

(a) Entries of the mean of $X(t)$ (b) Entries of the covariance matrix of $X(t)$

Figure 6.4 Evolution of the entries of the mean and the covariance matrix of $X(t) = [V_2(t), I_1(t), I_l(t)]^\top$.

with $m_{\Delta X}(0) = \mathbf{0}_3$ and $\Sigma_{\Delta X}(0) = \mathbf{0}_{3\times 3}$. Clearly, $m_{\Delta X}(t) = 0$ for all t; thus, it follows from (6.56) and (6.57) that

$$m_X(t) = x^\circ(t), \tag{6.63}$$

and

$$\Sigma_X(t) = \Sigma_{\Delta X}(t), \tag{6.64}$$

respectively. The trajectories followed by the entries of $m_X(t) = x^\circ(t)$ and the entries of $\Sigma_X(t)$, denoted by $[\Sigma_X(t)]_{i,j}$, $i = 1, 2, 3$, $j = 1, 2, 3$, are displayed in Fig. 6.4(a) and Fig. 6.4(b), respectively, for the case when $r = 1\ \Omega$, $\ell = 1$ H, $c = 1$ F, $v_s^\circ = 12$ V, $i_l^\circ = 4$ A, $a = -1$, and $b = 0.5$.

6.2.2 Gaussian Systems

Consider the system in (6.15) with

$$\begin{aligned} dV(t) &= \sqrt{Q(t)}dW(t) \\ &= \sqrt{Q(t)}\big(W(t+dt) - W(t)\big), \end{aligned} \tag{6.65}$$

where $W = \{W(t)\colon t \geq 0\}$ is assumed to be a standard Wiener vector process (see Section 2.1.5), i.e., the components of W,

$$W_i = \{W_i(t) : t \geq 0\},\ i = 1, 2, \ldots, m,$$

satisfy the following properties:

G1. $W_1, W_2, \ldots, W_m$ are independent.
G2. $W_i(0) = 0,\ i = 1, 2, \ldots, m.$

G3. W_i, $i = 1, 2, \ldots, m$, have independent increments, i.e., the distribution of $W_i(t) - W_i(s)$ depends on $t - s$ alone, and $W_i(t_j) - W_i(s_j)$, $j = 1, 2, \ldots, n$, are independent whenever the intervals $(s_j, t_j]$ are disjoint.

G4. $W_i(s+t) - W_i(s)$, $i = 1, 2, \ldots, m$, are normally distributed with zero mean and variance t for all $s, t \geq 0$; thus, $W_i(s+t) - W_i(s) = \sqrt{t} N_i(s)$, where $N_i(s)$ denotes a Gaussian random variable with zero mean and variance one.

G5. The sample paths of W_i are continuous.

Assume that $X(0) \in \mathbb{R}^n$ is a Gaussian random vector with mean $m_X(0) \in \mathbb{R}^n$ and covariance matrix $\Sigma_X(0) \in \mathbb{R}^{n \times n}$. Further assume that $X(0)$ and $W(t)$ are independent; thus, $X(t)$ and $W(t)$ are also independent. Therefore, $X(t)$ and $dW(t)$ are also independent and thus uncorrelated. Note that, except for $W_i(s+t) - W_i(s)$, $i = 1, 2, \ldots, m$, being Gaussian, Properties G1 – G4 are identical to Properties W1 – W4.

The setting here is clearly a special case of the setting discussed earlier; therefore we can use the techniques developed in Section 6.2.1 to characterize the mean and covariance functions of the stochastic process $X = \{X(t) : t \geq 0\}$. However, because of the additional structure of the Wiener process, the evolution of the pdf of $X(t)$, which we denote by $f_X(t, x)$, can be characterized by the following partial differential equation:

$$\frac{\partial f_X(t,x)}{\partial t} = -\sum_{i=1}^{n} \frac{\partial}{\partial x_i}\Big(a_i(t,x) f_X(t,x)\Big) + \frac{1}{2}\sum_{i=1}^{n}\sum_{j=1}^{n} d_{i,j}(t) \frac{\partial^2 f_X(t,x)}{\partial x_i \partial x_j}, \tag{6.66}$$

where $a_i(t, x)$ is the ith entry of the vector $A(t)x \in \mathbb{R}^n$, and $d_{i,j}(t)$ is the (i, j) entry of the matrix

$$D(t) = B(t)Q(t)B^\top(t) \in \mathbb{R}^{n \times n},$$

and with initial conditions as follows:

$$f_X(0,x) = \frac{1}{\sqrt{(2\pi)^n \mathbf{det}(\Sigma_X(0))}} e^{-\frac{(x-m_X(0))^\top \Sigma_X^{-1}(0)(x-m_X(0))}{2}}. \tag{6.67}$$

For the case when $n = m = 1$, and assuming $V(t) = W(t)$, we have that

$$dX(t) = \alpha(t)X(t)\,dt + \beta(t)dW(t), \tag{6.68}$$

where $\alpha(t)$ and $\beta(t)$ are scalars, and (6.66) reduces to

$$\frac{\partial f_X(t,x)}{\partial t} = -\alpha(t)\frac{\partial}{\partial x}\left(x f_X(t,x)\right) + \frac{1}{2}\beta^2(t)\frac{\partial^2 f_X(t,x)}{\partial x^2}. \tag{6.69}$$

Next, we show the steps to derive (6.69) and postpone the derivation of the expression for the general case in (6.66) to later when we discuss continuous-time nonlinear systems.

Consider a function $\psi\colon [0,\infty) \times \mathbb{R} \to \mathbb{R}$ that is continuously differentiable with respect to its first argument and twice continuously differentiable with respect to its second argument, and define $Y(t) = \psi\big(t, X(t)\big)$, where $X(t)$ satisfies (6.68). The differential formula for $Y(t)$, $dY(t) = d\psi\big(t, X(t)\big)$, is given by

$$\begin{aligned} dY(t) &= d\psi\big(t, X(t)\big) \\ &= \psi\big(t+dt, X(t+dt)\big) - \psi\big(t, X(t)\big) \\ &= \left(\frac{\partial\psi\big(t, X(t)\big)}{\partial t} + \frac{\partial\psi\big(t, X(t)\big)}{\partial x}\alpha(t)X(t) + \frac{1}{2}\frac{\partial^2\psi\big(t, X(t)\big)}{\partial x^2}\beta^2(t)\right)dt \\ &\quad + \frac{\partial\psi\big(t, X(t)\big)}{\partial x}\beta(t)dW(t). \end{aligned} \tag{6.70}$$

The expression above is a result of Itô's lemma and its derivation is fairly technical, and involves using results pertaining to convergence of a sequence of random variables. If we compare (6.70) with the expression that would result by applying the chain rule from calculus, one notices the extra term

$$\frac{1}{2}\frac{\partial^2\psi\big(t, X(t)\big)}{\partial x^2}\beta^2(t)dt;$$

next, we provide some heuristics to give the reader a feeling for where this term comes from. The idea is to expand $\psi\big(t+dt, X(t+dt)\big) = \psi\big(t+dt, X(t)+dX(t)\big)$ around $\big(t, X(t)\big)$. In doing so, we will keep the second-order term associated with the second argument of $\psi(\cdot,\cdot)$; this is necessary in light of (6.68) and the fact that $dW(t) = \sqrt{dt}N(t)$, where $N(t) = \big[N_1(t), N_2(t), \ldots, N_m(t)\big]^\top$, which makes the expectation of terms of the form $\big(dW(t)\big)^2$ of order dt instead of dt^2. Thus, for a sufficiently small $dt > 0$, we have that

$$\begin{aligned} dY(t) &= d\psi\big(t, X(t)\big) \\ &= \psi\big(t+dt, X(t+dt)\big) - \psi\big(t, X(t)\big) \\ &\approx \frac{\partial\psi\big(t, X(t)\big)}{\partial t}dt + \frac{\partial\psi\big(t, X(t)\big)}{\partial x}dX(t) + \frac{1}{2}\frac{\partial^2\psi\big(t, X(t)\big)}{\partial x^2}\big(dX(t)\big)^2 \\ &= \frac{\partial\psi\big(t, X(t)\big)}{\partial t}dt + \frac{\partial\psi\big(t, X(t)\big)}{\partial x}\big(\alpha(t)X(t)\,dt + \beta(t)dW(t)\big) \\ &\quad + \frac{1}{2}\frac{\partial^2\psi\big(t, X(t)\big)}{\partial x^2}\big(\alpha(t)X(t)\,dt + \beta(t)dW(t)\big)^2 \\ &\approx \left(\frac{\partial\psi\big(t, X(t)\big)}{\partial t} + \frac{\partial\psi\big(t, X(t)\big)}{\partial x}\alpha(t)X(t) + \frac{1}{2}\frac{\partial^2\psi\big(t, X(t)\big)}{\partial x^2}\beta^2(t)\right)dt \\ &\quad + \frac{\partial\psi\big(t, X(t)\big)}{\partial x}\beta(t)dW(t), \end{aligned} \tag{6.71}$$

where the last approximation results from dropping the terms

$$\frac{1}{2}\frac{\partial^2\psi\big(t, X(t)\big)}{\partial x^2}\big(\alpha(t)X(t)dt\big)^2,$$

and

$$\frac{\partial^2 \psi(t, X(t))}{\partial x^2}\alpha(t)\beta(t)X(t)dW(t)dt = \frac{\partial^2 \psi(t, X(t))}{\partial x^2}\alpha(t)\beta(t)X(t)N(t)dt^{1.5},$$

because their order is higher than dt, and approximating the term

$$\frac{1}{2}\frac{\partial^2 \psi(t, X(t))}{\partial x^2}\left(\beta(t)dW(t)\right)^2$$

by

$$\frac{1}{2}\frac{\partial^2 \psi(t, X(t))}{\partial x^2}\beta^2(t)dt,$$

in light of the fact that $\mathrm{E}\left[\left(dW(t)\right)^2\right] = dt$.

Now, by dividing both sides of (6.70) by dt, and taking expectations, we obtain

$$\begin{aligned}
&\mathrm{E}\left[\frac{\psi(t+dt, X(t+dt)) - \psi(t, X(t))}{dt}\right] \\
&= \mathrm{E}\left[\frac{\partial \psi(t, X(t))}{\partial t} + \frac{\partial \psi(t, X(t))}{\partial x}\alpha(t)X(t) + \frac{1}{2}\frac{\partial^2 \psi(t, X(t))}{\partial x^2}\beta^2(t)\right] \\
&\quad + \underbrace{\frac{1}{dt}\mathrm{E}\left[\frac{\partial \psi(t, X(t))}{\partial x}\beta(t)dW(t)\right]}_{=:\kappa(t)} \\
&= \mathrm{E}\left[\frac{\partial \psi(t, X(t))}{\partial t} + \frac{\partial \psi(t, X(t))}{\partial x}\alpha(t)X(t) + \frac{1}{2}\frac{\partial^2 \psi(t, X(t))}{\partial x^2}\beta^2(t)\right], \quad (6.72)
\end{aligned}$$

where the last equality follows from the fact that $X(t)$ and $dW(t)$ are independent as noted earlier; thus,

$$\begin{aligned}
\kappa(t) &= \mathrm{E}\left[\frac{\partial \psi(t, X(t))}{\partial x}\beta(t)\right]\mathrm{E}\left[dW(t)\right] \\
&= 0 \qquad (6.73)
\end{aligned}$$

because $\mathrm{E}\left[dW(t)\right] = 0$ for all $t \geq 0$. Then, by taking the limit on both sides of (6.72) as dt goes to zero and exchanging the limit and expectation operations (this can be done because $\psi(\cdot, \cdot)$ is continuously differentiable), we obtain

$$\frac{d}{dt}\mathrm{E}\left[\psi(t, X(t))\right] = \mathrm{E}\left[(L\psi)(t, X(t))\right], \qquad (6.74)$$

with

$$(L\psi)(t, x) = \frac{\partial \psi(t, x)}{\partial t} + \frac{\partial \psi(t, x)}{\partial x}\alpha(t)x + \frac{1}{2}\frac{\partial^2 \psi(t, x)}{\partial x^2}\beta^2(t). \qquad (6.75)$$

The operator $(L\psi)(t, x)$ is referred to as the generator of the stochastic process X, and the expression in (6.74) is known as Dynkin's formula.

Now, we will use Dynkin's formula to derive (6.69). On the one hand, considering the left-hand side of (6.74), we have that

$$
\begin{aligned}
\frac{d}{dt}\mathrm{E}\Big[\psi(t,X(t))\Big] &= \frac{d}{dt}\int_{-\infty}^{\infty}\psi(t,x)f_X(t,x)dx \\
&= \int_{-\infty}^{\infty}\frac{d}{dt}\Big(\psi(t,x)f_X(t,x)\Big)dx \\
&= \int_{-\infty}^{\infty}\left(\frac{\partial\psi(t,x)}{\partial t}f_X(t,x)+\psi(t,x)\frac{\partial f_X(t,x)}{\partial t}\right)dx \\
&= \int_{-\infty}^{\infty}\frac{\partial\psi(t,x)}{\partial t}f_X(t,x)dx+\int_{-\infty}^{\infty}\psi(t,x)\frac{\partial f_X(t,x)}{\partial t}dx. \qquad (6.76)
\end{aligned}
$$

On the other hand, considering the right-hand side of (6.74), we have that

$$
\begin{aligned}
\mathrm{E}\Big[(L\psi)(t,X(t))\Big] &= \mathrm{E}\left[\frac{\partial\psi(t,X(t))}{\partial t}+\frac{\partial\psi(t,X(t))}{\partial x}\alpha(t)x+\frac{1}{2}\frac{\partial^2\psi(t,X(t))}{\partial x^2}\beta^2(t)\right] \\
&= \int_{-\infty}^{\infty}\frac{\partial\psi(t,x)}{\partial t}f_X(t,x)dx+\underbrace{\int_{-\infty}^{\infty}\frac{\partial\psi(t,x)}{\partial x}\alpha(t)xf_X(t,x)dx}_{=:\,I_1} \\
&\quad+\underbrace{\int_{-\infty}^{\infty}\frac{1}{2}\frac{\partial^2\psi(t,x)}{\partial x^2}\beta^2(t)f_X(t,x)dx}_{=:\,I_2}. \qquad (6.77)
\end{aligned}
$$

For any $t \geq 0$, define $g_t(x) = \psi(t,x),\ x \in \mathbb{R}$. Assume $g_t(\cdot)$ has compact support, i.e., the set $\mathcal{X} = \{x \in \mathbb{R}\colon\ g_t(x) \neq 0\}$ is closed and bounded. Now, we integrate I_1 by parts (See Appendix A) as follows:

$$
\begin{aligned}
I_1 &= \int_{-\infty}^{\infty}\underbrace{\alpha(t)xf_X(t,x)}_{u(x)}\,\underbrace{\frac{\partial\psi(t,x)}{\partial x}dx}_{dv(x)} \\
&= u(x)v(x)\Big|_{-\infty}^{\infty}-\int_{-\infty}^{\infty}v(x)du(x) \\
&= \alpha(t)xf_X(t,x)\,\underbrace{\psi(t,x)}_{g_t(x)}\Big|_{-\infty}^{\infty}-\int_{-\infty}^{\infty}\psi(t,x)\frac{\partial}{\partial x}\Big(\alpha(t)xf_X(t,x)\Big)dx \\
&= -\int_{-\infty}^{\infty}\psi(t,x)\frac{\partial}{\partial x}\Big(\alpha(t)xf_X(t,x)\Big)dx, \qquad (6.78)
\end{aligned}
$$

where the last equality follows from the fact that $g_t(\cdot)$ has compact support; thus, $\lim_{x\to\infty} g_t(x) = \lim_{x\to-\infty} g_t(x) = 0$. Similarly, integrating I_2 by parts yields

$$I_2 = \int_{-\infty}^{\infty} \underbrace{\frac{1}{2}\beta^2(t) f_X(t,x)}_{u(x)} \underbrace{\frac{\partial^2 \psi(t,x)}{\partial x^2} dx}_{dv(x)}$$

$$= \underbrace{- \int_{-\infty}^{\infty} \frac{\partial \psi(t,x)}{\partial x} \frac{1}{2} \frac{\partial}{\partial x}\Big(\beta^2(t) f_X(t,x)\Big) dx}_{I_3} . \tag{6.79}$$

Finally, integrating I_3 by parts yields

$$I_3 = \int_{-\infty}^{\infty} \underbrace{\frac{1}{2} \frac{\partial}{\partial x}\Big(\beta^2(t) f_X(t,x)\Big)}_{u(x)} \underbrace{\frac{\partial \psi(t,x)}{\partial x} dx}_{dv(x)}$$

$$= - \int_{-\infty}^{\infty} \psi(t,x) \frac{1}{2} \frac{\partial^2}{\partial x^2}\Big(\beta^2(t) f_X(t,x)\Big) dx; \tag{6.80}$$

thus

$$I_2 = \int_{-\infty}^{\infty} \psi(t,x) \frac{1}{2} \frac{\partial^2}{\partial x^2}\Big(\beta^2(t) f_X(t,x)\Big) dx. \tag{6.81}$$

Then, by plugging the expressions in (6.78) and (6.81) into (6.77), we obtain

$$\begin{aligned} \mathrm{E}\big[(L\psi)(t,X(t))\big] = &\int_{-\infty}^{\infty} \frac{\partial \psi(t,x)}{\partial t} f_X(t,x) dx - \int_{-\infty}^{\infty} \psi(t,x) \frac{\partial}{\partial x}\Big(\alpha(t) x f_X(t,x)\Big) dx \\ &+ \int_{-\infty}^{\infty} \psi(t,x) \frac{1}{2} \frac{\partial^2}{\partial x^2}\Big(\beta^2(t) f_X(t,x)\Big) dx, \end{aligned} \tag{6.82}$$

which by subtracting from (6.76) yields

$$\begin{aligned} 0 &= \int_{-\infty}^{\infty} \left(\psi(t,x) \frac{\partial f_X(t,x)}{\partial t}\right) dx + \int_{-\infty}^{\infty} \psi(t,x) \frac{\partial}{\partial x}\Big(\alpha(t) x f_X(t,x)\Big) dx \\ &\quad - \int_{-\infty}^{\infty} \psi(t,x) \frac{1}{2} \frac{\partial^2}{\partial x^2}\Big(\beta^2(t) f_X(t,x)\Big) dx \\ &= \int_{-\infty}^{\infty} \psi(t,x) \left(\frac{\partial f_X(t,x)}{\partial t} + \frac{\partial}{\partial x}\Big(\alpha(t) x f_X(t,x)\Big) - \frac{1}{2} \frac{\partial^2}{\partial x^2}\Big(\beta^2(t) f_X(t,x)\Big)\right) dx; \end{aligned} \tag{6.83}$$

thus, the integrand of the integral in the last line above must be identically equal to zero; therefore,

$$\frac{\partial f_X(t,x)}{\partial t} = -\alpha(t) \frac{\partial}{\partial x}\Big(x f_X(t,x)\Big) + \frac{1}{2}\beta^2(t) \frac{\partial^2 f_X(t,x)}{\partial x^2} \tag{6.84}$$

as claimed in (6.69).

6.3 Continuous-Time Nonlinear Systems

Here we study systems of the form:

$$dX(t) = \alpha\big(t, X(t)\big)\, dt + \beta\big(t, X(t)\big)dV(t), \tag{6.85}$$

with $\alpha\colon [0,\infty) \times \mathbb{R}^n \to \mathbb{R}^n$, $\beta\colon [0,\infty) \times \mathbb{R}^n \to \mathbb{R}^{n\times m}$, and

$$V = \big\{V(t) : t \geq 0\big\},\ V(t) \in \mathbb{R}^m,$$

with $dV(t) = \sqrt{Q(t)}\big(W(t+dt) - W(t)\big) = \sqrt{Q(t)}dW(t)$, where $Q(t) \in \mathbb{R}^{m\times m}$ is symmetric and positive definite, and $W = \big\{W(t) : t \geq 0\big\}$, $W(t) \in \mathbb{R}^m$, is a standard Wiener vector process. Assume that $X(0) \in \mathbb{R}^n$ is Gaussian with mean $m_X(0) \in \mathbb{R}^n$ and covariance matrix $\Sigma_X(0) \in \mathbb{R}^{n\times n}$. Further, assume that $X(0)$ and $W(t)$ are independent for any $t \geq 0$; thus, $X(t)$ and $W(t)$ are also independent, and so are $X(t)$ and $dW(t)$. Then, we would like to characterize the mean and covariance functions of X and/or its distribution for which we will utilize Itô's chain rule and Dynkin's formula. Both of these tools were introduced earlier for a scalar linear system; here we generalize them to the multi-dimensional nonlinear system in (6.85).

Itô's Chain Rule. Consider a function $\psi\colon [0,\infty) \times \mathbb{R}^n \to \mathbb{R}$ that is continuously differentiable with respect to its first argument and twice continuously differentiable with respect to its second argument. Define $Y(t) = \psi\big(t, X(t)\big)$, where $X(t)$ satisfies (6.85); then, the differential formula for $Y(t)$ is given by

$$\begin{aligned} dY(t) &= d\psi\big(t, X(t)\big) \\ &= \psi\big(t+dt, X(t+dt)\big) - \psi\big(t, X(t)\big) \\ &= \left(\frac{\partial\psi\big(t, X(t)\big)}{\partial t} + \sum_{i=1}^{n} \frac{\partial\psi\big(t, X(t)\big)}{\partial x_i}\alpha_i\big(t, X(t)\big)\right. \\ &\quad \left. + \frac{1}{2}\sum_{i=1}^{n}\sum_{j=1}^{n} \frac{\partial^2\psi\big(t, X(t)\big)}{\partial x_i \partial x_j} d_{i,j}\big(t, X(t)\big)\right) dt \\ &\quad + \sum_{i=1}^{n} \frac{\partial\psi\big(t, X(t)\big)}{\partial x_i} \hat{u}_i^\top \beta\big(t, X(t)\big)\sqrt{Q(t)}dW(t), \end{aligned} \tag{6.86}$$

where $\alpha_i\big(t, X(t)\big)$ is the ith entry of the vector $\alpha\big(t, X(t)\big) \in \mathbb{R}^m$, $d_{i,j}\big(t, X(t)\big)$ is the (i,j) entry of the matrix $D\big(t, X(t)\big) = \beta\big(t, X(t)\big)Q(t)\beta^\top\big(t, X(t)\big) \in \mathbb{R}^{n\times n}$, and $\hat{u}_i$ is an n-dimensional vector with its ith entry being equal to one and all other entries being equal to zero.

Dynkin's Formula. If we divide both sides of (6.86) by dt and take expectations, we obtain

$$
\begin{aligned}
\mathrm{E}\left[\frac{d\psi(t,X(t))}{dt}\right] &= \mathrm{E}\left[\frac{\partial\psi(t,X(t))}{\partial t}+\sum_{i=1}^{n}\frac{\partial\psi(t,X(t))}{\partial x_i}\alpha_i(t,X(t))\right.\\
&\quad\left.+\frac{1}{2}\sum_{i=1}^{n}\sum_{j=1}^{n}\frac{\partial^2\psi(t,X(t))}{\partial x_i\partial x_j}d_{i,j}(t,X(t))\right]\\
&\quad+\frac{1}{dt}\sum_{i=1}^{n}\underbrace{\mathrm{E}\left[\frac{\partial\psi(t,X(t))}{\partial x_i}\hat{u}_i^{\top}\beta(t,X(t))\sqrt{Q(t)}dW(t)\right]}_{=\kappa_i(t)}\\
&= \mathrm{E}\left[\frac{\partial\psi(t,X(t))}{\partial t}+\sum_{i=1}^{n}\frac{\partial\psi(t,X(t))}{\partial x_i}\alpha_i(t,X(t))\right.\\
&\quad\left.+\frac{1}{2}\sum_{i=1}^{n}\sum_{j=1}^{n}\frac{\partial^2\psi(t,X(t))}{\partial x_i\partial x_j}d_{i,j}(t,X(t))\right],
\end{aligned}
\tag{6.87}
$$

where the last equality follows from the fact that $X(t)$ and $dW(t)$ are independent as noted earlier; thus, we have that

$$
\begin{aligned}
\kappa_i(t) &= \mathrm{E}\left[\frac{\partial\psi(t,X(t))}{\partial x}\hat{u}_i^{\top}\beta(t,X(t))\right]\sqrt{Q(t)}\mathrm{E}\big[dW(t)\big]\\
&= 0, \quad i=1,2,\dots,n,
\end{aligned}
\tag{6.88}
$$

because $\mathrm{E}\big[dW(t)\big]=\mathbf{0}_m$ for all $t\geq 0$. Then, by taking the limit on both sides of (6.87) as dt goes to zero and exchanging the limit and expectation operations (this can be done because $\psi(\cdot,\cdot)$ is continuously differentiable), we obtain

$$
\frac{d}{dt}\mathrm{E}\Big[\psi(t,X(t)\Big]=\mathrm{E}\Big[(L\psi)(t,X(t))\Big],
\tag{6.89}
$$

with

$$
\begin{aligned}
(L\psi)(t,x) &= \frac{\partial\psi(t,X(t))}{\partial t}+\sum_{i=1}^{n}\frac{\partial\psi(t,X(t))}{\partial x_i}\alpha_i(t,X(t))\\
&\quad+\frac{1}{2}\sum_{i=1}^{n}\sum_{j=1}^{n}\frac{\partial^2\psi(t,X(t))}{\partial x_i\partial x_j}d_{i,j}(t,X(t)).
\end{aligned}
\tag{6.90}
$$

6.3.1 Moments

Unlike the linear case, here we will not be able to write linear differential equations to describe the evolution of the mean and covariance matrix of $X(t)$, except for the case when $\alpha(t,\cdot)$ and $\beta(t,\cdot)$ are polynomials. To see why this is the case, assume $m=n=1$ and $Q=1$; then, (6.89–6.90) can be simplified, yielding

$$\frac{d}{dt}\mathrm{E}\Big[\psi\big(t, X(t)\big)\Big] = \mathrm{E}\Big[(L\psi)\big(t, X(t)\big)\Big], \tag{6.91}$$

with

$$(L\psi)(t,x) = \frac{\partial\psi(t,x)}{\partial t} + \frac{\partial\psi(t,x)}{\partial x}\alpha(t,x) + \frac{1}{2}\frac{\partial^2\psi(t,x)}{\partial x^2}\beta^2(t,x). \tag{6.92}$$

Now, set $\psi(t,x) = x$; then, (6.91) becomes

$$\frac{d}{dt}\mathrm{E}\big[X(t)\big] = \mathrm{E}\Big[\alpha\big(t, X(t)\big)\Big]. \tag{6.93}$$

Thus, unless $\alpha(t,x)$ is linear with x, it is in general not possible to obtain $\mathrm{E}\big[X(t)\big]$ by solving (6.93). Similarly, if one sets $\psi(t,x) = x^2$, (6.91) becomes

$$\frac{d}{dt}\mathrm{E}\Big[X^2(t)\Big] = 2\mathrm{E}\Big[X(t)\alpha\big(t, X(t)\big)\Big] + \mathrm{E}\Big[\beta^2\big(t, X(t)\big)\Big], \tag{6.94}$$

and unless both $\alpha(t,x)$ and $\beta(t,x)$ are linear with x, it is not possible to obtain $\mathrm{E}\big[X^2(t)\big]$ by solving (6.94).

One could attempt to linearize $\alpha(t,\cdot)$ and $\beta(t,\cdot)$ around the trajectory followed by

$$\frac{dx(t)}{dt} = \alpha\big(t, x(t)\big), \tag{6.95}$$

with $x(0) = m_X(0)$, and follow a procedure similar to that used in Section 5.3 when computing moments of discrete-time nonlinear systems. However, the problem here is that $W(t)$ is normally distributed; thus, the support of its pdf is not bounded. Therefore, the linearized system might not accurately capture the original system dynamics for values of $W(t)$ that are very large or very small.

6.3.2 Probability Distribution

The evolution of the pdf of $X(t)$, $f_X(t,x)$, is governed by the following partial differential equation:

$$\frac{\partial f_X(t,x)}{\partial t} = -\sum_{i=1}^{n}\frac{\partial}{\partial x_i}\Big(\alpha_i(t,x)f_X(t,x)\Big) + \frac{1}{2}\sum_{i=1}^{n}\sum_{j=1}^{n}\frac{\partial^2}{\partial x_i\partial x_j}\Big(d_{i,j}(t,x)f_X(t,x)\Big), \tag{6.96}$$

with $f_X(0,x)$ as in (6.67). We will establish (6.96) by using a similar approach to that used when establishing (6.69) for the one-dimensional linear case. As in that case, a key step of the derivation involves integration by parts (see Appendix A).

Consider Dynkin's formula in (6.89–6.90), where $\psi\colon [0,\infty) \times \mathbb{R}^n \to \mathbb{R}$ is assumed to be continuously differentiable with respect to its first argument and twice continuously differentiable with respect to its second argument. In addition, for every $t \geq 0$, assume $g_t(\cdot) := \psi(t,\cdot)$ is supported in some open bounded set $\mathcal{X}$ and let $\partial\mathcal{X}$ denote the boundary of $\mathcal{X}$. Clearly, for any $t \geq 0$, we have that

$$g_t(x) = 0, \quad x \in \partial\mathcal{X}. \tag{6.97}$$

Now, by using the left-hand side of (6.89), we have

$$\begin{aligned}
\frac{d}{dt}\mathrm{E}\Big[\psi(t, X(t))\Big] &= \frac{d}{dt}\int_{\mathcal{X}} \psi(t,x) f_X(t,x)\underbrace{dx_1 dx_2 \ldots dx_n}_{=:\,dx} \\
&= \int_{\mathcal{X}} \frac{\partial}{\partial t}\Big(\psi(t,x) f_X(t,x)\Big)dx \\
&= \int_{\mathcal{X}} \left(\frac{\partial\psi(t,x)}{\partial t} f_X(t,x) + \psi(t,x)\frac{\partial f_X(t,x)}{\partial t}\right)dx \\
&= \int_{\mathcal{X}} \frac{\partial\psi(t,x)}{\partial t} f_X(t,x)dx + \int_{\mathcal{X}} \psi(t,x)\frac{\partial f_X(t,x)}{\partial t}dx.
\end{aligned} \tag{6.98}$$

Also, by using the right-hand side of (6.89), we have

$$\begin{aligned}
\mathrm{E}\Big[(L\psi)(t, X(t))\Big] &= \mathrm{E}\Bigg[\frac{\partial\psi(t,X(t))}{\partial t} + \sum_{i=1}^{n} \frac{\partial\psi(t,X(t))}{\partial x_i}\alpha_i(t, X(t)) \\
&\quad + \frac{1}{2}\sum_{i=1}^{n}\sum_{j=1}^{n} \frac{\partial^2\psi(t,X(t))}{\partial x_i \partial x_j} d_{i,j}(t, X(t))\Bigg] \\
&= \int_{\mathcal{X}} \frac{\partial\psi(t,x)}{\partial t} f_X(t,x)dx + \underbrace{\int_{\mathcal{X}} \sum_{i=1}^{n} \frac{\partial\psi(t,x)}{\partial x_i}\alpha_i(t,x) f_X(t,x)dx}_{I_4} \\
&\quad + \frac{1}{2}\sum_{i=1}^{n} \underbrace{\int_{\mathcal{X}} \sum_{j=1}^{n} \frac{\partial^2\psi(t,x)}{\partial x_i \partial x_j} d_{i,j}(t,x) f_X(t,x)dx}_{I_5(i)}.
\end{aligned} \tag{6.99}$$

Now by using (A.29), we can integrate I_4 by parts as follows:

$$\begin{aligned}
I_4 &= \int_{\mathcal{X}} \sum_{i=1}^{n} \frac{\partial\psi(t,x)}{\partial x_i}\alpha_i(t,x) f_X(t,x)dx \\
&= \int_{\mathcal{X}} \langle \nabla\psi(t,x), \alpha(t,x) f_X(t,x)\rangle dx \\
&= \underbrace{\int_{\partial\mathcal{X}} \langle \psi(t,x)\alpha(t,x) f_X(t,x), n(x)\rangle dS}_{=\,0 \text{ by } (6.97)} - \int_{\mathcal{X}} \langle \psi(t,x)\nabla, \alpha(t,x) f_X(t,x)\rangle dx \\
&= -\int_{\mathcal{X}} \psi(t,x) \sum_{i=1}^{n} \frac{\partial}{\partial x_i}\Big(\alpha_i(t,x) f_X(t,x)\Big)dx.
\end{aligned} \tag{6.100}$$

Define $D^{(i)}(t,x) := \left[d_{i,1}(t,x), d_{i,2}(t,x), \ldots, d_{i,n}(t,x)\right]^\top$. Then, by using (A.29), we can integrate $I_5(i)$ by parts as follows:

$$\begin{aligned}
I_5(i) &= \int_{\mathcal{X}} \sum_{j=1}^{n} \frac{\partial^2 \psi(t,x)}{\partial x_i \partial x_j} d_{i,j}(t,x) f_X(t,x) dx \\
&= \int_{\mathcal{X}} \left\langle \nabla \frac{\partial \psi(t,x)}{\partial x_i}, D^{(i)}(t,x) f_X(t,x) \right\rangle dx \\
&= \underbrace{\int_{\partial \mathcal{X}} \left\langle \frac{\partial \psi(t,x)}{\partial x_i} D^{(i)}(t,x) f_X(t,x), n(x) \right\rangle dS}_{=0 \text{ by } (6.97)} \\
&\quad - \int_{\mathcal{X}} \left\langle \frac{\partial \psi(t,x)}{\partial x_i} \nabla, D^{(i)}(t,x) f_X(t,x) \right\rangle dx \\
&= - \int_{\mathcal{X}} \frac{\partial \psi(t,x)}{\partial x_i} \sum_{j=1}^{n} \frac{\partial}{\partial x_j} \Big(d_{i,j}(t,x) f_X(t,x) \Big) dx. \qquad (6.101)
\end{aligned}$$

Then, by using integration by parts again, it follows that

$$\begin{aligned}
\frac{1}{2} \sum_{i=1}^{n} I_5(i) &= \frac{1}{2} \sum_{i=1}^{n} \int_{\mathcal{X}} \sum_{j=1}^{n} \frac{\partial^2 \psi(t,x)}{\partial x_i \partial x_j} d_{i,j}(t,x) f_X(t,x) dx \\
&= -\frac{1}{2} \int_{\mathcal{X}} \sum_{i=1}^{n} \frac{\partial \psi(t,x)}{\partial x_i} \underbrace{\left(\sum_{j=1}^{n} \frac{\partial}{\partial x_j} \Big(d_{i,j}(t,x) f_X(t,x) \Big) \right)}_{=:\, \xi_i(t,x)} dx \\
&= -\frac{1}{2} \int_{\mathcal{X}} \Big\langle \nabla \psi(t,x), \underbrace{\left[\xi_1(t,x), \xi_2(t,x), \ldots, \xi_n(t,x)\right]^\top}_{=:\, \xi(t,x)} \Big\rangle dx \\
&= \underbrace{-\frac{1}{2} \int_{\partial \mathcal{X}} \Big\langle \psi(t,x) \xi(t,x), n(x) \Big\rangle dS}_{=0 \text{ by } (6.97)} + \frac{1}{2} \int_{\mathcal{X}} \Big\langle \psi(t,x) \nabla, \xi(t,x) \Big\rangle dx \\
&= \frac{1}{2} \int_{\mathcal{X}} \psi(t,x) \sum_{i=1}^{n} \frac{\partial \xi_i(t,x)}{\partial x_i} dx \\
&= \frac{1}{2} \int_{\mathcal{X}} \psi(t,x) \sum_{i=1}^{n} \sum_{j=1}^{n} \frac{\partial^2}{\partial x_i \partial x_j} \Big(d_{i,j}(t,x) f_X(t,x) \Big) dx. \qquad (6.102)
\end{aligned}$$

Now, by plugging the expressions in (6.100) and (6.102) into (6.99), we obtain

$$\begin{aligned}\mathrm{E}\big[(L\psi)(t,X(t))\big] &= \int_{\mathcal{X}} \frac{\partial\psi(t,x)}{\partial t} f_X(t,x)dx \\ &- \int_{\mathcal{X}} \psi(t,x)\sum_{i=1}^{n}\frac{\partial}{\partial x_i}\Big(\alpha_i(t,x)f_X(t,x)\Big)dx \\ &+ \frac{1}{2}\int_{\mathcal{X}} \psi(t,x)\sum_{i=1}^{n}\sum_{j=1}^{n}\frac{\partial^2}{\partial x_i\partial x_j}\Big(d_{i,j}(t,x)f_X(t,x)\Big)dx,\end{aligned} \tag{6.103}$$

and by subtracting from (6.98), we obtain

$$\begin{aligned}0 &= \int_{\mathcal{X}} \psi(t,x)\frac{\partial f_X(t,x)}{\partial t}dx + \int_{\mathcal{X}} \psi(t,x)\sum_{i=1}^{n}\frac{\partial}{\partial x_i}\Big(\alpha_i(t,x)f_X(t,x)\Big)dx \\ &- \frac{1}{2}\int_{\mathcal{X}} \psi(t,x)\sum_{i=1}^{n}\sum_{j=1}^{n}\frac{\partial^2}{\partial x_i\partial x_j}\Big(d_{i,j}(t,x)f_X(t,x)\Big)dx \\ &= \int_{\mathcal{X}} \psi(t,x)\Bigg(\frac{\partial f_X(t,x)}{\partial t} + \sum_{i=1}^{n}\frac{\partial}{\partial x_i}\Big(\alpha_i(t,x)f_X(t,x)\Big) \\ &- \frac{1}{2}\sum_{i=1}^{n}\sum_{j=1}^{n}\frac{\partial^2}{\partial x_i\partial x_j}\Big(d_{i,j}(t,x)f_X(t,x)\Big)\Bigg)dx;\end{aligned} \tag{6.104}$$

thus, the integrand of the integral in the last line above must be identically equal to zero; therefore,

$$\frac{\partial f_X(t,x)}{\partial t} = -\sum_{i=1}^{n}\frac{\partial}{\partial x_i}\Big(\alpha_i(t,x)f_X(t,x)\Big) + \frac{1}{2}\sum_{i=1}^{n}\sum_{j=1}^{n}\frac{\partial^2}{\partial x_i\partial x_j}\Big(d_{i,j}(t,x)f_X(t,x)\Big), \tag{6.105}$$

as claimed in (6.96).

6.4 Analysis of Microgrids under Sensor Measurement Uncertainty

In this section, we apply the techniques developed earlier in the chapter to analyze the dynamic behavior of inertia-less AC microgrids when the measurements provided by the sensors used by the control system are corrupted by random disturbances. The model adopted here is essentially the continuous-time counterpart of the model adopted in Section 5.4, augmented to take into account the effect of the aforementioned disturbances on the system dynamics; thus, many details already discussed in Section 5.4.1 are omitted.

6.4.1 System Model

Consider a three-phase microgrid comprising n buses ($n > 1$) indexed by the elements in the set $\mathcal{V} = \{1, 2, \ldots, n\}$, and l transmission lines

$(n-1 \leq l \leq n(n-1)/2)$ indexed by the elements in the set $\mathcal{L} = \{1, 2, \ldots, l\}$. In the remainder, we make the following assumptions (consistent with those made in Section 5.4.1):

A1. There is at most one transmission line connecting each pair of buses.
A2. The microgrid is balanced and operating in sinusoidal regime.
A3. Each transmission line $e \in \mathcal{L}$ linking a pair of buses, say bus i and bus j, is short and lossless; thus, it can be modeled by a series reactance, $X_{ij} > 0$.

Let $M \in \mathbb{R}^{n \times l}$ denote the node-to-edge incidence matrix of the directed graph describing the microgrid network topology, with the orientation on each edge of this graph in agreement with the (arbitrarily chosen) positive power flow direction on the corresponding transmission line. Let $p_i(t)$ denote the active power injected into the microgrid via bus i at time instant t, and let $V_i(t)$ and $\theta_i(t)$ respectively denote the magnitude and phase angle of the phasor associated with bus i's voltage at time instant t (measured relatively to a reference frame rotating at ω_0 rad/s corresponding to the microgrid nominal frequency). Define $p(t) = \left[p_1(t), p_2(t), \ldots, p_n(t)\right]^\top$ and $\theta(t) = \left[\theta_1(t), \theta_2(t), \ldots, \theta_n(t)\right]^\top$; then, we have that

$$p(t) = M\Gamma(t)\,\mathbf{sin}\big(M^\top \theta(t)\big), \tag{6.106}$$

where $\Gamma(t)$ is an $(l \times l)$-dimensional matrix whose e^{th} diagonal entry, denoted by $\gamma_e(t)$, is associated with transmission line e linking bus i and bus j, and defined as follows:

$$\gamma_e(t) = \frac{V_i(t)V_j(t)}{X_{ij}}, \tag{6.107}$$

and where $\mathbf{sin}(x) \in \mathbb{R}^l$, with $x = M^\top \theta(t) \in \mathbb{R}^l$, is defined as follows:

$$\mathbf{sin}(x) = \left[\sin(x_1), \sin(x_2), \ldots, \sin(x_l)\right]^\top.$$

Open-Loop Dynamics

Assume that connected to each bus i, $i = 1, 2, \ldots, n$, there is either a generating- or a load-type resource interfaced via a three-phase voltage source inverter whose controls will attempt to synthesize a sinusoidal voltage waveform at its output terminals. As a result, the inverter connected to bus i can be described by a per-phase equivalent circuit model comprising a controlled voltage source with terminal voltage

$$e_i(t) = \sqrt{2}E_i(t)\sin\big(\omega_0 t + \delta_i(t)\big), \tag{6.108}$$

in series with a reactance resulting from the inverter output filter. In the remainder, we make the following assumptions (consistent with those made in Section 5.4.1):

A4. The reactance of each voltage source inverter output filter is small when compared to the reactance values of the network transmission lines; thus, it can be neglected. Therefore, $V_i(t) \approx E_i(t)$ and $\theta_i(t) \approx \delta_i(t)$, $i = 1, 2, \ldots, n$.

A5. As a result of the action of the inverter outer voltage and inner current control loops holding $E_i(t)$, $i = 1, 2, \ldots, n$, constant throughout time, we have that $V_i(t) \approx V_i$, $i = 1, 2, \ldots, n$, where V_i is a positive constant.

Assume the phase angle of the inverter connected to bus i is regulated via a continuous-time frequency-droop control; then, because of Assumption A4, we have that bus i's voltage phase angle, $\theta_i(t)$, evolves according to

$$\frac{d\theta_i(t)}{dt} = \frac{1}{D_i}\big(u_i(t) - p_i(t)\big), \tag{6.109}$$

with D_i [s/rad] denoting the droop coefficient, and where $u_i(t)$ can be either set extraneously or by a control system. Define $u(t) = \big[u_1(t), u_2(t), \ldots, u_n(t)\big]^\top$; then, by combining (6.106–6.107) and (6.109), and assuming Assumption A5 holds, it follows that

$$\frac{d}{dt}\theta(t) = D^{-1}\Big(u(t) - M\Gamma\,\mathbf{sin}\big(M^\top\theta(t)\big)\Big), \tag{6.110}$$

where $D = \mathbf{diag}\big(D_1, D_2, \ldots, D_n\big)$, and $\Gamma = \mathbf{diag}\big(\gamma_1, \gamma_2, \ldots, \gamma_l\big)$, with

$$\gamma_e = \frac{V_i V_j}{X_{ij}}. \tag{6.111}$$

Note that if the system in (6.110) were operating in some steady state with $\sum_{i=1}^{n} u_i(t) = \mu$ for all $t \geq 0$, then,

$$\frac{d\theta_i(t)}{dt} = \frac{\mu}{\sum_{j=1}^{n} D_j} \qquad i = 1, 2, \ldots, n;$$

thus, since $\delta_i(t) \approx \theta_i(t)$, $i = 1, 2, \ldots, n$, it follows that the voltage synthesized by each voltage source inverter would have a frequency of

$$\omega_0 + \frac{\mu}{\sum_{j=1}^{n} D_j}.$$

Closed-Loop Dynamics

We would like to design a control scheme that regulates a subset of the $u_i(t)$'s so that the frequency of the voltage synthesized by each voltage source inverter is as close to the nominal value, ω_0, as possible. To this end, and without loss of generality, assume that

$$u_i(t) = \begin{cases} \xi_i(t), & i = 1, 2, \ldots, m, \\ r_i(t), & i = m+1, m+2, \ldots, n, \end{cases} \tag{6.112}$$

with $1 \leq m \leq n-1$, and where $\xi_i(t)$ is a known scalar extraneously set, whereas $r_i(t)$ is set by a closed-loop control system whose objective is to drive

$$\sum_{j=1}^{n} u_j(t) = \sum_{j=1}^{m} \xi_j(t) + \sum_{j=m+1}^{n} r_j(t)$$

to zero. Define the *average frequency error* as follows:

$$\overline{\Delta\omega}(t) := \frac{\sum_{i=1}^{n} D_i \frac{d\theta_i(t)}{dt}}{\sum_{i=1}^{n} D_i}. \tag{6.113}$$

Then, by using (6.109), we have that

$$\begin{aligned}\overline{\Delta\omega}(t) &= \frac{1}{n\overline{D}} \sum_{j=1}^{n} u_j(t) \\ &= \frac{1}{n\overline{D}} \left(\sum_{j=1}^{m} \xi_j(t) + \sum_{j=m+1}^{n} r_j(t) \right),\end{aligned} \tag{6.114}$$

where $\overline{D} := \frac{\sum_{i=1}^{n} D_i}{n}$. Then, driving the average frequency error to zero will also drive $\sum_{j=1}^{m} u_j(t)$ to zero. In an attempt to compute the average frequency error, a processor at each bus i, $i = m+1, m+2, \dots, n$, receives a measurement of $u_j(t)$, $j = 1, 2, \dots, n$, denoted by $u_{j,i}(t)$, corrupted by some additive disturbance, $y_{j,i}(t)$, i.e.,

$$u_{j,i}(t) = u_j(t) + y_{j,i}(t). \tag{6.115}$$

Each bus i, $i = m+1, m+2, \dots, n$, then computes an estimate of $\overline{\Delta\omega}(t)$, denoted by $\overline{\Delta\omega}_i(t)$, as follows:

$$\begin{aligned}\overline{\Delta\omega}_i(t) &= \frac{1}{n\overline{D}} \sum_{j=1}^{n} u_{j,i}(t) \\ &= \frac{1}{n\overline{D}} \left(\sum_{j=1}^{n} u_j + \sum_{j=1}^{n} y_{j,i}(t) \right) \\ &= \overline{\Delta\omega}(t) + \frac{1}{n\overline{D}} \mathbf{1}_n^\top y_i(t),\end{aligned} \tag{6.116}$$

where $y_i(t) = \left[y_{1,i}(t), y_{2,i}(t), \dots, y_{n,i}(t)\right]^\top$, and uses it to regulate the value of $r_i(t)$ via the following integral control scheme:

$$\begin{aligned}\frac{dz_i(t)}{dt} &= \alpha_i \overline{\Delta\omega}_i(t) \\ &= \alpha_i \overline{\Delta\omega}(t) + \frac{\alpha_i}{n\overline{D}} \mathbf{1}_n^\top y_i(t),\end{aligned} \tag{6.117}$$

$$r_i(t) = r_i^* + \beta_i z_i(t), \tag{6.118}$$

where α_i [s/rad] and β_i [p.u.] are constants that need to be appropriately chosen as discussed below. [In the model above, $y_{i,i}(t)$ attempts to capture measurement errors in the sensor used to acquire $u_i(t)$, whereas $y_{j,i}(t)$, $i \neq j$, attempts to capture both measurement errors in the sensor used to acquire $u_j(t)$ and any noise that corrupts the measurement when transmitted from bus j to bus i.]

Define $\xi(t) = \left[\xi_1(t), \xi_2(t), \ldots, \xi_m(t)\right]^\top$, $r(t) = \left[r_{m+1}(t), r_{m+2}(t), \ldots, r_n(t)\right]^\top$, $z(t) = \left[z_{m+1}(t), z_{m+2}(t), \ldots, z_n(t)\right]^\top$, $r^* = \left[r^*_{m+1}, r^*_{m+2}, \ldots, r^*_n\right]^\top$, and

$$y(t) = \left[y^\top_{m+1}(t), y^\top_{m+2}(t), \ldots, y^\top_n(t)\right]^\top \in \mathbb{R}^{n(n-m)};$$

then, by using (6.114), we can rewrite (6.117–6.118) as follows:

$$\frac{d}{dt}z(t) = \frac{1}{n\overline{D}}\alpha\left(\mathbf{1}^\top_m \xi(t) + \mathbf{1}^\top_{n-m} r(t)\right) + \frac{1}{n\overline{D}}\alpha H y(t), \tag{6.119}$$

$$r(t) = r^* + Bz(t), \tag{6.120}$$

with $\alpha = [\alpha_{m+1}, \alpha_{m+2}, \ldots, \alpha_n]^\top$, $B = \mathbf{diag}(\beta_{m+1}, \beta_{m+2}, \ldots, \beta_n)$, and

$$H = I_{n-m} \otimes \mathbf{1}^\top_n \in \mathbb{R}^{(n-m)\times n(n-m)},$$

where $\otimes$ denotes the Kronecker product (see Appendix A). Additionally, define $\mu(t) = \mathbf{1}^\top_m \xi(t) + \mathbf{1}^\top_{n-m} r^*$, then we can rewrite (6.119–6.120) more compactly as follows:

$$\frac{d}{dt}z(t) = \frac{1}{n\overline{D}}\alpha\beta^\top z(t) + \frac{1}{n\overline{D}}\alpha\mu(t) + \frac{1}{n\overline{D}}\alpha H y(t), \tag{6.121}$$

where $\beta = [\beta_{m+1}, \beta_{m+2}, \ldots, \beta_n]^\top$. Then, by using (6.110), (6.120), and (6.121), it follows that

$$\frac{d}{dt}\theta(t) = D^{-1}\left(K_\xi \xi(t) + K_r\left(r^* + Bz(t)\right) - M\Gamma\mathbf{sin}\left(M^\top\theta(t)\right)\right), \tag{6.122}$$

$$\frac{d}{dt}z(t) = \frac{1}{n\overline{D}}\alpha\beta^\top z(t) + \frac{1}{n\overline{D}}\alpha\mu(t) + \frac{1}{n\overline{D}}\alpha H y(t), \tag{6.123}$$

where

$$K_\xi = \begin{bmatrix} I_m \\ \mathbf{0}_{(n-m)\times m} \end{bmatrix}, \quad K_r = \begin{bmatrix} \mathbf{0}_{m\times(n-m)} \\ I_{n-m} \end{bmatrix}.$$

In the remainder, we will assume that

$$z(0) = \mathbf{0}_{n-m}, \tag{6.124}$$

while $\theta(0) \neq \mathbf{0}_n$ satisfies

$$K_\xi \xi(0) + K_r r^* = M\Gamma\mathbf{sin}\left(M^\top\theta(0)\right). \tag{6.125}$$

Such choice of $\theta(0)$ is attempting to capture the scenario in which the system is initially open-loop controlled and is operating in some steady state in which the system common frequency is ω_0, i.e., the system is initially operating at its nominal frequency.

Controller Gain Selection. Consider the system in (6.121) with the entries of $\alpha = [\alpha_{m+1}, \alpha_{m+2}, \ldots, \alpha_n]^\top$ and $\beta = [\beta_{m+1}, \beta_{m+2}, \ldots, \beta_n]^\top$ chosen so as to satisfy:

$$\begin{aligned} \alpha_i &< 0, \quad i = m+1, m+2, \ldots, n, \\ \beta_i &> 0, \quad i = m+1, m+2, \ldots, n. \end{aligned} \tag{6.126}$$

Then, the trajectories followed by (6.121) remain bounded as long as $\mu(t)$ and $y(t)$ remain bounded for all $t \geq 0$. To establish this result, define $\zeta(t) = \beta^\top z(t)$; then, multiplying both sides of (6.121) by $\beta^\top$ yields:

$$\frac{d\zeta(t)}{dt} = \frac{1}{n\overline{D}}\beta^\top \alpha\, \zeta(t) + \frac{1}{n\overline{D}}\beta^\top \alpha\, \mu(t) + \frac{1}{n\overline{D}}\beta^\top \alpha\, H y(t). \tag{6.127}$$

Now, it follows from (6.126) that

$$\frac{1}{n\overline{D}}\beta^\top \alpha < 0;$$

thus, the homogeneous part of (6.127) is asymptotically stable, therefore, $\zeta(t)$ will remain bounded for all $t \geq 0$ if $\mu(t)$ is bounded for all $t \geq 0$ and the entries of $y(t)$ are also bounded for all $t \geq 0$. Finally, since $\zeta(t) = \beta^\top z(t)$ and the entries of β are all positive, the entries of $z(t)$ must remain bounded for all $t \geq 0$.

Measurement Disturbance Model

Assume that the value taken by $y(t)$ in (6.123) is random and governed by a vector-valued stochastic process, $Y = \{Y(t) \colon t \geq 0\}$, $Y(t) \in \mathbb{R}^{n(n-m)}$, satisfying the following stochastic differential equation,

$$dY(t) = FY(t)\, dt + G dW(t), \tag{6.128}$$

where $Y(t) \in \mathbb{R}^{n(n-m)}$, $W = \{W(t) \colon t \geq 0\}$, $W(t) \in \mathbb{R}^{n(n-m)}$, is a standard Wiener vector process, $F \in \mathbb{R}^{n(n-m)\times n(n-m)}$, $G \in \mathbb{R}^{n(n-m)\times n(n-m)}$, and $Y(0)$ is a random $n(n-m)$-dimensional vector with mean $m_Y(0)$ and covariance matrix $\Sigma_Y(0) \in \mathbb{R}^{n(n-m)\times n(n-m)}$. In the remainder we will assume that $Y(0)$ and $W(t)$ are uncorrelated for any $t \geq 0$, i.e.,

$$\mathrm{E}\Big[\big(Y(0) - m_Y(0)\big) W^\top(t)\Big] = 0, \quad t \geq 0. \tag{6.129}$$

6.4.2 Average Frequency Error Statistical Characterization

Consider the closed-loop system (6.114) and (6.117–6.118). Because of the presence of the measurement disturbance, $y(t)$, neither the integral control state variable, $z(t)$, will converge to some constant vector value, nor the system average frequency error, $\overline{\Delta\omega}(t)$, will converge to zero. Thus, given the model in (6.128) describing the values taken by $y(t)$, we would like to characterize the statistics of the values taken by $z(t)$ and $\overline{\Delta\omega}(t)$. To this end, let $Z(t)$ and $\overline{\Delta\Omega}(t)$ denote the random vectors describing the values taken by $z(t)$ and $\overline{\Delta\omega}(t)$, respectively, and define the corresponding stochastic processes, $Z = \{Z(t) \colon t \geq 0\}$, and

$\overline{\Delta\Omega} = \{\overline{\Delta\Omega}(t) \colon t \geq 0\}$, respectively. Then, by using (6.114), (6.120), (6.121), and (6.128), we have that

$$\underbrace{\begin{bmatrix} dY(t) \\ dZ(t) \end{bmatrix}}_{= dX(t)} = \left(\underbrace{\begin{bmatrix} F & \mathbf{0}_{n(n-m)\times(n-m)} \\ \frac{1}{n\overline{D}}\alpha H & \frac{1}{n\overline{D}}\alpha\beta^\top \end{bmatrix}}_{= A} \underbrace{\begin{bmatrix} Y(t) \\ Z(t) \end{bmatrix}}_{= X(t)} + \underbrace{\begin{bmatrix} \mathbf{0}_{n(n-m)} \\ \frac{1}{n\overline{D}}\alpha \end{bmatrix}}_{= C} \mu(t) \right) dt$$
$$+ \underbrace{\begin{bmatrix} G \\ \mathbf{0}_{(n-m)\times n(n-m)} \end{bmatrix}}_{= B} dW(t), \tag{6.130}$$

$$\overline{\Delta\Omega}(t) = \frac{1}{n\overline{D}}\Big(\mu(t) + \beta^\top Z(t)\Big), \tag{6.131}$$

with

$$m_X(0) = \begin{bmatrix} m_Y^\top(0) \\ \mathbf{0}_{n-m} \end{bmatrix}, \quad \Sigma_X(0) = \begin{bmatrix} \Sigma_Y(0) & \mathbf{0}_{n(n-m)\times(n-m)} \\ \mathbf{0}_{(n-m)\times n(n-m)} & \mathbf{0}_{(n-m)\times(n-m)} \end{bmatrix}.$$

Note that $Z(0)$ and $W(t)$ are uncorrelated for any $t \geq 0$ because $Z(0) = \mathbf{0}_{n-m}$.

Now, we will use the tools in Section 6.2.1 to characterize the mean and covariance functions of the stochastic process $X = \{X(t) \colon t \geq 0\}$, which in turn can be used to characterize the mean and covariance functions of the stochastic process $\overline{\Delta\Omega} = \{\Delta\Omega(t) \colon t \geq 0\}$. [Note that because W is a Wiener vector process, we could potentially utilize the techniques in Section 6.2.2 to fully characterize the distribution of X and $\overline{\Delta\Omega}$ if $X(0)$ and $W(t)$ happened to be independent (we have only assumed uncorrelatedness of $Y(0)$ and $W(t)$).] Since (6.130) has a deterministic input, we need to use the formulae in (6.53–6.58). Thus, define $\Delta X(t) = X(t) - x^\circ(t)$, where $x^\circ(t)$ is governed by

$$\frac{d}{dt}x^\circ(t) = Ax^\circ(t) + C\mu(t), \quad t \geq 0, \tag{6.132}$$

with $x^\circ(0) = \left[\mathbf{0}_{n(n-m)}^\top, \mathbf{0}_{(n-m)}^\top\right]^\top$, and $\Delta X(t)$ satisfies

$$d\Delta X(t) = A\Delta X(t)\, dt + BdW(t), \tag{6.133}$$

with $m_{\Delta X}(0) = m_X(0)$ and $\Sigma_{\Delta X}(0) = \Sigma_X(0)$. By tailoring the expressions in (6.18) and (6.22) – (6.24) to the setting here, we obtain that the values taken by the mean of ΔX, $m_{\Delta X}(t)$, $t \geq 0$, are governed by

$$\frac{d}{dt}m_{\Delta X}(t) = Am_{\Delta X}(t), \quad t \geq 0, \tag{6.134}$$

with $m_{\Delta X}(0) = m_X(0)$, and

$$\frac{d}{dt}\Sigma_{\Delta X}(t) = A\Sigma_{\Delta X}(t) + \Sigma_{\Delta X}(t)A^\top + BB^\top, \quad t \geq 0, \tag{6.135}$$

$$C_{\Delta X}(t,s) = \begin{cases} e^{A(t-s)}\Sigma_{\Delta X}(s), & \text{if } t \geq s, \\ \Sigma_{\Delta X}(t)e^{A^\top(s-t)}, & \text{if } t < s, \end{cases} \tag{6.136}$$

with $\Sigma_{\Delta X}(0) = \Sigma_X(0)$. Now, we can use the expressions in (6.132–6.136) to characterize the mean and covariance functions of X and $\overline{\Delta\Omega}$.

Mean Function. The values taken by the mean function of X, $m_X(t)$, $t \geq 0$, and the mean function of $\overline{\Delta\Omega}$, $m_{\overline{\Delta\Omega}}(t)$, $t \geq 0$, are given by

$$m_X(t) = x^\circ(t) + m_{\Delta X}(t), \quad t \geq 0, \tag{6.137}$$

$$m_{\overline{\Delta\Omega}}(t) = \frac{1}{n\overline{D}}\Big(\mu(t) + K_z^\top m_X(t)\Big), \quad t \geq 0, \tag{6.138}$$

where $K_z = \left[\mathbf{0}_{n(n-m)}^\top, \beta^\top\right]^\top$.

The expression in (6.137) follows directly from (6.56), whereas the expression in (6.138) can be obtained from (6.131) as follows:

$$\begin{aligned} m_{\overline{\Delta\Omega}}(t) &= \mathrm{E}\left[\overline{\Delta\Omega}(t)\right] \\ &= \frac{1}{n\overline{D}}\Big(\mu(t) + \beta^\top m_Z(t)\Big) \\ &= \frac{1}{n\overline{D}}\Big(\mu(t) + K_z^\top m_X(t)\Big). \end{aligned} \tag{6.139}$$

Covariance Function. The values taken by the covariance function of X, $C_X(t,s), t,s \geq 0$, and the covariance function of $\overline{\Delta\Omega}$, $C_{\overline{\Delta\Omega}}(t,s)$, $t,s \geq 0$, are given by

$$C_X(t,s) = C_{\Delta X}(t,s), \quad t,s \geq 0, \tag{6.140}$$

$$C_{\overline{\Delta\Omega}}(t,s) = \frac{1}{n^2\overline{D}^2}K_z^\top C_X(t,s)K_z, \quad t,s \geq 0. \tag{6.141}$$

The expression in (6.140) follows directly from (6.57), whereas the expression in (6.141) can be obtained from (6.131) as follows:

$$\begin{aligned} C_{\overline{\Delta\Omega}}(t,s) &= \mathrm{E}\left[\Big(\overline{\Delta\Omega}(t) - m_{\Delta\Omega}(t)\Big)\Big(\overline{\Delta\Omega}(s) - m_{\Delta\Omega}(s)\Big)\right] \\ &= \frac{1}{n^2\overline{D}^2}\mathrm{E}\left[\beta^\top\Big(Z(t) - m_Z(t)\Big)\Big(Z(s) - m_Z(s)\Big)\beta\right] \\ &= \frac{1}{n^2\overline{D}^2}\mathrm{E}\left[K_z^\top\Big(X(t) - m_X(t)\Big)\Big(X(s) - m_X(s)\Big)^\top K_z\right] \\ &= \frac{1}{n^2\overline{D}^2}K_z^\top C_X(t,s)K_z. \end{aligned} \tag{6.142}$$

Example 6.3 (Three-bus microgrid average frequency error statistics) Consider a three-bus microgrid with a lossless network. Assume the generation and load resources connected to bus 1, bus 2, and bus 3 are interfaced with the network via droop-controlled voltage source inverters with droop coefficients $D_1 = 0.1$ s/rad, $D_2 = 0.25$ s/rad, and $D_3 = 1$ s/rad, respectively. Assume that connected to bus 1 there is a load with $u_1 = \xi_1 = -1$ p.u., connected to bus 2 there is a generator with $u_2 = \xi_2 = 0.75$ p.u., and connected to bus 3 there is a generator with $u_3 = r_3$ regulated with the control scheme in (6.117–6.118) with $r_3^* = 0.25$ p.u., $\alpha_3 = -1$ s/rad, and $\beta_3 = 0.5$ p.u.; thus,

$$\begin{aligned}
\frac{dz_3(t)}{dt} &= \alpha_3 \overline{\Delta\omega}(t) + \frac{\alpha_3}{D_1 + D_2 + D_3}\left(y_{1,3}(t) + y_{2,3}(t) + y_{3,3}(t)\right) \\
&= -0.3703 z_3(t) - 0.7407\left(y_{1,3}(t) + y_{2,3}(t) + y_{3,3}(t)\right), && (6.143) \\
r_3(t) &= r_3^* + \beta_3 z_3(t) \\
&= 0.25 + 0.5 z_3(t), && (6.144) \\
\overline{\Delta\omega}(t) &= \frac{\left(\xi_1 + \xi_2 + r_3(t)\right)}{D_1 + D_2 + D_3} \\
&= 0.3703 z_3(t), && (6.145)
\end{aligned}$$

with $z_3(0) = 0$. Define $y_3(t) = \left[y_{1,3}(t), y_{2,3}(t), y_{3,3}(t)\right]^\top$ and assume that the values it takes are governed by a stochastic process, $Y_3 = \{Y_3(t) \colon t \geq 0\}$, satisfying the following stochastic differential equation:

$$dY_3(t) = FY_3(t)\, dt + GdW(t), \tag{6.146}$$

where $Y_3(t) \in \mathbb{R}^3$, $W = \{W(t) \colon t \geq 0\}$, $W(t) \in \mathbb{R}^3$, is a standard Wiener vector process, $F = \mathbf{diag}(-1, -1.5, -2)$ and $G = \mathbf{diag}(1, 0.5, 0.2)$, and $Y_3(0) = \mathbf{0}_3$; thus, $Y_3(0)$ and $W(t)$ are uncorrelated for any $t \geq 0$. Since the values taken by $y_3(t)$ are governed by the stochastic process $Y_3 = \{Y_3(t) \colon t \geq 0\}$, the values taken by $z_3(t)$ are also governed by a stochastic process $Z_3 = \{Z_3(t) \colon t \geq 0\}$. Note that $Z_3(0)$ and $W(t)$ are also uncorrelated for any $t \geq 0$ because $Z_3(0) = \mathbf{0}_3$. Also, since the values taken by $z_3(t)$ are governed by the stochastic process $Z_3 = \{Z_3(t) \colon t \geq 0\}$, the values taken by $\overline{\Delta\omega}(t) = 0.3703 z_3(t)$ are also governed by a stochastic process $\overline{\Delta\Omega} = \{\overline{\Delta\Omega}(t) \colon t \geq 0\}$. Define $X(t) = [Y_3^\top(t), Z_3(t)]^\top$; then, by combining (6.143), (6.146), and (6.145), we obtain that

$$dX(t) = AX(t) + BdW(t), \tag{6.147}$$

$$\overline{\Delta\Omega}(t) = 0.3703 Z_3(t), \tag{6.148}$$

where

$$A = \begin{bmatrix} -1 & 0 & 0 & 0 \\ 0 & -1.5 & 0 & 0 \\ 0 & 0 & -2 & 0 \\ -0.7407 & -0.7407 & -0.7407 & -0.3703 \end{bmatrix}, \quad B = \begin{bmatrix} 1 & 0 & 0 \\ 0 & 0.5 & 0 \\ 0 & 0 & 0.2 \\ 0 & 0 & 0 \end{bmatrix}. \tag{6.149}$$

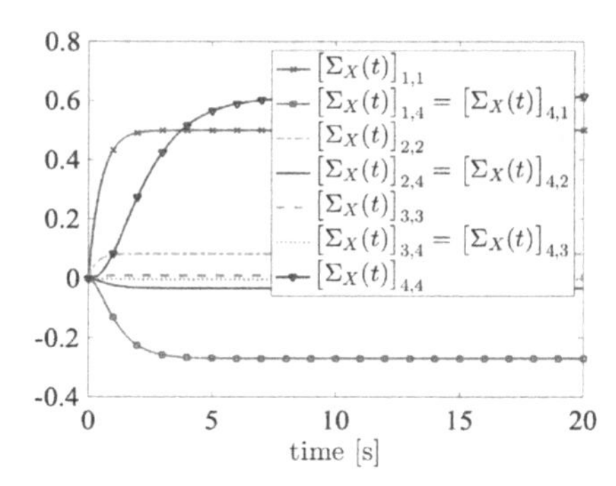

(a) Entries of the covariance matrix of $X(t)$

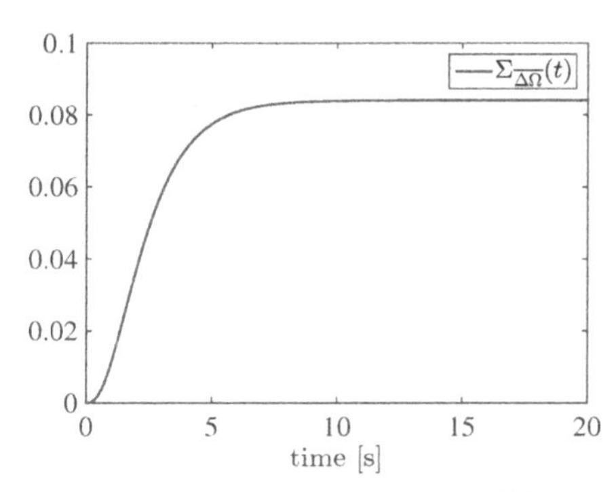

(b) Entries of the covariance of $\overline{\Delta\Omega}(t)$

Figure 6.5 Evolution of the non-zero entries of the covariance matrix of $X(t)$, and the covariance of $\overline{\Delta\Omega}(t)$ (note that because F and G are diagonal matrices, we have that $[\Sigma_X(t)]_{1,2} = [\Sigma_X(t)]_{2,1} = [\Sigma_X(t)]_{1,3} = [\Sigma_X(t)]_{3,1} = [\Sigma_X(t)]_{2,3} = [\Sigma_X(t)]_{3,2} = 0$).

Now, by tailoring the expressions in (6.137–6.138) to the setting here, and noting that $m_X(0) = \mathbf{0}_4$ and $\mu(t) = (\xi_1 + \xi_2 + r_3^*) = 0$ for all $t \geq 0$, it follows that

$$m_X(t) = \mathbf{0}_4, \tag{6.150}$$

$$m_{\overline{\Delta\Omega}}(t) = 0. \tag{6.151}$$

In addition, $C_X(t,s) \in \mathbb{R}^4$ can be computed for any $t, s \geq 0$ by using (6.140). The result can then be plugged into (6.141) to compute the values taken by the covariance of $\overline{\Delta\Omega}$, $C_{\overline{\Delta\Omega}}(t,s)$, $t, s \geq 0$. Figure 6.5(a) shows the evolution of the non-zero entries of $\Sigma_X(t) = C_X(t,t)$, whereas Fig. 6.5(b) shows the evolution of $\Sigma_{\overline{\Delta\Omega}}(t) = C_{\overline{\Delta\Omega}}(t,t)$.

6.5 Notes and References

The expressions describing the evolution of the mean, covariance, and correlation functions of a linear system are standard; see, for example, [10, 40]. The use of the Kronecker product for rewriting matrix equations is discussed in [41]. The material on stochastic differential equations is mostly standard and can be found in [42, 43]. The heuristic characterization of a white noise process as the time derivative of a Wiener process is from [42]. The derivation of the partial differential equation describing the evolution of the pdf of the state of a stochastic differential equation is in the spirit of that in [44]. The microgrid model used in Section 6.4 has been adapted from the model in [39].

7 Static Systems: Set-Theoretic Input Uncertainty

In this chapter, we consider the same class of systems as that considered in Chapter 3, i.e., systems in which the relation between the inputs and the states is described by some (possibly nonlinear) mapping. The values that the inputs take are also assumed to be uncertain, but unlike the model in Chapter 3, here we assume they are known to belong to particular classes of closed, convex sets, namely, ellipsoids and zonotopes. Then, the objective is to characterize the set containing all possible values that the states can take. We refer the reader to Section 2.2 for a review of the set-theoretic notions used throughout this chapter.

7.1 Introduction

Consider again the linear DC circuit discussed in Section 3.1 whose schematic is reproduced in Fig. 7.1, and recall the relation between the input variables, v_s and i_l, and the state variables, v_2 and i_1, given in (3.1):

$$\begin{bmatrix} v_2 \\ i_1 \end{bmatrix} = \begin{bmatrix} \frac{r_2}{r_1+r_2} & -\frac{r_1 r_2}{r_1+r_2} \\ \frac{1}{r_1+r_2} & \frac{r_2}{r_1+r_2} \end{bmatrix} \begin{bmatrix} v_s \\ i_l \end{bmatrix}, \tag{7.1}$$

which is just a particular instance of the general class of linear static systems described as follows:

$$x = Hw, \tag{7.2}$$

where $w \in \mathbb{R}^m$ is referred to as the input vector, $x \in \mathbb{R}^n$ is referred to as the state vector, and $H \in \mathbb{R}^{n \times m}$.

In Chapter 3, we developed tools to study the case where the possible values that w can take were described by a random vector with known statistics, and the objective was to characterize the statistics of the random vector describing the possible values taken by x. In this chapter, we will consider a set-theoretic model to capture the uncertainty in the value that w can take, i.e., the value that w takes is known to belong to some set, $\mathcal{W}$. For example, in the context of the circuit in Fig. 7.1, the value that v_s can take is known to lie within the interval $[9, 11]$ V, whereas the value of i_l is known to belong to the interval $[4, 6]$ A. The objective is to characterize the set of possible values that $x = [v_2, i_1]^\top$ can take.

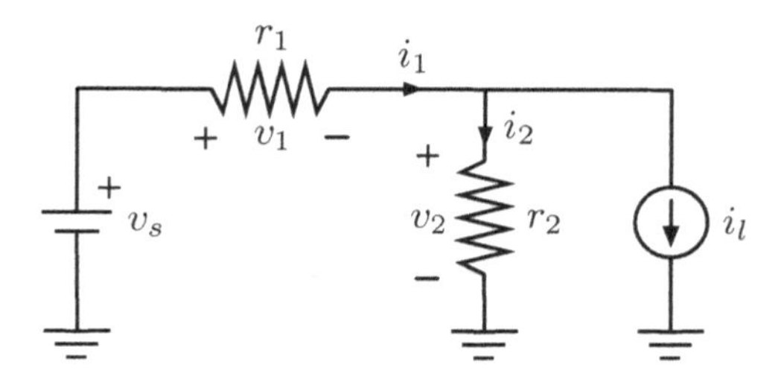

Figure 7.1 Linear DC circuit.

While the discussion above is centered around the linear case, in this chapter, we will consider a more general setting of the form:

$$x = h(w), \tag{7.3}$$

where $x \in \mathbb{R}^n$, $w \in \mathcal{W} \subseteq \mathbb{R}^m$, and $h(\cdot)$ is some function mapping elements in $\mathbb{R}^m$ into elements in $\mathbb{R}^n$. While in general the set $\mathcal{W}$ describing the possible values that the input can take can have any shape, in this chapter, we will restrict our analysis to two particular classes of closed, convex sets as follows:

S1. The set $\mathcal{W}$ is described by an ellipsoid, i.e.,

$$\mathcal{E} = \Big\{w \in \mathbb{R}^m \colon (w - w_0)^\top E^{-1}(w - w_0) \leq 1\Big\}, \tag{7.4}$$

where $w_0 \in \mathbb{R}^m$ is the center of the ellipsoid, and $E \in \mathbb{R}^{m \times m}$ is a symmetric positive definite matrix referred to as the shape matrix.

S2. The set $\mathcal{W}$ is described by a zonotope, i.e.,

$$\mathcal{Z} = \left\{z \in \mathbb{R}^m \colon z = z_0 + \sum_{j=1}^{s} \alpha_j e_j, \ -1 \leq \alpha_j \leq 1\right\}. \tag{7.5}$$

where $z_0 \in \mathbb{R}^m$ is the center of the zonotope, and $e_1, e_2, \ldots, e_s \in \mathbb{R}^m$ are referred to as the generators of the zonotope.

Then, for both S1 and S2, the objective is to characterize the set $\mathcal{X}$ containing the possible values that x can take. In this regard, when the mapping between the input w and the state x is linear, we will be able to provide an exact characterization of $\mathcal{X}$, whereas, for the case when the input-to-state mapping is nonlinear, we will resort to linearization to approximately compute $\mathcal{X}$.

The computation of the set $\mathcal{X}$ containing all possible values that x can take (or an approximation thereof) is useful because it allows us to determine whether or not the system state will always be within some set reflecting some requirements on the value that x can take for the system function to be fulfilled satisfactorily. For example, in power system applications, the so-called power flow equations can be used to map power injections to voltage magnitudes and line flows. If some of these power injections are associated with renewable-based generation resources, they can be modeled as unknown quantities belonging to some set. Then, one can use the techniques developed in this chapter to determine whether or not bus voltage magnitudes and line flows will remain within allowed limits.

7.2 Ellipsoid-Based Input Set Description

In this section, we consider Scenario S1, i.e., the case when the set describing the possible values that the system input can take is an ellipsoid. For the linear system case, we will show that the set containing all the possible values that the state can take can be computed exactly. For the nonlinear case, we will provide a linearization-based method to compute an approximation to the actual set that contains all possible values that the state can take.

7.2.1 Linear Setting

Consider a linear static system of the form

$$x = Hw, \tag{7.6}$$

where $x \in \mathbb{R}^n$, $H \in \mathbb{R}^{n \times m}$ is full rank, and $w \in \mathcal{W} \subseteq \mathbb{R}^m$, where $\mathcal{W}$ is an ellipsoid with center $w_0 \in \mathbb{R}^m$ and shape matrix $E \in \mathbb{R}^{m \times m}$, i.e.,

$$\mathcal{W} = \Big\{ w \in \mathbb{R}^m \colon (w - w_0)^\top E^{-1}(w - w_0) \leq 1 \Big\}; \tag{7.7}$$

thus, the support function of $\mathcal{W}$, denoted by $S_{\mathcal{W}}(\cdot)$, is given by (see (2.70))

$$S_{\mathcal{W}}(\eta) = \eta^\top w_0 + \sqrt{\eta^\top E \eta}, \quad \eta \in \mathbb{R}^m. \tag{7.8}$$

Then, the goal is to obtain the set that contains all possible values that x can take; we denote this set by $\mathcal{X}$. We need to consider two cases:

E1. The dimension of x is smaller than or equal to the dimension of w, i.e., $m \geq n$.

E2. The dimension of x is greater than the dimension of w, i.e., $m < n$.

Case E1. We will first consider the case when $m = n = 1$, i.e., x and w are scalars, and $H = a$ is also a scalar such that $a \neq 0$. Now, assume that $w \in \mathcal{W}$, where

$$\mathcal{W} = \Big\{ w \in \mathbb{R} \colon w_0 - \sqrt{e} \leq w \leq w_0 + \sqrt{e} \Big\}, \tag{7.9}$$

with $e > 0$. Clearly, $\mathcal{W}$ is a closed interval in $\mathbb{R}$, but by rewriting it as

$$\mathcal{W} = \Big\{ w \in \mathbb{R} \colon e^{-1}(w - w_0)^2 \leq 1 \Big\}, \tag{7.10}$$

and noting (7.4), one can also see $\mathcal{W}$ as a one-dimensional ellipsoid with center w_0, and shape matrix $E = e$. Then, the set $\mathcal{X}$ that contains all possible values of x such that $x = aw$, $w \in \mathcal{W}$, must also be a closed interval in $\mathbb{R}$. In order to compute the end points of $\mathcal{X}$, $\underline{x}$, and $\overline{x}$, we need to map the end points of $\mathcal{W}$, $\underline{w} = w_0 - \sqrt{e}$ and $\overline{w} = w_0 + \sqrt{e}$, via the relation between w and x. For $a > 0$, we have that

$$\begin{aligned}
\underline{x} &= a\underline{w} \\
&= a(w_0 - \sqrt{e}) \\
&= aw_0 - |a|\sqrt{e}, \\
\overline{x} &= a\overline{w} \\
&= a(w_0 + \sqrt{e}) \\
&= aw_0 + |a|\sqrt{e},
\end{aligned} \tag{7.11}$$

whereas for $a < 0$, we have that

$$\begin{aligned}
\underline{x} &= a\underline{w} \\
&= a(w_0 + \sqrt{e}) \\
&= aw_0 - |a|\sqrt{e}, \\
\overline{x} &= a\overline{w} \\
&= a(w_0 - \sqrt{e}) \\
&= aw_0 + |a|\sqrt{e};
\end{aligned} \tag{7.12}$$

from where it follows that

$$\begin{aligned}
\mathcal{X} &= \left\{x \in \mathbb{R} \colon aw_0 - |a|\sqrt{e} \le x \le aw_0 + |a|\sqrt{e}\right\} \\
&= \left\{x \in \mathbb{R} \colon |x - aw_0| \le |a|\sqrt{e}\right\},
\end{aligned} \tag{7.13}$$

which can be rewritten as

$$\mathcal{X} = \left\{x \in \mathbb{R} \colon f^{-1}(x - x_0)^2 \le 1\right\}, \tag{7.14}$$

where $x_0 = aw_0$, and $f = a^2 e$. Thus, the set $\mathcal{X}$ is also a one-dimensional ellipsoid with center $x_0 = aw_0$, and shape matrix $F = f = a^2 e$.

For the general case, i.e., $m \ge n > 1$, we have that

$$\mathcal{X} = \left\{x \in \mathbb{R}^n \colon (x - x_0)^\top F^{-1}(x - x_0) \le 1\right\}, \tag{7.15}$$

where

$$\begin{aligned}
x_0 &= Hw_0, \\
F &= HEH^\top.
\end{aligned} \tag{7.16}$$

The result in (7.15)–(7.16) can be obtained by computing the support function of $\mathcal{X}$, denoted by $S_{\mathcal{X}}(\cdot)$, as follows.

First, by using the definition of the support function in (2.55), we have that

$$S_{\mathcal{X}}(\eta) = \eta^\top x^*(\eta), \quad \eta \in \mathbb{R}^n, \tag{7.17}$$

where

$$x^*(\eta) := \arg\max_{x \in \mathcal{X}} \eta^\top x. \tag{7.18}$$

Then, since $x = Hw$, $w \in \mathcal{W}$, with $\mathcal{W}$ as defined in (7.7), we have that

$$S_{\mathcal{X}}(\eta) = \eta^\top H w^*(\eta), \tag{7.19}$$

where

$$w^*(\eta) := \arg\max_{w \in \mathcal{W}} \eta^\top H w. \tag{7.20}$$

From the geometry of the maximization problem in (7.20), one can see that $w^*(\eta)$ is a point on the boundary of $\mathcal{W}$; thus, we can obtain $w^*(\eta)$ from local extremum points of the following optimization problem:

$$\begin{aligned} &\underset{w}{\text{maximize}} && \eta^\top H w \\ &\text{subject to} && (w - w_0)^\top E^{-1}(w - w_0) = 1. \end{aligned} \tag{7.21}$$

By introducing a Lagrange multiplier, λ, we can reformulate (7.21) as an unconstrained optimization problem as follows:

$$\underset{w,\lambda}{\text{maximize}}\ f(w, \lambda), \tag{7.22}$$

where

$$f(w, \lambda) = \eta^\top H w - \lambda\big((w - w_0)^\top E^{-1}(w - w_0) - 1\big).$$

Now, in order to obtain the solution to (7.22), we compute the values of w and λ for which the gradient of $f(\cdot, \cdot)$ is equal to zero. Thus, we have that

$$\begin{aligned} \frac{\partial f(w, \lambda)}{\partial w} &= \eta^\top H - 2\lambda(w - w_0)^\top E^{-1} \\ &= 0, \end{aligned} \tag{7.23}$$

$$\begin{aligned} \frac{\partial f(w, \lambda)}{\partial \lambda} &= (w - w_0)^\top E^{-1}(w - w_0) - 1 \\ &= 0. \end{aligned} \tag{7.24}$$

It is easy to see from (7.23) that:

$$\begin{aligned} w &= w_0 + \frac{1}{2\lambda} E^\top H^\top \eta \\ &= w_0 + \frac{1}{2\lambda} E H^\top \eta, \end{aligned} \tag{7.25}$$

and by plugging this expression into (7.24), and solving for λ, we obtain two solutions, $\lambda^{(1)}$ and $\lambda^{(2)}$, as follows:

$$\begin{aligned} \lambda^{(1)} &= \frac{1}{2}\sqrt{\eta^\top H E H^\top \eta}, \\ \lambda^{(2)} &= -\frac{1}{2}\sqrt{\eta^\top H E H^\top \eta}, \end{aligned} \tag{7.26}$$

which by substituting in (7.25) and solving for w yield two solutions, $w^{(1)}$ and $w^{(2)}$, respectively, as follows:

$$w^{(1)} = w_0 + \frac{1}{\sqrt{\eta^\top HEH^\top \eta}} EH^\top \eta,$$
$$w^{(2)} = w_0 - \frac{1}{\sqrt{\eta^\top HEH^\top \eta}} EH^\top \eta. \tag{7.27}$$

Evaluating $f(w, \lambda)$ for $w = w^{(1)}$ and $\lambda = \lambda^{(1)}$ yields

$$f(w^{(1)}, \lambda^{(1)}) = \eta^\top Hw_0 + \sqrt{\eta^\top HEH^\top \eta},$$

whereas evaluating $f(w, \lambda)$ for $w = w^{(2)}$ and $\lambda = \lambda^{(2)}$ yields

$$f(w^{(2)}, \lambda^{(2)}) = \eta^\top Hw_0 - \sqrt{\eta^\top HEH^\top \eta};$$

thus, the pair

$$\left(w_0 + \frac{1}{\sqrt{\eta^\top HEH^\top \eta}} EH^\top \eta, \frac{1}{2}\sqrt{\eta^\top HEH^\top \eta} \right)$$

is a maximizer of (7.22). Returning to (7.19), we have that $w^*(\eta) = w^{(1)}$, which results in

$$\begin{aligned} S_{\mathcal{X}}(\eta) &= \eta^\top Hw^*(\eta) \\ &= \eta^\top Hw_0 + \sqrt{\eta^\top HEH^\top \eta}. \end{aligned} \tag{7.28}$$

Then, by defining $x_0 = Hw_0$ and $F = HEH^\top$, we obtain that

$$S_{\mathcal{X}}(\eta) = \eta^\top x_0 + \sqrt{\eta^\top F \eta}. \tag{7.29}$$

Since H is full rank and E is symmetric and positive definite, F is also symmetric and positive definite; therefore, (7.29) corresponds to the support function of an n-dimensional ellipsoid with center x_0 and shape matrix F.

Example 7.1 (Linear circuit with ellipsoidal input and state sets) Consider the linear DC circuit in Fig. 7.1 and assume that $r_1 = r_2 = 1\ \Omega$; then it follows from (7.1–7.2) that

$$\underbrace{\begin{bmatrix} v_2 \\ i_1 \end{bmatrix}}_{=:\,x} = \underbrace{\begin{bmatrix} \frac{1}{2} & -\frac{1}{2} \\ \frac{1}{2} & \frac{1}{2} \end{bmatrix}}_{=:\,H} \underbrace{\begin{bmatrix} v_s \\ i_l \end{bmatrix}}_{=:\,w}. \tag{7.30}$$

Assume that

$$w \in \mathcal{W} = \left\{ w \in \mathbb{R}^2 \colon (w - w_0)^\top E^{-1} (w - w_0) \leq 1 \right\}, \tag{7.31}$$

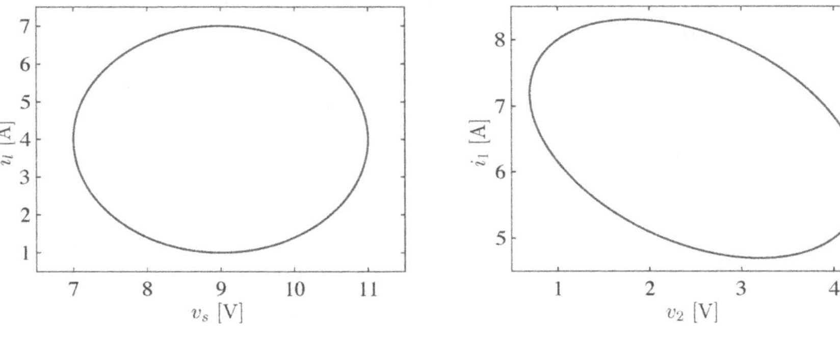

(a) Input-bounding ellipsoid $\mathcal{W}$. (b) State-bounding ellipsoid $\mathcal{X}$.

Figure 7.2 Bounding ellipsoids for the linear DC circuit in Example 7.1.

where $w_0 = [9 \text{ V}, 4 \text{ A}]^\top$, and

$$E = \begin{bmatrix} 4 & 0 \\ 0 & 9 \end{bmatrix}; \tag{7.32}$$

see Fig. 7.2(a) for a graphical depiction of the ellipsoid $\mathcal{W}$. In this case, we have that $m = n = 2$; thus, by using (7.15–7.16), we obtain that

$$\mathcal{X} = \Big\{ x \in \mathbb{R}^2 \colon (x - x_0)^{-1} F^{-1} (x - x_0) \leq 1 \Big\}, \tag{7.33}$$

where $x_0 = H w_0 = [2.5 \text{ V}, \; 6.5 \text{ A}]$, and

$$F = HEH^\top = \begin{bmatrix} 3.25 & -1.25 \\ -1.25 & 3.25 \end{bmatrix} \tag{7.34}$$

(see Fig. 7.2(b) for a graphical depiction).

Case E2. Here, we have that $H \in \mathbb{R}^{n \times m}$, $n > m$. The support function of $\mathcal{X}$ can be obtained in the same way as in Case E1, and it results in an identical expression to that in (7.29):

$$S_{\mathcal{X}}(\eta) = \eta^\top x_0 + \sqrt{\eta^\top F \eta}, \tag{7.35}$$

where $x_0 = H w_0$, and $F = HEH^\top$. However, although E is symmetric and positive definite, F is not full rank because $n > m$. Therefore, F is positive semidefinite, which implies $\mathcal{X}$ is not an ellipsoid. To show this, assume that there exist some $x_0' \in \mathbb{R}^n$ and $F' \in \mathbb{R}^{n \times n}$ that is symmetric and positive definite such that

$$S_{\mathcal{X}}(\eta) = \eta^\top x_0' + \sqrt{\eta^\top F' \eta}, \quad \eta \in \mathbb{R}^n \tag{7.36}$$

(this would mean that F' is a valid shape matrix and thus $\mathcal{X}$ would be an ellipsoid); we will show next that such F' cannot exist. By using (7.35), we have that $S_{\mathcal{X}}(\eta) = \eta^\top x_0 + \sqrt{\eta^\top F \eta}$ and $S_{\mathcal{X}}(-\eta) = -\eta^\top x_0 + \sqrt{\eta^\top F \eta}$, from where it follows that

$$S_{\mathcal{X}}(\eta) - S_{\mathcal{X}}(-\eta) = 2\eta^\top x_0, \tag{7.37}$$

for any $\eta \in \mathbb{R}^n$. Similarly, by using the expression in (7.36), we obtain that $S_{\mathcal{X}}(\eta) = \eta^\top x_0' + \sqrt{\eta^\top F' \eta}$ and $S_{\mathcal{X}}(-\eta) = -\eta^\top x_0' + \sqrt{\eta^\top F' \eta}$, from where it follows that

$$S_{\mathcal{X}}(\eta) - S_{\mathcal{X}}(-\eta) = 2\eta^\top x_0', \tag{7.38}$$

for any $\eta \in \mathbb{R}^n$. Then, by equating the right-hand side of (7.37) and (7.38), it follows that

$$\eta^\top (x_0 - x_0') = 0, \tag{7.39}$$

for any $\eta \in \mathbb{R}^n$; thus, $x_0 = x_0'$. Now, let η_0 denote an eigenvector of F associated with a zero eigenvalue; then, on the one hand, by using (7.35), we have that

$$S_{\mathcal{X}}(\eta_0) = \eta_0^\top x_0. \tag{7.40}$$

On the other hand, by using (7.36) and the fact that $x_0' = x_0$, we obtain that

$$\begin{aligned} S_{\mathcal{X}}(\eta_0) &= \eta_0^\top x_0' + \sqrt{\eta_0^\top F' \eta_0} \\ &= \eta_0^\top x_0 + \sqrt{\eta_0^\top F' \eta_0}. \end{aligned} \tag{7.41}$$

But since the matrix F' is positive definite, we have that $\sqrt{\eta_0^\top F' \eta_0} > 0$; therefore, $S_{\mathcal{X}}(\eta_0) > \eta_0^\top x_0$, which contradicts (7.40).

In this case, it is obvious that we cannot use the expression in (7.29) to describe $\mathcal{X}$. However, recall the alternative description of a closed convex set $\mathcal{X} \subseteq \mathbb{R}^n$ given in (2.61):

$$\mathcal{X} = \left\{ x \in \mathbb{R}^n : \eta^\top x \leq S_{\mathcal{X}}(\eta), \quad \eta \in \mathcal{B} \right\}, \tag{7.42}$$

where $\mathcal{B} = \{\eta \in \mathbb{R}^n : \eta^\top \eta = 1\}$. Then, by using (7.35), it follows that

$$\begin{aligned} \mathcal{X} &= \left\{ x \in \mathbb{R}^n : \eta^\top x \leq S_{\mathcal{X}}(\eta), \quad \eta^\top \eta = 1 \right\} \\ &= \left\{ x \in \mathbb{R}^n : \eta^\top x \leq \eta^\top x_0 + \sqrt{\eta^\top F \eta}, \quad \eta^\top \eta = 1 \right\}. \end{aligned} \tag{7.43}$$

The interpretation of the description in (7.43) is as follows. For each η, the inequality $\eta^\top x \leq \eta^\top x_0 + \sqrt{\eta^\top F \eta}$ defines a half space, which is determined by the hyperplane

$$\mathcal{H}(\eta) = \left\{ x \in \mathbb{R}^n : \eta^\top x = \eta^\top x_0 + \sqrt{\eta^\top F \eta} \right\},$$

whose normal vector is η. As discussed in Section 2.2.2, $\mathcal{H}(\eta)$ touches the boundary of $\mathcal{X}$. Thus, as η varies, $\mathcal{H}(\eta)$ will sweep the boundary of $\mathcal{X}$. The following example illustrates these ideas.

Example 7.2 (Linear circuit with ellipsoidal input set and non-ellipsoidal state set) Consider again the circuit in Fig. 7.1 and assume that $r_1 = r_2 = 0.5\ \Omega$, $i_l = 0$, and $v_s \in [9.5, 10.5]$ V. Then, by letting $x = [v_2, i_1]^\top$, and $w = v_s$, we have that

$$x = Hw, \tag{7.44}$$

where

$$H = \begin{bmatrix} \frac{1}{2} \\ 1 \end{bmatrix}, \tag{7.45}$$

$$w \in \mathcal{W} = \left\{ w \in \mathbb{R}\colon E^{-1}(w - w_0)^2 \leq 1 \right\}, \tag{7.46}$$

with $E = \frac{1}{4}$, and $w_0 = 10$. We are interested in finding the set $\mathcal{X}$ that contains all possible values that x can take.

On the one hand, one can compute $\mathcal{X}$ as follows. Since $v_2 = \frac{1}{2}w$ and $i_1 = w$, we have that $i_1 = 2v_2$, where $v_2 \in [4.75, 5.25]$; thus, we can write

$$\mathcal{X} = \left\{ \begin{bmatrix} v_2 \\ i_1 \end{bmatrix} : i_1 = 2v_2,\ 4.75 \leq v_2 \leq 5.25 \right\}, \tag{7.47}$$

i.e., $\mathcal{X}$ is a line segment as depicted in Fig. 7.3. On the other hand, by tailoring (7.35–7.43) to the setting in (7.44–7.46), we have that

$$\begin{aligned} S_{\mathcal{X}}(\eta) &= \eta^\top H w_0 + \sqrt{\eta^\top H E H^\top \eta} \\ &= [\eta_1, \eta_2] \begin{bmatrix} 5 \\ 10 \end{bmatrix} + \frac{1}{2} \left| [\eta_1, \eta_2] \begin{bmatrix} \frac{1}{2} \\ 1 \end{bmatrix} \right|, \end{aligned}$$

and

$$\mathcal{X} = \left\{ \begin{bmatrix} v_2 \\ i_1 \end{bmatrix} : \eta_1 v_2 + \eta_2 i_1 \leq 5\eta_1 + 10\eta_2 + 0.5\left|0.5\eta_1 + \eta_2\right|,\ \eta_1^2 + \eta_2^2 = 1 \right\}. \tag{7.48}$$

The characterizations of $\mathcal{X}$ in (7.47) and (7.48) are equivalent. In fact, as discussed earlier, the constraint in (7.48) parametrizes all half spaces in $\mathbb{R}^2$ containing $\mathcal{X}$ with the lines (hyperplanes in $\mathbb{R}^2$) defining the boundary of each such half space touching the boundary of $\mathcal{X}$. To visualize this, consider the following values of η and the corresponding conditions that, according to (7.48), v_2 and i_1 must satisfy:

$$\begin{aligned} \eta &= [1, 0]^\top : \ v_2 \leq 5.25, \\ \eta &= [-1, 0]^\top : \ v_2 \geq 4.75, \\ \eta &= [0, 1]^\top : \ i_1 \leq 10.5, \\ \eta &= [0, -1]^\top : \ i_1 \geq 9.5, \end{aligned}$$

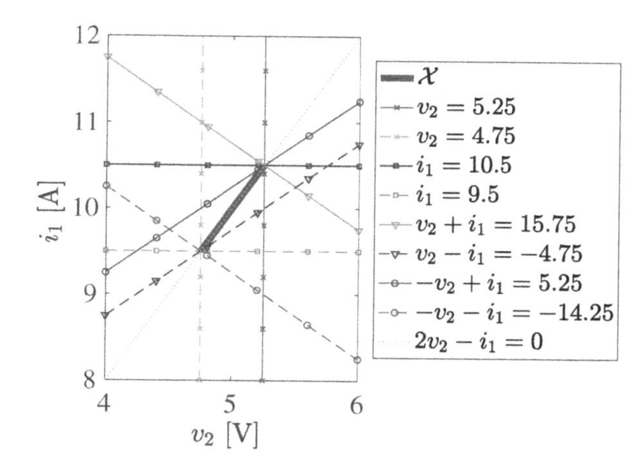

Figure 7.3 Set $\mathcal{X}$ for the linear DC circuit in Example 7.2, and boundaries of several half spaces containing its boundary.

$$\eta = \frac{1}{\sqrt{2}}[1, 1]^\top : \ v_2 + i_1 \leq 15.75,$$
$$\eta = \frac{1}{\sqrt{2}}[1, -1]^\top : \ v_2 - i_1 \leq -4.75,$$
$$\eta = \frac{1}{\sqrt{2}}[-1, 1]^\top : \ -v_2 + i_1 \leq 5.25,$$
$$\eta = \frac{1}{\sqrt{2}}[-1, -1]^\top : \ -v_2 - i_1 \leq -14.25,$$
$$\eta = \frac{1}{\sqrt{5}}[2, -1]^\top : \ 2v_2 - i_1 \leq 0;$$

these define several half spaces whose boundaries are depicted in Fig. 7.3 along with $\mathcal{X}$. One can easily check that such half spaces contain $\mathcal{X}$ while, as shown in Fig. 7.3, the lines defining the corresponding boundaries touch the boundary of $\mathcal{X}$.

Deterministic Inputs

In the developments above, we have not considered the effect of deterministic inputs, but these can be easily handled as follows. Consider a system of the form:

$$x = Hw + Lu, \tag{7.49}$$

where $H \in \mathbb{R}^{n \times m}$ is as in (7.6), $w \in \mathcal{W}$, with $\mathcal{W}$ as in (7.7), and $L \in \mathbb{R}^{n \times l}$ and $u \in \mathbb{R}^l$ are known. Define $z = x - Lu$, then we have that

$$z = Hw, \quad w \in \mathcal{W}, \tag{7.50}$$

which is of the form in (7.6–7.7). Thus,

$$z \in \mathcal{Z} = \left\{ z \in \mathbb{R}^n : (z - Hw_0)^\top \left(HEH^\top\right)^{-1} (z - Hw_0) \leq 1 \right\}, \tag{7.51}$$

if $m \geq n$ (Case E1 above), and

$$z \in \mathcal{Z} = \left\{ z \in \mathbb{R}^n : \eta^\top z \leq \eta^\top H w_0 + \sqrt{\eta^\top H E H^\top \eta}, \quad \eta^\top \eta = 1 \right\}, \tag{7.52}$$

if $m < n$ (Case E2 above). Then, since $x = z + Lu$, it follows that

$$\mathcal{X} = \{x^\circ\} + \mathcal{Z}, \tag{7.53}$$

where $x^\circ = Lu$.

7.2.2 Nonlinear Setting

Consider a system of the form

$$x = h(w), \tag{7.54}$$

where $x \in \mathbb{R}^n$, $w \in \mathbb{R}^m$ belongs to an ellipsoid, $\mathcal{W}$, with center $w_0 \in \mathbb{R}^m$, and shape matrix $E \in \mathbb{R}^{m \times m}$, i.e.,

$$\mathcal{W} = \left\{ w \in \mathbb{R}^m : (w - w_0)^\top E^{-1} (w - w_0) \leq 1 \right\}, \tag{7.55}$$

and $h \colon \mathbb{R}^m \to \mathbb{R}^n$. Assume that the Jacobian of $h(w)$, denoted by $\frac{\partial h(w)}{\partial w}$, is full rank for all $w \in \mathcal{W}$. The objective is then to obtain a mathematical description of the set $\mathcal{X}$ containing all possible values that x can take. Unlike the linear case, obtaining an exact characterization of $\mathcal{X}$ is in general difficult; thus, instead, we will settle for an approximate characterization of $\mathcal{X}$. This approximation is obtained by relying on a linear model derived using the same linearization technique as in Section 3.2.2, which we recall next.

Let the pair w^*, x^* satisfy $x^* = h(w^*)$, and let Δx denote a small variation in x around x^* that arises from a small variation in w around w^*, which we similarly denote by Δw; then, we have that

$$x^* + \Delta x = h(w^* + \Delta w). \tag{7.56}$$

Now, by using the Taylor series expansion of $h(\cdot)$ around w^* and only keeping the first-order term, we obtain that

$$\Delta x \approx H(w^*) \Delta w, \tag{7.57}$$

where

$$H(w^*) = \left. \frac{\partial h(w)}{\partial w} \right|_{w = w^*} \in \mathbb{R}^{n \times m}.$$

Here, we also need to consider two cases as follows:

F1. The dimension of x is smaller than or equal to the dimension of w, i.e., $m \geq n$.

F2. The dimension of x is greater than the dimension of w, i.e., $m < n$.

Case F1. Let $\Delta w = w - w_0$ and $\Delta x = x - h(w_0)$; then it follows from (7.55) and (7.57) that

$$\begin{aligned} \Delta x &\approx H(w_0)\Delta w, \\ \Delta w &\in \Delta\mathcal{W} = \{\Delta w \colon \Delta w^\top E^{-1}\Delta w \leq 1\}. \end{aligned} \tag{7.58}$$

where $H(w_0) = \frac{\partial h(w)}{\partial w}\Big|_{w=w_0} \in \mathbb{R}^{n\times m}$, $m \geq n$. Now, define $\widetilde{\Delta x} := H(w_0)\Delta w$; then it follows from the developments for Case E1 in Section 7.2.1 that

$$\widetilde{\Delta x} \in \widetilde{\Delta\mathcal{X}} = \left\{\widetilde{\Delta x} \in \mathbb{R}^n \colon \left(\widetilde{\Delta x}\right)^\top F^{-1}(w_0)\widetilde{\Delta x} \leq 1\right\}, \tag{7.59}$$

where $F(w_0) = H(w_0)EH^\top(w_0)$. Finally, since $\Delta x = x - x_0 \approx \widetilde{\Delta x}$, where $x_0 = h(w_0)$, we have that $x \approx h(w_0) + \widetilde{\Delta x}$, and it follows that

$$\begin{aligned} x \in \mathcal{X} &\approx \{x_0\} + \widetilde{\Delta\mathcal{X}} \\ &= \left\{x \in \mathbb{R}^n \colon \left(x - h(w_0)\right)^\top F^{-1}(w_0)\left(x - h(w_0)\right) \leq 1\right\}. \end{aligned} \tag{7.60}$$

Case F2. As in Case F1, $\Delta w = w - w_0$ and $\Delta x = x - h(w_0)$; then,

$$\begin{aligned} \Delta x &\approx H(w_0)\Delta w, \\ \Delta w &\in \Delta\mathcal{W} = \{\Delta w \colon \Delta w^\top E^{-1}\Delta w \leq 1\}, \end{aligned} \tag{7.61}$$

where $H(w_0) = \frac{\partial h(w)}{\partial w}\Big|_{w=w_0} \in \mathbb{R}^{n\times m}$, $m < n$. Now, define $\widetilde{\Delta x} := H(w_0)\Delta w$; then, from the developments for Case E2 in Section 7.2.1, we obtain that

$$\widetilde{\Delta x} \in \widetilde{\Delta\mathcal{X}} = \left\{\widetilde{\Delta x} \in \mathbb{R}^n \colon \eta^\top \widetilde{\Delta x} \leq \sqrt{\eta^\top H(w_0)EH^\top(w_0)\eta}, \quad \eta^\top\eta = 1\right\}. \tag{7.62}$$

Finally, since $\Delta x = x - x_0 \approx \widetilde{\Delta x}$, where $x_0 = h(w_0)$, we have that

$$x \approx h(w_0) + \widetilde{\Delta x},$$

and it follows that

$$\begin{aligned} x \in \mathcal{X} &\approx \{x_0\} + \widetilde{\Delta\mathcal{X}} \\ &= \left\{x \in \mathbb{R}^n \colon \eta^\top x \leq \eta^\top h(w_0) + \sqrt{\eta^\top H(w_0)EH^\top(w_0)\eta}, \quad \eta^\top\eta = 1\right\}. \end{aligned} \tag{7.63}$$

Example 7.3 (Nonlinear resistor) Consider a nonlinear resistor with the relation between the voltage across its terminals, v, and the current flowing through it, i, given by

$$v = ri^{1/3}, \tag{7.64}$$

with $r > 0$. Assume that we connect an ideal voltage source across the nonlinear resistor terminals with its terminal voltage, v_s, taking values in the interval $[9, 11]$ V. We are interested in obtaining the interval that contains the possible values that i can take, which we denote by $\mathcal{I}$. Then, by inverting the relation in (7.64), we obtain

$$i = \left(\frac{v_s}{r}\right)^3, \tag{7.65}$$

with $v_s \in \{v_s \colon (v_s - 10)^2 \leq 1\}$. One can directly compute the end points of the interval containing all possible values that i can take by plugging the end points of the interval $[9, 11]$ V into (7.65); this calculation yields

$$\left(\frac{9}{r}\right)^3 \leq i \leq \left(\frac{11}{r}\right)^3. \tag{7.66}$$

Alternatively, we can use the procedure outlined in (7.58–7.60) to obtain an approximation to the interval containing all possible values that i can take, which we denote by $\widetilde{\mathcal{I}}$. To this end, by letting $h(v_s) = \left(\frac{v_s}{r}\right)^3$, we have that

$$\begin{aligned} H(v_s) &:= \frac{dh(v_s)}{dv_s} \\ &= \frac{3v_s^2}{r^3}; \end{aligned} \tag{7.67}$$

thus, by using (7.60), and noting that in this case $E = 1$, we obtain that

$$\begin{aligned} i \in \mathcal{I} &\approx \{h(10)\} + \left\{\widetilde{\Delta i} \colon H^{-2}(10)\left(\widetilde{\Delta i}\right)^2 \leq 1\right\} \\ &= \left\{i \colon H^{-2}(10)\left(i - h(10)\right)^2 \leq 1\right\} \\ &= \left\{i \colon \frac{r^6}{300^2}\left(i - \frac{1000}{r^3}\right)^2 \leq 1\right\} \\ &=: \widetilde{\mathcal{I}}. \end{aligned} \tag{7.68}$$

Now, let $r = 1$ Ω; then, by using (7.66) and (7.68), we obtain

$$\begin{aligned} \mathcal{I} &= \{i \colon (i - 1030)^3 \leq 90{,}601\}, \\ \widetilde{\mathcal{I}} &= \{i \colon (i - 1000)^2 \leq 90{,}000\}. \end{aligned} \tag{7.69}$$

Similarly, for $r = 2$ Ω, we obtain

$$\begin{aligned} \mathcal{I} &= \{i \colon (i - 128.75)^3 \leq 1{,}415.64\}, \\ \widetilde{\mathcal{I}} &= \{i \colon (i - 125)^2 \leq 1{,}406.25\}. \end{aligned} \tag{7.70}$$

Example 7.4 (Nonlinear circuit with ellipsoidal input set) Consider the circuit in Fig. 7.4. Both voltage and current sources are linear and their terminal behavior is the same as those in Example 7.1. The element interconnecting them is a nonlinear resistor similar to that in Example 7.3; thus the relation between the voltage across its terminals, v_1, and the current flowing through it, i_1, is given by

$$v_1 = r_1 i_1^{1/3}. \tag{7.71}$$

Assume that $r_1 = 1$ Ω, $v_s = 10$ V, and 3 A $\leq i_l \leq$ 5 A; thus,

$$i_l \in \left\{i_l \colon (i_l - 4)^2 \leq 1\right\}.$$

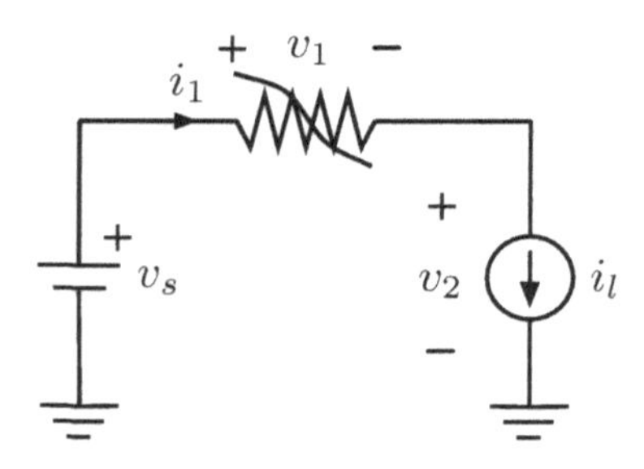

Figure 7.4 Nonlinear DC circuit.

Then, by using Kirchhoff's voltage law, we can obtain the mapping between i_l, and $x = [v_2, i_1]^\top$ as follows:

$$\begin{aligned} v_2 &= h_1(i_l) \\ &= \begin{cases} 10 - i_l^{1/3}, & \text{if } i_l \geq 0, \\ 10 + (-i_l)^{1/3}, & \text{if } i_l < 0; \end{cases} \end{aligned} \tag{7.72}$$

thus, we have that

$$\begin{aligned} \begin{bmatrix} v_2 \\ i_1 \end{bmatrix} &= \begin{bmatrix} h_1(i_l) \\ h_2(i_l) \end{bmatrix} \\ &= \begin{bmatrix} 10 - i_l^{1/3} \\ i_l \end{bmatrix}. \end{aligned} \tag{7.73}$$

Now, by taking the partial derivative of $h_1(\cdot)$ and $h_2(\cdot)$ with respect to i_l, we obtain

$$\begin{aligned} \frac{\partial h_1(i_l)}{\partial i_l} &= \frac{-1}{3\sqrt[3]{i_l^2}}, \\ \frac{\partial h_2(i_l)}{\partial i_l} &= 1. \end{aligned} \tag{7.74}$$

Then, by tailoring (7.61) to the setting here, we obtain

$$\begin{aligned} \begin{bmatrix} \Delta v_2 \\ \Delta i_1 \end{bmatrix} &\approx \begin{bmatrix} \left.\frac{\partial h_1(i_l)}{\partial i_l}\right|_{i_l=4} \\ \left.\frac{\partial h_2(i_l)}{\partial i_l}\right|_{i_l=4} \end{bmatrix} \Delta i_l \\ &= \begin{bmatrix} -\frac{1}{3\sqrt[3]{16}} \\ 1 \end{bmatrix} \Delta i_l \\ &=: \begin{bmatrix} \Delta \widetilde{v}_2 \\ \Delta \widetilde{i}_1 \end{bmatrix}, \end{aligned} \tag{7.75}$$

and by using (7.63), noting that in this case $E = 1$, we obtain that

$$\mathcal{X} \approx \left\{ \begin{bmatrix} v_2 \\ i_1 \end{bmatrix} : \eta_1 v_2 + \eta_2 i_1 \leq (10 - 4^{1/3})\eta_1 \right.$$
$$\left. + 4\eta_2 + \left| -\frac{\eta_1}{3\sqrt[3]{16}} + \eta_2 \right|, \quad \eta_1^2 + \eta_2^2 = 1 \right\}. \qquad (7.76)$$

Example 7.5 (Two-bus power system with ellipsoidal input set) Consider the two-bus power system in Fig. 7.5 and assume it is operating in quasi-steady-state sinusoidal regime; thus, it can be described by the standard power flow model (see Appendix B). The network admittance matrix in this case is as follows:

$$\overline{Y} = G + jB, \qquad (7.77)$$

where

$$G = \frac{1}{R_{12}^2 + X_{12}^2}\begin{bmatrix} R_{12} & -R_{12} \\ -R_{12} & R_{12} \end{bmatrix}, \quad B = \frac{1}{R_{12}^2 + X_{12}^2}\begin{bmatrix} -X_{12} & X_{12} \\ X_{12} & -X_{12} \end{bmatrix}, \qquad (7.78)$$

with $R_{12} > 0$ and $X_{12} > 0$ being respectively the series resistance and reactance of the transmission line between buses 1 and 2.

Assume that the voltage magnitude and phase angle of the voltage source on the left are known constants, denoted by V_0 and δ, respectively. Also, assume that the active and reactive power withdrawn from the network at bus 2, denoted by P_2^D and Q_2^D, respectively, are known. Finally, assume that the active and reactive power injected into the network at bus 2, denoted by p_2^G and q_2^G, respectively, are uncertain but known to belong to an ellipsoid with center $w_0 = [P_2^G, Q_2^G]^\top$ and shape matrix $E \in \mathbb{R}^{2\times 2}$. Let v_2 and θ_2 denote the magnitude and phase angle of the phasor associated with the voltage at bus 2. Define $x = [\theta_2, v_2]^\top$, $w = \left[p_2^G, q_2^G\right]^\top$, $u = \left[P_2^D, Q_2^D\right]^\top$; then, we have that

$$w = g(x) + u, \qquad (7.79)$$

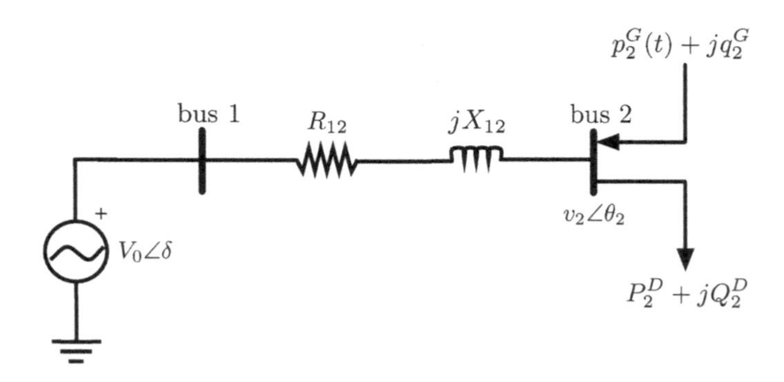

Figure 7.5 Two-bus power system.

where $g(x) = \big[g_1(x), g_2(x)\big]^\top$, with $g_1(\cdot)$ and $g_2(\cdot)$ defined as follows:

$$g_1(x) = \frac{v_2}{R_{12}^2 + X_{12}^2}\left(R_{12}v_2 + \Big(-R_{12}\cos(\theta_2 - \delta) + X_{12}\sin(\theta_2 - \delta)\Big)V_0\right),$$

$$g_2(x) = \frac{v_2}{R_{12}^2 + X_{12}^2}\left(X_{12}v_2 - \Big(R_{12}\sin(\theta_2 - \delta) + X_{12}\cos(\theta_2 - \delta)\Big)V_0\right), \quad (7.80)$$

and

$$w \in \mathcal{W} = \Big\{w \in \mathbb{R}^2 \colon (w - w_0)^\top E^{-1}(w - w_0) \leq 1\Big\}. \quad (7.81)$$

The objective here is to approximately characterize the set $\mathcal{X}$ containing all possible values that x can take. Note that (7.79–7.81) do not match the form of the setting in (7.54–7.55). However, we can use a linearization approach similar to the one used for handling (7.54–7.55). Let x_0 denote the value of x that satisfies (7.79) for $w = w_0$. Let $\Delta x = x - x_0$, then, we have that

$$w_0 + \Delta w = g(x_0 + \Delta x) + u, \quad (7.82)$$

where

$$\Delta w \in \Delta\mathcal{W} = \Big\{\Delta w \in \mathbb{R}^2 \colon (\Delta w)^\top E^{-1}\Delta w \leq 1\Big\}. \quad (7.83)$$

Now, by using the Taylor series expansion of $g(\cdot)$ around x_0 and only keeping the first-order term, we obtain that

$$\Delta w \approx G(x_0)\Delta x, \quad (7.84)$$

where $G(x_0) = \left.\frac{\partial g(x)}{\partial x}\right|_{x=x_0} \in \mathbb{R}^{2\times 2}$. Assume that $G(x_0)$ is invertible; then it follows from the inverse function theorem (see Appendix A.3) that

$$\Delta x \approx H(x_0)\Delta w, \quad (7.85)$$

where $\Delta w \in \Delta\mathcal{W}$ and $H(x_0) = G^{-1}(x_0)$. Now define $\widetilde{\Delta x} := H(x_0)\Delta w$; then, we have that

$$\widetilde{\Delta x} \in \widetilde{\Delta\mathcal{X}} = \Big\{\widetilde{\Delta x} \in \mathbb{R}^2 \colon \big(\widetilde{\Delta x}\big)^\top F^{-1}(x_0)\widetilde{\Delta x} \leq 1\Big\}, \quad (7.86)$$

where $F(x_0) = H(x_0)EH^\top(x_0)$. Finally, since $\Delta x = x - x_0 \approx \widetilde{\Delta x}$, we have that

$$\begin{aligned} x \in \mathcal{X} &\approx \{x_0\} + \widetilde{\Delta\mathcal{X}} \\ &= \Big\{x \in \mathbb{R}^2 \colon \big(x - x_0\big)^\top F^{-1}(x_0)\big(x - x_0\big) \leq 1\Big\}. \end{aligned} \quad (7.87)$$

Letting $w_0 = [0.3 \text{ p.u.}, 0 \text{ p.u.}]^\top$ and using the parameter values in Table 7.1 yields $x_0 = [-0.0011 \text{ rad}, 0.9564 \text{ p.u.}]^\top$ and

Table 7.1 Two-bus System Model Parameter Values

V_0 [p.u.]	δ [rad]	R_{12} [p.u.]	X_{12} [p.u.]	P_2^D [p.u.]	Q_2^D [p.u.]
0.97	0	0.01	0.02	0.8	0.5

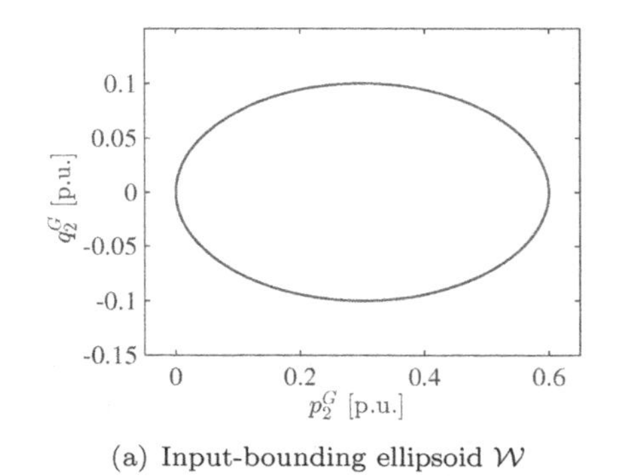

(a) Input-bounding ellipsoid $\mathcal{W}$

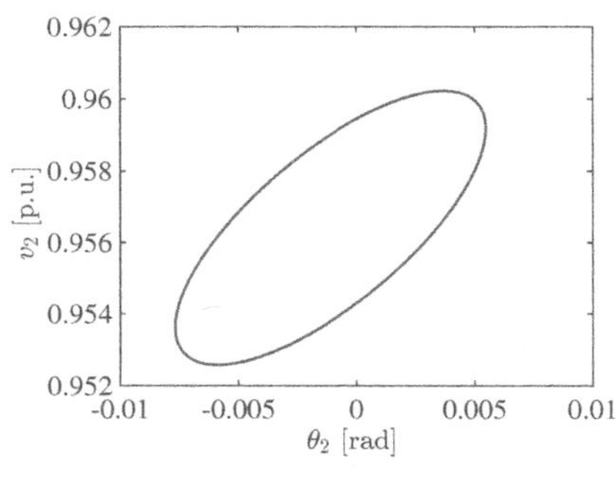

(b) State-bounding ellipsoid $\mathcal{X}$

Figure 7.6 Bounding ellipsoids for the two-bus power system in Example 7.5.

$$\begin{aligned} H(x_0) &= G^{-1}(x_0) \\ &= \begin{bmatrix} 37.0883 & 18.8142 \\ -18.5942 & 37.7329 \end{bmatrix}^{-1} \\ &= \begin{bmatrix} 0.0216 & -0.0108 \\ 0.0106 & 0.0212 \end{bmatrix}. \end{aligned} \tag{7.88}$$

Assume that p_2^G can vary ± 0.3 p.u. around $P_2^G = 0.3$ p.u. and q_2^G can vary ± 0.1 p.u. around $Q_2^G = 0$ p.u., then, it follows that

$$E = 10^{-2} \begin{bmatrix} 9 & 0 \\ 0 & 1 \end{bmatrix}; \tag{7.89}$$

see Fig. 7.6(a) for a graphical depiction of the resulting ellipsoid. Finally, we have that

$$\begin{aligned} F(x_0) &= H(x_0) E H^\top(x_0) \\ &= 10^{-4} \begin{bmatrix} 0.4303 & 0.1836 \\ 0.1836 & 0.1466 \end{bmatrix}; \end{aligned} \tag{7.90}$$

see Fig. 7.6(b) for a graphical depiction of the resulting ellipsoid.

7.3 Zonotope-Based Input Set Description

In this section, we consider Scenario S2, i.e., the set containing all possible values that the input can take is a zonotope. For the linear case, we develop techniques to compute the exact set containing all possible values the state can take, while for the nonlinear case, we develop techniques to compute an approximation thereof.

7.3.1 Linear Setting

Consider a linear static system of the form

$$x = Hw, \tag{7.91}$$

where $H \in \mathbb{R}^{n\times m}$ is full rank, $x \in \mathbb{R}^n$, and $w \in \mathcal{W} \subseteq \mathbb{R}^m$, where $\mathcal{W}$ is a zonotope with center $w_0 \in \mathbb{R}^m$, and generators $e_1, e_2, \ldots, e_s \in \mathbb{R}^m$, i.e.,

$$\mathcal{W} = \left\{ w \in \mathbb{R}^m \colon w = w_0 + \sum_{j=1}^{s} \alpha_j e_j, \; -1 \leq \alpha_j \leq 1 \right\}; \tag{7.92}$$

thus, the support function of $\mathcal{W}$, $S_{\mathcal{W}}(\cdot)$, is given by (see (2.83))

$$S_{\mathcal{W}}(\eta) = \eta^\top w_0 + \sum_{j=1}^{s} |\eta^\top e_j|, \quad \eta \in \mathbb{R}^m. \tag{7.93}$$

As in Scenario S1, the goal here is to characterize the set $\mathcal{X}$ containing all possible values that the state, x, can take.

In this case, since any $w \in \mathcal{W}$ can be written as a linear combination of the e_j's, i.e., $w = w_0 + \sum_{j=1}^{s} \alpha_j e_j, \; -1 \leq \alpha_j \leq 1$, we have that:

$$\begin{aligned} x &= Hw \\ &= H\left(w_0 + \sum_{j=1}^{s} \alpha_j e_j\right) \\ &= Hw_0 + \sum_{j=1}^{s} \alpha_i H e_j. \end{aligned} \tag{7.94}$$

Now, by defining $x_0 = Hw_0$ and $f_j = He_j, \; j = 1, 2, \ldots, s$, we have that the set $\mathcal{X}$ containing all possible values that x can take can be described as follows:

$$\mathcal{X} = \left\{ x \in \mathbb{R}^n \colon x = x_0 + \sum_{j=1}^{s} \alpha_j f_j, \quad -1 \leq \alpha_j \leq 1 \right\}, \tag{7.95}$$

which is also a zonotope with center $x_0 \in \mathbb{R}^n$ and generators $f_1, f_2, \ldots, f_s \in \mathbb{R}^n$. In light of (7.93) and (7.95), we have that the support function of $\mathcal{X}$, denoted by $S_{\mathcal{X}}(\cdot)$, is given by

$$S_{\mathcal{X}}(\eta) = \eta^\top x_0 + \sum_{j=1}^{s} |\eta^\top f_j|, \quad \eta \in \mathbb{R}^n. \tag{7.96}$$

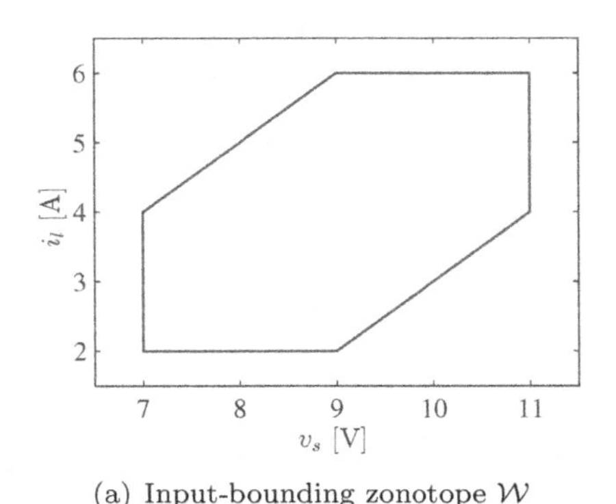

(a) Input-bounding zonotope $\mathcal{W}$

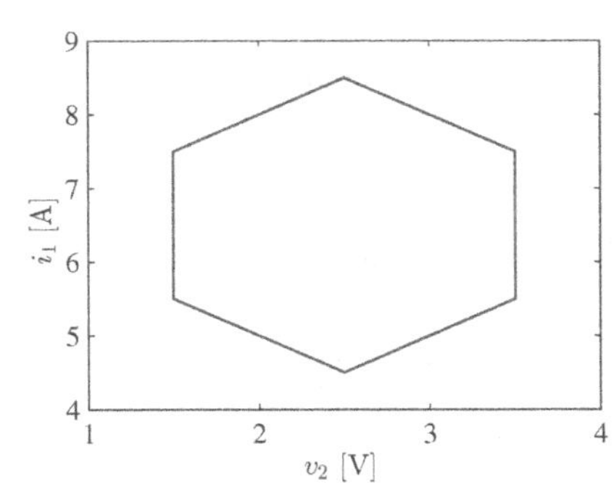

(b) State-bounding zonotope $\mathcal{X}$

Figure 7.7 Bounding zonotopes for the linear DC circuit in Example 7.6.

Example 7.6 (Linear circuit with zonotope-based input) Consider again the circuit in Fig. 7.1 and, as in Example 7.1, assume that $r_1 = r_2 = 1\ \Omega$. Then, recalling from (7.30), we have that

$$\underbrace{\begin{bmatrix} v_2 \\ i_1 \end{bmatrix}}_{=:\,x} = \underbrace{\begin{bmatrix} \frac{1}{2} & -\frac{1}{2} \\ \frac{1}{2} & \frac{1}{2} \end{bmatrix}}_{=:\,H} \underbrace{\begin{bmatrix} v_s \\ i_l \end{bmatrix}}_{=:\,w}. \tag{7.97}$$

Assume that w is contained in a zonotope $\mathcal{W}$ with center $w_0 = [9, 4]^\top$, and generators $e_1 = [1, 0]^\top$, $e_2 = [0, 1]^\top$, and $e_3 = [1, 1]^\top$, i.e.,

$$w \in \mathcal{W} = \left\{ w \in \mathbb{R}^2 \colon w = w_0 + \sum_{j=1}^{3} \alpha_j e_j, \quad -1 \leq \alpha_j \leq 1 \right\}; \tag{7.98}$$

see Fig. 7.7(a) for a graphical representation. Now, by tailoring (7.95) to the setting here, it follows that x also belongs to a zonotope $\mathcal{X}$ with center

$$x_0 = H w_0 = [2.5, 6.5]^\top,$$

and generators as follows: $f_1 = He_1 = [0.5, 0.5]^\top$, $f_2 = He_2 = [-0.5, 0.5]^\top$, and $f_3 = He_3 = [0, 1]^\top$, i.e.,

$$\mathcal{X} = \left\{ x \in \mathbb{R}^n \colon x = x_0 + \sum_{j=1}^{3} \alpha_j f_j, \quad -1 \leq \alpha_j \leq 1 \right\}; \tag{7.99}$$

see Fig. 7.7(b) for its depiction.

Deterministic Inputs

Consider a system of the form:

$$x = Hw + Lu, \tag{7.100}$$

where $H \in \mathbb{R}^{n \times m}$ is as in (7.91), $w \in \mathcal{W}$, with $\mathcal{W}$ as in (7.92), and $L \in \mathbb{R}^{n \times l}$ and $u \in \mathbb{R}^l$ are known. Define $z = x - Lu$, then, we have that

$$z = Hw, \quad w \in \mathcal{W}, \tag{7.101}$$

which is of the form in (7.91–7.92). Thus,

$$z \in \mathcal{Z} = \left\{ z \in \mathbb{R}^n : z = Hw_0 + \sum_{j=1}^{s} \alpha H e_j, \quad -1 \leq \alpha_j \leq 1 \right\}. \tag{7.102}$$

Then, since $x = z + Lu$, it follows that

$$\mathcal{X} = \{x^\circ\} + \mathcal{Z}, \tag{7.103}$$

where $x^\circ = Lu$.

7.3.2 Nonlinear Setting

Consider a nonlinear static system of the form

$$x = h(w), \tag{7.104}$$

where $x \in \mathbb{R}^n$, $w \in \mathbb{R}^m$ belongs to a zonotope, $\mathcal{W}$, with center $w_0 \in \mathbb{R}^m$, and generators $e_1, e_2, \dots, e_s \in \mathbb{R}^m$, i.e.,

$$\mathcal{W} = \left\{ w \in \mathbb{R}^m : w = w_0 + \sum_{j=1}^{s} \alpha_j e_j, \ -1 \leq \alpha_j \leq 1 \right\}, \tag{7.105}$$

and $h \colon \mathbb{R}^m \to \mathbb{R}^n$. Assume the Jacobian of $h(w)$, which we denote by $\frac{\partial h(w)}{\partial w}$, is full rank for all $w \in \mathcal{W}$. The objective here is to obtain a set $\widetilde{\mathcal{X}}$ approximating the set $\mathcal{X}$ containing all possible values that x can take. As in Section 7.2.2, we will leverage the Taylor series expansion of $h(\cdot)$ around w_0 – the center of the zonotope $\mathcal{W}$.

Let $\Delta w = w - w_0$ and $\Delta x = x - h(w_0)$; then it follows from (7.57) and (7.105) that

$$\Delta x \approx H(w_0)\Delta w,$$

$$\Delta w \in \Delta\mathcal{W} = \left\{ \Delta w \colon \Delta w = \sum_{j=1}^{s} \alpha_j e_j, \ -1 \leq \alpha_j \leq 1 \right\}, \tag{7.106}$$

where $H(w_0) = \left.\frac{\partial h(w)}{\partial w}\right|_{w=w_0} \in \mathbb{R}^{n \times m}$. Now, define $\widetilde{\Delta x} := H(w_0)\Delta w$; then it follows from the developments in Section 7.3.1 that

$$\widetilde{\Delta x} \in \widetilde{\Delta \mathcal{X}} = \left\{ \widetilde{\Delta x} \in \mathbb{R}^n : \widetilde{\Delta x} = \sum_{j=1}^{s} \alpha_j f_j, \ -1 \leq \alpha_j \leq 1 \right\}, \tag{7.107}$$

where $f_j = H(w_0)e_j$, $j = 1, 2, \dots, s$. Finally, since $\Delta x = x - x_0 \approx \widetilde{\Delta x}$, where $x_0 = h(w_0)$, we have that $x \approx h(w_0) + \widetilde{\Delta x}$; thus it follows that

$$x \in \mathcal{X} \approx \{x_0\} + \widetilde{\Delta\mathcal{X}}$$
$$= \left\{ x \in \mathbb{R}^n : x = h(w_0) + \sum_{j=1}^{s} \alpha_j f_j, \quad -1 \leq \alpha_j \leq 1 \right\}. \tag{7.108}$$

Example 7.7 (Nonlinear circuit with zonotope input set) Consider again the circuit in Fig. 7.4. Assume that the value of $w = [v_s, i_l]^\top$ is uncertain but known to take values in the zonotope

$$\mathcal{W} = \left\{ w \in \mathbb{R}^2 : w = w_0 + \sum_{j=1}^{3} \alpha_j e_j, \quad -1 \leq \alpha_j \leq 1 \right\}, \tag{7.109}$$

where $w_0 = [10, 4]^\top$, $e_1 = [1, 0]^\top$, $e_2 = [0, 1]^\top$, and $e_3 = [1, -1]^\top$; see Fig. 7.8(a) for a graphical depiction. Then, since

$$\begin{bmatrix} v_2 \\ i_1 \end{bmatrix} = \begin{bmatrix} h_1(v_s, i_l) \\ h_2(v_s, i_l) \end{bmatrix} = \begin{bmatrix} v_s - i_l^{1/3} \\ i_l \end{bmatrix}, \tag{7.110}$$

we have that

$$\begin{bmatrix} \Delta v_2 \\ \Delta i_1 \end{bmatrix} \approx \begin{bmatrix} \left.\frac{\partial h_1(v_s,i_l)}{\partial v_s}\right|_{v_s=10,i_l=4} & \left.\frac{\partial h_1(v_s,i_l)}{\partial i_l}\right|_{v_s=10,i_l=4} \\ \left.\frac{\partial h_2(v_s,i_l)}{\partial v_s}\right|_{v_s=10,i_l=4} & \left.\frac{\partial h_2(v_s,i_l)}{\partial i_l}\right|_{v_s=10,i_l=4} \end{bmatrix} \begin{bmatrix} \Delta v_s \\ \Delta i_l \end{bmatrix}$$
$$= \begin{bmatrix} 1 & -\frac{1}{3\sqrt[3]{16}} \\ 0 & 1 \end{bmatrix} \begin{bmatrix} \Delta v_s \\ \Delta i_l \end{bmatrix}$$
$$=: \begin{bmatrix} \widetilde{\Delta v_2} \\ \widetilde{\Delta i_1} \end{bmatrix}, \tag{7.111}$$

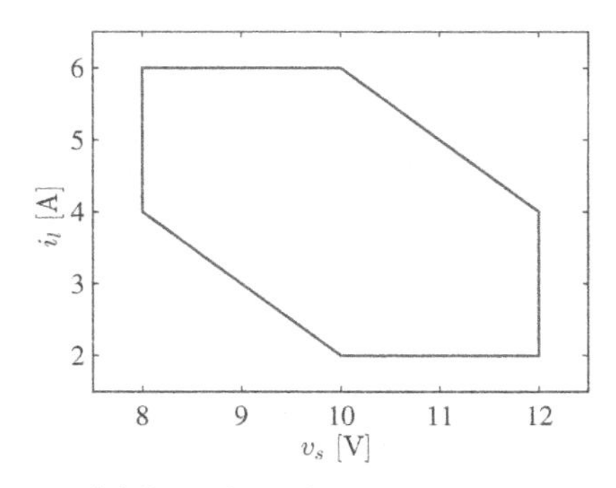

(a) Input-bounding zonotope $\mathcal{W}$

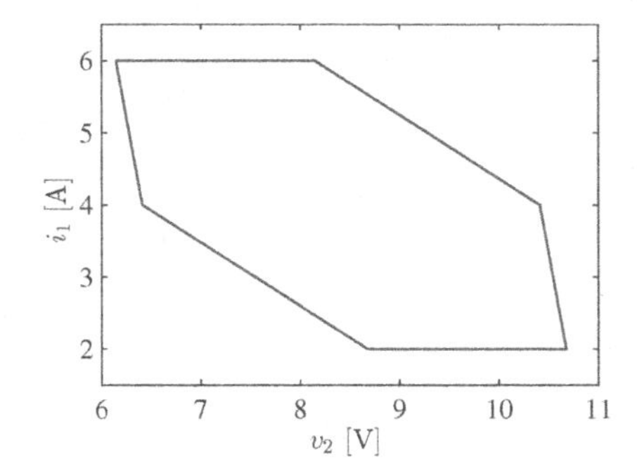

(b) Zonotope approximation to the state-bounding set $\mathcal{X}$

Figure 7.8 Bounding zonotopes for nonlinear DC circuit in Example 7.7.

where

$$[\Delta v_s, \Delta i_l]^\top \in \Delta\mathcal{W} = \left\{ \Delta w \in \mathbb{R}^2 : \Delta w = \sum_{j=1}^{3} \alpha_j e_j, \ -1 \leq \alpha_j \leq 1 \right\}. \quad (7.112)$$

Now, by using (7.107), we obtain that

$$\left[\widetilde{\Delta v}_2, \widetilde{\Delta i}_1\right]^\top \in \widetilde{\Delta\mathcal{X}} = \left\{ \widetilde{\Delta x} \in \mathbb{R}^2 : \widetilde{\Delta x} = \sum_{j=1}^{3} \alpha_j f_j, \ -1 \leq \alpha_j \leq 1 \right\}, \quad (7.113)$$

where

$$\begin{aligned} f_1 &= \begin{bmatrix} 1 & -\frac{1}{3\sqrt[3]{16}} \\ 0 & 1 \end{bmatrix} e_1 \\ &= \begin{bmatrix} 1 \\ 0 \end{bmatrix}, \\ f_2 &= \begin{bmatrix} 1 & -\frac{1}{3\sqrt[3]{16}} \\ 0 & 1 \end{bmatrix} e_2 \\ &= \begin{bmatrix} -\frac{1}{3\sqrt[3]{16}} \\ 1 \end{bmatrix}, \\ f_3 &= \begin{bmatrix} 1 & -\frac{1}{3\sqrt[3]{16}} \\ 0 & 1 \end{bmatrix} e_3 \\ &= \begin{bmatrix} 1 + \frac{1}{3\sqrt[3]{16}} \\ -1 \end{bmatrix}. \end{aligned} \quad (7.114)$$

Finally, by using (7.108). we obtain that

$$\begin{aligned} [v_2, i_1]^\top \in \mathcal{X} &\approx \{h(10,4)\} + \widetilde{\Delta\mathcal{X}} \\ &= \left\{ x \in \mathbb{R}^2 : x = h(10,4) + \sum_{j=1}^{3} \alpha_j f_j, \ -1 \leq \alpha_j \leq 1 \right\}, \end{aligned} \quad (7.115)$$

with $h(10,4) = [10 - 4^{1/3}, 4]^\top$. The set $\{h(10,4)\} + \widetilde{\Delta\mathcal{X}}$ is depicted in Fig. 7.8(b).

Example 7.8 (Two-bus power system with zonotope input set) Consider again the two-bus power system in Example 7.5 and recall the relation between the active and reactive power injection vector $w = [p_2^g, q_2^g]^\top$ and the voltage magnitude and angle vector $x = [\theta_2, v_2]^\top$ given in (7.79):

$$w = g(x) + u. \quad (7.116)$$

Assume that w belongs to a zonotope $\mathcal{W}$ with center $w_0 \in \mathbb{R}^2$ and generators $e_1, e_2, e_3 \in \mathbb{R}^2$, i.e.,

$$\mathcal{W} = \left\{ w \in \mathbb{R}^2 \colon w = w_0 + \sum_{j=1}^{3} \alpha_j e_j, \quad -1 \leq \alpha_j \leq 1 \right\}. \tag{7.117}$$

As in Example 7.5, the objective is to approximately characterize the set $\mathcal{X}$ containing all possible values that x can take. Let x_0 denote the value of x that satisfies (7.116) for $w = w_0$. Define $\Delta x = x - x_0$, then by following a procedure similar to that in (7.82–7.85), we obtain that

$$\Delta x \approx H(x_0)\Delta w, \tag{7.118}$$

where

$$H(x_0) = \left(\frac{\partial g(x)}{\partial x}\bigg|_{x=x_0} \right)^{-1} \tag{7.119}$$

and

$$\Delta w \in \Delta\mathcal{W} = \left\{ \Delta w \in \mathbb{R}^2 \colon \Delta w = \sum_{j=1}^{3} \alpha_j e_j, \quad -1 \leq \alpha_j \leq 1 \right\}. \tag{7.120}$$

Now, let $\widetilde{\Delta x} = H(x_0)\Delta w$; then we have that

$$\widetilde{\Delta x} \in \widetilde{\Delta \mathcal{X}} = \left\{ \widetilde{\Delta x} \in \mathbb{R}^2 \colon \widetilde{\Delta x} = \sum_{j=1}^{3} \alpha_j f_j, \quad -1 \leq \alpha_j \leq 1 \right\}, \tag{7.121}$$

where $f_j = H(x_0)e_j$, $j = 1, 2, 3$. Finally, since $\Delta x = x - x_0 \approx \widetilde{\Delta x}$, we have that

$$\begin{aligned} x \in \mathcal{X} &\approx \{x_0\} + \widetilde{\Delta \mathcal{X}} \\ &= \left\{ x \in \mathbb{R}^2 \colon x = x_0 + \sum_{j=1}^{3} \alpha_j f_j, \quad -1 \leq \alpha_j \leq 1 \right\}. \end{aligned} \tag{7.122}$$

Assume $w_0 = [0.3 \text{ p.u.}, 0 \text{ p.u.}]^\top$, $e_1 = [0.01, 0]^\top$, $e_2 = [0.01, 0.01]^\top$, and $e_3 = [0.005, 0.01]^\top$; see Fig. 7.9(a) for a depiction of the resulting $\mathcal{W}$. Then, by using the parameter values in Table 7.1, we have that

$$x_0 = [-0.0011 \text{ rad}, 0.9564 \text{ p.u.}]^\top, \tag{7.123}$$

$$H(x_0) = \begin{bmatrix} 0.0216 & -0.0108 \\ 0.0106 & 0.0212 \end{bmatrix}, \tag{7.124}$$

which yields

$$\begin{aligned} f_1 &= 10^{-3}[0.216, 0.106]^\top, \\ f_2 &= 10^{-3}[0.108, 0.318]^\top, \\ f_3 &= 10^{-3}[0, 0.265]^\top; \end{aligned}$$

see Fig. 7.9(b) for a graphical representation of the resulting zonotope.

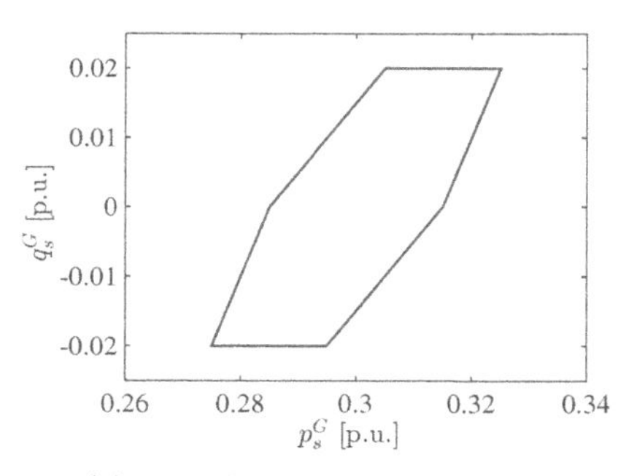

(a) Input-bounding zonotope $\mathcal{W}$

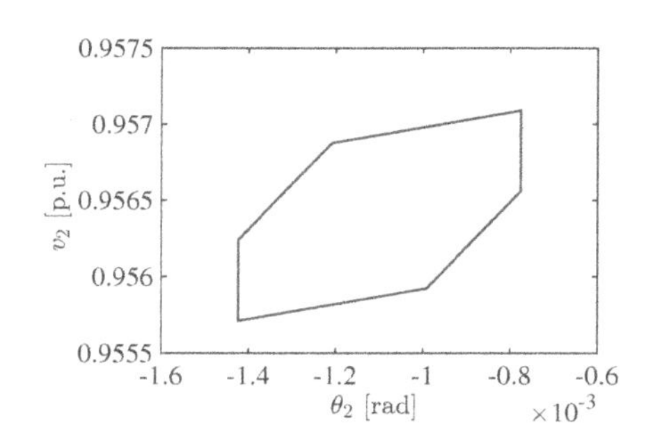

(b) Zonotope approximation to the state-bounding set $\mathcal{X}$

Figure 7.9 Bounding ellipsoids for the two-bus power system in Example 7.8.

7.4 Performance Requirements Verification

Consider

$$x = h(w), \tag{7.125}$$

with $h\colon \mathbb{R}^m \to \mathbb{R}^n$, and where $w \in \mathbb{R}^m$ belongs to an ellipsoid as described in (7.55), or a zonotope as described in (7.105). We can then utilize the tools developed earlier to compute the set $\mathcal{X}$ containing all possible values that x can take, or an approximation thereof, $\widetilde{\mathcal{X}}$. Computing such a set is relevant in applications where performance requirements impose the state vector, $x = [x_1, x_2, \dots, x_n]^\top$, to remain within a certain region of the state space at all times as described by some set $\mathcal{R} \subseteq \mathbb{R}^n$.

While in general the set $\mathcal{R}$ can be of any form, in most practical applications, performance requirements will dictate that for the system to perform its function, each x_i, $i = 1, 2, \dots, n$, must remain above some minimum value, $x_i^m \in \mathbb{R}$, and below some maximum value, $x_i^M \in \mathbb{R}$. Thus, the set $\mathcal{R}$ can be described as the Cartesian product of n real-valued intervals as follows:

$$\mathcal{R} = \mathcal{R}_1 \times \mathcal{R}_2 \times \cdots \times \mathcal{R}_n, \tag{7.126}$$

where

$$\mathcal{R}_i = \left\{ x_i \in \mathbb{R}\colon x_i^m < x_i < x_i^M \right\}. \tag{7.127}$$

Then, in order to verify that the system meets its performance requirements, one needs to check if the set $\mathcal{X}$ is fully contained in the set $\mathcal{R}$.

For the case when the system is linear, i.e., $x = Hw$, we have developed techniques that provide an exact characterization of $\mathcal{X}$; see Section 7.2.1 for the case when $\mathcal{W}$ is an ellipsoid, and Section 7.3.1 for the case when $\mathcal{W}$ is a zonotope. In both cases, we can easily check whether or not $\mathcal{X} \subset \mathcal{R}$, as follows. Let $\hat{u}_i$ denote an n-dimensional vector whose i^{th} entry is equal to 1 and all other entries are equal to zero. Then, we have that $\mathcal{X} \subset \mathcal{R}$, if

$$S_{\mathcal{X}}(\hat{u}_i) < x_i^M \tag{7.128}$$

and

$$S_{\mathcal{X}}(-\hat{u}_i) < -x_i^m, \tag{7.129}$$

for all $i = 1, 2, \ldots, n$, where

$$S_{\mathcal{X}}(\eta) = \eta^\top H w_0 + \sqrt{\eta^\top H E H^\top \eta}, \quad \eta \in \mathbb{R}^n,$$

if $\mathcal{W}$ is an ellipsoid with center $w_0 \in \mathbb{R}^m$ and shape matrix $E \in \mathbb{R}^{m \times m}$, and

$$S_{\mathcal{X}}(\eta) = \eta^\top H w_0 + \sum_{j=1}^{s} |\eta^\top H e_j|, \quad \eta \in \mathbb{R}^n,$$

if $\mathcal{W}$ is a zonotope with center $w_0 \in \mathbb{R}^m$ and generators $e_1, e_2, \ldots, e_s \in \mathbb{R}^m$. The justification for the criterion in (7.128–7.129) is as follows. Recalling the definition of support function of a convex set in (2.55), we have that

$$\begin{aligned} S_{\mathcal{X}}(\hat{u}_i) &= \max_{x \in \mathcal{X}} \hat{u}_i^\top x \\ &= \max_{x \in \mathcal{X}} x_i, \end{aligned} \tag{7.130}$$

i.e., $S_{\mathcal{X}}(\hat{u}_i)$ is the maximum value that the ith entry of x can take for any $x \in \mathcal{X}$. Similarly,

$$\begin{aligned} S_{\mathcal{X}}(-\hat{u}_i) &= \max_{x \in \mathcal{X}} \; -\hat{u}_i^\top x \\ &= \max_{x \in \mathcal{X}} \; -x_i && (7.131) \\ &= -\min_{x \in \mathcal{X}} x_i, && (7.132) \end{aligned}$$

i.e., $-S_{\mathcal{X}}(-\hat{u}_i)$ is the minimum value that the ith entry of x can take for any $x \in \mathcal{X}$. Thus, if (7.128–7.129) hold, we ensure that $x_i^m < x_i < x_i^M$ for all $i = 1, 2, \ldots, n$, which implies that $x \in \mathcal{R}$ for all $x \in \mathcal{X}$; thus, $\mathcal{X} \subset \mathcal{R}$.

7.5 Application to Power Flow Analysis

In this section, we consider the set-theoretic counterpart of the problem tackled in Section 3.5. Namely, given a set-theoretic model for the extraneous, uncontrolled power injections into a power network, we are interested in computing the set containing all possible values that active power flows along the transmission lines take. We will adopt the power flow model used in Section 3.5, the main elements of which are summarized below (the reader is referred to Appendix B for its detailed derivation).

7.5.1 Power Flow Model

Consider a three-phase power network comprising n buses ($n > 1$) indexed by the elements in $\mathcal{V} = \{1, 2, \dots, n\}$, and l transmission lines $\left(n - 1 \leq l \leq n(n-1)/2\right)$ indexed by the elements in the set $\mathcal{L} = \{1, 2, \dots, l\}$. Here, we make the following assumptions (consistent with those made in Section 3.5):

A1. The power system is balanced and operating in sinusoidal regime.
A2. There is at most one transmission line connecting each pair of buses.
A3. Each transmission line $e \in \mathcal{L}$ linking a pair of buses, say bus i and bus j, is short and lossless; thus it can be modeled by a series reactance, $X_{ij} > 0$.
A4. The voltage magnitude at each bus $i \in \mathcal{V}$, denoted by V_i, is fixed by some control mechanism.

Let p_i denote the active power injected into the power network via bus $i, i = 1, 2, \dots, n$, and let ϕ_e denote the active power flowing on transmission line $e, e = 1, 2, \dots, l$. Assume that

$$p_i = \xi_i, \quad i = 1, 2, \dots, m,$$

$1 \leq m \leq n - 1$, where ξ_i is extraneously set and a priori unknown. Further assume that

$$p_i = r_i + \alpha_i \left(\sum_{j=1}^{m} \xi_j + \sum_{j=m+1}^{n} r_j \right), \quad i = m+1, m+2, \dots, n, \tag{7.133}$$

where r_i is some nominal setpoint, and α_i is a nonpositive scalar such that $\sum_{i=1}^{n} \alpha_i = -1$. Define $r = [r_{m+1}, r_{m+2}, \dots, r_n]^\top$, $\alpha = [\alpha_{m+1}, \alpha_{m+2}, \dots, \alpha_n]^\top$, $\xi = [\xi_1, \xi_2, \dots, \xi_m]^\top$, and $\phi = [\phi_1, \phi_2, \dots, \phi_l]^\top$. Let $M \in \mathbb{R}^{n \times l}$ denote the node-to-edge incidence matrix of the directed graph describing the network topology, with the orientation on each edge of this graph in agreement with the (arbitrarily chosen) positive power flow direction on the corresponding transmission line. Then, under some suitable assumptions on the system operating conditions (see Appendix B for the details), we have that

$$\begin{bmatrix} I_m \\ \alpha \mathbf{1}_m^\top \end{bmatrix} \xi + \begin{bmatrix} \mathbf{0}_{m \times (n-m)} \\ I_{n-m} + \alpha \mathbf{1}_{n-m}^\top \end{bmatrix} r = M\phi, \tag{7.134}$$

$$\mathbf{0}_{l-n+1} = N \mathbf{arcsin}\,(\Gamma^{-1}\phi), \tag{7.135}$$

where $N \in \mathbb{R}^{(l-n+1) \times l}$ is full rank and such that $MN^\top = \mathbf{0}_{n,l-n+1}$,

$$\Gamma = \mathbf{diag}\,(\gamma_1, \gamma_2, \dots, \gamma_l),$$

with

$$\gamma_e = \frac{V_i V_j}{X_{ij}},$$

and, for $y = [y_1, y_2, \dots, y_l]$, $\mathbf{arcsin}(y)$ is defined as:

$$\mathbf{arcsin}\,(y) = [\arcsin(y_1), \arcsin(y_2), \dots, \arcsin(y_l)]^\top. \tag{7.136}$$

7.5.2 Ellipsoidal-Based Description of Possible Extraneous Power Injection Values

Assume that the extraneous power injection vector, ξ, is uncertain but known to take values in an ellipsoidal-shaped set $\mathcal{P}$ defined as follows:

$$\mathcal{P} = \Big\{\xi \in \mathbb{R}^m : (\xi - \xi_0)^\top E^{-1}(\xi - \xi_0) \leq 1\Big\}, \tag{7.137}$$

where $\xi_0 \in \mathbb{R}^m$, and $E \in \mathbb{R}^{m\times m}$ is a symmetric positive definite matrix. The goal is to obtain the set of all possible values that the transmission line flow vector ϕ can take; we denote this set by $\mathcal{F}$.

Tree Networks

In this case, $N = 0$; thus, the expressions in (7.134–7.135) reduce to

$$\begin{bmatrix} I_m \\ \alpha \mathbf{1}_m^\top \end{bmatrix} \xi + \begin{bmatrix} \mathbf{0}_{m\times(n-m)} \\ I_{n-m} + \alpha \mathbf{1}_{n-m}^\top \end{bmatrix} r = M\phi. \tag{7.138}$$

Let $\widetilde{M} \in \mathbb{R}^{(n-1)\times(n-1)}$ denote the matrix that results from removing the last row in M, and let $\widetilde{I}_{n-m} \in \mathbb{R}^{(n-m-1)\times(n-m)}$ denote the matrix that results from removing the last row in I_{n-m}. Also, let $\widetilde{\alpha} \in \mathbb{R}^{n-m-1}$ denote the vector that results from removing the last entry in α. Then, by manipulating (7.134–7.135), we can write

$$\phi = H\xi + Lr, \tag{7.139}$$

where

$$H = \widetilde{M}^{-1} \begin{bmatrix} I_m \\ \widetilde{\alpha} \mathbf{1}_m^\top \end{bmatrix} \in \mathbb{R}^{(n-1)\times m} \tag{7.140}$$

is a full-column rank matrix, and

$$L = \widetilde{M}^{-1} \begin{bmatrix} \mathbf{0}_{m\times(n-m)} \\ \widetilde{I}_{n-m} + \widetilde{\alpha} \mathbf{1}_{n-m}^\top \end{bmatrix} \in \mathbb{R}^{(n-1)\times(n-m)}. \tag{7.141}$$

Now, given $\xi \in \mathcal{P}$, with $\mathcal{P}$ as defined in (7.137), the objective is to obtain the set $\mathcal{F}$ containing all possible values that the vector ϕ can take. We need to distinguish two cases: (i) $m = n - 1$ and (ii) $1 \leq m < n - 1$.

For the case when $m = n - 1$, one can see that $H = \widetilde{M}^{-1} \in \mathbb{R}^{(n-1)\times(n-1)}$ and $L = 0$; thus, the setting above reduces to Case E1 in Section 7.2.1; therefore, the set $\mathcal{F}$ is an ellipsoid. Then, by using the expressions in (7.15)–(7.16), we obtain:

$$\mathcal{F} = \Big\{\phi \in \mathbb{R}^{n-1} : (\phi - \phi_0)^\top F^{-1}(\phi - \phi_0) \leq 1\Big\}, \tag{7.142}$$

with $F = HEH^\top$ and $\phi_0 = H\xi_0$.

For $1 \leq m < n - 1$, we first define $\widetilde{\phi} = \phi - Lr$ and rewrite (7.139) as follows:

$$\widetilde{\phi} = H\xi, \tag{7.143}$$

which is of the form of the setting in Case $E2$, and as discussed in Section 7.2.1, the set $\widetilde{\mathcal{F}}$ containing all possible values that $\widetilde{\phi}$ can take is not an ellipsoid. However, by using the expression in (7.43), we can describe it as follows:

$$\widetilde{\mathcal{F}} = \left\{ \widetilde{\phi} \in \mathbb{R}^{n-1} \colon \eta^\top \widetilde{\phi} \leq \eta^\top \widetilde{\phi}_0 + \sqrt{\eta^\top F \eta}, \quad \eta^\top \eta = 1 \right\}, \tag{7.144}$$

with $F = HEH^\top$ and $\widetilde{\phi}_0 = H\xi_0$. Finally, the set $\mathcal{F}$ can be obtained as the Minkowski sum of $\widetilde{\mathcal{F}}$ and the singleton $\{Lr\}$, yielding

$$\mathcal{F} = \left\{ \phi \in \mathbb{R}^{n-1} : \eta^\top \phi \leq \eta^\top \phi_0 + \sqrt{\eta^\top F \eta}, \ \eta^\top \eta = 1 \right\}, \tag{7.145}$$

with $\phi_0 = \widetilde{\phi}_0 + Lr = H\xi_0 + Lr$.

Example 7.9 (Three-node tree network with two uncertain power injections) Recall the three-node lossless power network of Example 3.11, the topology of which is described by the graph in Fig. 7.10. As in that example, we will assume that the power injections into nodes 1 and 2, denoted respectively by ξ_1 and ξ_2, are set extraneously; thus, recalling (3.138), we have

$$\phi = H\xi, \tag{7.146}$$

where $\xi = [\xi_1, \xi_2]^\top$, $\phi = [\phi_1, \phi_2]^\top$, and

$$H = \begin{bmatrix} 1 & 0 \\ 1 & 1 \end{bmatrix}. \tag{7.147}$$

Assume that the extraneous power injection vector, ξ, is uncertain but known to take values in the ellipsoid

$$\mathcal{P} = \left\{ \xi : (\xi - \xi_0)^\top E^{-1} (\xi - \xi_0) \leq 1 \right\}, \tag{7.148}$$

where

$$E = \begin{bmatrix} 0.01 & 0 \\ 0 & 0.01 \end{bmatrix}, \tag{7.149}$$

and $\xi_0 = [0.2, -0.5]^\top$; see Fig. 7.11(a) for a graphical representation. Then, by using (7.142), we can obtain the ellipsoidal set $\mathcal{F}$ containing all possible values that ϕ takes; the calculation yields

$$\mathcal{F} = \left\{ \phi \in \mathbb{R}^2 : (\phi - \phi_0)^\top F^{-1} (\phi - \phi_0) \leq 1 \right\}, \tag{7.150}$$

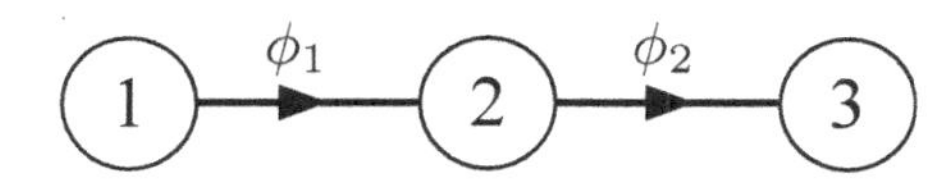

Figure 7.10 Graph describing the topology of a three-node tree power network.

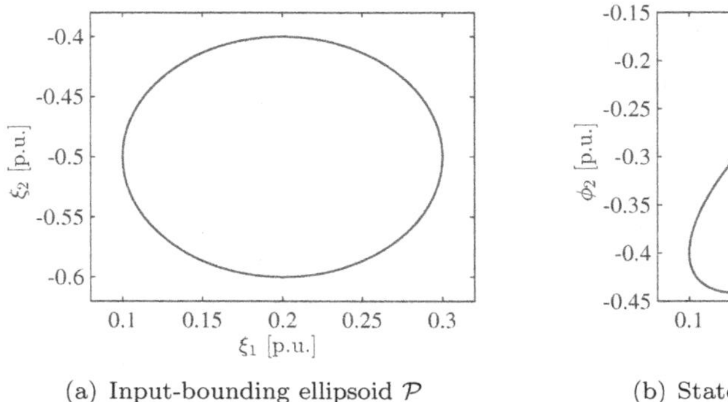

(a) Input-bounding ellipsoid $\mathcal{P}$ (b) State-bounding ellipsoid $\mathcal{F}$

Figure 7.11 Bounding ellipsoids for the three-bus tree network in Example 7.9.

with

$$F = HEH^\top = \begin{bmatrix} 0.01 & 0.01 \\ 0.01 & 0.02 \end{bmatrix}, \tag{7.151}$$

and

$$\phi_0 = H\xi_0 \tag{7.152}$$

$$= \begin{bmatrix} 0.2 \\ -0.3 \end{bmatrix}. \tag{7.153}$$

The ellipsoid $\mathcal{F}$ is depicted in Fig. 7.11(b).

Example 7.10 (Three-node tree network with one uncertain power injection) Consider the same power network as in Example 7.9. In this case, we will assume that ξ_1 is uncertain but known to take values in the interval $[0.1, 0.3]$ and the active power injected into bus 2 and bus 3, p_2 and p_3, respectively, are set as follows

$$p_2 = \alpha_2(\xi_1 + r_2 + r_3),$$
$$p_3 = \alpha_3(\xi_1 + r_2 + r_3), \tag{7.154}$$

with $\alpha_2 = \alpha_3 = -1/2$, and $r_2 = r_3 = 0$. Then, by recalling (3.143), it follows that

$$\underbrace{\begin{bmatrix} \phi_1 \\ \phi_2 \end{bmatrix}}_{=\phi} = \underbrace{\begin{bmatrix} 1 \\ \frac{1}{2} \end{bmatrix}}_{=H} \xi_1, \tag{7.155}$$

with

$$\xi_1 \in \mathcal{P} = \left\{\xi_1 \in \mathbb{R} : E^{-1}(\xi_1 - 0.2)^2 \leq 1\right\}, \tag{7.156}$$

where $E = 0.01$.

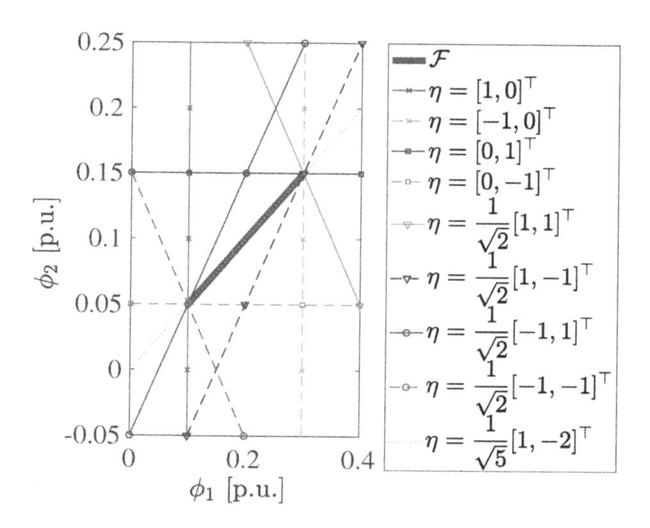

Figure 7.12 Set $\mathcal{F}$ for three-node network in Example 7.10 and boundaries of several half spaces containing its boundary.

As we did in Example 7.2, we can directly calculate the set $\mathcal{F}$ containing the possible values that ϕ can take by noting that $\phi_2 = \frac{1}{2}\phi_1$, where

$$\phi_1 = \xi_1 \in [0.1, 0.3];$$

thus, we can write

$$\mathcal{F} = \left\{ \begin{bmatrix} \phi_1 \\ \phi_2 \end{bmatrix} : \phi_2 = \frac{1}{2}\phi_1, \ 0.1 \leq \phi_1 \leq 0.3 \right\}, \tag{7.157}$$

i.e., $\mathcal{F}$ is a line segment as shown in Fig. 7.12. Alternatively, since in this case $m = 1$ and $n = 3$, we can use the representation given in (7.145) to describe $\mathcal{F}$, which in this case yields

$$\begin{aligned} \mathcal{F} &= \left\{ \phi \in \mathbb{R}^2 : \eta^\top \phi \leq \eta^\top H \xi_0 + \sqrt{\eta^\top H E H^\top \eta}, \quad \eta^\top \eta = 1 \right\} \\ &= \left\{ \begin{bmatrix} \phi_1 \\ \phi_2 \end{bmatrix} : \eta_1 \phi_1 + \eta_2 \phi_2 \leq 0.2\eta_1 + 0.1\eta_2 + 0.1 \left| \eta_1 + 0.5\eta_2 \right|, \quad \eta_1^2 + \eta_2^2 = 1 \right\}. \end{aligned} \tag{7.158}$$

Figure 7.12 shows the boundary of the half space defined by the constraint in (7.158) for different values of η. One can easily check that each such half space contains $\mathcal{F}$, while, as shown in Fig. 7.12, its boundary touches the boundary of the set $\mathcal{F}$.

Loopy Networks

In this case $N \neq 0$, thus the power flow model in (7.134–7.135) is nonlinear and we need to resort to the techniques in Section 7.2.2. Let ϕ^* denote the value of ϕ

that satisfies (7.134–7.135) for $\xi = \xi_0$, and let $\Delta\phi$ denote the change in ϕ around ϕ^* that arises from a change in ξ around ξ_0, which is similarly denoted by $\Delta\xi$. Then, as already shown in Section 3.5 for the probabilistic counterpart of the problem of interest here, we have that

$$\Delta\phi \approx H(\phi^*)\Delta\xi, \tag{7.159}$$

where

$$H(\phi^*) = \begin{bmatrix} \widetilde{M} \\ J(\phi^*) \end{bmatrix}^{-1} \begin{bmatrix} I_m \\ \tilde{\alpha}\mathbf{1}_m^\top \\ \mathbf{0}_{(l-n+1)\times m} \end{bmatrix} \in \mathbb{R}^{l\times m}, \tag{7.160}$$

with

$$J(\phi^*) = N\frac{\partial \mathbf{arcsin}(y)}{\partial y}\bigg|_{y=\Gamma^{-1}\phi^*}\Gamma^{-1} \in \mathbb{R}^{(l-n+1)\times l}. \tag{7.161}$$

In this case, the dimension of $\Delta\phi$ is strictly greater than the dimension of ξ; thus, the setting here is essentially that of Case F2. Then, we can tailor the expression in (7.63) to obtain an approximation of the set $\mathcal{F}$ as follows:

$$\phi \in \mathcal{F} \approx \Big\{\phi \in \mathbb{R}^l : \eta^\top\phi \le \eta^\top\phi^* + \sqrt{\eta^\top F(\phi^*)\eta},\ \eta^\top\eta = 1\Big\}, \tag{7.162}$$

where

$$F(\phi^*) = H(\phi^*)EH^\top(\phi^*). \tag{7.163}$$

Example 7.11 (Three-node cycle network with two uncertain power injections) Consider the same power network in Example 3.14, the topology of which is reproduced in Fig. 7.13. As in that example, we will assume that the power injections into nodes 1 and 2, denoted respectively by ξ_1 and ξ_2, are set extraneously, with the extraneous power injection vector, $\xi = [\xi_1, \xi_2]^\top$, taking values in the ellipsoidal set

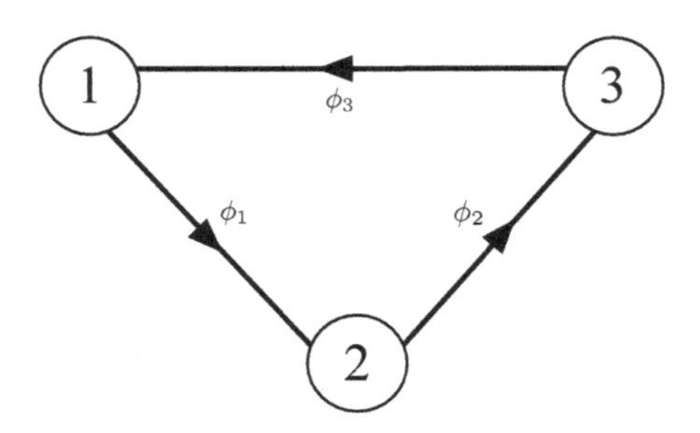

Figure 7.13 Graph describing the topology of a three-node cycle power network.

$$\mathcal{P} = \left\{ \xi \in \mathbb{R}^2 : (\xi - \xi_0)^\top E^{-1} (\xi - \xi_0) \le 1 \right\}, \tag{7.164}$$

where

$$E = \begin{bmatrix} 0.01 & 0 \\ 0 & 0.01 \end{bmatrix}, \tag{7.165}$$

and $\xi_0 = [0.2, -0.5]^\top$. Here, we are interested in approximately characterizing the set $\mathcal{F}$ containing all possible values that the transmission line flow vector $\phi = [\phi_1, \phi_2, \phi_3]^\top$ can take.

Assuming the same model parameters as those used in Example 3.14, we obtain

$$\Delta\phi \approx H(\phi^*)\Delta\xi, \tag{7.166}$$

$$\Delta\xi \in \Delta\mathcal{P} = \left\{ \Delta\xi : (\Delta\xi)^\top E^{-1} \Delta\xi \le 1 \right\}, \tag{7.167}$$

where $\Delta\xi = \xi - \xi_0$, $\Delta\phi = \phi - \phi^*$, with $\phi^* = [0.2337, -0.2667, 0.0337]^\top$, and

$$H(\phi^*) = \begin{bmatrix} 0.3263 & -0.3383 \\ 0.3263 & 0.6617 \\ -0.6737 & -0.3383 \end{bmatrix}. \tag{7.168}$$

Since the number of columns of $H(\phi^*)$ is smaller than the number of rows, this setting corresponds to that of Case F2; thus, the set $\Delta\mathcal{F}$ containing all possible values that $\widetilde{\Delta\phi} := H(\phi^*)\Delta\xi$, $\Delta\xi \in \Delta\mathcal{P}$, can take is not an ellipsoid. Thus, we need to use the representation given in (7.162) to approximately describe the set $\mathcal{F}$, which yields

$$\mathcal{F} \approx \{\phi^*\} + \Delta\mathcal{F} = \left\{ \phi \in \mathbb{R}^3 : \eta^\top \phi \le \eta^\top \phi^* + \sqrt{\eta^\top F(\phi^*) \eta},\ \eta^\top \eta = 1 \right\}, \tag{7.169}$$

with

$$\begin{aligned} F(\phi^*) &= H(\phi^*) E H^\top(\phi^*) \\ &= 10^{-4} \begin{bmatrix} 22.09 & -11.74 & -10.54 \\ -11.74 & 54.43 & -44.37 \\ -10.54 & -44.39 & 56.83 \end{bmatrix}. \end{aligned} \tag{7.170}$$

7.5.3 Zonotope-Based Description of Possible Extraneous Power Injection Values

Assume that the extraneous power injection vector, ξ, takes values in a zonotope, $\mathcal{P}$, as follows:

$$\mathcal{P} = \left\{ \xi \in \mathbb{R}^m : \xi = \xi_0 + \sum_{j=1}^{s} \alpha_j e_j,\ -1 \le \alpha_j \le 1 \right\}, \tag{7.171}$$

where $\xi_0 \in \mathbb{R}^m$, and $e_1, e_2, \ldots, e_s \in \mathbb{R}^m$. As in the ellipsoidal-based extraneous power injection vector description, the goal here is to characterize the set $\mathcal{F}$ containing all possible values that the transmission line flow vector, ϕ, can take.

Tree Networks

Recall from (7.139) that, in this case, the relation between ϕ and ξ is given by

$$\phi = H\xi + Lr, \tag{7.172}$$

which can be rewritten as

$$\widetilde{\phi} = H\xi, \tag{7.173}$$

where $\widetilde{\phi} = \phi - Lr$. Now, we can tailor the expression in (7.95) to obtain a zonotope $\widetilde{\mathcal{F}}$ containing all possible values that $\widetilde{\phi}$ can take as follows:

$$\widetilde{\mathcal{F}} = \left\{ \widetilde{\phi} \in \mathbb{R}^l : \widetilde{\phi} = \widetilde{\phi}_0 + \sum_{j=1}^{s} \alpha_j f_j, \quad -1 \leq \alpha_j \leq 1 \right\}, \tag{7.174}$$

where $\widetilde{\phi}_0 = H\xi_0$ and $f_j = He_j$. Finally, since $\phi = \widetilde{\phi} + Lr$, we can obtain the set $\mathcal{F}$ containing all possible values that ϕ can take as the Minkowski sum of $\widetilde{\mathcal{F}}$ and the singleton $\{Lr\}$, yielding

$$\mathcal{F} = \left\{ \phi \in \mathbb{R}^l : \phi = \phi_0 + \sum_{j=1}^{s} \alpha_j f_j, \quad -1 \leq \alpha_j \leq 1 \right\}, \tag{7.175}$$

where $\phi_0 = \widetilde{\phi}_0 + Lr = H\xi_0 + Lr$.

Example 7.12 (Three-node tree network with two uncertain power injections) Consider the power network in Example 7.9 as described by (7.146–7.147). Here assume that the extraneous power injection vector, $\xi = [\xi_1, \xi_2]^\top$, is uncertain but known to take values in the following zonotope:

$$\mathcal{P} = \left\{ \xi \in \mathbb{R}^2 : \xi = \xi_0 + \sum_{j=1}^{2} \alpha_j e_j, \quad -1 \leq \alpha_j \leq 1 \right\}, \tag{7.176}$$

where $\xi_0 = [0.2, -0.5]^\top$, $e_1 = [0.05, 0.05]^\top$, and $e_2 = [-0.05, 0.05]^\top$; see Fig. 7.14(a) for a graphical depiction. Then, by using (7.175) together with (7.147), we obtain that the zonotope $\mathcal{F}$ containing all possible values that the transmission line flow vector, $\phi = [\phi_1, \phi_2]^\top$, can take is given by

$$\mathcal{F} = \left\{ \phi \in \mathbb{R}^2 : \phi = \phi_0 + \sum_{j=1}^{2} \alpha_j f_j, \quad -1 \leq \alpha_j \leq 1 \right\}, \tag{7.177}$$

where $\phi_0 = [0.2, -0.3]^\top$, $f_1 = [0.05, 0.01]^\top$, and $f_2 = [-0.05, 0]^\top$; see Fig. 7.14(b) for a graphical depiction.

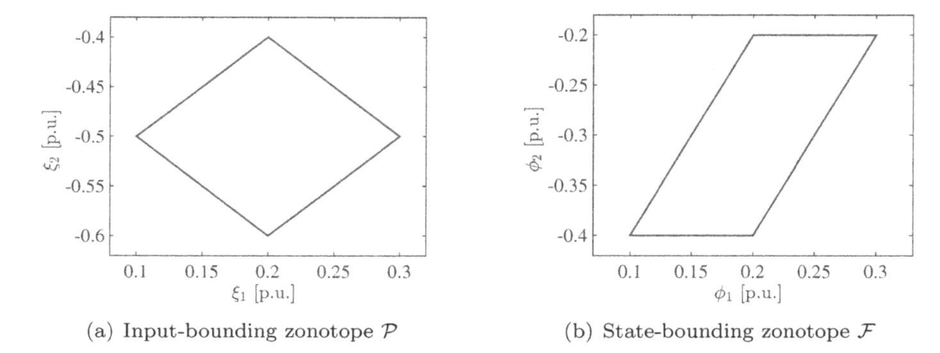

(a) Input-bounding zonotope $\mathcal{P}$ (b) State-bounding zonotope $\mathcal{F}$

Figure 7.14 Bounding zonotope for the three-bus tree network in Example 7.12.

Loopy Networks

Recall from (7.159) that in this case, we can approximately relate $\Delta\xi = \xi - \xi_0$ and $\Delta\phi = \phi - \phi^*$, where ϕ^* is the value of ϕ that satisfies (7.134–7.135) for $\xi = \xi_0$ as follows:

$$\Delta\phi \approx H(\phi^*)\Delta\xi, \tag{7.178}$$

with $H(\phi^*)$ as defined in (7.160). Then, by tailoring the expression in (7.108) to the setting here, we obtain

$$\phi \in \mathcal{F} \approx \left\{ \phi \in \mathbb{R}^l \colon \phi = \phi_0 + \sum_{j=1}^{s} \alpha_j f_j, \quad -1 \leq \alpha_j \leq 1 \right\} \tag{7.179}$$

where $\phi_0 = \phi^*$, and $f_j = H(\phi^*)e_j,\ j = 1, 2, \ldots, s$.

Example 7.13 (Three-node cycle network with two uncertain power injections) Consider the power network in Example 7.11 as described by (7.166) and (7.168). Assume that the extraneous power injection vector, $\xi = [\xi_1, \xi_2]^\top$, is uncertain but known to take values in the following zonotope

$$\mathcal{P} = \left\{ \xi \in \mathbb{R}^2 \colon \xi = \xi_0 + \sum_{j=1}^{2} \alpha_j e_j, \quad -1 \leq \alpha_j \leq 1 \right\}, \tag{7.180}$$

where $\xi_0 = [0.2, -0.5]^\top$, $e_1 = [0.05, 0.05]^\top$, and $e_2 = [-0.05, 0.05]^\top$. Then, by using (7.179) together with (7.168), we obtain

$$\phi \in \mathcal{F} \approx \left\{ \phi \in \mathbb{R}^3 \colon \phi = \phi_0 + \sum_{j=1}^{2} \alpha_j f_j, \quad -1 \leq \alpha_j \leq 1 \right\}, \tag{7.181}$$

where $\phi_0 = [0.2337, -0.2667, 0.0337]^\top$, $f_1 = [-0.0006, 0.0494, -0.0506]^\top$, and $f_2 = [-0.0332, 0.0168, 0.0168]^\top$.

7.6 Notes and References

The material on analysis of static linear systems subject to inputs belonging to an ellipsoidal set is based on [10]. The zonotope parametrization used in this chapter is based on that used in [45]. The use of ellipsoids for analyzing power systems under power injection uncertainty was first proposed in [46, 47]. Unlike the framework in Section 3.5.1, which adopts a simplified power flow model formulation, these two papers use the full nonlinear power flow model (see Section B.3) as a starting point and then use linearization to approximate the relation between state variables, i.e., bus voltage magnitudes and phase angles, and unknown inputs, i.e., active and reactive power injections (the method is illustrated in Example 7.5, which was also featured in [46]). A zonotope-based framework for analyzing power systems under power injection uncertainty was proposed in [47]. Interval analysis techniques have also been proposed to analyze the impact of uncertainty on the solution to the power flow model; see, e.g., [48].

8 Discrete-Time Systems: Set-Theoretic Input Uncertainty

In this chapter, we extend the set-theoretic input uncertainty framework for static systems developed in Chapter 7 to the case where the relation between the input and the state is governed by a discrete-time state-space model. Here, we will assume that the values that the input can take belong to an ellipsoid. The objective is then to characterize the set containing all possible values that the state can take. Fundamental concepts from set theory and discrete-time linear dynamical systems used throughout this chapter are reviewed in Sections 2.2.2 and 2.3.1, respectively.

8.1 Introduction

Consider again the bank accounts example in Section 5.1, and recall the relation between the amount of money in Jane's savings account at the beginning of months k and $k+1$, denoted by s_k and s_{k+1}, respectively, and the amount of money transferred from her checking account to her savings account at the beginning of month k, denoted by d_k:

$$s_{k+1} = \alpha s_k + d_k. \tag{8.1}$$

Due to unexpected new expenditures, there are months in which Jane needs to transfer money from her savings account to her checking account; thus, in this case, d_k can also be negative. Because of the bank terms on Jane's savings account, s_k must remain above some minimum amount, $\underline{s}$; otherwise, the bank will assess a fee of β [$/month] for every month that s_k is smaller than $\underline{s}$ (this fee is charged to her checking account, so it does not get reflected in her savings account). From analyzing her expenses, Jane has determined that the money transfers into the savings account never exceed some value $\overline{d} > 0$ and are never smaller than some value $\underline{d} < 0$, i.e., $d_k \in [\underline{d}, \overline{d}]$ for all $k \geq 0$. To check whether or not she will be assessed the fee, she needs to check whether or not $s_k \leq \underline{s}$ for all $k \geq 0$. To this end, she will need to characterize the real interval containing all possible values that s_k can take.

The system in (8.1) belongs to the class of linear systems, which can be compactly described in state-space form as follows:

$$x_{k+1} = G_k x_k + H_k w_k, \tag{8.2}$$

where $x_k \in \mathbb{R}^n$, $A_k \in \mathbb{R}^{n \times n}$, $B_k \in \mathbb{R}^{n \times m}$, and $w_k \in \mathbb{R}^m$ is unknown but it belongs to some closed and bounded set $\mathcal{W}_k$. While in general the set $\mathcal{W}_k$ can have any shape, in this chapter we will restrict our analysis to the case when $\mathcal{W}_k$ is an ellipsoid, i.e.,

$$\mathcal{W}_k = \{w \in \mathbb{R}^m \colon (w - \overline{w}_k)^\top E_k^{-1}(w - \overline{w}_k) \leq 1\}, \tag{8.3}$$

where $\overline{w}_k \in \mathbb{R}^m$ and E_k is an $(m \times m)$-dimensional symmetric positive definite matrix. We will also provide techniques for analyzing the nonlinear counterpart of (8.2):

$$x_{k+1} = h_k(x_k, w_k), \tag{8.4}$$

where $w_k \in \mathcal{W}_k$, with $\mathcal{W}_k$ as in (8.3). The objective then is to characterize the set containing all possible values that x_k can take. Such set is referred to as the *reachable set* of the system and, in general, is *not* an ellipsoid (even in the linear case).

The computation of the reachable set of (8.2) (or (8.4)) is useful in situations in which we would like to verify whether or not the possible system trajectories remain at all times within some set $\mathcal{R}$ prescribed by performance requirements on the particular application. For example, this is important in settings when excursions outside the set $\mathcal{R}$ could result in unsafe operation or trigger some protection schemes resulting in the shutdown of the system.

8.2 Discrete-Time Linear Systems

Consider a system of the form:

$$x_{k+1} = G_k x_k + H_k w_k, \quad k = 0, 1, \ldots, \tag{8.5}$$

where $x_k \in \mathbb{R}^n$, $G_k \in \mathbb{R}^{n \times n}$ is invertible, $H_k \in \mathbb{R}^{n \times m}$ such that $H_k \neq \mathbf{0}_{n \times m}$, $w_k \in \mathbb{R}^m$ is uncertain but known to belong to a convex set $\mathcal{W}_k \subseteq \mathbb{R}^m$, and x_0 is also uncertain but known to belong to a convex set $\mathcal{X}_0 \subseteq \mathbb{R}^n$. For any $k \geq 0$, we are interested in characterizing the *reachable set* (i.e., the set containing all possible values that x_k can take), which we denote by $\mathcal{X}_k$.

Recall from (2.108) that

$$x_k = \Phi_{k,0} x_0 + \sum_{\ell=0}^{k-1} \Phi_{k,\ell+1} H_\ell w_\ell, \quad k \geq 1, \tag{8.6}$$

where

$$\Phi_{k,\ell} = \begin{cases} G_{k-1} G_{k-2} \cdots G_\ell, & \text{if } k > \ell \geq 0, \\ I_n, & \text{if } k = \ell. \end{cases} \tag{8.7}$$

Then, the reachable set at instant $k \geq 1$, $\mathcal{X}_k$, can be described as follows:

$$\mathcal{X}_k = \left\{ x_k \in \mathbb{R}^n : x_k = \Phi_{k,0} x_0 + \sum_{\ell=0}^{k-1} \Phi_{k,\ell+1} H_\ell w_\ell, \ x_0 \in \mathcal{X}_0, \ w_\ell \in \mathcal{W}_\ell \right\}, \quad (8.8)$$

or alternatively as the Minkowski sum of the following $k+1$ sets:

$$\Phi_{k,0}\mathcal{X}_0, \quad \Phi_{k,1}H_0\mathcal{W}_0, \quad \Phi_{k,2}H_1\mathcal{W}_1, \quad \dots \quad \Phi_{k,k}H_{k-1}\mathcal{W}_{k-1}.$$

Also, in light of (8.5), we can recursively define the reachable set as follows:

$$\mathcal{X}_{k+1} = \left\{ x \in \mathbb{R}^n : x = G_k y + H_k z, \ y \in \mathcal{X}_k, \ z \in \mathcal{W}_k \right\}, \quad k = 0, 1, \dots, \quad (8.9)$$

which can be equivalently written as follows:

$$\mathcal{X}_{k+1} = G_k \mathcal{X}_k + H_k \mathcal{W}_k, \quad (8.10)$$

where $\mathcal{X}_0$ and $\mathcal{W}_k$, $k \geq 0$, are given.

Recall the notion of support function of a convex set $\mathcal{X} \subseteq \mathbb{R}^n$, which we denote by $S_{\mathcal{X}}(\cdot)$, given in (2.55):

$$S_{\mathcal{X}}(\eta) = \max_{x \in \mathcal{X}} \eta^\top x, \quad \eta \in \mathbb{R}^n, \quad (8.11)$$

and the following two properties given in (2.57) and (2.59), respectively:

P1. The support function of the Minkowski sum of $\mathcal{X} \subseteq \mathbb{R}^n$ and $\mathcal{Y} \subseteq \mathbb{R}^n$, denoted by $S_{\mathcal{X}+\mathcal{Y}}(\cdot)$, is given by

$$S_{\mathcal{X}+\mathcal{Y}}(\eta) = S_{\mathcal{X}}(\eta) + S_{\mathcal{Y}}(\eta), \quad \eta \in \mathbb{R}^n. \quad (8.12)$$

P2. The support function of $H\mathcal{W} \subseteq \mathbb{R}^n$, $H \in \mathbb{R}^{n \times m}$ and $\mathcal{W} \subseteq \mathbb{R}^m$, denoted by $S_{H\mathcal{W}}(\cdot)$, is given by

$$S_{H\mathcal{W}}(\eta) = S_{\mathcal{W}}(H^\top \eta), \quad \eta \in \mathbb{R}^n. \quad (8.13)$$

Then, by using Properties P1 and P2 together with (8.10), we have that

$$S_{\mathcal{X}_{k+1}}(\eta) = S_{\mathcal{X}_k}(G_k^\top \eta) + S_{\mathcal{W}_k}(H_k^\top \eta), \quad \eta \in \mathbb{R}^n. \quad (8.14)$$

Also, since $\mathcal{X}_k$ can be written as the Minkowski sum of

$$\Phi_{k,0}\mathcal{X}_0, \ \Phi_{k,\ell+1}H_\ell\mathcal{W}_\ell, \ \ell = 0, 1, \dots, k-1,$$

it follows from Properties P1 and P2 that

$$S_{\mathcal{X}_k}(\eta) = S_{\mathcal{X}_0}\left(\Phi_{k,0}^\top \eta\right) + \sum_{\ell=0}^{k-1} S_{\mathcal{W}_\ell}\left(H_\ell^\top \Phi_{k,\ell+1}^\top \eta\right), \quad \eta \in \mathbb{R}^n. \quad (8.15)$$

8.2.1 Ellipsoidal-Based Input Description

Consider the system in (8.5) and assume that the sets $\mathcal{W}_k$, $k \geq 0$, and $\mathcal{X}_0$ are ellipsoids with center $\overline{w}_k \in \mathbb{R}^m$ and $\overline{x}_0 \in \mathbb{R}^n$, respectively, and shape matrix $E_k \in \mathbb{R}^{m \times m}$ and $F_0 \in \mathbb{R}^{n \times n}$, respectively; i.e.,

$$\mathcal{W}_k = \left\{ w \in \mathbb{R}^m : (w - \overline{w}_k)^\top E_k^{-1}(w - \overline{w}_k) \leq 1 \right\}, \quad (8.16)$$

$$\mathcal{X}_0 = \{x \in \mathbb{R}^n : (x - \overline{x}_0)^\top F_0^{-1}(x - \overline{x}_0) \leq 1\}. \tag{8.17}$$

Then the support functions of $\mathcal{W}_k$ and $\mathcal{X}_0$ are given by:

$$S_{\mathcal{W}_k}(\eta) = \eta^\top \overline{w}_k + \sqrt{\eta^\top E_k \eta}, \quad \eta \in \mathbb{R}^m, \tag{8.18}$$

$$S_{\mathcal{X}_0}(\eta) = \eta^\top \overline{x}_0 + \sqrt{\eta^\top F_0 \eta}, \quad \eta \in \mathbb{R}^n, \tag{8.19}$$

respectively. Thus, by plugging the expressions in (8.18–8.19) into (8.15), we obtain the following expression for the support function of the set $\mathcal{X}_k$:

$$\begin{aligned} S_{\mathcal{X}_k}(\eta) = \eta^\top & \left(\Phi_{k,0}\overline{x}_0 + \sum_{\ell=0}^{k-1} \Phi_{k,\ell+1} H_\ell \overline{w}_\ell \right) \\ & + \sqrt{\eta^\top \Phi_{k,0} F_0 \Phi_{k,0}^\top \eta} + \sum_{\ell=0}^{k-1} \sqrt{\eta^\top \Phi_{k,\ell+1} H_\ell E_\ell H_\ell^\top \Phi_{k,\ell+1}^\top \eta}. \end{aligned} \tag{8.20}$$

Therefore, in general, the set $\mathcal{X}_k$ is not an ellipsoid because the expression in (8.20) does not conform to the expression for the support function of an ellipsoid. The goal is then to outer bound the set $\mathcal{X}_k$, $k \geq 0$, by one or more ellipsoids that are optimal in some sense, e.g., their volume is minimum. To this end, consider ellipsoids of the form

$$\mathcal{X}_k^+ = \{x \in \mathbb{R}^n : (x - \overline{x}_k)^\top F_k^{-1}(x - \overline{x}_k) \leq 1\}, \quad k = 0, 1, \ldots, \tag{8.21}$$

with the center $\overline{x}_k \in \mathbb{R}^n$ and shape matrix $F_k \in \mathbb{R}^{n \times n}$ defined recursively as follows:

$$\overline{x}_{k+1} = G_k \overline{x}_k + H_k \overline{w}_k, \quad k \geq 0, \tag{8.22}$$

$$F_{k+1} = (1 + \gamma_k) G_k F_k G_k^\top + \left(1 + \frac{1}{\gamma_k}\right) H_k E_k H_k^\top, \quad k \geq 0, \tag{8.23}$$

where $\gamma_k > 0$, and $\overline{x}_0$ and F_0 are given; then $\mathcal{X}_k \subseteq \mathcal{X}_k^+$ for all $k \geq 0$. To establish this, we use induction, for which we need the following two facts.

F1. The support function of $\mathcal{X}_{k+1}^+$ satisfies the following relation:

$$\begin{aligned} S_{\mathcal{X}_{k+1}^+}(\eta) &= \eta^\top \overline{x}_{k+1} + \sqrt{\eta^\top F_{k+1} \eta} \\ &= \eta^\top (G_k \overline{x}_k + H_k \overline{w}_k) \\ &\quad + \sqrt{\eta^\top \left((1 + \gamma_k) G_k F_k G_k^\top + \left(1 + \frac{1}{\gamma_k}\right) H_k E_k H_k^\top \right) \eta}; \end{aligned} \tag{8.24}$$

this follows directly from (8.23) and the definition of the support function of an ellipsoid.

F2. For any $\gamma > 0$, the following relation holds:

$$(a+b)^2 \leq (1+\gamma)a^2 + \left(1+\frac{1}{\gamma}\right)b^2. \tag{8.25}$$

This can be established by noting that

$$\left(\sqrt{\gamma}a - \frac{1}{\sqrt{\gamma}}b\right)^2 = \gamma a^2 + \frac{1}{\gamma}b^2 - 2ab \geq 0;$$

thus,

$$2ab \leq \gamma a^2 + \frac{1}{\gamma}b^2,$$

and by adding $a^2 + b^2$ on both sides of the inequality, we obtain

$$a^2 + b^2 + 2ab \leq (1+\gamma)a^2 + \left(1+\frac{1}{\gamma}\right)b^2,$$

and (8.25) follows.

For $k = 0$, it clearly holds that $\mathcal{X}_0 \subseteq \mathcal{X}_0^+$ because $\mathcal{X}_0^+ = \mathcal{X}_0$, thus by using the result in (8.25), we have that

$$\begin{aligned}
S_{\mathcal{X}_1}(\eta) &= S_{\mathcal{X}_0}(G_0^\top \eta) + S_{\mathcal{W}_0}(H_0^\top \eta) \\
&= \eta^\top (G_0 \overline{x}_0 + H_0 \overline{w}_0) + \sqrt{\eta^\top G_0 F_0 G_0^\top \eta} + \sqrt{\eta^\top H_0 E_0 H_0^\top \eta} \\
&\leq \eta^\top (G_0 \overline{x}_0 + H_0 \overline{w}_0) \\
&\quad + \sqrt{\eta^\top \left((1+\gamma_0) G_0 F_0 G_0^\top + \left(1 + \frac{1}{\gamma_0}\right) H_0 E_0 H_0^\top \right) \eta} \\
&= \eta^\top \overline{x}_1 + \sqrt{\eta^\top F_1 \eta} \\
&= S_{\mathcal{X}_1^+}(\eta),
\end{aligned} \tag{8.26}$$

for some $\gamma_0 > 0$. Now, recall that for any two closed and convex sets $\mathcal{X}, \mathcal{Y} \in \mathbb{R}^n$, $\mathcal{X} \subseteq \mathcal{Y}$ if and only if $S_{\mathcal{X}}(\eta) \leq S_{\mathcal{Y}}(\eta)$ for all $\eta \in \mathbb{R}^n$; therefore, we conclude that $\mathcal{X}_1 \subseteq \mathcal{X}_1^+$. Now, we need to verify that F_1 is a valid shape matrix, for which we need to check that it is symmetric and positive definite. To this end, since F_0 is symmetric and positive definite and G_0 is invertible, it follows that $(1+\gamma_0)G_0 F_0 G_0^\top$ is also symmetric and positive definite. Also, because E_0 is symmetric and positive definite, it follows that $(1+1/\gamma_0)H_0 E_0 H_0^\top$ is symmetric and positive (semi)definite. But the sum of a symmetric positive definite matrix and a symmetric positive (semi)definite matrix is always a positive definite matrix; thus, F_1 is symmetric and positive definite and therefore a valid shape matrix.

For $k > 0$, assume that $\mathcal{X}_k \subseteq \mathcal{X}_k^+$, then, by using the result in (8.25), we have that

$$\begin{aligned}
S_{\mathcal{X}_{k+1}}(\eta) &= S_{\mathcal{X}_k}(G_k^\top \eta) + S_{\mathcal{W}_k}(H_k^\top \eta) \\
&\leq S_{\mathcal{X}_k^+}(G_k^\top \eta) + S_{\mathcal{W}_k}(H_k^\top \eta) \\
&= \eta^\top (G_k \overline{x}_k + H_k \overline{w}_k) + \sqrt{\eta^\top G_k F_k G_k^\top \eta} + \sqrt{\eta^\top H_k E_k H_k^\top \eta} \\
&\leq \eta^\top (G_k \overline{x}_k + H_k \overline{w}_k) \\
&\quad + \sqrt{\eta^\top \left((1+\gamma_k) G_k F_k G_k^\top + \left(1 + \frac{1}{\gamma_k}\right) H_k E_k H_k^\top \right) \eta} \\
&= \eta^\top \overline{x}_{k+1} + \sqrt{\eta^\top F_{k+1} \eta} \\
&= S_{\mathcal{X}_{k+1}^+}(\eta),
\end{aligned} \tag{8.27}$$

for some $\gamma_k > 0$; thus, $\mathcal{X}_{k+1} \subseteq \mathcal{X}_{k+1}^+$. Now, because F_k is symmetric and positive definite and G_k is invertible, it follows that the matrix $(1 + \gamma_k)G_k F_k G_k^\top$ is also symmetric and positive definite. Similarly, the matrix $(1 + 1/\gamma_k) H_k E_k H_k^\top$ is symmetric and positive (semi)definite because E_k is symmetric and positive definite. Thus, F_{k+1} is the sum of a symmetric positive definite matrix and a symmetric positive (semi)definite matrix; therefore it is symmetric and positive definite, which means it is a valid shape matrix.

From the analysis above, one can see that imposing G_k, $k \geq 0$, to be invertible is sufficient to ensure that F_k, $k \geq 0$, is positive definite. In reality, what is needed is that either $G_k F_k G_k^\top$ is positive definite or $H_k E_k H_k^\top$ is positive definite. Thus, the condition on G_k being invertible can be relaxed if $H_k \in \mathbb{R}^{n \times m}$, $k \geq 0$, is full rank and $n \leq m$; this would ensure that $H_k E_k H_k^\top$, $k \geq 0$, is positive definite.

8.2.2 Choice of Parameter γ_k

One possibility is to simply set γ_k to some positive constant for all $k \geq 0$. However, while this approach is computationally simple because (8.23) is linear in F_k and guarantees that $\mathcal{X}_k \subseteq \mathcal{X}_k^+$ for all $k \geq 0$, we do not know whether $\mathcal{X}_k^+$ "tightly" approximates the reachable set $\mathcal{X}_k$ or it is a loose outer bound. As an alternative, one can choose γ_k so that $\mathcal{X}_k^+$ is optimal in some sense, e.g., its volume is minimum, or the sum of its squared principal semi-axes is minimum; or it is tight in the sense that its boundary touches the boundary of $\mathcal{X}_k$. Next, we explore all these alternatives.

Positive Constant

Consider the system in (8.5) with $\mathcal{W}_k$ and $\mathcal{X}_0$ as given in (8.16) and (8.17), respectively. Let $\mathcal{X}_k^+ \subseteq \mathbb{R}^n$ denote an ellipsoid with center $\overline{x}_k \in \mathbb{R}^n$ and shape matrix $F_k \in \mathbb{R}^{n \times n}$ recursively defined as in (8.22) and (8.23), respectively, with

$$\gamma_k = \gamma > 0, \quad k \geq 0, \tag{8.28}$$

where γ is a positive constant. While this choice guarantees that $\mathcal{X}_k \subseteq \mathcal{X}_k^+$ for all $k \geq 0$, we do not know how "tightly" $\mathcal{X}_k^+$ approximates the reachable set $\mathcal{X}_k$, or even if it will remain bounded as time evolves. It turns out that if the system is time-invariant, i.e., $G_k = G \in \mathbb{R}^{n\times n}$ and $H_k = H \in \mathbb{R}^{n\times m}$, for all $k \geq 0$, then one needs to impose additional constraints on γ to ensure that $\mathcal{X}_k^+$ does not grow unbounded with k. Specifically, let $\lambda_1, \lambda_2, \ldots, \lambda_n$ denote the eigenvalues of G. Assume that all the eigenvalues of G are within the unit circle, i.e., $|\lambda_i| < 1, i = 1, 2, \ldots, n$; this in turn is a sufficient condition for the system trajectories to not grow unbounded. Then, in addition to (8.28), one needs to enforce that

$$\gamma < \frac{1}{\max\{|\lambda_1|^2, |\lambda_2|^2, \ldots, |\lambda_n|^2\}} - 1. \tag{8.29}$$

To see this, we set $G_k = G$, $H_k = H$, and $\gamma_k = \gamma$ in the recursion in (8.23) and rewrite it in vector form using the Kronecker product (see Appendix A) as follows:

$$\mathbf{vec}(F_{k+1}) = (1+\gamma)(G \otimes G)\mathbf{vec}(F_k) + \left(1 + \frac{1}{\gamma}\right)\mathbf{vec}(HE_kH^\top), \tag{8.30}$$

where $\mathbf{vec}(F_k)$ and $\mathbf{vec}(HE_kH^\top)$ are n^2-dimensional vectors obtained by respectively stacking the columns of F_k and $HE_kH^\top$ from left to right.[1]

Now, for the homogeneous part of (8.30) to be asymptotically stable, the eigenvalues of the matrix

$$A := (1+\gamma)(G \otimes G)$$

must be within the unit circle. This in turn is a sufficient condition to ensure that the entries of F_k remain bounded because the entries of E_k are bounded for all $k \geq 0$. Let $\sigma(G)$ denote the spectrum of the matrix G; then the spectrum of the matrix A, $\sigma(A)$, is given by

$$\sigma(A) = \{\lambda \colon \lambda = (1+\gamma)\lambda_i\lambda_j,\ \lambda_i \in \sigma(G), \lambda_j \in \sigma(G)\}. \tag{8.31}$$

Now, since $|\lambda_i| < 1,\ i = 1, 2, \ldots, n$, it follows from (8.31) that, by enforcing

$$\gamma < \frac{1}{\max\{|\lambda_1|^2, |\lambda_2|^2, \ldots, |\lambda_n|^2\}} - 1, \tag{8.32}$$

the magnitude of the eigenvalues of the matrix A will be strictly smaller than one; thus, the homogeneous part of (8.30) is asymptotically stable. Therefore the outer-bounding ellipsoid that results from (8.30) with such choice of γ will not grow unbounded as time evolves.

Note that it is possible that even if some of the eigenvalues of G have magnitude one, the system trajectories might still be bounded if the input is bounded; this in turn would mean that the reachable set, $\mathcal{X}_k$, $k \geq 0$, would not grow unbounded as

[1] Given $A = [a_{i,j}] \in \mathbb{R}^{n\times m}$, we can associate a vector $\mathbf{vec}(A) \in \mathbb{R}^{nm}$ defined as follows:

$$\mathbf{vec}(A) = [a_{1,1}, a_{2,1}, \ldots, a_{n,1}, a_{1,2}, a_{2,2}, \ldots, \ldots, a_{n,2}, \ldots, a_{1,m}, a_{2,m}, \ldots, a_{n,m}]^\top.$$

time progresses. However, the bounding ellipsoid $\mathcal{X}_k^+$, $k \geq 0$, with shape matrix defined by (8.30) would grow unbounded as time progresses because according to (8.31) one or more eigenvalues of the matrix A would have magnitude greater than one. In summary, earlier analysis showed that as long as γ_k is positive for all $k \geq 0$, the recursions in (8.22–8.23) would define an outer-bounding ellipsoid for $\mathcal{X}_k$. However, the analysis here shows that for linear time-invariant systems, such outer-bounding ellipsoids might be very loose if one does not choose the parameter γ_k carefully.

Example 8.1 (Positive constant) Consider the following one-dimensional linear time-invariant system

$$x_{k+1} = \alpha x_k + \beta w_k, \quad k \geq 0, \tag{8.33}$$

where $x_k \in \mathbb{R}$, $w_k \in [w_k^m, w_k^M]$, $w_k^m < w_k^M$, $-1 < \alpha < 1$, and $\beta \neq 0$. Further assume that $x_0 \in [x_0^m, x_0^M]$, $x_0^m < x_0^M$.

One can easily check that the condition $w_k \in \left[w_k^m, w_k^M\right]$ can be equivalently written as follows:

$$w_k \in \mathcal{W}_k = \left\{w \in \mathbb{R} \colon E_k^{-1}\left(w_k - \overline{w}_k\right)^2 \leq 1\right\}, \tag{8.34}$$

where

$$E_k = \left(\frac{w_k^M - w_k^m}{2}\right)^2, \qquad \overline{w}_k = \frac{w_k^m + w_k^M}{2}.$$

Similarly, one can easily check that the condition $x_0 \in \left[x_0^m, x_0^M\right]$ can be equivalently written as follows:

$$x_0 \in \mathcal{X}_0 = \left\{x \in \mathbb{R} \colon F_0^{-1}\left(x_0 - \overline{x}_0\right)^2 \leq 1\right\}, \tag{8.35}$$

where

$$F_0 = \left(\frac{x_0^M - x_0^m}{2}\right)^2, \qquad \overline{x}_0 = \frac{x_0^m + x_0^M}{2}.$$

In this case, the reachable set, $\mathcal{X}_k$, is a real interval; thus, the objective is to compute a real interval, $\mathcal{X}_k^+$, such that $\mathcal{X}_k \subseteq \mathcal{X}_k^+$.

By tailoring the expressions in (8.22–8.23) and (8.28) to the setting here, it follows that

$$\mathcal{X}_k \subseteq \mathcal{X}_k^+ = \left\{x \in \mathbb{R} \colon F_k^{-1}\left(x - \overline{x}_k\right)^2 \leq 1\right\},$$

with $\overline{x}_k \in \mathbb{R}$ and $F_k \in \mathbb{R}$ determined by

$$\overline{x}_{k+1} = \alpha \overline{x}_k + \beta \overline{w}_k, \tag{8.36}$$

$$F_{k+1} = (1 + \gamma)\alpha^2 F_k + \left(1 + \frac{1}{\gamma}\right)\beta^2 E_k, \tag{8.37}$$

where $\gamma > 0$. Thus, since E_k is bounded for all $k \geq 0$, for F_k to remain bounded, in addition to being greater than zero, γ must be chosen such that

$$\gamma < \frac{1}{\alpha^2} - 1,$$

which is clearly a special case of the result in (8.32).

As we did with the sets $\mathcal{W}_k$ and $\mathcal{X}_0$, we can write $\mathcal{X}_k$ as follows:

$$\mathcal{X}_k = \left\{x_k \in \mathbb{R}\colon (F_k^*)^{-1}\left(x_k - \overline{x}_k^*\right)^2 \leq 1\right\}, \tag{8.38}$$

where F_k^* is some positive real number, and $\overline{x}_k^*$ is some real number. Now, on the one hand, we have that

$$\begin{aligned} S_{\mathcal{X}_{k+1}}(\eta) &= \eta\overline{x}_{k+1}^* + \sqrt{\eta^2 F_{k+1}^*} \\ &= \eta\overline{x}_{k+1}^* + |\eta|\sqrt{F_{k+1}^*}, \end{aligned} \tag{8.39}$$

for any $\eta \in \mathbb{R}$. On the other hand, it follows from (8.14) that

$$\begin{aligned} S_{\mathcal{X}_{k+1}}(\eta) &= S_{\mathcal{X}_k}(\alpha\eta) + S_{\mathcal{W}_k}(\beta\eta) \\ &= \eta(\alpha\overline{x}_k^* + \beta\overline{w}_k) + \sqrt{\eta^2\alpha^2 F_k^*} + \sqrt{\eta^2\beta^2 E_k} \\ &= \eta(\alpha\overline{x}_k^* + \beta\overline{w}_k) + |\eta|\left(|\alpha|\sqrt{F_k^*} + |\beta|\sqrt{E_k}\right), \end{aligned} \tag{8.40}$$

for any $\eta \in \mathbb{R}$. Then, by equating the right-hand side of (8.39) and (8.40), we obtain that

$$\overline{x}_{k+1}^* = \alpha\overline{x}_k^* + \beta\overline{w}_k, \tag{8.41}$$

$$F_{k+1}^* = \left(|\alpha|\sqrt{F_k^*} + |\beta|\sqrt{E_k}\right)^2, \tag{8.42}$$

with $\overline{x}_0^* = \overline{x}_0$ and $F_0^* = F_0$. We can now show by induction that indeed $\mathcal{X}_k \subseteq \mathcal{X}_k^+$ for all $k \geq 0$. First note that since $\overline{x}_0^* = \overline{x}_0$, it follows from (8.36) and (8.41) that $\overline{x}_k^* = \overline{x}_k$ for all $k \geq 0$. Now, since $F_0^* = F_0$, it follows that $\mathcal{X}_0 = \mathcal{X}_0^+$. Then, by using (8.25), we obtain that

$$\begin{aligned} F_1^* &= \left(|\alpha|\sqrt{F_0^*} + |\beta|\sqrt{E_0}\right)^2 \\ &\leq (1+\gamma)\alpha^2 F_0^* + \left(1 + \frac{1}{\gamma}\right)\beta^2 E_0 \\ &= (1+\gamma)\alpha^2 F_0 + \left(1 + \frac{1}{\gamma}\right)\beta^2 E_0 \\ &= F_1, \end{aligned} \tag{8.43}$$

for some positive γ; therefore $\mathcal{X}_1 \subseteq \mathcal{X}_1^+$. Now assume that $\mathcal{X}_k \subseteq \mathcal{X}_k^+$; thus, $F_k^* \leq F_k$. Then, by using (8.25) again, we obtain that

$$\begin{aligned}
F_{k+1}^* &= \left(|\alpha|\sqrt{F_k^*} + |\beta|\sqrt{E_k}\right)^2 \\
&\le (1+\gamma)\alpha^2 F_k^* + \left(1 + \frac{1}{\gamma}\right)\beta^2 E_k \\
&\le (1+\gamma)\alpha^2 F_k + \left(1 + \frac{1}{\gamma}\right)\beta^2 E_k \\
&= F_{k+1},
\end{aligned} \tag{8.44}$$

for some positive γ; therefore $\mathcal{X}_{k+1} \subseteq \mathcal{X}_{k+1}^+$.

Minimum Volume Ellipsoid

Consider the system in (8.5) with $\mathcal{W}_k$ and $\mathcal{X}_0$ as given in (8.16) and (8.17), respectively. Let $\mathcal{X}_k^+ \subseteq \mathbb{R}^n$ denote an ellipsoid with center $\overline{x}_k \in \mathbb{R}^n$ and shape matrix $F_k \in \mathbb{R}^{n\times n}$ recursively defined as in (8.22) and (8.23), respectively, with γ_k being the unique solution to the equation

$$\sum_{i=1}^{n} \frac{\psi_{k,i}}{\psi_{k,i} + \gamma_k} = \frac{n\gamma_k}{1+\gamma_k}, \tag{8.45}$$

where $\psi_{k,i}$, $i = 1, 2, \dots, n$, are the roots of the polynomial

$$p_k(\psi) = \mathbf{det}(Q_k - \psi P_k), \tag{8.46}$$

with $P_k = G_k F_k G_k^\top$ and $Q_k = H_k E_k H_k^\top$. Then, $\mathcal{X}_k^+$ is optimal in the sense that its volume is minimum among all the outer-bounding ellipsoids within the class recursively defined by (8.22–8.23). To show this, recall from Section 2.2 that the volume of an ellipsoid $\mathcal{E} \subseteq \mathbb{R}^n$ with shape matrix $E \in \mathbb{R}^{n\times n}$ is given by

$$\mathbf{vol}(\mathcal{E}) = \frac{\pi^{n/2}\sqrt{\mathbf{det}(E)}}{\Gamma(\frac{n}{2}+1)},$$

where $\Gamma(\cdot)$ is Euler's gamma function. Thus, by choosing γ_k so as to minimize $\mathbf{det}(F_{k+1})$ (F_0 is given) for all $k \ge 0$, we ensure that

$$\mathbf{vol}(\mathcal{X}_{k+1}^+) - \mathbf{vol}(\mathcal{X}_k^+)$$

is minimum for all $k \ge 0$. Now, we proceed by induction.

For $k = 0$, $\mathcal{X}_0^+ = \mathcal{X}_0$, thus clearly $\mathbf{vol}(\mathcal{X}_0^+)$ is minimum. For $k = 1$, within the class of ellipsoids recursively defined by (8.22) and (8.23), we can find $\mathcal{X}_1^+$ with minimum volume such that $\mathcal{X}_1 \subseteq \mathcal{X}_1^+$ by finding $\gamma_0 > 0$ that results in the determinant of

$$F_1 = (1+\gamma_0)G_0 F_0 G_0^\top + \left(1 + \frac{1}{\gamma_0}\right) H_0 E_0 H_0^\top$$

being minimum; this can be formulated as the following optimization problem:

$$\underset{\gamma_0>0}{\text{minimize}} \quad \underbrace{\mathbf{det}\left((1+\gamma_0)G_0F_0G_0^\top + \left(1+\frac{1}{\gamma_0}\right)H_0E_0H_0^\top\right)}_{=:\,\varphi(\gamma_0)}. \tag{8.47}$$

Define the following two matrices: $P_0 := G_0F_0G_0^\top$ and $Q_0 := H_0E_0H_0^\top$. Then, since F_0 is symmetric and positive definite and G_0 is invertible, P_0 is symmetric and positive definite. Also, since E_0 is symmetric and positive definite, Q_0 is symmetric and positive (semi)definite. Thus, P_0 and Q_0 can be diagonalized simultaneously (see Appendix A.2), i.e., there exists some nonsingular matrix $V_0 \in \mathbb{R}^{n\times n}$ such that

$$P_0 = V_0V_0^\top, \qquad Q_0 = V_0\Psi_0V_0^\top,$$

with

$$\Psi_0 = \mathbf{diag}(\psi_{0,1}, \psi_{0,2}, \ldots, \psi_{0,n}), \tag{8.48}$$

where $\psi_{0,i} \geq 0,\ i = 1, 2, \ldots, n$, are the roots of the polynomial

$$p_0(\psi) = \mathbf{det}(Q_0 - \psi P_0).$$

Then, we have that

$$\begin{aligned}
\varphi(\gamma_0) &= \mathbf{det}\left((1+\gamma_0)\underbrace{G_0F_0G_0^\top}_{=\,V_0V_0^\top} + \left(1+\frac{1}{\gamma_0}\right)\underbrace{H_0E_0H_0^\top}_{=\,V_0\Psi_0V_0^\top}\right)\\
&= \mathbf{det}\left(V_0\left((1+\gamma_0)I_n + \left(1+\frac{1}{\gamma_0}\right)\Psi_0\right)V_0^\top\right)\\
&= \underbrace{\mathbf{det}(V_0)}_{=:\,c_0}\mathbf{det}\left((1+\gamma_0)I_n + \left(1+\frac{1}{\gamma_0}\right)\Psi_0\right)\underbrace{\mathbf{det}(V_0^\top)}_{=:\,c_0}\\
&= c_0^2\,\mathbf{det}\left((1+\gamma_0)I_n + \left(1+\frac{1}{\gamma_0}\right)\Psi_0\right)\\
&= c_0^2\,\mathbf{det}\left((1+\gamma_0)\left(I_n + \frac{1}{\gamma_0}\Psi_0\right)\right)\\
&= c_0^2\,(1+\gamma_0)^n\mathbf{det}\left(I_n + \frac{1}{\gamma_0}\Psi_0\right).
\end{aligned} \tag{8.49}$$

Now, for any invertible matrix $A(p) \in \mathbb{R}^{n\times n}$, recall the formula in (A.12):

$$\frac{d}{dp}\mathbf{det}(A(p)) = \mathbf{det}(A(p))\,\mathbf{tr}\left(A^{-1}(p)\frac{d}{dp}A(p)\right),$$

which we can use then to differentiate $\varphi(\gamma_0)$ as follows:

$$\begin{aligned}
\frac{d\varphi(\gamma_0)}{d\gamma_0} &= c_o^2 \left(n(1+\gamma_0)^{n-1}\mathbf{det}\Big(I_n + \frac{1}{\gamma_0}\Psi_0\Big) + (1+\gamma_0)^n \frac{d}{d\gamma_0}\mathbf{det}\Big(I_n + \frac{1}{\gamma_0}\Psi_0\Big)\right) \\
&= c_o^2\, n(1+\gamma_0)^{n-1}\mathbf{det}\Big(I_n + \frac{1}{\gamma_0}\Psi_0\Big) \\
&\quad - c_o^2\,(1+\gamma_0)^n\mathbf{det}\Big(I_n + \frac{1}{\gamma_0}\Psi_0\Big)\mathbf{tr}\left(\Big(I_n + \frac{1}{\gamma_0}\Psi_0\Big)^{-1}\frac{1}{\gamma_0^2}\Psi_0\right) \\
&= c_o^2\,(1+\gamma_0)^{n-1}\mathbf{det}\Big(I_n + \frac{1}{\gamma_0}\Psi_0\Big) \\
&\quad \times \left[n - (1+\gamma_0)\mathbf{tr}\left(\Big(I_n + \frac{1}{\gamma_0}\Psi_0\Big)^{-1}\frac{1}{\gamma_0^2}\Psi_0\right)\right].
\end{aligned} \tag{8.50}$$

Now, since

$$(1+\gamma_0)^{n-1}\mathbf{det}\Big(I_n + \frac{1}{\gamma_0}\Psi_0\Big) > 0$$

for any $\gamma_0 > 0$, it follows that $\frac{d\varphi(\gamma_0)}{d\gamma_0} = 0$ if and only if

$$n - \frac{1+\gamma_0}{\gamma_0}\mathbf{tr}\left(\Big(I_n + \frac{1}{\gamma_0}\Psi_0\Big)^{-1}\frac{1}{\gamma_0}\Psi_0\right) = 0. \tag{8.51}$$

By using (8.48), one can check that

$$\Big(I_n + \frac{1}{\gamma_0}\Psi_0\Big)^{-1}\frac{1}{\gamma_0}\Psi_0 = \mathbf{diag}\left(\frac{\psi_{0,1}}{\psi_{0,1}+\gamma_0}, \frac{\psi_{0,2}}{\psi_{0,2}+\gamma_0}, \ldots, \frac{\psi_{0,n}}{\psi_{0,n}+\gamma_0}\right); \tag{8.52}$$

thus,

$$\mathbf{tr}\left(\Big(I_n + \frac{1}{\gamma_0}\Psi_0\Big)^{-1}\frac{1}{\gamma_0}\Psi_0\right) = \sum_{i=1}^{n}\frac{\psi_{0,i}}{\psi_{0,i}+\gamma_0}, \tag{8.53}$$

which by substituting in (8.51) and rearranging yields

$$\sum_{i=1}^{n}\frac{\psi_{0,i}}{\psi_{0,i}+\gamma_0} = \frac{n\gamma_0}{1+\gamma_0}. \tag{8.54}$$

Now, to see that (8.54) has a unique root, define the following two functions:

$$\alpha_0(\gamma_0) := \sum_{i=1}^{n}\frac{\psi_{0,i}}{\psi_{0,i}+\gamma_0}, \qquad \beta_0(\gamma_0) := \frac{n\gamma_0}{1+\gamma_0};$$

one can easily check that $\frac{d\alpha_0(\gamma_0)}{d\gamma_0} < 0$ for all $\gamma_0 \geq 0$ and $\frac{d\beta_0(\gamma_0)}{d\gamma_0} > 0$; thus, $\alpha_0(\cdot)$ is a strictly monotonically decreasing function in $[0, \infty)$ and $\beta_0(\cdot)$ is a strictly monotonically increasing function in $[0, \infty)$. Then, since

$$\alpha_0(0) = n, \qquad \lim_{\gamma_0 \to \infty}\alpha_0(\gamma_0) = 0,$$

and

$$\beta_0(0) = 0, \qquad \lim_{\gamma_0 \to \infty} \beta_0(\gamma_0) = n,$$

and because of the monotonicity properties of $\alpha_0(\cdot)$ and $\beta_0(\cdot)$, we conclude that there exists a unique $\gamma_0 > 0$ such that $\alpha_0(\gamma_0) = \beta_0(\gamma_0)$.

For $k > 0$, assume that we are given some ellipsoid $\mathcal{X}_k^+$ with center $\overline{x}_k$ and shape matrix F_k that has minimum volume among those within the class of ellipsoids recursively defined by (8.22–8.23). Then, in order to find $\mathcal{X}_{k+1}^+$ with minimum volume in the class of ellipsoids recursively defined by (8.22–8.23), we need to find γ_k that results in the determinant of

$$F_{k+1} = (1+\gamma_k)G_k F_k G_k^\top + \left(1 + \frac{1}{\gamma_k}\right) H_k E_k H_k^\top$$

being minimum. Define

$$\varphi_k(\gamma_k) := \mathbf{det}\left((1+\gamma_k)P_k + \left(1 + \frac{1}{\gamma_k}\right) Q_k \right),$$

where $P_k = G_k F_k G_k^\top$ and $Q_k = H_k E_k H_k^\top$. Because F_k is symmetric and positive definite and G_k is invertible, P_k is also symmetric and positive definite. Also, because E_k is symmetric and positive definite, Q_k is symmetric and positive (semi)definite. Thus, P_k and Q_k are simultaneously diagonalizable, i.e., there exists some invertible matrix $V_k \in \mathbb{R}^{n \times n}$ so that

$$P_k = V_k V_k^\top, \qquad Q_k = V_k \Psi_k V_k^\top,$$

with

$$\Psi_k = \mathbf{diag}(\psi_{k,1}, \psi_{k,2}, \ldots, \psi_{k,n}), \tag{8.55}$$

where $\psi_{k,i} \geq 0,\ i = 1, 2, \ldots, n$, are the roots of the polynomial

$$p_k(\psi) = \mathbf{det}(Q_k - \psi P_k) = 0.$$

Then, we have that

$$\varphi_k(\gamma_k) = c_k^2\, \mathbf{det}\left((1+\gamma_k) I_n + \left(1 + \frac{1}{\gamma_k}\right) \Psi_k \right),$$

where $c_k = \mathbf{det}(V_k)$; thus,

$$\frac{d\varphi_k(\gamma_k)}{d\gamma_k} = \underbrace{c_k^2\ (1+\gamma_k)^{n-1} \mathbf{det}\left(I_n + \frac{1}{\gamma_k}\Psi_k \right)}_{>0 \text{ for any } \gamma_k > 0} \times \left[n - (1+\gamma_k)\mathbf{tr}\left(\left(I_n + \frac{1}{\gamma_k}\Psi_k \right)^{-1} \frac{1}{\gamma_k^2}\Psi_0 \right) \right]. \tag{8.56}$$

Therefore, $\frac{d\varphi_k(\gamma_k)}{d\gamma_k} = 0$ if and only if

$$n - \frac{1+\gamma_k}{\gamma_k}\mathbf{tr}\left(\left(I_n + \frac{1}{\gamma_k}\Psi_k\right)^{-1}\frac{1}{\gamma_k}\Psi_k\right) = 0. \tag{8.57}$$

Now, one can check that

$$\mathbf{tr}\left(\left(I_n + \frac{1}{\gamma_k}\Psi_k\right)^{-1}\frac{1}{\gamma_k}\Psi_k\right) = \sum_{i=1}^{n}\frac{\psi_{k,i}}{\psi_{k,i}+\gamma_k}, \tag{8.58}$$

which by substituting in (8.57) and rearranging yields

$$\underbrace{\sum_{i=1}^{n}\frac{\psi_{k,i}}{\psi_{k,i}+\gamma_k}}_{=:\,\alpha_k(\gamma_k)} = \underbrace{\frac{n\gamma_k}{1+\gamma_k}}_{=:\,\beta_k(\gamma_k)}. \tag{8.59}$$

Finally, one can check that (8.59) has a unique root by verifying that $\alpha_k(\cdot)$ $\big(\beta_k(\cdot)\big)$ is a strictly monotonically increasing (decreasing) function in $[0,\infty)$, and

$$\alpha_k(0) = n, \qquad \lim_{\gamma_k\to\infty}\alpha_k(\gamma_k) = 0,$$

$$\beta_k(0) = 0, \qquad \lim_{\gamma_k\to\infty}\beta_k(\gamma_k) = n;$$

thus, it must exist a unique $\gamma_k > 0$ such that $\alpha_k(\gamma_k) = \beta_k(\gamma_k)$.

Example 8.2 (Minimum volume ellipsoid) Consider again the one-dimensional system in Example 8.1 as described by (8.33–8.35). Tailoring the expressions in (8.22–8.23) and (8.45–8.46) to the setting here yields

$$\mathcal{X}_k \subseteq \mathcal{X}_k^+ = \left\{x \in \mathbb{R}^n : F_k^{-1}\big(x - \overline{x}_k\big)^2 \leq 1\right\},$$

where $\overline{x}_k \in \mathbb{R}$ and $F_k \in \mathbb{R}$ are determined by

$$\overline{x}_{k+1} = \alpha\overline{x}_k + \beta\overline{w}_k, \tag{8.60}$$

$$F_{k+1} = (1+\gamma_k)\alpha^2 F_k + \left(1 + \frac{1}{\gamma_k}\right)\beta^2 E_k, \tag{8.61}$$

with γ_k satisfying

$$\frac{\psi_k}{\psi_k + \gamma_k} = \frac{\gamma_k}{1+\gamma_k}, \tag{8.62}$$

where ψ_k is such that

$$\beta^2 E_k - \psi_k\alpha^2 F_k = 0. \tag{8.63}$$

Then, solving for ψ_k in (8.63) and plugging the result into (8.62) yields

$$\frac{\beta^2 E_k}{\beta^2 E_k + \gamma_k\alpha^2 F_k} = \frac{\gamma_k}{1+\gamma_k}, \tag{8.64}$$

and after some manipulation, we obtain that

$$\gamma_k = \left|\frac{\beta}{\alpha}\right| \sqrt{\frac{E_k}{F_k}}. \tag{8.65}$$

In this case, it turns out that $\mathcal{X}_k^+ = \mathcal{X}_k$ for all $k \geq 0$. To see this, first recall from Example 8.1 that

$$\mathcal{X}_k = \left\{x_k \in \mathbb{R} \colon (F_k^*)^{-1}\left(x_k - \overline{x}_k^*\right)^2 \leq 1\right\}, \quad k \geq 0,$$

where F_k^* and $\overline{x}_k^*$ are determined by

$$\overline{x}_{k+1}^* = \alpha \overline{x}_k^* + \beta \overline{w}_k, \tag{8.66}$$

$$F_{k+1}^* = \left(|\alpha|\sqrt{F_k^*} + |\beta|\sqrt{E_k}\right)^2, \tag{8.67}$$

with $\overline{x}_0^* = \overline{x}_0$ and $F_0^* = F_0$. Then, by plugging the expression for γ_k in (8.65) into (8.61), we obtain that

$$\begin{aligned} F_{k+1} &= \left(1 + \left|\frac{\beta}{\alpha}\right|\sqrt{\frac{E_k}{F_k}}\right)\alpha^2 F_k + \left(1 + \left|\frac{\alpha}{\beta}\right|\sqrt{\frac{F_k}{E_k}}\right)\beta^2 E_k \\ &= \left(|\alpha|\sqrt{F_k} + |\beta|\sqrt{E_k}\right)^2, \end{aligned} \tag{8.68}$$

but this recursion coincides with that in (8.67); thus clearly $F_k^* = F_k$ for all $k \geq 0$. Also, the recursion in (8.66) coincides with that in (8.60). Thus, clearly $\overline{x}_k^* = \overline{x}_k$ for all $k \geq 0$. Therefore, $\mathcal{X}_k^+ = \mathcal{X}_k$, for all $k \geq 0$. This result should not be surprising because in order to obtain (8.60–8.61), we need to minimize the volume of $\mathcal{X}_k^+$ such that $\mathcal{X}_k \subseteq \mathcal{X}_k^+$, but in this case $\mathcal{X}_k^+$ is just a real interval, so $\mathcal{X}_k^+$ is the minimum-length interval containing $\mathcal{X}_k$; thus, $\mathcal{X}_k^+$ has to be identically equal to $\mathcal{X}_k$.

Ellipsoid with Minimum Sum of Squares of the Semi-Axis Lengths

Consider the system in (8.5) with $\mathcal{W}_k$ and $\mathcal{X}_0$ as given in (8.16) and (8.17), respectively. Let $\mathcal{X}_k^+ \subseteq \mathbb{R}^n$ denote an ellipsoid with center $\overline{x}_k \in \mathbb{R}^n$ and shape matrix $F_k \in \mathbb{R}^{n\times n}$ recursively defined as in (8.22) and (8.23), respectively, for given $\overline{x}_0$ and F_0, with γ_k chosen as follows:

$$\gamma_k = \sqrt{\frac{\mathbf{tr}\left(H_k E_k H_k^\top\right)}{\mathbf{tr}\left(G_k F_k G_k^\top\right)}}. \tag{8.69}$$

This choice of γ_k makes the sum of squares of the semi-axis lengths of the resulting outer-bounding ellipsoid $\mathcal{X}_k^+$ minimum among all outer-bounding ellipsoids within the class recursively defined by (8.22) and (8.23). In this regard, recall from the discussion on ellipsoids in Section 2.2.2 that the eigenvalues of the matrix F_k are the squares of the semi-axis lengths; thus, in order to establish the result, we need to show that the trace of F_k is minimum for all k, for which we use induction.

For $k = 0$, $\mathcal{X}_0^+ = \mathcal{X}_0$, thus clearly the sum of squares of the semi-axis lengths of the ellipsoid $\mathcal{X}_0^+$ is minimum. For $k = 1$, with $\mathcal{X}_1^+$ chosen in the class of ellipsoids recursively defined by (8.22–8.23), we have that

$$\begin{aligned}\mathbf{tr}(F_1) &= \mathbf{tr}\left((1+\gamma_0)G_0F_0G_0^\top + \left(1+\frac{1}{\gamma_0}\right)H_0E_0H_0^\top\right)\\ &= (1+\gamma_0)\mathbf{tr}\big(G_0F_0G_0^\top\big) + \left(1+\frac{1}{\gamma_0}\right)\mathbf{tr}\big(H_0E_kH_0^\top\big)\\ &=: \phi_0(\gamma_0). \end{aligned} \tag{8.70}$$

Now, in order to obtain the value of γ_0 that minimizes $\mathbf{tr}(F_1)$, we differentiate $\phi_0(\gamma_0)$ with respect to γ_0, which yields

$$\frac{d\phi_0(\gamma_0)}{d\gamma_0} = \mathbf{tr}\big(G_0F_0G_0^\top\big) - \frac{1}{\gamma_0^2}\mathbf{tr}\big(H_0E_0H_0^\top\big). \tag{8.71}$$

Note that $\mathbf{tr}\big(G_0F_0G_0^\top\big) > 0$ because $G_0F_0G_0^\top$ is positive definite. Also, because $H_0E_0H_0^\top$ is positive (semi)definite, we have that $\mathbf{tr}\big(H_0E_0H_0^\top\big) > 0$ (even if $H_0E_0H_0^\top$ is positive semidefinite, its trace cannot be identically equal to zero because at least one eigenvalue of $H_0E_0H_0^\top$ has to be positive because $H_0 \neq \mathbf{0}_{n\times m}$). Then, by equating (8.71) to zero and solving for γ_0, we obtain

$$\gamma_0^{(1)} = \sqrt{\frac{\mathbf{tr}\big(H_0E_0H_0^\top\big)}{\mathbf{tr}\big(G_0F_0G_0^\top\big)}}, \quad \gamma_0^{(2)} = -\sqrt{\frac{\mathbf{tr}\big(H_0E_0H_0^\top\big)}{\mathbf{tr}\big(G_0F_0G_0^\top\big)}},$$

but recall from (8.23) that γ_k must be greater than zero for all $k \geq 0$; thus, only the positive solution is valid. Now, in order to check that $\gamma_0^{(1)}$ is indeed a minimum, we differentiate $\phi_0(\gamma_0)$ twice with respect to γ_0, which yields

$$\frac{d^2\phi_0(\gamma_0)}{d\gamma_0^2} = \frac{2}{\gamma_0^3}\mathbf{tr}\big(H_0E_0H_0^\top\big), \tag{8.72}$$

and by evaluating for $\gamma_0^{(1)}$, we obtain:

$$\begin{aligned}\left.\frac{d^2\phi_0(\gamma_0)}{d\gamma_0^2}\right|_{\gamma_0=\gamma_0^{(1)}} &= 2\sqrt{\frac{\Big(\mathbf{tr}\big(G_0F_0G_0^\top\big)\Big)^3}{\mathbf{tr}\big(H_0E_0H_0^\top\big)}}\\ &> 0; \end{aligned} \tag{8.73}$$

thus, the sum of squares of the semi-axis lengths of the ellipsoid $\mathcal{X}_1^+$ is minimum among those within the class of ellipsoids recursively defined by (8.22) and (8.23).

For $k > 0$, assume we are given some ellipsoid $\mathcal{X}_k^+$ with center $\overline{x}_k$ and shape matrix F_k with sum of squares of the lengths of the semi-axes being minimum among those ellipsoids within the class recursively defined by (8.22–8.23). Then, in order to find the ellipsoid $\mathcal{X}_{k+1}^+$ with sum of squares of the lengths of the semi-axes being minimum in the class of ellipsoids recursively defined by (8.22–8.23), we need to find the value of γ_k that minimizes the trace of F_{k+1} as defined in (8.23). To do so, we follow a similar procedure to that used for computing the optimal γ_0 above, i.e., we define

$$\phi_k(\gamma_k) := (1+\gamma_k)\mathbf{tr}(G_k F_k G_k^\top) + \left(1+\frac{1}{\gamma_k}\right)\mathbf{tr}(H_k E_k H_k^\top),$$

compute its first derivative, equate the resulting expression to zero, and solve for γ_k; this procedure yields again two solutions as follows:

$$\gamma_k^{(1)} = \sqrt{\frac{\mathbf{tr}(H_k E_k H_k^\top)}{\mathbf{tr}(G_k F_k G_k^\top)}}, \quad \gamma_k^{(2)} = -\sqrt{\frac{\mathbf{tr}(H_k E_k H_k^\top)}{\mathbf{tr}(G_k F_k G_k^\top)}}, \tag{8.74}$$

(as for $k=0$, one can check that $\mathbf{tr}(G_k F_k G_k^\top) > 0$ and $\mathbf{tr}(H_k E_k H_k^\top) > 0$), and as before, only $\gamma_k^{(1)}$ is valid because of the constraint that $\gamma_k > 0$ for all k. The final step is to check that $\gamma_k^{(1)}$ is indeed a minimum, which as before, we can accomplish by computing the second derivative of $\phi_k(\gamma_k)$ and verifying that the resulting expression is positive for $\gamma_k = \gamma_k^{(1)}$; we omit the details as the expressions are identical to those derived for the case when $k=0$ if one replaces 0 by k in all subindices of (8.72) and (8.73). Then, we can conclude that the sum of squares of the semi-axis lengths of the ellipsoid $\mathcal{X}_k^+$ is minimum among those within the class of ellipsoids recursively defined by (8.22) and (8.23).

Example 8.3 (Ellipsoid with minimum sum of squares of the semi-axis lengths) Consider again the one-dimensional system in Examples 8.1 and 8.2 as described by (8.33–8.35). By tailoring the expressions in (8.22–8.23) and (8.69) to the setting here, it follows that

$$\mathcal{X}_k \subseteq \mathcal{X}_k^+ = \left\{x \in \mathbb{R} \colon F_k^{-1}(x-\overline{x}_k)^2 \leq 1\right\},$$

where $\overline{x}_k \in \mathbb{R}$ and $F_k \in \mathbb{R}$ are determined by

$$\overline{x}_{k+1} = \alpha\overline{x}_k + \beta\overline{w}_k, \tag{8.75}$$

$$F_{k+1} = (1+\gamma_k)\alpha^2 F_k + \left(1+\frac{1}{\gamma_k}\right)\beta^2 E_k, \tag{8.76}$$

with

$$\gamma_k = \sqrt{\frac{\beta^2 E_k}{\alpha^2 F_k}} = \left|\frac{\beta}{\alpha}\right|\sqrt{\frac{E_k}{F_k}}. \tag{8.77}$$

Note that the expression in (8.77) matches that in (8.65). This should not be surprising given that, in obtaining (8.77), we are minimizing $\mathbf{tr}(F_k)$, whereas in obtaining (8.65), we are minimizing $\mathbf{det}(F_k)$, but in this case, $\mathbf{tr}(F_k) = \mathbf{det}(F_k)$ since F_k is a scalar.

Tight Ellipsoid

Consider the system in (8.5) with $\mathcal{W}_k$ and $\mathcal{X}_0$ as given in (8.16) and (8.17), respectively. Here, we will additionally assume that the matrix $H_k \in \mathbb{R}^{n\times m}$ is such that $n \leq m$ and $\mathbf{rank}(H_k) = n$ for all $k \geq 0$. Let $\mathcal{X}_k^+ \subseteq \mathbb{R}^n$ denote an ellipsoid with center $\overline{x}_k \in \mathbb{R}^n$ and shape matrix $F_k \in \mathbb{R}^{n\times n}$ recursively defined as in (8.22) and (8.23), respectively, for given $\overline{x}_0$ and F_0, with γ_k chosen as follows:

$$\gamma_k = \sqrt{\frac{\eta_k^\top G_k^{-1} H_k E_k H_k^\top \left(G_k^\top\right)^{-1} \eta_k}{\eta_k^\top F_k \eta_k}}, \tag{8.78}$$

where η_k is recursively defined by

$$\eta_{k+1} = (G_k^\top)^{-1} \eta_k, \tag{8.79}$$

for given η_0. Then, the outer-bounding ellipsoid $\mathcal{X}_k^+$ is tight to $\mathcal{X}_k$ in the sense that the boundaries of $\mathcal{X}_k^+$ and $\mathcal{X}_k$ touch each other at a point x_k^* defined as follows:

$$x_k^* = \overline{x}_k + \frac{1}{\sqrt{\eta_k^\top F_k \eta_k}} F_k \eta_k. \tag{8.80}$$

The additional conditions on H_k ensure that the matrix $H_k E_k H_k^\top$ is positive definite for all $k \geq 0$, which in turn ensures that $\gamma_k > 0$ for all $k \geq 0$.

To establish the result in (8.80) we need to show that $S_{\mathcal{X}_k}(\eta_k) = S_{\mathcal{X}_k^+}(\eta_k)$ for all $k \geq 0$. To this end, we use induction, for which we will need the following two facts (which were established in Section 2.2):

G1. Given two closed and convex sets $\mathcal{X}$ and $\mathcal{Y}$, if $S_\mathcal{X}(\eta) = S_\mathcal{Y}(\eta)$ for some η, then the boundaries of $\mathcal{X}$ and $\mathcal{Y}$ touch each other at the point

$$\begin{aligned} x^*(\eta) &= \arg \underbrace{\max_{x\in\mathcal{X}} \eta^\top x}_{= S_\mathcal{X}(\eta)} \\ &= \arg \underbrace{\max_{x\in\mathcal{Y}} \eta^\top x}_{= S_\mathcal{Y}(\eta)}. \end{aligned} \tag{8.81}$$

G2. For an ellipsoid

$$\mathcal{E} = \left\{x \in \mathbb{R}^n \colon (x - \overline{x})^\top E^{-1} (x - \overline{x}) \leq 1\right\},$$

we have that

$$\begin{aligned} x^*(\eta) &= \arg \max_{x\in\mathcal{E}} \eta^\top x \\ &= \overline{x} + \frac{1}{\sqrt{\eta^\top E \eta}} E\eta, \end{aligned} \tag{8.82}$$

for any $\eta \in \mathbb{R}^n$.

For $k = 0$, $\mathcal{X}_0 = \mathcal{X}_0^+$, thus it trivially follows that the boundaries of $\mathcal{X}_0^+$ and $\mathcal{X}_0$ touch each other at all points, and in particular they touch at

$$x_0^* = \overline{x}_0 + \frac{1}{\sqrt{\eta_0^\top F_0 \eta_0}} F_0 \eta_0.$$

On the one hand, by using (8.14) and (8.79) and the fact that $S_{\mathcal{X}_0}(\eta_0) = S_{\mathcal{X}_0^+}(\eta_0)$, we have that

$$\begin{aligned} S_{\mathcal{X}_1}(\eta_1) &= S_{\mathcal{X}_0}(G_0^\top \eta_1) + S_{\mathcal{W}_0}(H_0^\top \eta_1) \\ &= S_{\mathcal{X}_0}(\eta_0) + S_{\mathcal{W}_0}(H_0^\top (G_0^\top)^{-1} \eta_0) \\ &= S_{\mathcal{X}_0^+}(\eta_0) + S_{\mathcal{W}_0}(H_0^\top (G_0^\top)^{-1} \eta_0) \\ &= \eta_0^\top (\overline{x}_0 + G_0^{-1} H_0 \overline{w}_0) \\ &\quad + \sqrt{\eta_0^\top F_0 \eta_0} + \sqrt{\eta_0^\top G_0^{-1} H_0 E_0 H_0^\top (G_0^\top)^{-1} \eta_0}. \end{aligned} \tag{8.83}$$

On the other hand, by using (8.22) and (8.23), we have that

$$\begin{aligned} S_{\mathcal{X}_1^+}(\eta_1) &= \eta_1^\top \overline{x}_1 + \sqrt{\eta_1^\top F_1 \eta_1} \\ &= \eta_1^\top (G_0 \overline{x}_0 + H_0 \overline{w}_0) \\ &\quad + \sqrt{\eta_1^\top \left((1+\gamma_0) G_0 F_0 G_0^\top + \left(1 + \frac{1}{\gamma_0}\right) H_0 E_0 H_0^\top \right) \eta_1} \\ &= \eta_0^\top (\overline{x}_0 + G_0^{-1} H_0 \overline{w}_0) \\ &\quad + \sqrt{\underbrace{\eta_0^\top \left((1+\gamma_0) F_0 + \left(1 + \frac{1}{\gamma_0}\right) G_0^{-1} H_0 E_0 H_0^\top (G_0^\top)^{-1} \right) \eta_0}_{=: R_0}}. \end{aligned} \tag{8.84}$$

Then, by using (8.78), one can check that

$$R_0 = \left(\sqrt{\eta_0^\top F_0 \eta_0} + \sqrt{\eta_0^\top G_0^{-1} H_0 E_0 H_0^\top (G_0^\top)^{-1} \eta_0} \right)^2, \tag{8.85}$$

and by plugging into (8.84), we obtain

$$\begin{aligned} S_{\mathcal{X}_1^+}(\eta_1) &= \eta_0^\top (\overline{x}_0 + G_0^{-1} H_0 \overline{w}_0) \\ &\quad + \sqrt{\eta_0^\top F_0 \eta_0} + \sqrt{\eta_0^\top G_0^{-1} H_0 E_0 H_0^\top (G_0^\top)^{-1} \eta_0}, \end{aligned} \tag{8.86}$$

which matches the expression for $S_{\mathcal{X}_1}(\eta_1)$ in (8.83); thus, $S_{\mathcal{X}_1}(\eta_1) = S_{\mathcal{X}_1^+}(\eta_1)$. Now, in light of (8.81) and (8.82), we have that the boundaries of $\mathcal{X}_1^+$ and $\mathcal{X}_1$ touch each other at

$$x_1^* = \overline{x}_1 + \frac{1}{\sqrt{\eta_1^\top F_1 \eta_1}} F_1 \eta_1.$$

For $k > 0$, assume that $S_{\mathcal{X}_k}(\eta_k) = S_{\mathcal{X}_k^+}(\eta_k)$; thus, by (8.81) and (8.82), the boundaries of $\mathcal{X}_k$ and $\mathcal{X}_k^+$ touch each other at

$$x_k^* = \overline{x}_k + \frac{1}{\sqrt{\eta_k^\top F_k \eta_k}} F_k \eta_k. \tag{8.87}$$

On the one hand, by using (8.14) and (8.79) and taking into account that $S_{\mathcal{X}_k}(\eta_k) = S_{\mathcal{X}_k^+}(\eta_k)$, we have that

$$\begin{aligned} S_{\mathcal{X}_{k+1}}(\eta_{k+1}) &= S_{\mathcal{X}_k}(G_k^\top \eta_{k+1}) + S_{\mathcal{W}_k}(H_k^\top \eta_{k+1}) \\ &= S_{\mathcal{X}_k}(\eta_k) + S_{\mathcal{W}_k}(H_k^\top (G_k^\top)^{-1} \eta_k) \\ &= S_{\mathcal{X}_k^+}(\eta_k) + S_{\mathcal{W}_k}(H_k^\top (G_k^\top)^{-1} \eta_k) \\ &= \eta_k^\top (\overline{x}_k + G_k^{-1} H_k \overline{w}_k) \\ &\quad + \sqrt{\eta_k^\top F_k \eta_k} + \sqrt{\eta_k^\top G_k^{-1} H_k E_k H_k^\top (G_k^\top)^{-1} \eta_k}. \end{aligned} \tag{8.88}$$

On the other hand, by using (8.22) and (8.23), we have that

$$\begin{aligned} S_{\mathcal{X}_{k+1}^+}(\eta_{k+1}) &= \eta_{k+1}^\top \overline{x}_{k+1} + \sqrt{\eta_{k+1}^\top F_{k+1} \eta_{k+1}} \\ &= \eta_{k+1}^\top (G_k \overline{x}_k + H_k \overline{w}_k) \\ &\quad + \sqrt{\eta_{k+1}^\top \left((1+\gamma_k) G_k F_k G_k^\top + \left(1 + \frac{1}{\gamma_k}\right) H_k E_k H_k^\top \right) \eta_{k+1}} \\ &= \eta_k^\top (\overline{x}_k + G_k^{-1} H_k \overline{w}_k) \\ &\quad + \sqrt{\underbrace{\eta_k^\top \left((1+\gamma_k) F_k + \left(1 + \frac{1}{\gamma_k}\right) G_k^{-1} H_k E_k H_k^\top (G_k^\top)^{-1} \right) \eta_k}_{=:\, R_k}}. \end{aligned} \tag{8.89}$$

Then, by using (8.78), one can check that

$$R_k = \left(\sqrt{\eta_k^\top F_k \eta_k} + \sqrt{\eta_k^\top G_k^{-1} H_k E_k H_k^\top (G_k^\top)^{-1} \eta_k} \right)^2, \tag{8.90}$$

and by plugging into (8.89), we obtain that

$$\begin{aligned} S_{\mathcal{X}_{k+1}^+}(\eta_{k+1}) &= \eta_k^\top (\overline{x}_k + G_k^{-1} H_k \overline{w}_k) \\ &\quad + \sqrt{\eta_k^\top F_k \eta_k} + \sqrt{\eta_k^\top G_k^{-1} H_k E_k H_k^\top (G_k^\top)^{-1} \eta_k}, \end{aligned} \tag{8.91}$$

which matches the expression for $S_{\mathcal{X}_{k+1}}(\eta_{k+1})$ in (8.88); thus, we have that $S_{\mathcal{X}_{k+1}}(\eta_{k+1}) = S_{\mathcal{X}_{k+1}^+}(\eta_{k+1})$. Finally, by (8.81) and (8.82), we have that the boundaries of $\mathcal{X}_{k+1}^+$ and $\mathcal{X}_{k+1}$ touch each other at

$$x_{k+1}^* = \overline{x}_{k+1} + \frac{1}{\sqrt{\eta_{k+1}^\top F_{k+1} \eta_{k+1}}} F_{k+1} \eta_{k+1}.$$

Example 8.4 (Tight ellipsoid) Consider again the one-dimensional system in Examples 8.1–8.3 as described by (8.33–8.35). By tailoring the expressions in (8.22–8.23) and (8.78–8.79) to the setting here, it follows that

$$\mathcal{X}_k \subseteq \mathcal{X}_k^+ = \left\{x \in \mathbb{R} \colon F_k^{-1}\left(x - \overline{x}_k\right)^2 \leq 1\right\},$$

where $\overline{x}_k \in \mathbb{R}$ and $F_k \in \mathbb{R}$ are determined by

$$\overline{x}_{k+1} = \alpha \overline{x}_k + \beta \overline{w}_k, \tag{8.92}$$

$$F_{k+1} = \left(1 + \gamma_k\right)\alpha^2 F_k + \left(1 + \frac{1}{\gamma_k}\right)\beta^2 E_k, \tag{8.93}$$

where

$$\begin{aligned}\gamma_k &= \sqrt{\frac{\beta^2 E_k \left(\alpha^{-1}\eta_k\right)^2}{F_k \eta_k^2}} \\ &= \left|\frac{\beta}{\alpha}\right| \sqrt{\frac{E_k}{F_k}}.\end{aligned} \tag{8.94}$$

Note that the expression for γ_k above coincides with that obtained in Examples 8.2–8.3, and, as in those examples, it yields the exact $\mathcal{X}_k$. Also note that in this case, η_k is a scalar, and while the theory determines that its evolution is governed by

$$\eta_{k+1} = \alpha^{-1}\eta_k, \tag{8.95}$$

for given $\gamma_0 \in \mathbb{R}$, in practice, it does not play any role in the computation of γ_k.

Reachable Set Exact and Approximate Characterization

Tight ellipsoids can be utilized to describe the reachable set $\mathcal{X}_k$ exactly. Let $\mathcal{X}_k^+(\xi)$ denote an ellipsoid with center $\overline{x}_k \in \mathbb{R}^n$ and shape matrix $F_k \in \mathbb{R}^{n \times n}$ recursively defined as in (8.22) and (8.23), where γ_k is chosen as in (8.78–8.79) with $\eta_0 = \xi \in \mathcal{B} = \left\{\nu \in \mathbb{R}^n \colon \nu^\top \nu = 1\right\}$. Then, we have that

$$\mathcal{X}_k = \bigcap_{\xi \in \mathcal{B}} \mathcal{X}_k^+(\xi). \tag{8.96}$$

While the formula above implies that in order to exactly characterize the reachable set it is necessary to compute an infinite number of tight bounding ellipsoids, in practice, highly accurate approximations can be obtained with just a few of such tight bounding ellipsoids. More generally, approximations to the reachable set can be obtained as the intersection of a family of bounding ellipsoids obtained by using (8.22) and (8.23) with γ_k chosen to be constant. Also, note that the intersection of a minimum volume ellipsoid and an ellipsoid with smallest sum of squares of the semi-axis lengths also results in an outer-bounding set.

Discussion on Computational Burden

The computation of a minimum-volume ellipsoid $\mathcal{X}_k^+$ for all k involves finding a matrix V_k that simultaneously diagonalizes $P_k = G_k F_k G_k^\top$ and $Q_k = H_k E_k H_k^\top$ and then numerically solving (8.45) to compute the value of γ_k; thus, it may be computationally expensive for large systems. The computation of ellipsoids with minimum sum of squares of the semi-axis lengths involves computing products of matrices, namely $G_k F_k G_k^\top$ and $H_k E_k H_k^\top$; thus, it is relatively efficient in terms of computational cost as compared to obtaining ellipsoids with minimum volume. The computation of "tight" ellipsoids involves computing the inverse of G_k and products of matrices and vectors of the form $\eta_k^\top G_k^{-1} H_k E_k H_k^\top \left(G_k^\top\right)^{-1} \eta_k$ and $\eta_k^\top F_k \eta_k$; thus, it is more computationally expensive than computing ellipsoids with minimum sum of squares of the semi-axis lengths, but still less computationally expensive than computing ellipsoids with minimum volume.

8.2.3 Deterministic Inputs

In (8.5) we have not considered deterministic inputs; however, these can be easily incorporated as detailed next. Consider the system

$$x_{k+1} = G_k x_k + H_k w_k + L_k u_k, \tag{8.97}$$

where $L_k \in \mathbb{R}^{n \times l}$ and u_k are known for all $k \geq 0$, and G_k, H_k, x_k, and w_k are as in (8.5). Define $z_k = x_k - x_k^\circ$, where x_k° is governed by the following recursion:

$$x_{k+1}^\circ = G_k x_k^\circ + L_k u_k, \tag{8.98}$$

with $x_0^\circ = \mathbf{0}_n$. Then, it follows that

$$\begin{aligned} z_{k+1} &= x_{k+1} - x_{k+1}^\circ \\ &= G_k x_k + H_k w_k + L_k u_k - \left(G_k x_k^\circ + L_k u_k\right) \\ &= G_k z_k + H_k w_k, \end{aligned} \tag{8.99}$$

where $w_k \in \mathcal{W}_k$ and $z_0 \in \mathcal{Z}_0 = \mathcal{X}_0$; this system is now in the form of the system in (8.5). Let $\mathcal{Z}_k$ denote the reachable set of (8.99); then, since $x_k = x_k^\circ + z_k$, the reachable set of (8.97), $\mathcal{X}_k$, can be obtained as follows:

$$\mathcal{X}_k = \left\{x_k^\circ\right\} + \mathcal{Z}_k. \tag{8.100}$$

Now, we can use the techniques discussed earlier to find an ellipsoid $\mathcal{Z}_k^+$ containing the reachable set $\mathcal{Z}_k$. Let $\overline{z}_k$ and F_k denote the center and shape matrix of $\mathcal{Z}_k^+$. Then, it follows from (8.100) that

$$\begin{aligned} \mathcal{X}_k &\subseteq \mathcal{X}_k^+ \\ &:= \left\{x_k^\circ\right\} + \mathcal{Z}_k^+, \end{aligned} \tag{8.101}$$

where $\mathcal{X}_k^+$ is an ellipsoid with the same shape matrix as that of $\mathcal{Z}_k^+$ and center $\overline{x}_k = x_k^\circ + \overline{z}_k$.

8.3 Discrete-Time Nonlinear Systems

In this section, we study systems of the form

$$x_{k+1} = h_k(x_k, w_k), \tag{8.102}$$

where $x_k \in \mathbb{R}^n$, $w_k \in \mathbb{R}^m$, and $h_k \colon \mathbb{R}^n \times \mathbb{R}^m \to \mathbb{R}^n$. Assume that $w_k \in \mathcal{W}_k$ and $x_0 \in \mathcal{X}_0$, where the sets $\mathcal{W}_k$, $k \geq 0$, and $\mathcal{X}_0$ are ellipsoids with center $\overline{w}_k \in \mathbb{R}^m$ and $\overline{x}_0 \in \mathbb{R}^n$, respectively, and shape matrix $E_k \in \mathbb{R}^{m\times m}$ and $F_0 \in \mathbb{R}^{n\times n}$, respectively, i.e.,

$$\mathcal{W}_k = \big\{w \in \mathbb{R}^m \colon (w - \overline{w}_k)^\top E_k^{-1}(w - \overline{w}_k) \leq 1\big\}, \tag{8.103}$$

$$\mathcal{X}_0 = \big\{x \in \mathbb{R}^n \colon (x - \overline{x}_0)^\top F_0^{-1}(x - \overline{x}_0) \leq 1\big\}. \tag{8.104}$$

As in the linear case, we would like to characterize the set $\mathcal{X}_k$ containing all possible values that x_k can take. However, because of the nonlinear dependence of x_{k+1} on x_k and w_k, it is usually difficult, if not impossible, to exactly characterize $\mathcal{X}_k$. Instead, we will settle for obtaining an approximate characterization of $\mathcal{X}_k$ by resorting to linearization.

Let $\overline{x}_k$, $k \geq 0$, denote the trajectory followed by the system

$$x_{k+1} = h_k\big(x_k, w_k\big), \tag{8.105}$$

for $w_k = \overline{w}_k$, $k \geq 0$, and $x_0 = \overline{x}_0$. Let Δx_k denote a small variation of x_k around $\overline{x}_k$ that results from a small variation in w_k around $\overline{w}_k$, which we denote by Δw_k. Then, since $\Delta w_k = w_k - \overline{w}_k$, $\overline{w}_k \in \mathcal{W}_k$, and $\Delta x_0 = x_0 - \overline{x}_0$, it follows that

$$\Delta w_k \in \Delta\mathcal{W}_k = \big\{\Delta w \in \mathbb{R}^m \colon (\Delta w)^\top E_k^{-1}(\Delta w) \leq 1\big\}, \tag{8.106}$$

$$\Delta x_0 \in \Delta\mathcal{X}_0 = \big\{\Delta x \in \mathbb{R}^n \colon (\Delta x)^\top F_0^{-1}(\Delta x) \leq 1\big\}. \tag{8.107}$$

Thus,

$$\overline{x}_{k+1} + \Delta x_{k+1} = h_k\big(\overline{x}_k + \Delta x_k, \overline{w}_k + \Delta w_k\big), \tag{8.108}$$

where $\Delta w_k \in \Delta\mathcal{W}_k$, and $\Delta x_0 \in \Delta\mathcal{X}_0$. Now, by linearizing $h_k(\cdot,\cdot)$ around $\overline{x}_k$ and $\overline{w}_k$, we have that Δx_k can be approximated by some $\widetilde{\Delta x}_k$ the evolution of which is given by

$$\widetilde{\Delta x}_{k+1} = G_k \widetilde{\Delta x}_k + H_k \Delta w_k, \tag{8.109}$$

with $\Delta w_k \in \Delta\mathcal{W}_k$, and $\Delta x_0 \in \Delta\mathcal{X}_0$, and

$$G_k = \left.\frac{\partial h_k(x,w)}{\partial x}\right|_{x=\overline{x}_k, w=\overline{w}_k} \in \mathbb{R}^{n\times n},$$

$$H_k = \left.\frac{\partial h_k(x,w)}{\partial w}\right|_{x=\overline{x}_k, w=\overline{w}_k} \in \mathbb{R}^{n\times m}.$$

For the system in (8.109) to provide a good approximation, Δw_k needs to be sufficiently small. In the context of the system in (8.102), this translates into $\mathcal{W}_k$ being sufficiently small. In addition, $\mathcal{X}_0$ also needs to be sufficiently small.

We can now use the same techniques we developed for the linear case (see Section 8.2.1) to find an ellipsoid $\widetilde{\Delta\mathcal{X}}_k^+$ containing the reachable set of (8.109), $\widetilde{\Delta\mathcal{X}}_k$, for all $k \geq 0$. In this case because both $\Delta\mathcal{W}_k$ and $\widetilde{\Delta\mathcal{X}}_0$ are centered around zero, we have that

$$\widetilde{\Delta\mathcal{X}}_k^+ = \{\widetilde{\Delta x} \in \mathbb{R}^n : \widetilde{\Delta x}^\top \widetilde{F}_k^{-1} \widetilde{\Delta x} \leq 1\},$$

where $\widetilde{F}_k$ is recursively defined as follows:

$$\widetilde{F}_{k+1} = (1 + \gamma_k) G_k \widetilde{F}_k G_k^\top + \left(1 + \frac{1}{\gamma_k}\right) H_k E_k H_k^\top, \tag{8.110}$$

with $\widetilde{F}_0 = F_0$ and $\gamma_k > 0$ for all $k \geq 0$. As with the linear case, we can set γ_k to some positive constant γ, or choose it so that $\widetilde{\Delta\mathcal{X}}_k^+$ is optimal in some sense as for the linear case discussed earlier (see Section 8.2.2). For example, by choosing γ_k to be the unique solution of

$$\sum_{i=1}^{n} \frac{\psi_{k,i}}{\psi_{k,i} + \gamma_k} = \frac{n\gamma_k}{1 + \gamma_k}, \tag{8.111}$$

with $\psi_{k,i}$, $i = 1, 2, \ldots, n$, such that $\mathbf{det}(Q_k - \psi_{k,i} P_k) = 0$, where $P_k := G_k \widetilde{F}_k G_k^\top$ and $Q_k := H_k E_k H_k^\top$, the resulting ellipsoid $\widetilde{\Delta\mathcal{X}}_k^+$ has minimum volume. Also, if we choose γ_k as follows:

$$\gamma_k = \sqrt{\frac{\mathbf{tr}\left(H_k E_k H_k^\top\right)}{\mathbf{tr}\left(G_k \widetilde{F}_k G_k^\top\right)}}, \tag{8.112}$$

the sum of squares of the lengths of the semi-axes of the resulting ellipsoid $\widetilde{\Delta\mathcal{X}}_k^+$ is minimum. Finally, if we choose γ_k as follows:

$$\gamma_k = \sqrt{\frac{\eta_k^\top G_k^{-1} H_k E_k H_k^\top \left(G_k^\top\right)^{-1} \eta_k}{\eta_k^\top \widetilde{F}_k \eta_k}}, \tag{8.113}$$

where η_k is recursively defined by

$$\eta_{k+1} = (G_k^\top)^{-1} \eta_k, \tag{8.114}$$

for given η_0, the resulting ellipsoid, $\widetilde{\Delta\mathcal{X}}_k^+$, and $\widetilde{\Delta\mathcal{X}}_k$ touch each other at the following point:

$$\widetilde{\Delta x}_k^* = \frac{1}{\sqrt{\eta_k^\top \widetilde{F}_k \eta_k}} \widetilde{F}_k \eta_k. \tag{8.115}$$

Then, since $x_k \approx \overline{x}_k + \widetilde{\Delta x}_k$, we have that

$$\begin{aligned} x_k \in \mathcal{X}_k \approx \widetilde{\mathcal{X}}_k &:= \{\overline{x}_k\} + \widetilde{\Delta\mathcal{X}}_k \\ &\subseteq \{\overline{x}_k\} + \widetilde{\Delta\mathcal{X}}_k^+ \\ &=: \widetilde{\mathcal{X}}_k^+. \end{aligned} \tag{8.116}$$

Note that we cannot guarantee that $\mathcal{X}_k \subseteq \widetilde{\mathcal{X}}_k^+$ because the set $\widetilde{\mathcal{X}}_k$ is not guaranteed to contain the reachable set $\mathcal{X}_k$. However, in many practical applications it will be the case that the set $\widetilde{\mathcal{X}}_k$ will provide a good approximation to $\mathcal{X}_k$.

8.4 Performance Requirements Verification

Computing reachable sets are relevant in many applications where performance requirements impose the state vector, $x_k = [x_{k,1}, x_{k,2}, \ldots, x_{k,n}]^\top$, to remain within some set $\mathcal{R}_k \subseteq \mathbb{R}^n$ for all $k \geq 0$. While the set $\mathcal{R}_k$ can have any shape, in most practical applications, performance requirements will impose each entry of the state vector x_k to remain within some particular range. As such, the set $\mathcal{R}_k$ can be described as the Cartesian product of n real intervals as follows:

$$\mathcal{R}_k = \mathcal{R}_{k,1} \times \mathcal{R}_{k,2} \times \cdots \times \mathcal{R}_{k,n}, \tag{8.117}$$

with

$$\mathcal{R}_{k,i} = \{x_i \in \mathbb{R} \colon x_{k,i}^m < x_i < x_{k,i}^M\}, \tag{8.118}$$

where $x_{k,i}^m, x_{k,i}^M \in \mathbb{R}$ respectively denote the minimum and maximum values that $x_{k,i}$ can take. Thus, in order to check that the system meets its performance requirements, one needs to verify that the reachable set, $\mathcal{X}_k$, is contained in the set $\mathcal{R}_k$ for all $k \geq 0$. Since in general we will not be able to obtain the exact characterization of the reachable set, we will need to utilize the ellipsoids obtained using the techniques discussed earlier.

Linear Case. Consider the system in (8.5) with $\mathcal{W}_k$ and $\mathcal{X}_0$ as given in (8.16) and (8.17), respectively. Assume we are given a finite number of bounding ellipsoids, $\mathcal{X}_{k,1}^+, \mathcal{X}_{k,2}^+, \ldots, \mathcal{X}_{k,q}^+$, with centers $\overline{x}_k^{(j)}$, $j = 1, 2, \ldots, q$, and shape matrices $F_k^{(j)} \in \mathbb{R}^{n \times n}$, $j = 1, 2, \ldots, q$, respectively. Let $\hat{u}_i$ denote an n-dimensional vector whose i^{th} entry is equal to one and all others are equal to zero, and define:

$$\begin{aligned} M_{k,i} &= \min_j \left\{ \hat{u}_i^\top \overline{x}_k^{(j)} + \sqrt{\hat{u}_i^\top F_k^{(j)} \hat{u}_i} \right\}, \\ m_{k,i} &= \max_j \left\{ \hat{u}_i^\top \overline{x}_k^{(j)} - \sqrt{\hat{u}_i^\top F_k^{(j)} \hat{u}_i} \right\}; \end{aligned} \tag{8.119}$$

then, we have that $\mathcal{X}_k \subset \mathcal{R}_k$ if

$$\begin{aligned} M_{k,i} &< x_{k,i}^M, \\ m_{k,i} &> x_{k,i}^m, \end{aligned} \tag{8.120}$$

for all $i = 1, 2, \ldots, n$. To establish this result, first recall that for any closed and convex set $\mathcal{X} \in \mathbb{R}^n$,

$$\begin{aligned} S_{\mathcal{X}}(\hat{u}_i) &= \max_{x \in \mathcal{X}} \hat{u}_i^\top x \\ &= \max_{x \in \mathcal{X}} x_i, \end{aligned} \tag{8.121}$$

i.e., $S_{\mathcal{X}}(\hat{u}_i)$ is the maximum value that the ith entry of $x \in \mathcal{X}$ can take. Similarly,

$$\begin{aligned} S_{\mathcal{X}}(-\hat{u}_i) &= \max_{x \in \mathcal{X}} -\hat{u}_i^\top x \\ &= \max_{x \in \mathcal{X}} -x_i \\ &= -\min_{x \in \mathcal{X}} x_i, \end{aligned} \tag{8.122}$$

i.e., $-S_{\mathcal{X}}(-\hat{u}_i)$ is the minimum value that the ith entry of $x \in \mathcal{X}$ can take. Now, since

$$\begin{aligned} S_{\mathcal{X}_{k,j}}(\hat{u}_i) &= \hat{u}_i^\top \overline{x}_k^{(j)} + \sqrt{\hat{u}_i^\top F_k^{(j)} \hat{u}_i}, \\ S_{\mathcal{X}_{k,j}}(-\hat{u}_i) &= -\hat{u}_i^\top \overline{x}_k^{(j)} + \sqrt{\hat{u}_i^\top F_k^{(j)} \hat{u}_i}, \end{aligned} \tag{8.123}$$

and the set $\cap_{i=1}^q \mathcal{X}_{k,i}^+$ is convex, it follows that

$$\begin{aligned} S_{\cap_{j=1}^q \mathcal{X}_{k,j}^+}(\hat{u}_i) &= \max_{x \in \cap_{j=1}^q \mathcal{X}_{k,j}^+} x_i \\ &= \min_j \underbrace{\max_{x \in \mathcal{X}_{k,j}^+} x_i}_{=S_{\mathcal{X}_{k,j}^+}(\hat{u}_i)} \\ &= \min_j \left\{ \hat{u}_i^\top \overline{x}_k^{(j)} + \sqrt{\hat{u}_i^\top F_k^{(j)} \hat{u}_i} \right\} \\ &= M_{k,i}, \end{aligned} \tag{8.124}$$

and

$$\begin{aligned} S_{\cap_{j=1}^q \mathcal{X}_{k,j}^+}(-\hat{u}_i) &= \max_{x \in \cap_{j=1}^q \mathcal{X}_{k,j}^+} -x_i \\ &= \min_j \underbrace{\max_{x \in \mathcal{X}_{k,j}^+} -x_i}_{=S_{\mathcal{X}_{k,j}^+}(-\hat{u}_i)} \\ &= \min_j \left\{ -\hat{u}_i^\top \overline{x}_k^{(j)} + \sqrt{\hat{u}_i^\top F_k^{(j)} \hat{u}_i} \right\} \\ &= -\max_j \left\{ \hat{u}_i^\top \overline{x}_k^{(j)} - \sqrt{\hat{u}_i^\top F_k^{(j)} \hat{u}_i} \right\} \\ &= -m_{k,i}. \end{aligned} \tag{8.125}$$

Then, since $\mathcal{X}_k \subseteq \cap_{i=1}^q \mathcal{X}_{k,i}^+$, we have that $\mathcal{X}_k \subset \mathcal{R}_k$ if $M_{k,i} < x_{k,i}^M$ and $m_{k,i} > x_{k,i}^m$.

Nonlinear Case. Consider the system in (8.102) with $\mathcal{W}_k$ and $\mathcal{X}_0$ as given in (8.103–8.104), respectively. Assume we are given a family of ellipsoids, $\widetilde{\mathcal{X}}_{k,1}^+, \widetilde{\mathcal{X}}_{k,2}^+, \dots, \widetilde{\mathcal{X}}_{k,q}^+$, obtained by using the linearization approach described earlier. As noted, these ellipsoids are only guaranteed to outer bound the reachable set of the linearized system, but not necessarily the reachable set of the original nonlinear system in (8.102). Thus, even if the criteria in (8.120) are satisfied, we cannot say with certainty whether or not $\mathcal{X}_k \subset \mathcal{R}_k$.

8.5 Analysis of Microgrids under Power Injection Uncertainty

In this section, we utilize the techniques developed earlier in the chapter to analyze inertia-less AC microgrids when the power injected into a subset of the buses is uncertain but known to belong to an ellipsoid. Except for the uncertainty model, the setting here is identical to that in Section 5.4; thus, the reader is referred to Section 5.4.1 for a detailed derivation of the model used here.

8.5.1 System Model

Consider a three-phase microgrid comprising n buses ($n > 1$) indexed by the elements in $\mathcal{V} = \{1, 2, \ldots, n\}$, and l transmission lines $\big(n-1 \leq l \leq n(n-1)/2\big)$ indexed by the elements in $\mathcal{L} = \{1, 2, \ldots, l\}$. In subsequent developments, we make the following assumptions (identical to those made in Section 5.4.1):

A1. The microgrid is balanced and operating in sinusoidal regime.

A2. There is at most one transmission line connecting each pair of buses.

A3. Each transmission line $e \in \mathcal{L}$ linking a pair of buses, say bus i and bus j, is short and lossless; thus, it can be modeled by a series reactance, $X_{ij} > 0$.

A4. Connected to each bus there is either a generating- or a load-type resource interfaced via a three-phase voltage source inverter whose controls will attempt to synthesize a sinusoidal voltage waveform at its output terminals.

A5. The reactance of each voltage source inverter output filter is small when compared to the reactance values of the network transmission lines; thus, it can be neglected.

A6. The phase angle of the inverter connected to each bus is regulated via a discrete-time frequency-droop control scheme that gets updated every T_s [s].

A7. The inverter outer voltage and inner current control loops hold the inverter output voltage magnitude constant throughout time.

Because of Assumptions A1–A2, we can define a directed graph describing the microgrid network topology, with each node of this graph corresponding to an element in the bus set, $\mathcal{V}$, and each edge of this graph corresponding to an element in the transmission line set, $\mathcal{L}$, with its orientation in agreement with the (arbitrarily chosen) positive power flow direction on the corresponding transmission line. Let $p_i(t)$ denote the active power injected into the microgrid via bus i at time instant t, and let $V_i(t)$ and $\theta_i(t)$ respectively denote the magnitude and phase angle of the phasor associated with bus i's (sinusoidal) voltage at time instant t. Because of Assumption A5 and Assumption A6, we have that

$$\theta_{i,k+1} = \theta_{i,k} + \frac{T_s}{D_i}\big(u_{i,k} - p_{i,k}\big), \quad i = 1, 2, \ldots, n,$$

where D_i [s/rad] denotes the droop coefficient of bus i's inverter, $p_{i,k} := p_i(kT_s)$, $\theta_{i,k} := \theta_i(kT_s)$, $k \geq 0$, and where $u_{i,k}$ can be either extraneously set (thus, from the system point of view acts as a disturbance) or set by a control system. Also, because of Assumption A5 and Assumption A7, we have that

$$V_{i,k} \approx V_i, \; i = 1, 2, \ldots, n,$$

where $V_{i,k} := V_i(kT_s)$, $k \geq 0$, and V_i is a positive scalar. Define the following two vectors: $\theta_k = \big[\theta_{1,k}, \theta_{2,k}, \ldots, \theta_{n,k}\big]^\top$ and $u_k = \big[u_{1,k}, u_{2,k}, \ldots, u_{n,k}\big]^\top$; then, it follows that

$$\theta_{k+1} = \theta_k + T_s D^{-1}\Big(u_k - M\Gamma\mathbf{sin}\big(M^\top\theta_k\big)\Big), \tag{8.126}$$

where $M \in \mathbb{R}^{n\times l}$ denotes the node-to-edge incidence matrix of the directed graph describing the microgrid network topology, $D = \mathbf{diag}\big(D_1, D_2, \ldots, D_n\big)$, $\Gamma = \mathbf{diag}\big(\gamma_1, \gamma_2, \ldots, \gamma_l\big)$, with

$$\gamma_e = \frac{V_i V_j}{X_{ij}}, \tag{8.127}$$

and $\mathbf{sin}(x)$, $x = [x_1, x_2, \ldots, x_l]^\top$, defined as follows:

$$\mathbf{sin}(x) = \big[\sin(x_1), \sin(x_2), \ldots, \sin(x_l)\big]^\top.$$

As in Section 5.4.1, assume that

$$u_{i,k} = \begin{cases} \xi_{i,k}, & i = 1, 2, \ldots, m, \\ r_{i,k}, & i = m+1, m+2, \ldots, n, \end{cases} \tag{8.128}$$

$1 \leq m \leq n-1$, where $\xi_{i,k}$ is extraneously set and a priori unknown, and $r_{i,k}$ is regulated via a closed-loop control system as follows:

$$\begin{aligned} z_{i,k+1} &= z_{i,k} + \alpha_i \overline{\Delta\omega}_k, \\ r_{i,k} &= r_i^* + \beta_i z_{i,k}, \end{aligned} \tag{8.129}$$

where $\overline{\Delta\omega}_k$ is referred to as the *average frequency error* at instant k, and is defined as follows:

$$\overline{\Delta\omega}_k = \frac{\sum_{i=1}^m \xi_{i,k} + \sum_{i=m+1}^n r_{i,k}}{\sum_{i=1}^n D_i}, \tag{8.130}$$

and α_i [s/rad] and β_i [p.u.] are constants chosen so that

$$\begin{aligned} \alpha_i &< 0, \quad i = m+1, m+2, \ldots, n, \\ \beta_i &> 0, \quad i = m+1, m+2, \ldots, n, \\ \sum_{j=1}^n \alpha_j\beta_j &> -2\sum_{j=1}^n D_i. \end{aligned} \tag{8.131}$$

Such choice ensures that if $\xi_{i,k}$, $i = 1, 2 \ldots, m$, are bounded for all $k \geq 0$, so are the trajectories followed by (8.129–8.130) (see discussion on controller gain selection in Section 5.4.1). Define the following vectors: $\xi_k = \left[\xi_{1,k}, \xi_{2,k}, \ldots, \xi_{m,k}\right]^\top$, $r^* = \left[r^*_{m+1}, r^*_{m+2}, \ldots, r^*_n\right]^\top$, and $z_k = \left[z_{m+1,k}, z_{m+2,k}, \ldots, z_{n,k}\right]^\top$. Then, by combining (8.126) and (8.129–8.130), we obtain that

$$\theta_{k+1} = \theta_k + T_s D^{-1}\Big(K_\xi \xi_k + K_r\big(r^* + Bz_k\big) - M\Gamma\mathbf{sin}\big(M^\top \theta_k\big)\Big), \qquad (8.132)$$

$$z_{k+1} = \left(I_{n-m} + \frac{1}{n\overline{D}}\alpha\beta^\top\right)z_k + \frac{1}{n\overline{D}}\alpha\Big(\mathbf{1}_m^\top \xi_k + \mathbf{1}_{n-m}^\top r^*\Big), \qquad (8.133)$$

with $\overline{D} := \frac{\sum_{i=1}^n D_i}{n}$, $B = \mathbf{diag}(\beta_{m+1}, \beta_{m+2}, \ldots, \beta_n)$, $\beta = \left[\beta_{m+1}, \beta_{m+2}, \ldots, \beta_n\right]^\top$, and

$$K_\xi = \begin{bmatrix} I_m \\ \mathbf{0}_{(n-m)\times m}\end{bmatrix}, \quad K_r = \begin{bmatrix}\mathbf{0}_{m\times(n-m)} \\ I_{n-m}\end{bmatrix}, \qquad (8.134)$$

with initial conditions as follows:

$$z_0 = \mathbf{0}_{n-m}, \qquad (8.135)$$

and θ_0 satisfying

$$K_\xi \xi^0 + K_r r^* = M\Gamma\mathbf{sin}\big(M^\top \theta_0\big), \qquad (8.136)$$

for some constant $\xi^0 = \left[\xi^0_1, \xi^0_2, \ldots, \xi^0_m\right]^\top$.

In the remainder, we will assume that the disturbance ξ_k, $k \geq 0$, is of the form

$$\xi_k = \xi^0 + w_k, \quad k \geq 0, \qquad (8.137)$$

where w_k is known to belong to an ellipsoid $\mathcal{W}_k$ with center $\overline{w}_k \in \mathbb{R}^m$ and shape matrix $E_k \in \mathbb{R}^{m\times m}$, i.e.,

$$\mathcal{W}_k = \Big\{w \in \mathbb{R}^m \colon (w - \overline{w}_k)^\top E_k^{-1}(w - \overline{w}_k) \leq 1\Big\}. \qquad (8.138)$$

8.5.2 Characterization of the Set Containing the Average Frequency Error

Here we are interested in (i) characterizing the set containing all possible values that the control internal state vector, z_k, can take and (ii) computing upper and lower bounds on the values that the average frequency error, $\overline{\Delta\omega}_k$, can take. First, in light of (8.136), we have that $\mathbf{1}_m^\top \xi^0 + \mathbf{1}_{n-m}^\top r^* = 0$; then, using (8.137–8.138) together with (8.130) and (8.133), yields

$$z_{k+1} = \left(I_{n-m} + \frac{1}{n\overline{D}}\alpha\beta^\top\right)z_k + \frac{1}{n\overline{D}}\alpha\mathbf{1}_m^\top w_k, \qquad (8.139)$$

$$\overline{\Delta\omega}_k = \frac{1}{n\overline{D}}\big(\mathbf{1}_m^\top w_k + \beta^\top z_k\big), \qquad (8.140)$$

where $w_k \in \mathcal{W}_k$, with $\mathcal{W}_k$ as defined in (8.138), and $z_0 = \mathbf{0}_{n-m}$.

Let $\mathcal{Z}_k$ denote the set containing all possible values that z_k can take. Then, in order to apply the techniques developed earlier to the setting here, we need the matrix

$$G := I_{n-m} + \frac{1}{n\overline{D}}\alpha\beta^\top \tag{8.141}$$

to be full rank. This indeed will be the case as long as $\frac{1}{n\overline{D}}\alpha^\top\beta \neq -1$; otherwise, one can check that the matrix G would have an eigenvalue at zero. Thus, in the remainder, we assume that α and β are chosen so as to satisfy (8.131) and such that $\alpha^\top\beta \neq -n\overline{D}$. Also note that the initial conditions here, $z_0 = \mathbf{0}_{n-m}$, do not technically conform to those in (8.17); however, we can always define

$$\mathcal{Z}_0 = \left\{z \in \mathbb{R}^{n-m} \colon (z - \overline{z}_0)^\top F_0^{-1}(z - \overline{z}_0) \leq 1\right\}, \tag{8.142}$$

where $\overline{z}_0 = z_0 = \mathbf{0}_{n-m}$, and $F_0 = \varepsilon I_{n-m}$, where ε is a small positive constant. Then, by assuming that $z_0 \in \mathcal{Z}_0$, the setting here fully conforms to that in (8.5) and (8.16–8.17).

Let $\mathcal{Z}_k \subseteq \mathbb{R}^{n-m}$ denote the set of all possible values that z_k can take; then, we can use the techniques developed earlier in the chapter to find an ellipsoid $\mathcal{Z}_k^+$ such that $\mathcal{Z}_k \subseteq \mathcal{Z}_k^+$. To this end, let $\overline{z}_k \in \mathbb{R}^{n-m}$ and $F_k \in \mathbb{R}^{(n-m)\times(n-m)}$ denote the center and shape matrix of $\mathcal{Z}_k^+$; then, direct application of the formulae in (8.22–8.23) to the model in (8.138), (8.139), and (8.142) yields

$$\overline{z}_{k+1} = G\overline{z}_z + \frac{1}{n\overline{D}}\alpha\mathbf{1}_m^\top\overline{w}_k, \tag{8.143}$$

$$F_{k+1} = (1+\gamma_k)GF_kG^\top + \left(1 + \frac{1}{\gamma_k}\right)\frac{1}{n^2\overline{D}^2}\alpha\mathbf{1}_m^\top E_k\mathbf{1}_m\alpha^\top, \tag{8.144}$$

where $\overline{z}_0 = \mathbf{0}_{n-m}$, $F_0 = \varepsilon I_{n-m}$, and $\gamma_k > 0$ for all $k \geq 0$.

Let $\overline{\Delta\Omega}_k$ denote the set containing all possible values that $\overline{\Delta\omega}_k$ can take; then, by using (8.140), it follows that

$$\overline{\Delta\Omega}_k = \left\{y \colon y = \frac{1}{n\overline{D}}(\mathbf{1}_m^\top w + \beta^\top z),\ w \in \mathcal{W}_k,\ z \in \mathcal{Z}_k\right\}, \tag{8.145}$$

and

$$\begin{aligned}\overline{\Delta\Omega}_k \subseteq \mathcal{Y}_k &= \left\{y \colon y = \frac{1}{n\overline{D}}(\mathbf{1}_m^\top w + \beta^\top z),\ w \in \mathcal{W}_k,\ z \in \mathcal{Z}_k^+\right\}\\ &= \left\{y \colon \overline{y}_k - \widetilde{\Delta y}_k \leq y \leq \overline{y}_k + \widetilde{\Delta y}_k\right\},\end{aligned} \tag{8.146}$$

where

$$\overline{y}_k = \frac{\mathbf{1}_m^\top\overline{w}_k + \beta^\top\overline{z}_k}{n\overline{D}}, \quad \widetilde{\Delta y}_k = \frac{\sqrt{\mathbf{1}_m^\top E_k\mathbf{1}_m} + \sqrt{\beta^\top F_k\beta}}{n\overline{D}}.$$

To establish (8.146) we make use of the properties of support functions for convex sets in (8.12–8.13). First note that since $\mathcal{Z}_k \subseteq \mathcal{Z}_k^+$, we have that $S_{\mathcal{Z}_k}(\eta) \leq S_{\mathcal{Z}_k^+}(\eta)$ for all $\eta \in \mathbb{R}^{n-m}$, and

$$
\begin{aligned}
S_{\overline{\Delta\Omega}_k}(\mu) &= S_{\mathcal{W}_k}\left(\frac{1}{n\overline{D}}\mathbf{1}_m\mu\right) + S_{\mathcal{Z}_k}\left(\frac{1}{n\overline{D}}\beta\mu\right) \\
&\leq S_{\mathcal{W}_k}\left(\frac{1}{n\overline{D}}\mathbf{1}_m\mu\right) + S_{\mathcal{Z}_k^+}\left(\frac{1}{n\overline{D}}\beta\mu\right) \\
&= S_{\mathcal{Y}_k}(\mu), \quad \mu \in \mathbb{R};
\end{aligned} \tag{8.147}
$$

thus, $\overline{\Delta\Omega}_k \subseteq \mathcal{Y}_k$. Now since $\mathcal{W}_k$ and $\mathcal{Z}_k^+$ are ellipsoids with center $\overline{w}_k$ and $\overline{z}_k$, respectively, and shape matrix E_k and F_k, respectively, it follows that

$$
\begin{aligned}
S_{\mathcal{Y}_k}(\mu) &= S_{\mathcal{W}_k}\left(\frac{1}{n\overline{D}}\mathbf{1}_m\mu\right) + S_{\mathcal{Z}_k^+}\left(\frac{1}{n\overline{D}}\beta\mu\right) \\
&= \frac{\mu(\mathbf{1}_m^\top\overline{w}_k + \beta^\top\overline{z}_k)}{n\overline{D}} + \frac{|\mu|\left(\sqrt{\mathbf{1}_m^\top E_k \mathbf{1}_m} + \sqrt{\beta^\top F_k \beta}\right)}{n\overline{D}}, \quad \mu \in \mathbb{R}.
\end{aligned} \tag{8.148}
$$

Thus,

$$
\begin{aligned}
\mathcal{Y}_k &= \left\{y \colon \mu y \leq S_{\mathcal{Y}_k}(\mu), \quad |\mu| = 1\right\} \\
&= \left\{y \colon \mu y \leq \frac{\mu(\mathbf{1}_m^\top\overline{w}_k + \beta^\top\overline{z}_k)}{n\overline{D}} + \frac{|\mu|\left(\sqrt{\mathbf{1}_m^\top E_k \mathbf{1}_m} + \sqrt{\beta^\top F_k \beta}\right)}{n\overline{D}}, \quad |\mu| = 1\right\} \\
&= \left\{y \colon \overline{y}_k - \widetilde{\Delta y}_k \leq y \leq \overline{y}_k + \widetilde{\Delta y}_k\right\},
\end{aligned} \tag{8.149}
$$

where

$$
\overline{y}_k = \frac{\mathbf{1}_m^\top\overline{w}_k + \beta^\top\overline{z}_k}{n\overline{D}}, \quad \widetilde{\Delta y}_k = \frac{\sqrt{\mathbf{1}_m^\top E_k \mathbf{1}_m} + \sqrt{\beta^\top F_k \beta}}{n\overline{D}},
$$

as claimed in (8.146).

Choice of Parameter γ_k. For the case when $n - m = 1$, i.e., there is only one controllable resource, the system in (8.139) reduces to

$$
z_{k+1} = \left(1 + \frac{\alpha\beta}{n\overline{D}}\right)z_k + \frac{\alpha}{n\overline{D}}\mathbf{1}_{n-1}^\top w_k, \tag{8.150}
$$

where α and β are constants such that $\frac{\alpha\beta}{n\overline{D}} \neq -1$ and $-2 < \frac{\alpha\beta}{n\overline{D}} < 0$, $z_k \in \mathbb{R}$, and $w_k \in \mathbb{R}^{n-1}$. Clearly,

$$
\left|1 + \frac{\alpha\beta}{n\overline{D}}\right| < 1;
$$

thus, according to the discussion in Section 8.2.2, we can set $\gamma_k = \gamma$, with γ satisfying (8.32), which in this case yields

$$
\gamma < \frac{n^2\overline{D}^2}{(n\overline{D} + \alpha\beta)^2} - 1.
$$

We can also choose γ_k according to the criteria in (8.45) (minimum volume ellipsoid), (8.69) (ellipsoid with minimum sum of squares of the semi-axis lengths), and (8.78–8.79) (tight ellipsoid); however, all these choices are identical because the system is one-dimensional, and in fact, they yield the exact characterization of the set $\mathcal{Z}_k$ containing all possible values that z_k can take, which in this case is just a real interval (see the analysis in Examples 8.2–8.4).

For the case when $n - m > 1$, the matrix G has at least one eigenvalue at one; thus, according to the discussion in Section 8.2.2, we cannot set γ_k equal to a positive constant; otherwise, the corresponding bounding ellipsoid would grow unbounded. In this case, we can still use the criteria in (8.45) (minimum volume ellipsoid) and (8.69) (ellipsoid with minimum sum of squares of the semi-axis lengths); however, we cannot use the criterion in (8.78–8.79) (tight ellipsoid) because the matrix

$$\frac{1}{n\overline{D}}\alpha \mathbf{1}_m^\top$$

is not full row rank.

Exact Computation of Upper and Lower Bounds on $\overline{\Delta\omega}_k$. The procedure discussed above would provide approximate bounds on the possible values taken by $\overline{\Delta\omega}_k$; however, because of the structure of the system in (8.139), it is actually possible to obtain the exact bounds on the values that $\overline{\Delta\omega}_k$ can take. To this end, define $\zeta_k = \beta^\top z_k$; then, multiplying by $\beta^\top$ both sides of (8.139) yields

$$\zeta_{k+1} = \left(1 + \frac{1}{n\overline{D}}\beta^\top \alpha\right)\zeta_k + \frac{1}{n\overline{D}}\beta^\top \alpha \mathbf{1}_m^\top w_k, \tag{8.151}$$

where $w_k \in \mathcal{W}_k$, with $\mathcal{W}_k$ as defined in (8.138), and

$$\zeta_0 \in Z_0 = \{\zeta \in \mathbb{R} \colon (\zeta - \overline{\zeta}_0)^2 \leq f_0\},$$

with $\overline{\zeta}_0 = 0$ and $f_0 = \varepsilon\|\beta\|_2^2$, $\varepsilon > 0$, where we have taken into account the modified initial conditions of (8.139) as defined by (8.142). Also, by using (8.140), it follows that

$$\overline{\Delta\omega}_k = \frac{1}{n\overline{D}}(\mathbf{1}_m^\top w_k + \zeta_k). \tag{8.152}$$

Let Z_k denote the real interval containing all possible values that ζ_k can take, and let

$$Z_k^+ = \left\{\zeta \in \mathbb{R} \colon (\zeta - \overline{\zeta}_k)^2 \leq f_k\right\}, \tag{8.153}$$

with $\overline{\zeta}_k$ and f_k defined as follows:

$$\overline{\zeta}_{k+1} = a\,\overline{\zeta}_k + b\,\mathbf{1}_m^\top \overline{w}_k, \tag{8.154}$$

$$f_{k+1} = (1+\gamma_k)a^2 f_k + (1+\gamma_k^{-1})b^2\, e_k, \tag{8.155}$$

where $a = 1 + \frac{1}{n\overline{D}}\beta^\top\alpha$, $b = \frac{1}{n\overline{D}}\beta^\top\alpha$, $e_k = \mathbf{1}_m^\top E_k \mathbf{1}_m$, and $\gamma_k > 0$ for all $k \geq 0$. The expressions in (8.153–8.155) are just a special case of the expressions in (8.21–8.23) when $n = 1$; thus, $Z_k \subseteq Z_k^+$. Now, we can choose γ_k according to the criteria in (8.45) (minimum volume ellipsoid), (8.69) (ellipsoid with minimum sum of squares of the semi-axis lengths), and (8.78–8.79) (tight ellipsoid); however, as shown earlier in Examples 8.2–8.4, all these choices are identical and in this case yield

$$\gamma_k = \left|\frac{b}{a}\right| \sqrt{\frac{e_k}{f_k}}, \tag{8.156}$$

and in fact $Z_k^+ = Z_k$ for all $k \geq 0$. Finally, let $\overline{\Delta\Omega}_k$ denote the set containing all possible values that $\overline{\Delta\omega}_k$ can take; then, by using (8.152), it follows that

$$\overline{\Delta\Omega}_k = \left\{\overline{\Delta\omega} \colon \overline{y}_k - \Delta y_k \leq \overline{\Delta\omega} \leq \overline{y}_k + \Delta y_k\right\}, \tag{8.157}$$

where

$$\overline{y}_k = \frac{\mathbf{1}_m^\top \overline{w}_k + \overline{\zeta}_k}{n\overline{D}}, \quad \Delta y_k = \frac{\sqrt{e_k} + \sqrt{f_k}}{n\overline{D}}.$$

The derivation of the expression in (8.157) is very similar to the derivation of the expression in (8.146) and therefore omitted.

Example 8.5 (Three-bus microgrid average frequency error characterization) Consider a three-bus microgrid with a lossless network. Assume the generation and load resources connected to bus 1, bus 2, and bus 3 are interfaced with the network via droop-controlled voltage source inverters with droop coefficients $D_1 = 0.1$ s/rad, $D_2 = 0.25$ s/rad, and $D_3 = 1$ s/rad, respectively. Connected to bus 1 there is a load with $u_{1,k} = \xi_{1,k} = \xi_1^0 + w_{1,k}$, where $\xi_1^0 = -1$ p.u., and

$$w_{1,k} \in \mathcal{W}_k = \left\{w \in \mathbb{R} \colon (w - \overline{w}_{1,k})^2 \leq e\right\},$$

with $\overline{w}_{1,k} = -0.25$ p.u. for all $k \geq 0$, and $e = 0.1$. Additionally, there is a generator connected to each bus 2 and bus 3. Assume that the power produced by both these generators can be regulated using the control scheme in (8.129–8.130) with $\alpha_2 = -0.5$ s/rad, $\beta_2 = 1$ p.u., and $r_2^* = 0.75$ p.u. for the generator connected to bus 2, and $\alpha_3 = -1$ s/rad, $\beta_3 = 0.5$ p.u., and $r_3^* = 0.25$ p.u., for the generator connected to bus 3. Thus, in this case, we have that $n = 3$, $\overline{D} = 0.45$ s/rad, $\alpha = [-0.5, -1]^\top$ s/rad, $\beta = [1, 0.5]^\top$ p.u., which by plugging into (8.151) and (8.152) yields

$$\begin{aligned} \zeta_{k+1} &= 0.2593\zeta_k - 0.7407 w_{1,k}, \\ \overline{\Delta\omega}_k &= 0.7407(w_{1,k} + \zeta_k). \end{aligned} \tag{8.158}$$

The intervals bounding the possible values that ζ_k and $\overline{\Delta\omega}_k$ can take, denoted by Z_k and $\overline{\Delta\Omega}_k$, respectively, can then be computed using the formulas in (8.154–8.155); the results are displayed in Fig. 8.1 for $f_0 = 10^{-6}$.

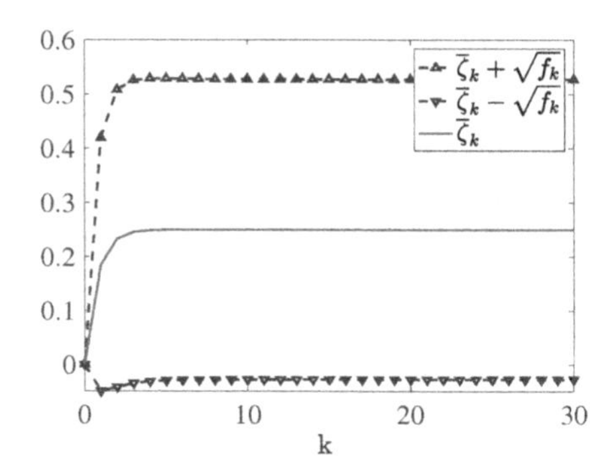

(a) Evolution of the intervals bounding the possible values taken by ζ_k.

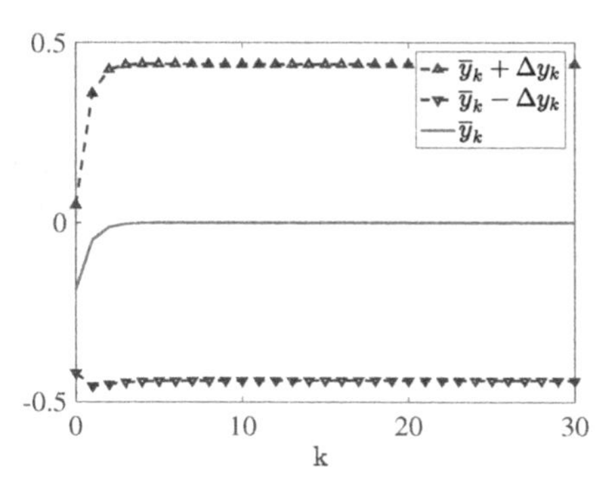

(b) Evolution of the intervals bounding the possible values taken by $\overline{\Delta\omega_k}$.

Figure 8.1 Evolution of the intervals bounding the possible values taken by ζ_k and $\overline{\Delta\omega_k}$ in Example 8.5.

8.5.3 Characterization of Set Containing the Bus Phase Angles

Consider the system in (8.132–8.133) with $\xi_k = \xi^0 + w_k$, $w_k \in \mathcal{W}_k$, with $\mathcal{W}_k$ as defined in (8.138), $z_0 = \mathbf{0}_{n-m}$, and θ_0 satisfying (8.136). Here we will utilize the techniques developed in Section 8.3 to approximately characterize the set containing all possible values that $x_k = \left[\theta_k^\top, z_k^\top\right]^\top$ can take.

Let $\overline{x}_k = \left[\overline{\theta}_k^\top, \overline{z}_k^\top\right]^\top$, $k \geq 0$, denote the trajectory followed by (8.132–8.133) for $\xi_k = \xi^0 + \overline{w}_k$, $k \geq 0$, and $x_0 = \left[\theta_0^\top, \mathbf{0}_{n-m}^\top\right]^\top$. Define $\Delta w_k = w_k - \overline{w}_k$ and

$$\begin{aligned}\Delta x_k &= \left[\Delta\theta_k^\top, \Delta z_k^\top\right]^\top \\ &= \left[\theta_k^\top, z_k^\top\right]^\top - \left[\overline{\theta}_k^\top, \overline{z}_k^\top\right]^\top.\end{aligned}$$

Then, by linearizing the right-hand side of (8.132–8.133) around $(\overline{\theta}_k, \overline{z}_k, \overline{w}_k)$, the vector Δx_k can be approximated by another vector, $\widetilde{\Delta x}_k = \left[\widetilde{\Delta\theta}_k^\top, \widetilde{\Delta z}_k^\top\right]^\top$, whose evolution is governed by

$$\begin{aligned}\begin{bmatrix}\widetilde{\Delta\theta}_{k+1} \\ \widetilde{\Delta z}_{k+1}\end{bmatrix} &= \underbrace{\begin{bmatrix} I_n - T_s D^{-1} J(\overline{\theta}_k) & T_s D^{-1} K_r B \\ \mathbf{0}_{(n-m)\times n} & I_{n-m} + \frac{1}{nD}\alpha\beta^\top \end{bmatrix}}_{=:\,G_k} \begin{bmatrix}\widetilde{\Delta\theta}_k \\ \widetilde{\Delta z}_k\end{bmatrix} \\ &\quad + \underbrace{\begin{bmatrix} T_s D^{-1} K_\xi \\ \frac{1}{nD}\alpha \mathbf{1}_m^\top \end{bmatrix}}_{=:\,H} \Delta w_k, \end{aligned} \tag{8.159}$$

with

$$J(\overline{\theta}_k) = M\Gamma\,\mathbf{diag}(\phi_{1,k}, \phi_{2,k}, \ldots, \phi_{l,k})M^\top, \tag{8.160}$$

where

$$\phi_{e,k} = \cos\left(\overline{\theta}_{i,k} - \overline{\theta}_{j,k}\right),$$

$$\Delta w_k \in \Delta\mathcal{W}_k = \left\{\Delta w \colon \Delta w^\top E_k^{-1} \Delta w \leq 1\right\}, \tag{8.161}$$

$\widetilde{\Delta\theta}_0 = \mathbf{0}_n^\top$, and $\widetilde{\Delta z}_0 = \mathbf{0}_{n-m}^\top$. Clearly the system in (8.159–8.161) is in the form of (8.106), (8.107), and (8.109) except for the initial conditions. To resolve this issue, define

$$\Delta\mathcal{X}_0 = \left\{\Delta x \in \mathbb{R}^{2n-m} \colon \Delta x^\top F_0^{-1} \Delta x \leq 1\right\}, \tag{8.162}$$

with $F_0 = \varepsilon I_{2n-m}$, where ε is a small positive constant. By assuming that

$$\Delta x_0 = \left[\widetilde{\Delta\theta}_0^\top, \widetilde{\Delta z}_0^\top\right]^\top \in \Delta\mathcal{X}_0,$$

the setting in (8.159), (8.161), and (8.162) fully conforms to that in (8.106), (8.107), and (8.109).

Let $\widetilde{\Delta\mathcal{X}}_k$ denote the reachable set of (8.159), (8.161), and (8.162). Then, we can follow the procedure in Section 8.3 to find an ellipsoid

$$\widetilde{\Delta\mathcal{X}}_k^+ = \left\{\widetilde{\Delta x} \in \mathbb{R}^{2n-m} \colon \widetilde{\Delta x}^\top \widetilde{F}_k^{-1} \widetilde{\Delta x} \leq 1\right\},$$

where $\widetilde{F}_k \in \mathbb{R}^{(2n-m)\times(2n-m)}$ is recursively defined as follows:

$$\widetilde{F}_{k+1} = (1+\gamma_k) G_k \widetilde{F}_k G_k^\top + \left(1 + \frac{1}{\gamma_k}\right) H E_k H^\top, \tag{8.163}$$

with $\widetilde{F}_0 = F_0 = \varepsilon I_{2n-m}$ and $\gamma_k > 0$ for all $k \geq 0$; this ellipsoid is guaranteed to contain $\widetilde{\Delta\mathcal{X}}_k$. Let $\mathcal{X}_k$ denote the set containing all possible values that $x_k = \left[\theta_k^\top, z_k^\top\right]^\top$ can take. Then, since $x_k \approx \overline{x}_k + \widetilde{\Delta x}_k$, we can approximate $\mathcal{X}_k$ by another set $\widetilde{\mathcal{X}}_k$ as follows:

$$\begin{aligned} x_k \in \mathcal{X}_k &\approx \widetilde{\mathcal{X}}_k \\ &:= \{\overline{x}_k\} + \widetilde{\Delta\mathcal{X}}_k \\ &\subseteq \{\overline{x}_k\} + \widetilde{\Delta\mathcal{X}}_k^+. \end{aligned} \tag{8.164}$$

Choice of Parameter γ_k. The matrix G_k has at least one eigenvalue at one. To see this, note that $\mathbf{rank}(M) = n-1$; thus $J(\overline{\theta}_k)$ is rank deficient, which means that it has at least one eigenvalue at zero. Thus, the upper-left block matrix in G_k, $I_n - T_s D^{-1} J(\overline{\theta}_k)$, has at least one eigenvalue at one, which means that the matrix G_k has also at least one eigenvalue at one (this follows from the block structure of the matrix G_k). Therefore, even if $G_k = G$, where G is some constant matrix, we cannot set γ_k to a positive constant for all $k \geq 0$ because the resulting bounding ellipsoid would grow unbounded. In general, we can still use the criteria in (8.45) (minimum volume ellipsoid) and (8.69) (ellipsoid with minimum sum of squares of the semi-axis lengths). However, we cannot use the criterion in (8.78–8.79) (tight ellipsoid) because of the dimensions of the matrix H.

8.6 Notes and References

The derivation of a generic bounding ellipsoid using Holder's inequality is due to Schweppe [10]. The results on ellipsoids with minimum volume and minimum sum of squares of the semi-axis lengths are due to Chernousko [17], although the derivations here are slightly different from those in [17]. The results on tight ellipsoids are due to Kurzhanskiy and Varaiya [49], although the derivation here is slightly different from that in [49]. The microgrid model used in Section 8.5 has been adapted from the model in [39].

9 Continuous-Time Systems: Set-Theoretic Input Uncertainty

In this chapter, we extend the framework developed in Chapter 8 to settings in which the relation between the input vector and the state vector is governed by a continuous-time state-space model. As in Chapter 8, we assume that the values that the input vector can take belong to an ellipsoid. Then the objective is to characterize the set containing all possible values that the state vector can take. Fundamental concepts from set theory and continuous-time linear dynamical systems used throughout this chapter are reviewed in Section 2.2.2 and Section 2.3.1, respectively. The material presented here heavily builds on Chapter 8; thus, the reader is encouraged to read that chapter before embarking on the reading of this chapter.

9.1 Introduction

Consider the circuit in Fig. 9.1, which we featured in Section 6.1, and recall the relation between the inputs, $v_s(t)$ and $i_l(t)$, and the states, $v_2(t)$ and $i_1(t)$, given in (6.1):

$$\frac{d}{dt}\begin{bmatrix} v_2(t) \\ i_1(t) \end{bmatrix} = \begin{bmatrix} 0 & \frac{1}{c} \\ -\frac{1}{\ell} & -\frac{r}{\ell} \end{bmatrix}\begin{bmatrix} v_2(t) \\ i_1(t) \end{bmatrix} + \begin{bmatrix} 0 & -\frac{1}{c} \\ \frac{1}{\ell} & 0 \end{bmatrix}\begin{bmatrix} v_s(t) \\ i_l(t) \end{bmatrix}. \tag{9.1}$$

In Chapter 6, we studied the case when the values that the input vector, $w(t)=\left[v_s(t), i_l(t)\right]^\top$, takes were not a priori known but described by some stochastic processes with known first and second moments or known probability distribution. Here, we consider a set-theoretic model to capture the uncertainty in the value that the input vector can take, i.e., the value that $w(t)$ takes is unknown but it belongs to some closed and bounded set, $\mathcal{W}(t)$.

For the circuit to perform its intended function, at any time t, the voltage across the terminals of the current source, $v_2(t)$, must be within the interval $[v_2^m, v_2^M]$. To check whether or not this is the case, and since the evolution of $v_2(t)$ depends on $i_1(t)$, we first need to characterize the set containing all possible values that the state vector, $x(t) = \left[v_2(t), i_1(t)\right]^\top$, can take. This set, which we denote by $\mathcal{X}(t)$, is referred to as the reachable set of (9.1) when $w(t)$ takes values

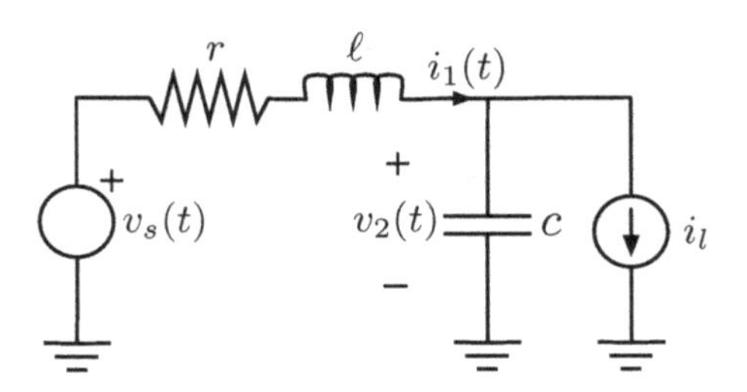

Figure 9.1 AC linear circuit.

in $\mathcal{W}(t)$. Once the reachable set is calculated, we can determine whether or not $v_2(t) \in [v_2^m, v_2^M]$ for all $t \geq 0$ by checking whether or not the reachable set, $\mathcal{X}(t)$, is fully contained in the set $\mathcal{R} = \{x \in \mathbb{R}^2 \colon v_2^m \leq x_1 \leq v_2^M,\ x_2 \in \mathbb{R}\}$, which is defined by the performance requirements described above.

As discussed in Chapter 6, the system in (9.1) belongs to the class of linear systems, which can be compactly described in state-space form as follows:

$$\frac{d}{dt}x(t) = A(t)x(t) + B(t)w(t), \tag{9.2}$$

where $x(t) \in \mathbb{R}^n$, $w(t) \in \mathcal{W}(t) \subseteq \mathbb{R}^m$, $A(t) \in \mathbb{R}^{n\times n}$ and $B(t) \in \mathbb{R}^{n\times m}$ [in the specific example in (9.1), all the model parameters are time invariant; thus both $A(t)$ and $B(t)$ do not depend on t]. While in general the set $\mathcal{W}(t)$ describing the possible values that the input can take can have any shape, in this chapter, we will restrict our analysis to the case when $\mathcal{W}(t)$ is an ellipsoid. We will then provide tools to characterize the reachable set of (9.2), which in general will not be an ellipsoid. We will also provide techniques for characterizing the reachable set of the nonlinear counterpart of (9.2):

$$\frac{d}{dt}x(t) = f\big(t, x(t), w(t)\big), \tag{9.3}$$

where $x(t) \in \mathbb{R}^n$, $w(t) \in \mathcal{W}(t) \subseteq \mathbb{R}^m$, and $f\colon [0,\infty) \times \mathbb{R}^n \times \mathbb{R}^m \to \mathbb{R}^n$.

The computation of the reachable set of (9.2) [or (9.3)] is useful because it allows the system analyst to determine whether or not the system trajectories will remain within some set, $\mathcal{R} \subseteq \mathbb{R}^n$, defined by performance requirements, for all possible realizations of the system inputs. For example, in the context of a power system, the techniques developed in this chapter can be used to analyze the performance of certain control functions for different levels of renewable-based generation, which acts as an unknown disturbance to the total amount of power injected into the system, and can be described by a set-theoretic model.

9.2 Continuous-Time Linear Systems

Consider a system of the form:

$$\frac{d}{dt}x(t) = A(t)x(t) + B(t)w(t), \tag{9.4}$$

where $x(t) \in \mathbb{R}^n$, $w(t) \in \mathbb{R}^m$, $A(t) \in \mathbb{R}^{n\times n}$, and $B(t) \in \mathbb{R}^{n\times m}$ such that $B(t) \neq \mathbf{0}_{n\times m}$. Assume that $w(t)$, $t \geq 0$, is uncertain but known to belong to an ellipsoid $\mathcal{W}(t)$ with center $\overline{w}(t) \in \mathbb{R}^m$ and shape matrix $E(t) \in \mathbb{R}^{m\times m}$, i.e.,

$$\mathcal{W}(t) = \Big\{w \in \mathbb{R}^m\colon \big(w - \overline{w}(t)\big)^\top E^{-1}(t)\big(w - \overline{w}(t)\big) \leq 1\Big\}. \tag{9.5}$$

Similarly, assume that $x(0)$ is uncertain but known to belong to an ellipsoid $\mathcal{X}(0)$ with center $\overline{x}(0) \in \mathbb{R}^n$ and shape matrix $F(0) \in \mathbb{R}^{n\times n}$, i.e.,

$$\mathcal{X}(0) = \Big\{x \in \mathbb{R}^n\colon \big(x - \overline{x}(0)\big)^\top F^{-1}(0)\big(x - \overline{x}(0)\big) \leq 1\Big\}. \tag{9.6}$$

Recall that $E(t)$, $t \geq 0$, and $F(0)$ are symmetric and positive definite matrices. Also, recall that the support functions of $\mathcal{W}(t)$ and $\mathcal{X}(0)$, denoted by $S_{\mathcal{W}(t)}(\eta)$ and $S_{\mathcal{X}(0)}(\eta)$, respectively, are given by

$$S_{\mathcal{W}(t)}(\eta) = \eta^\top \overline{w}(t) + \sqrt{\eta^\top E(t)\eta}, \quad \eta \in \mathbb{R}^m, \tag{9.7}$$

$$S_{\mathcal{X}(0)}(\eta) = \eta^\top \overline{x}(0) + \sqrt{\eta^\top F(0)\eta}, \quad \eta \in \mathbb{R}^n. \tag{9.8}$$

The objective here is to characterize, exactly or approximately, the reachable set of (9.4–9.6), i.e., the set $\mathcal{X}(t)$ containing all possible values that $x(t)$ can take for all $t \geq 0$.

As in the discrete-time case discussed in Chapter 8, even if $\mathcal{W}(t)$ and $\mathcal{X}(0)$ are ellipsoids, the reachable set, $\mathcal{X}(t)$, is not an ellipsoid in general; thus, we would like to obtain an ellipsoid $\mathcal{X}^+(t)$ that contains the set $\mathcal{X}(t)$ for all $t \geq 0$. To this end, consider ellipsoids of the form

$$\mathcal{X}^+(t) = \Big\{x \in \mathbb{R}^n\colon \big(x - \overline{x}(t)\big)^\top F^{-1}(t)\big(x - \overline{x}(t)\big) \leq 1\Big\}, \tag{9.9}$$

with the evolution of $\overline{x}(t) \in \mathbb{R}^n$ and $F(t) \in \mathbb{R}^{n\times n}$ governed by

$$\frac{d}{dt}\overline{x}(t) = A(t)\overline{x}(t) + B(t)\overline{w}(t), \tag{9.10}$$

$$\frac{d}{dt}F(t) = A(t)F(t) + F(t)A^\top(t) + \beta(t)F(t) + \frac{1}{\beta(t)}B(t)E(t)B^\top(t), \tag{9.11}$$

where $\beta(t) > 0$, and $\overline{x}(0)$ and $F(0)$ are given. Then, $\mathcal{X}(t) \subset \mathcal{X}^+(t)$ for all $t \geq 0$. In order to establish this result, we will discretize the differential equation in (9.4), follow similar steps to those followed in the discrete-time case, and then obtain the result by taking limits.

First, note that, for small $dt > 0$, we have that

$$\frac{d}{dt}x(t) \approx \frac{1}{dt}\big(x(t + dt) - x(t)\big);$$

thus, by using (9.4), we can write

$$x(t+dt) \approx \underbrace{\left(I_n + A(t)dt\right)}_{=:\,G(t)} x(t) + \underbrace{B(t)dt}_{=:\,H(t)} w(t). \tag{9.12}$$

Next, for $\mathcal{X}, \mathcal{Y} \subseteq \mathbb{R}^n$, recall the expression for the support function of the set $\mathcal{X} + \mathcal{Y}$, $S_{\mathcal{X}+\mathcal{Y}}(\cdot)$, given in (2.57):

$$S_{\mathcal{X}+\mathcal{Y}}(\eta) = S_{\mathcal{X}}(\eta) + S_{\mathcal{Y}}(\eta), \quad \eta \in \mathbb{R}^n. \tag{9.13}$$

Also, given $H \in \mathbb{R}^{n\times m}$ and $\mathcal{W} \subseteq \mathbb{R}^m$, recall the expression for the support function of the set $H\mathcal{W}$ given in (2.59):

$$S_{H\mathcal{W}}(\eta) = S_{\mathcal{W}}(H^\top \eta), \quad \eta \in \mathbb{R}^n. \tag{9.14}$$

Assume that $\mathcal{X}(t) \subseteq \mathcal{X}^+(t) = \left\{x \in \mathbb{R}^n : \left(x - \overline{x}(t)\right)^\top F^{-1}(t)\left(x - \overline{x}(t)\right) \leq 1\right\}$; thus, $S_{\mathcal{X}(t)}(\eta) \leq S_{\mathcal{X}^+(t)}(\eta)$ for all $\eta \in \mathbb{R}^n$. Then, by using (9.13–9.14), and the fact that

$$(a+b)^2 \leq (1+\gamma)a^2 + \left(1 + \frac{1}{\gamma}\right)b^2, \tag{9.15}$$

for $\gamma > 0$ and any $a, b \in \mathbb{R}$ (see Section 8.2.1), it follows that

$$\begin{aligned}
S_{\mathcal{X}(t+dt)}(\eta) &= S_{\mathcal{X}(t)}\left(G^\top(t)\eta\right) + S_{\mathcal{W}(t)}\left(H^\top(t)\eta\right) \\
&\leq S_{\mathcal{X}^+(t)}\left(G^\top(t)\eta\right) + S_{\mathcal{W}(t)}\left(H^\top(t)\eta\right) \\
&= \eta^\top G(t)\overline{x}(t) + \sqrt{\eta^\top G(t)F(t)G^\top(t)\eta} \\
&\quad + \eta^\top H(t)\overline{w}(t) + \sqrt{\eta^\top H(t)E(t)H^\top(t)\eta} \\
&\leq \eta^\top \left(G(t)\overline{x}(t) + H(t)\overline{w}(t)\right) \\
&\quad + \sqrt{\eta^\top \left[\left(1+\gamma(t)\right)G(t)F(t)G^\top(t) + \left(1 + \frac{1}{\gamma(t)}\right)H(t)E(t)H^\top(t)\right]\eta},
\end{aligned} \tag{9.16}$$

where

$$\gamma(t) = \beta(t)dt, \quad \beta(t) > 0. \tag{9.17}$$

Note that since $G(t) = I_n + A(t)dt$ and $H(t) = B(t)dt$, choosing $\gamma(t)$ as in (9.17) makes the second inequality in (9.16) tight as dt approaches zero. Now, since $F(t)$ is symmetric and positive definite and $G(t)$ is invertible for sufficiently small dt, it follows that $\left(1+\gamma(t)\right)G(t)F(t)G^\top(t)$ is also symmetric and positive definite. Similarly, since $E(t)$ is positive definite, it follows that $\left(1+1/\gamma(t)\right)H(t)E(t)H^\top(t)$ is symmetric and positive (semi)definite. But the sum of a symmetric positive definite matrix and a symmetric positive (semi)definite matrix is always symmetric and positive definite; thus, the matrix

$$\Big(1+\gamma(t)\Big)G(t)F(t)G^\top(t)+\left(1+\frac{1}{\gamma(t)}\right)H(t)E(t)H^\top(t) \tag{9.18}$$

is symmetric and positive definite. Next, define

$$\mathcal{X}^+(t+dt):=\left\{x\in\mathbb{R}^n\colon\big(x-\overline{x}(t+dt)\big)^\top F^{-1}(t+dt)\big(x-\overline{x}(t+dt)\big)\le 1\right\},$$

where

$$\overline{x}(t+dt):=G(t)\overline{x}(t)+H(t)\overline{w}(t), \tag{9.19}$$

$$F(t+dt):=\Big(1+\gamma(t)\Big)G(t)F(t)G^\top(t)+\left(1+\frac{1}{\gamma(t)}\right)H(t)E(t)H^\top(t); \tag{9.20}$$

then, by using (9.16), we obtain that $S_{\mathcal{X}(t+dt)}(\eta)\le S_{\mathcal{X}^+(t+dt)}(\eta)$ for all $\eta\in\mathbb{R}^n$, from where it follows that $\mathcal{X}(t+dt)\subseteq\mathcal{X}^+(t+dt)$. Now, plugging the expression for $G(t)$ and $H(t)$ into (9.19) and rearranging terms yields

$$\frac{1}{dt}\big(\overline{x}(t+dt)-\overline{x}(t)\big)=A(t)\overline{x}(t)+B(t)\overline{w}(t),$$

and by taking the limit as dt goes to zero on both sides, we obtain

$$\frac{d}{dt}\overline{x}(t)=A(t)\overline{x}(t)+B(t)\overline{w}(t), \tag{9.21}$$

an expression that matches that in (9.10). Also, by plugging the expressions for $G(t)$ and $H(t)$ into (9.20) and taking into account that $\gamma(t)=\beta(t)dt$, we obtain

$$\begin{aligned}F(t+dt)&=\Big(1+\beta(t)dt\Big)\Big(I_n+A(t)dt\Big)F(t)\Big(I_n+A(t)dt\Big)^\top\\&\quad+\left(1+\frac{1}{\beta(t)dt}\right)(dt)^2B(t)E(t)B^\top(t)\\&=F(t)+dt\left(A(t)F(t)+F(t)A^\top(t)+\beta(t)F(t)+\frac{1}{\beta(t)}B(t)E(t)B^\top(t)\right)\\&\quad+(dt)^2\Big(A(t)F(t)A^\top(t)+\beta(t)\big(A(t)F(t)+F(t)A^\top(t)\big)\\&\quad+B(t)E(t)B^\top(t)\Big)+(dt)^3\beta(t)A(t)F(t)A^\top(t),\end{aligned} \tag{9.22}$$

which can be rewritten as

$$\begin{aligned}\frac{1}{dt}\Big(F(t+dt)-F(t)\Big)&=A(t)F(t)+F(t)A^\top(t)+\beta(t)F(t)\\&\quad+\frac{1}{\beta(t)}B(t)E(t)B^\top(t)\\&\quad+dt\Big(A(t)F(t)A^\top(t)+\beta(t)\big(A(t)F(t)+F(t)A^\top(t)\big)+B(t)E(t)B^\top(t)\Big)\\&\quad+(dt)^2\beta(t)A(t)F(t)A^\top(t).\end{aligned} \tag{9.23}$$

Then, by taking the limit as dt goes to zero on both sides of (9.23), we obtain

$$\frac{d}{dt}F(t) = A(t)F(t) + F(t)A^\top(t) + \beta(t)F(t) + \frac{1}{\beta(t)}B(t)E(t)B^\top(t), \qquad (9.24)$$

an expression that matches that in (9.11).

9.2.1 Choice of Parameter $\beta(t)$

One possibility is to set $\beta(t)$ to some positive constant for all $t \geq 0$, which simplifies the matrix differential equation in (9.11); however, there is no guarantee on whether the outer-bounding ellipsoid that results, $\mathcal{X}^+(t)$, will be optimal in any sense. An alternative is to choose $\beta(t)$ so that the ellipsoid $\mathcal{X}^+(t)$ is optimal in some sense. Recall that in the discrete-time case in Chapter 8, given $\mathcal{X}_k^+$, as defined by some symmetric and positive definite matrix F_k, we chose γ_k (the discrete-time counterpart of the parameter $\beta(t)$) so as to minimize the volume, or the sum of squares of the semi-axis lengths, of $\mathcal{X}_{k+1}^+$ by means of minimizing $\mathbf{det}(F_{k+1})$ or $\mathbf{tr}(F_{k+1})$. These, in turn, respectively ensured that

$$\Delta\mathbf{det}(F_k) := \mathbf{det}(F_{k+1}) - \mathbf{det}(F_k), \qquad (9.25)$$

and

$$\Delta\mathbf{tr}(F_k) := \mathbf{tr}(F_{k+1}) - \mathbf{tr}(F_k), \qquad (9.26)$$

were minimum. Thus, the natural extension of (9.25) to the continuous-time case is to choose $\beta(t)$ so as to minimize

$$\frac{d}{dt}\mathbf{det}\big(F(t)\big), \qquad (9.27)$$

while the natural extension of (9.26) to the continuous-time case is to choose $\beta(t)$ so as to minimize

$$\frac{d}{dt}\mathbf{tr}\big(F(t)\big). \qquad (9.28)$$

The criterion in (9.27) will result in an ellipsoid that is optimal within the class of outer-bounding ellipsoids defined by (9.10) and (9.11) in the sense of having minimum volume. Similarly, the criterion in (9.28) will result in an ellipsoid that is optimal within the class of outer-bounding ellipsoids defined by (9.10) and (9.11) in the sense of having minimum sum of squares of the semi-axis lengths. Next, we will explore all these alternative choices.

Positive Constant

Consider the system in (9.4–9.6). Let $\mathcal{X}^+(t) \subseteq \mathbb{R}^n$ denote an ellipsoid with center $\overline{x}(t) \in \mathbb{R}^n$ and shape matrix $F(t) \in \mathbb{R}^{n \times n}$ whose evolution with time is determined by (9.10–9.11), for given $\overline{x}(0)$ and $F(0)$, with $\beta(t)$ chosen as follows:

$$\beta(t) = \beta, \qquad (9.29)$$

where β is a positive scalar. This choice simplifies the matrix differential equation in (9.11), and in the case when the system is time invariant, i.e., $A(t) = A \in \mathbb{R}^{n\times n}$ and $B(t) = B \in \mathbb{R}^{n\times m}$, it also makes it time invariant:

$$\frac{d}{dt}F(t) = AF(t) + F(t)A^\top + \beta F(t) + \frac{1}{\beta}BE(t)B^\top; \tag{9.30}$$

however, it is not clear a priori how tight the resulting outer-bounding ellipsoid will be or even if it will remain bounded at all times.

For the case when A is Hurwitz, i.e., the real part of all the eigenvalues of A is strictly negative, which ensures that the system trajectories will remain bounded because the inputs are also bounded, in addition to enforcing the condition in (9.29), we need to impose the following additional condition on β:

$$\frac{\beta}{2} < \min\{|\mathbf{real}(\lambda_1)|, |\mathbf{real}(\lambda_2)|, \ldots, |\mathbf{real}(\lambda_n)|\}, \tag{9.31}$$

where $\lambda_1, \lambda_2, \ldots, \lambda_n$ denote the eigenvalues of the matrix A. Then, by choosing β in such a way, we ensure that the entries of $F(t)$ remain bounded for all $t \geq 0$, thus yielding a useful outer-bounding ellipsoid. To see this, we rewrite (9.30) in vector form as follows:

$$\frac{d}{dt}\mathbf{vec}(F(t)) = \underbrace{[(I_n \otimes A) + (A \otimes I_n) + \beta I_{n^2}]}_{=:\,\widetilde{A}}\mathbf{vec}(F(t)) + \frac{1}{\beta}\mathbf{vec}(BE(t)B^\top), \tag{9.32}$$

where "$\otimes$" denotes the Kronecker product, and $\mathbf{vec}(F(t))$ and $\mathbf{vec}(BE(t)B^\top)$ are n^2-dimensional vectors obtained by stacking the columns of the matrices $F(t)$ and $BE(t)B^\top$, respectively, from left to right.[1] Now, for the trajectories of (9.32) to remain bounded, a sufficient condition is that the matrix $\widetilde{A}$ is Hurwitz. Let $\sigma(A)$ denote the spectrum of A, then, the spectrum of $\widetilde{A}$, denoted by $\sigma(\widetilde{A})$, is given by

$$\sigma(\widetilde{A}) = \{\lambda\colon \lambda = \lambda_i + \lambda_j + \beta,\ \lambda_i \in \sigma(A),\ \lambda_j \in \sigma(A)\}. \tag{9.33}$$

Thus, by choosing β so that

$$\frac{\beta}{2} < \min\Big\{|\mathbf{real}(\lambda_1)|, |\mathbf{real}(\lambda_2)|, \ldots, |\mathbf{real}(\lambda_n)|\Big\}, \tag{9.34}$$

we ensure that $\tilde{A}$ is Hurwitz; therefore, the entries of $F(t)$ will remain bounded at all times.

[1] Given $A = [a_{i,j}] \in \mathbb{R}^{n\times m}$, we can associate a vector $\mathbf{vec}(A) \in \mathbb{R}^{nm}$ defined as follows:

$$\mathbf{vec}(A) = [a_{1,1}, a_{2,1}, \ldots, a_{n,1}, a_{1,2}, a_{2,2}, \ldots, a_{n,2}, \ldots, a_{1,m}, a_{2,m}, \ldots, a_{n,m}]^\top.$$

Example 9.1 (Positive constant) Consider the following one-dimensional LTI system:

$$\frac{d}{dt}x(t) = ax(t) + bw(t), \tag{9.35}$$

where $x(t) \in \mathbb{R}$, $w(t) \in [w^m(t), w^M(t)]$, $w^m(t) < w^M(t)$, $a < 0$, and $b \neq 0$. Also assume that $x(0) \in [x^m(0), x^M(0)]$, $x^m(0) < x^M(0)$.

First note that condition $w(t) \in [w^m(t), w^M(t)]$ can be rewritten as follows:

$$w(t) \in \mathcal{W}(t) = \left\{w(t) \in \mathbb{R} \colon E^{-1}(t)\big(w(t) - \overline{w}(t)\big)^2 \leq 1\right\}, \tag{9.36}$$

where

$$E(t) = \left(\frac{w^M(t) - w^m(t)}{2}\right)^2, \qquad \overline{w}(t) = \frac{w^m(t) + w^M(t)}{2}.$$

Similarly, note that condition $x(0) \in [x^m(0), x^M(0)]$ can be rewritten as follows:

$$x(0) \in \mathcal{X}(0) = \left\{x(0) \in \mathbb{R} \colon F^{-1}(0)\big(x(0) - \overline{x}(0)\big)^2 \leq 1\right\}, \tag{9.37}$$

where

$$F(0) = \left(\frac{x^M(0) - x^m(0)}{2}\right)^2, \qquad \overline{x}(0) = \frac{x^m(0) + x^M(0)}{2}.$$

Then, the objective is to compute a real interval, $\mathcal{X}^+(t)$, such that $\mathcal{X}(t) \subseteq \mathcal{X}^+(t)$, where $\mathcal{X}(t)$ denotes the reachable set of (9.35), which in this case is just a real interval.

Tailoring the expressions in (9.10–9.11) and (9.29) to the setting here yields

$$\mathcal{X}(t) \subseteq \mathcal{X}^+(t) = \left\{x \in \mathbb{R} \colon F^{-1}(t)\big(x - \overline{x}(t)\big)^2 \leq 1\right\},$$

where $F(t) \in \mathbb{R}$ and $\overline{x}(t) \in \mathbb{R}$ are determined by

$$\begin{aligned}\frac{d}{dt}\overline{x}(t) &= a\overline{x}(t) + b\overline{w}(t),\\ \frac{d}{dt}F(t) &= (2a + \beta)F(t) + \frac{1}{\beta}b^2E(t),\end{aligned} \tag{9.38}$$

with $\beta > 0$. Thus, since $E(t)$ is bounded for all $t \geq 0$, for $F(t)$ to remain bounded for all $t \geq 0$, the parameter β must be chosen so that

$$\beta < -2a,$$

which is clearly a special case of the result in (9.31). Note that if a were positive, and because of the constraint $\beta > 0$ in (9.38), $F(t)$ would grow unbounded as t goes to infinity, but this should not come as a surprise since, in such a case, there would be some trajectories of (9.35) that would also grow unbounded as t goes to infinity.

In this case, we can compute the exact reachable set, $\mathcal{X}(t)$, as follows. First, note that $\mathcal{X}(t)$ can be written as follows:

$$\mathcal{X}(t) = \left\{x \in \mathbb{R} \colon \left(F^*(t)\right)^{-1}(x - \overline{x}^*(t))^2 \leq 1\right\}, \tag{9.39}$$

where $\overline{x}^*$ is a real number and $F^*(t)$ is a positive real number. Now, for $dt > 0$ small enough, we have that

$$x(t + dt) \approx (1 + adt)x(t) + bdtw(t). \tag{9.40}$$

Then, on the one hand, we have that

$$S_{\mathcal{X}(t+dt)}(\eta) = \eta\overline{x}^*(t + dt) + |\eta|\sqrt{F^*(t + dt)}, \quad \eta \in \mathbb{R}. \tag{9.41}$$

On the other hand, by using (9.13–9.14), it follows that

$$\begin{aligned} S_{\mathcal{X}(t+dt)}(\eta) &\approx S_{\mathcal{X}(t)}\big((1 + dta)\eta\big) + S_{\mathcal{W}(t)}\big(dtb\eta\big) \\ &= \eta\big((1 + adt)\overline{x}^*(t) + bdt\overline{w}(t)\big) + |\eta|\left(|1 + adt|\sqrt{F^*(t)} + |bdt|\sqrt{E(t)}\right). \end{aligned} \tag{9.42}$$

Now, equating the right-hand sides of (9.41) and (9.42) yields

$$\begin{aligned} \overline{x}^*(t + dt) &\approx (1 + adt)\overline{x}^*(t) + bdt\overline{w}(t), \\ F^*(t + dt) &\approx \left(|1 + adt|\sqrt{F^*(t)} + |bdt|\sqrt{E(t)}\right)^2, \end{aligned} \tag{9.43}$$

which by rearranging and dividing throughout by dt results in

$$\begin{aligned} \frac{\overline{x}^*(t + dt) - \overline{x}^*(t)}{dt} &\approx a\overline{x}^*(t) + b\overline{w}(t), \\ \frac{F^*(t + dt) - F^*(t)}{dt} &\approx (2a + a^2dt)F^*(t) + 2\big|(1 + adt)b\big|\sqrt{F^*(t)E(t)} + b^2dtE(t). \end{aligned} \tag{9.44}$$

Finally, by taking the limit as dt goes to zero on both sides of the expressions in (9.44), we obtain that

$$\begin{aligned} \frac{d}{dt}\overline{x}^*(t) &= a\overline{x}^*(t) + b\overline{w}(t), \\ \frac{d}{dt}F^*(t) &= 2aF^*(t) + 2|b|\sqrt{F^*(t)E(t)}. \end{aligned} \tag{9.45}$$

Minimum Volume Ellipsoid

Consider the system in (9.4–9.6). Let $\mathcal{X}^+(t) \subseteq \mathbb{R}^n$ denote an ellipsoid with center $\overline{x}(t) \in \mathbb{R}^n$ and shape matrix $F(t) \in \mathbb{R}^{n \times n}$ whose evolution with time is governed by (9.10–9.11), for given $\overline{x}(0)$ and $F(0)$, with $\beta(t)$ chosen as follows:

$$\beta(t) = \sqrt{\frac{\mathbf{tr}\Big(F^{-1}(t)B(t)E(t)B^\top(t)\Big)}{n}}. \tag{9.46}$$

Then, $\mathcal{X}^+(t)$ is optimal in the sense that its volume is minimum among all the outer-bounding ellipsoids within the class defined by (9.10) and (9.11). To see this, first note that for $\epsilon > 0$ small enough, we have that

$$\mathbf{det}\big(F(t+\epsilon)\big) \approx \mathbf{det}\big(F(t)\big) + \epsilon \frac{d}{dt}\mathbf{det}\big(F(t)\big). \tag{9.47}$$

In light of (9.11), it is clear that $\frac{d}{dt}\mathbf{det}\big(F(t)\big)$ depends on $\beta(t)$; thus, in order to minimize the determinant of $F(t+\epsilon)$, given the matrix $F(t)$ has minimum determinant among those matrices defined by (9.11), we need to find the value of $\beta(t)$ that minimizes $\frac{d}{dt}\mathbf{det}\big(F(t)\big)$. Then, by minimizing $\mathbf{det}\big(F(t)\big)$, we are minimizing the volume of $\mathcal{X}^+(t)$ because, as discussed in Section 2.2, the volume of an ellipsoid $\mathcal{E} \subseteq \mathbb{R}^n$ with shape matrix $E \in \mathbb{R}^{n\times n}$ is given by

$$\mathbf{vol}(\mathcal{E}) = \frac{\pi^{n/2}\sqrt{\mathbf{det}(E)}}{\Gamma\left(\frac{n}{2}+1\right)},$$

where $\Gamma(\cdot)$ is Euler's gamma function.

By using Jacobi's formula as given in (A.12), we have that

$$\begin{aligned}\frac{d}{dt}\mathbf{det}\big(F(t)\big) &= \mathbf{det}\big(F(t)\big)\mathbf{tr}\left(F^{-1}(t)\frac{d}{dt}F(t)\right)\\ &= \mathbf{det}\big(F(t)\big)\mathbf{tr}\Bigg(F^{-1}(t)A(t)F(t) + A^\top(t) + \beta(t)I_n\\ &\qquad + \frac{1}{\beta(t)}F^{-1}(t)B(t)E(t)B^\top(t)\Bigg)\end{aligned} \tag{9.48}$$

Define

$$\varphi_t(z) = \mathbf{tr}\left(F^{-1}(t)A(t)F(t) + A^\top(t) + zI_n + \frac{1}{z}F^{-1}(t)B(t)E(t)B^\top(t)\right); \tag{9.49}$$

then, since $\mathbf{det}(F(t)) > 0$ for any $t \geq 0$, finding the value of $\beta(t)$ that minimizes $\frac{d}{dt}\mathbf{det}(F(t))$ is equivalent to finding the value of z that minimizes $\varphi_t(z)$. Thus, we take the derivative of $\varphi_t(z)$ with respect to z, which yields

$$\frac{d\varphi_t(z)}{dz} = \mathbf{tr}\left(I_n - \frac{1}{z^2}F^{-1}(t)B(t)E(t)B^\top(t)\right), \tag{9.50}$$

and by equating to zero and solving for z, we obtain

$$z^{(1)} = \sqrt{\frac{\mathbf{tr}\Big(F^{-1}(t)B(t)E(t)B^\top(t)\Big)}{n}}, \quad z^{(2)} = -\sqrt{\frac{\mathbf{tr}\Big(F^{-1}(t)B(t)E(t)B^\top(t)\Big)}{n}},$$

but recall from (9.11) that $\beta(t) > 0$ for all t; thus, only $z^{(1)}$ is a potentially valid solution. [Note that $\mathbf{tr}\big(F^{-1}(t)B(t)E(t)B^\top(t)\big) > 0$ because $E(t)$ and $F(t)$ are symmetric and positive definite and $B(t) \neq \mathbf{0}_{n\times m}$.] Now, in order to check that $z^{(1)}$ is indeed a minimum, we differentiate $\frac{d\varphi_t(z)}{dz}$, which yields

$$\frac{d^2\varphi_t(z)}{dz^2} = \frac{2}{z^3}\mathbf{tr}\Big(F^{-1}(t)B(t)E(t)B^\top(t)\Big), \tag{9.51}$$

from where it follows that

$$\begin{aligned}\left.\frac{d^2\varphi_t(z)}{dz^2}\right|_{z=z^{(1)}} &= \frac{2n^{3/2}}{\sqrt{\mathbf{tr}\Big(F^{-1}(t)B(t)E(t)B^\top(t)\Big)}} \\ &> 0;\end{aligned} \tag{9.52}$$

thus,

$$\beta(t) = \sqrt{\frac{\mathbf{tr}\Big(F^{-1}(t)B(t)E(t)B^\top(t)\Big)}{n}} \tag{9.53}$$

minimizes $\frac{d}{dt}\mathbf{det}\big(F(t)\big)$ as claimed.

Example 9.2 (Minimum volume ellipsoid) Consider again the system analyzed in Example 9.1 [see (9.35–9.37) for its description]. By tailoring the expressions in (9.10–9.11) and (9.46) to the setting here, it follows that

$$\mathcal{X}(t) \subseteq \mathcal{X}^+(t) = \Big\{x \in \mathbb{R}\colon F^{-1}(t)\big(x - \overline{x}(t)\big)^2 \leq 1\Big\},$$

where $\overline{x}(t) \in \mathbb{R}$ and $F(t) \in \mathbb{R}$ are determined by

$$\frac{d}{dt}\overline{x}(t) = a\overline{x}(t) + b\overline{w}(t), \tag{9.54}$$

$$\frac{d}{dt}F(t) = \big(2a + \beta(t)\big)F(t) + \frac{1}{\beta(t)}b^2E(t), \tag{9.55}$$

with

$$\begin{aligned}\beta(t) &= \sqrt{\mathbf{tr}\Big(b^2F^{-1}(t)E(t)\Big)} \\ &= |b|\sqrt{\frac{E(t)}{F(t)}}.\end{aligned} \tag{9.56}$$

One can check that the expressions in (9.54–9.56) are identical to those in (9.45) exactly characterizing $\mathcal{X}(t)$; thus, in this case $\mathcal{X}(t) = \mathcal{X}^+(t)$. This result should be expected, as in order to obtain (9.54–9.56), we are minimizing the volume of $\mathcal{X}^+(t)$ so that $\mathcal{X}(t) \subseteq \mathcal{X}^+(t)$, but in this case, because $\mathcal{X}^+(t)$ is just a real interval, minimizing the volume of $\mathcal{X}^+(t)$ really means minimizing its length while containing $\mathcal{X}(t)$; thus, clearly $\mathcal{X}^+(t)$ must be identically equal to $\mathcal{X}(t)$.

Ellipsoid with Minimum Sum of Squares of the Semi-Axis Lengths

Consider the system in (9.4–9.6). Let $\mathcal{X}^+(t) \subseteq \mathbb{R}^n$ denote an ellipsoid with center $\overline{x}(t) \in \mathbb{R}^n$ and shape matrix $F(t) \in \mathbb{R}^{n\times n}$ whose evolution with time is governed by (9.10–9.11), for given $\overline{x}(0)$ and $F(0)$, and $\beta(t)$ chosen as follows:

$$\beta(t) = \sqrt{\frac{\mathbf{tr}\Big(B(t)E(t)B^\top(t)\Big)}{\mathbf{tr}\big(F(t)\big)}}. \tag{9.57}$$

Then, the outer-bounding ellipsoid $\mathcal{X}^+(t)$ is optimal in the sense that the sum of squares of its semi-axis lengths is minimum among all the outer-bounding ellipsoids within the class defined by (9.10) and (9.11).

To see this, first note that for $\epsilon > 0$ small enough, we have that

$$\mathbf{tr}\big(F(t+\epsilon)\big) \approx \mathbf{tr}\big(F(t)\big) + \epsilon\frac{d}{dt}\mathbf{tr}\big(F(t)\big). \tag{9.58}$$

Then, in light of (9.11), it is clear that $\frac{d}{dt}\mathbf{tr}\big(F(t)\big)$ depends on $\beta(t)$; thus, in order to minimize the trace of $F(t+\epsilon)$, given the matrix $F(t)$ has minimum trace among those matrices defined by (9.11), we need to find the value of $\beta(t)$ that minimizes $\frac{d}{dt}\mathbf{tr}\big(F(t)\big)$. Thus, by minimizing $\mathbf{tr}\big(F(t)\big)$, we are minimizing the sum of squares of the semi-axis lengths of $\mathcal{X}^+(t)$ because, as discussed in Section 2.2.2, the eigenvalues of the shape matrix of an ellipsoid are the squares of its semi-axis lengths.

Define

$$\phi_t(z) := \mathbf{tr}\left(A(t)F(t) + F(t)A^\top(t) + zF(t) + \frac{1}{z}B(t)E(t)B^\top(t)\right); \tag{9.59}$$

then, by differentiating, we obtain

$$\frac{d\phi_t(z)}{dz} = \mathbf{tr}\left(F(t) - \frac{1}{z^2}B(t)E(t)B^\top(t)\right), \tag{9.60}$$

which by equating to zero and solving for z yields the following two solutions:

$$z^{(1)} = \sqrt{\frac{\mathbf{tr}\Big(B(t)E(t)B^\top(t)\Big)}{\mathbf{tr}\big(F(t)\big)}},\; z^{(2)} = -\sqrt{\frac{\mathbf{tr}\Big(B(t)E(t)B^\top(t)\Big)}{\mathbf{tr}\big(F(t)\big)}}.$$

But recall from (9.11) that $\beta(t) > 0$ for all t; thus, only $z^{(1)}$ is a potentially valid solution. [Note that $z^{(1)}$ is strictly positive because of the fact that $E(t)$ and $F(t)$ are positive definite and $B(t) \neq \mathbf{0}_{n\times m}$; thus, $\mathbf{tr}\big(B(t)E(t)B^\top(t)\big) > 0$ and $\mathbf{tr}\big(F(t)\big) > 0$.] Then, by differentiating $\frac{d\phi_t(z)}{dz}$ and evaluating at $z = z^{(1)}$, we obtain

$$\begin{aligned}\left.\frac{d^2\phi_t(z)}{dz^2}\right|_{z=z^{(1)}} &= 2\sqrt{\frac{\Big(\mathbf{tr}\big(F(t)\big)\Big)^3}{\mathbf{tr}\big(B(t)E(t)B^\top(t)\big)}}\\ &> 0;\end{aligned} \tag{9.61}$$

thus,

$$\beta(t) = \sqrt{\frac{\mathbf{tr}\Big(B(t)E(t)B^\top(t)\Big)}{\mathbf{tr}\big(F(t)\big)}} \tag{9.62}$$

minimizes $\frac{d}{dt}\mathbf{tr}\big(F(t)\big)$ as claimed.

Example 9.3 (Ellipsoid with minimum sum of squares of the semi-axis lengths) Consider again the system analyzed in Examples 9.1–9.2 [see (9.35–9.37) for its description]. By tailoring the expressions in (9.10–9.11) and (9.57) to the setting here, it follows that

$$\mathcal{X}(t) \subseteq \mathcal{X}^+(t) = \Big\{x \in \mathbb{R}\colon\, F^{-1}(t)\big(x - \overline{x}(t)\big)^2 \leq 1\Big\},$$

where $\overline{x}(t) \in \mathbb{R}$ and $F(t) \in \mathbb{R}$ are determined by

$$\frac{d}{dt}\overline{x}(t) = a\overline{x}(t) + b\overline{w}(t), \tag{9.63}$$

$$\frac{d}{dt}F(t) = \big(2a + \beta(t)\big)F(t) + \frac{1}{\beta(t)}b^2E(t), \tag{9.64}$$

with

$$\begin{aligned}\beta(t) &= \sqrt{\frac{\mathbf{tr}\big(b^2E(t)\big)}{\mathbf{tr}\big(F(t)\big)}} \\ &= |b|\sqrt{\frac{E(t)}{F(t)}}.\end{aligned} \tag{9.65}$$

Note that the expression for $\beta(t)$ in (9.65) matches that in (9.56). This is because in obtaining (9.65), we are minimizing $\frac{d}{dt}\mathbf{tr}\big(F(t)\big)$, whereas in obtaining (9.56), we are minimizing $\frac{d}{dt}\mathbf{det}\big(F(t)\big)$, but in this case $\mathbf{tr}\big(F(t)\big) = \mathbf{det}\big(F(t)\big)$; hence the solution must be the same in both cases, and as argued in Example 9.2 the resulting $\mathcal{X}^+(t)$ is identically equal to $\mathcal{X}(t)$ (see Example 9.1 for its characterization).

Tight Ellipsoid

Consider the system in (9.4–9.6). Here we also assume that $B(t) \in \mathbb{R}^{n\times m}$ is such that $n \leq m$ and $\mathbf{rank}\big(B(t)\big) = n$ for all $t \geq 0$. Let $\mathcal{X}^+(t) \subseteq \mathbb{R}^n$ denote an ellipsoid with center $\overline{x}(t) \in \mathbb{R}^n$ and shape matrix $F(t) \in \mathbb{R}^{n\times n}$ whose evolution with time is governed by (9.10–9.11), for given $\overline{x}(0)$ and $F(0)$, with $\beta(t)$ chosen as follows:

$$\beta(t) = \sqrt{\frac{\eta^\top(t)B(t)E(t)B^\top(t)\eta(t)}{\eta^\top(t)F(t)\eta(t)}}, \tag{9.66}$$

where $\eta(t)$ evolves according to

$$\frac{d}{dt}\eta(t) = -A^\top(t)\eta(t), \tag{9.67}$$

for given $\eta(0)$. Then, the outer-bounding ellipsoid $\mathcal{X}^+(t)$ is tight to $\mathcal{X}(t)$ in the sense that the boundaries of $\mathcal{X}^+(t)$ and $\mathcal{X}(t)$ touch each other at a point $x^*(t)$ defined as follows:

$$x^*(t) = \overline{x}(t) + \frac{1}{\sqrt{\eta(t)^\top F(t)\eta(t)}} F(t)\eta(t). \tag{9.68}$$

To establish this result, we need to show that $S_{\mathcal{X}(t)}\big(\eta(t)\big) = S_{\mathcal{X}^+(t)}\big(\eta(t)\big)$ for all $t \geq 0$, for which we will need the following two facts (which were established in Section 2.2):

F1. Given two closed and convex sets $\mathcal{X}, \mathcal{Y} \in \mathbb{R}^n$, if $S_{\mathcal{X}}(\eta) = S_{\mathcal{Y}}(\eta)$ for some $\eta \in \mathbb{R}^n$, then the boundaries of $\mathcal{X}$ and $\mathcal{Y}$ touch each other at the following point:

$$\begin{aligned} x^*(\eta) &= \underbrace{\arg\max_{x \in \mathcal{X}} \eta^\top x}_{= S_{\mathcal{X}}(\eta)} \\ &= \underbrace{\arg\max_{x \in \mathcal{Y}} \eta^\top x}_{= S_{\mathcal{Y}}(\eta)}. \end{aligned} \tag{9.69}$$

F2. For an ellipsoid

$$\mathcal{E} = \big\{ x \in \mathbb{R}^n \colon (x - \overline{x})^\top E^{-1} (x - \overline{x}) \leq 1 \big\},$$

we have that

$$\begin{aligned} x^*(\eta) &= \arg\max_{x \in \mathcal{E}} \eta^\top x \\ &= \overline{x} + \frac{1}{\sqrt{\eta^\top E \eta}} E\eta, \quad \eta \in \mathbb{R}^n. \end{aligned} \tag{9.70}$$

Now, for $dt > 0$ small enough, we have that

$$x(t + dt) = \big(I_n + A(t)dt\big)x(t) + dt\, B(t)w(t), \tag{9.71}$$

$$\begin{aligned} \eta(t + dt) &= \big(I_n - A^\top(t)dt\big)\eta(t) \\ &\approx \big(I_n + A^\top(t)dt\big)^{-1}\eta(t); \end{aligned} \tag{9.72}$$

thus,

$$\begin{aligned} S_{\mathcal{X}(t+dt)}\big(\eta(t+dt)\big) &\approx S_{\mathcal{X}(t)}\Big(\big(I_n + A(t)dt\big)^\top \eta(t+dt)\Big) \\ &\quad + dt S_{\mathcal{W}(t)}\big(B^\top(t)\eta(t+dt)\big) \\ &= S_{\mathcal{X}(t)}\Big(\big(I_n + A^\top(t)dt\big)\big(I_n + A^\top(t)dt\big)^{-1}\eta(t)\Big) \\ &\quad + dt S_{\mathcal{W}(t)}\Big(B^\top(t)\big(I_n + A^\top(t)dt\big)^{-1}\eta(t)\Big) \\ &= S_{\mathcal{X}(t)}\big(\eta(t)\big) \\ &\quad + dt S_{\mathcal{W}(t)}\Big(B^\top(t)\big(I_n + A^\top(t)dt\big)^{-1}\eta(t)\Big), \end{aligned} \tag{9.73}$$

which by rearranging and dividing throughout by dt yields

$$\frac{S_{\mathcal{X}(t+dt)}\big(\eta(t+dt)\big) - S_{\mathcal{X}(t)}\big(\eta(t)\big)}{dt} = S_{\mathcal{W}(t)}\Big(B^\top(t)\big(I_n + A^\top(t)dt\big)^{-1}\eta(t)\Big). \tag{9.74}$$

Then, taking the limits on both sides of (9.74) as dt goes to zero yields

$$\frac{d}{dt}S_{\mathcal{X}(t)}\big(\eta(t)\big) = S_{\mathcal{W}(t)}\big(B^\top(t)\eta(t)\big). \tag{9.75}$$

Now, recall that $S_{\mathcal{X}^+(t)}\big(\eta(t)\big) = \eta^\top(t)\overline{x}(t) + \sqrt{\eta^\top(t)F(t)\eta(t)}$; thus,

$$\begin{aligned}
\frac{d}{dt}S_{\mathcal{X}^+(t)}\big(\eta(t)\big) &= \frac{d\eta^\top(t)}{dt}\overline{x}(t) + \eta^\top(t)\frac{d\overline{x}(t)}{dt}\\
&\quad + \frac{1}{2\sqrt{\eta^\top(t)F(t)\eta(t)}}\left(2\frac{d\eta^\top(t)}{dt}F(t)\eta(t) + \eta^\top(t)\frac{dF(t)}{dt}\eta(t)\right)\\
&= -\eta^\top(t)A(t)\overline{x}(t) + \eta^\top(t)\big(A(t)\overline{x}(t) + B(t)\overline{w}(t)\big)\\
&\quad + \frac{1}{2\sqrt{\eta^\top(t)F(t)\eta(t)}}\Big(-2\eta^\top(t)A(t)F(t)\eta(t) + \eta^\top(t)A(t)F(t)\eta(t)\\
&\qquad + \eta^\top(t)F(t)A^\top(t)\eta(t) + \beta(t)\eta^\top(t)F(t)\eta(t)\\
&\qquad + \frac{1}{\beta(t)}\eta^\top(t)B(t)E(t)B^\top(t)\eta(t)\Big)\\
&= \eta^\top(t)B(t)\overline{w}(t)\\
&\quad + \frac{1}{2\sqrt{\eta^\top(t)F(t)\eta(t)}}\Big(\beta(t)\eta^\top(t)F(t)\eta(t)\\
&\qquad + \frac{1}{\beta(t)}\eta^\top(t)B(t)E(t)B^\top(t)\eta(t)\Big)\\
&= \eta^\top(t)B(t)\overline{w}(t) + \sqrt{\eta^\top(t)B(t)E(t)B(t)^\top\eta(t)}\\
&= S_{\mathcal{W}(t)}\big(B(t)^\top\eta(t)\big).
\end{aligned} \tag{9.76}$$

Then, it follows from (9.75) and (9.76) that

$$\frac{d}{dt}S_{\mathcal{X}(t)}\big(\eta(t)\big) = \frac{d}{dt}S_{\mathcal{X}^+(t)}\big(\eta(t)\big), \quad t \geq 0. \tag{9.77}$$

Now, since $\mathcal{X}(0) = \mathcal{X}^+(0)$, we have that $S_{\mathcal{X}(0)}\big(\eta(0)\big) = S_{\mathcal{X}^+(0)}\big(\eta(0)\big)$; then, in light of (9.77), it follows that

$$S_{\mathcal{X}(t)}\big(\eta(t)\big) = S_{\mathcal{X}^+(t)}\big(\eta(t)\big), \quad t \geq 0.$$

Finally, by (9.69) and (9.70), we have that the boundaries of $\mathcal{X}^+(t)$ and $\mathcal{X}(t)$

touch each other at

$$x^*(t) = \overline{x}(t) + \frac{1}{\sqrt{\eta^\top(t)F(t)\eta(t)}}F(t)\eta(t).$$

Example 9.4 (Tight ellipsoid) Consider again the system analyzed in Examples 9.1–9.3 [see (9.35– 9.37) for its description]. By tailoring the expressions in (9.10–9.11) and (9.66–9.67) to the setting here, it follows that

$$\mathcal{X}(t) \subseteq \mathcal{X}^+(t) = \left\{x \in \mathbb{R}^n : F^{-1}(t)\big(x - \overline{x}(t)\big)^2 \leq 1\right\},$$

where $\overline{x}(t) \in \mathbb{R}$ and $F(t) \in \mathbb{R}$ are determined by

$$\frac{d}{dt}\overline{x}(t) = a\overline{x}(t) + b\overline{w}(t), \tag{9.78}$$

$$\frac{d}{dt}F(t) = \big(2a + \beta(t)\big)F(t) + \frac{1}{\beta(t)}b^2E(t), \tag{9.79}$$

where

$$\begin{aligned}\beta(t) &= \sqrt{\frac{E(t)\big(b\eta(t)\big)^2}{F(t)\eta^2(t)}} \\ &= |b|\sqrt{\frac{E(t)}{F(t)}}.\end{aligned} \tag{9.80}$$

Note that, according to (9.67), the evolution of $\eta(t)$ is determined by

$$\frac{d}{dt}\eta(t) = -a\eta(t), \tag{9.81}$$

for a given $\eta(0)$; however, in this case, $\eta(t)$ does not play any role in determining $\beta(t)$. Again, the solution obtained here coincides with that obtained in Examples 9.2–9.3, and as argued earlier, the resulting $\mathcal{X}^+(t)$ is identically equal to $\mathcal{X}(t)$ (see Example 9.1 for its characterization).

Reachable Set Exact and Approximate Characterization

Tight ellipsoids can be utilized to exactly characterize the reachable set $\mathcal{X}(t)$. Let $\mathcal{X}_\xi^+(t)$ denote an ellipsoid with center $\overline{x}(t) \in \mathbb{R}^n$ and shape matrix $F(t) \in \mathbb{R}^{n\times n}$ defined by (9.10–9.11), where $\beta(t)$ is chosen as in (9.66–9.67) with

$$\eta(0) = \xi \in \mathcal{B} = \left\{\nu \in \mathbb{R}^n : \nu^\top \nu = 1\right\}.$$

Then, we have that

$$\mathcal{X}(t) = \bigcap_{\xi \in \mathcal{B}} \mathcal{X}_\xi^+(t). \tag{9.82}$$

While the formula above implies that in order to exactly characterize the reachable set it is necessary to compute an infinite number of tight bounding ellipsoids, in practice, highly accurate approximation can be obtained with just a few such ellipsoids. More generally, approximations to the reachable set can be obtained as the intersection of a family of bounding ellipsoids obtained by using (9.10–9.11) with $\beta(t)$ chosen to be constant.

9.2.2 Deterministic Inputs

As we did in Chapter 8 for the discrete-time case, we will now show how to handle deterministic inputs. Consider the system

$$\frac{d}{dt}x(t) = A(t)x(t) + B(t)w(t) + C(t)u(t), \tag{9.83}$$

where $C(t) \in \mathbb{R}^{n \times l}$ and $u(t)$ are known for all $t \geq 0$, and with $A(t)$, $B(t)$, $x(t)$, and $w(t)$ as in (9.4). Define $z(t) = x(t) - x^{\circ}(t)$, where $x^{\circ}(t)$ is governed by

$$\frac{d}{dt}x^{\circ}(t) = A(t)x^{\circ}(t) + C(t)u(t), \tag{9.84}$$

with $x(0)^{\circ} = \mathbf{0}_n$. Then, it follows that

$$\begin{aligned}\frac{d}{dt}z(t) &= \frac{d}{dt}\big(x(t) - x^{\circ}(t)\big) \\ &= A(t)x(t) + B(t)w(t) + C(t)u(t) - \big(A(t)x^{\circ}(t) + C(t)u(t)\big) \\ &= A(t)z(t) + B(t)w(t),\end{aligned} \tag{9.85}$$

where $w(t) \in \mathcal{W}(t)$ and $z(0) \in \mathcal{Z}(0) = \mathcal{X}(0)$, which is now in the form of (9.4). Let $\mathcal{Z}(t)$ denote the reachable set of (9.85); then, since $x(t) = x^{\circ}(t) + z(t)$, the reachable set of (9.83), $\mathcal{X}(t)$ can be obtained as follows:

$$\mathcal{X}(t) = \big\{x^{\circ}(t)\big\} + \mathcal{Z}(t). \tag{9.86}$$

Now, we can use the techniques discussed earlier to find an ellipsoid $\mathcal{Z}^{+}(t)$ containing the reachable set $\mathcal{Z}(t)$ for all $t \geq 0$. Let $\overline{z}(t)$ and $F(t)$ denote the center and shape matrix of $\mathcal{Z}^{+}(t)$. Then, it follows from (9.86) that

$$\begin{aligned}\mathcal{X}(t) &\subseteq \big\{x^{\circ}(t)\big\} + \mathcal{Z}^{+}(t) \\ &:= \mathcal{X}^{+}(t);\end{aligned} \tag{9.87}$$

thus, the ellipsoid $\mathcal{X}^{+}(t)$ has the same shape matrix as that of $\mathcal{Z}^{+}(t)$ and center $\overline{x}(t) = x^{\circ}(t) + \overline{z}(t)$.

Example 9.5 (AC linear circuit) Consider the circuit in Fig. 9.1, where the value of the voltage source magnitude, $v(t)$, is constant, i.e., $v(t) = V_s$, $V_s > 0$, for all $t \geq 0$, and the value of the load current, $i_l(t)$, is not a priori known except for some upper and lower values, i.e., $i_l(t) \in \big[i_l^m, i_l^M\big]$, $0 < i_l^m < i_l^M$, for all $t \geq 0$.

Table 9.1 AC linear circuit model parameter values

r [Ω]	ℓ [μH]	c [μF]	V_s [V]	i_l^m [A]	i_l^M [A]
0.3	50	100	5	1.9	2.1

As we did in Example 9.1, the constraint $i_l(t) \in \left[i_l^m, i_l^M\right]$ can be rewritten as follows:

$$i_l(t) \in \left\{i \in \mathbb{R} \colon E^{-1}(i - \bar{i}) \leq 1\right\}, \tag{9.88}$$

where

$$E = \left(\frac{i_l^M - i_l^m}{2}\right)^2, \quad \bar{i} = \frac{i_l^M + i_l^m}{2}.$$

Then, by letting $x(t) = \left[v_2(t), i_1(t)\right]^\top$, $w(t) = i_l(t)$, $u(t) = V_s$, and following the notation in (9.83), we have

$$\frac{d}{dt}x(t) = A(t)x(t) + B(t)w(t) + C(t)u(t) \tag{9.89}$$

where

$$A(t) = \begin{bmatrix} 0 & \frac{1}{c} \\ -\frac{1}{\ell} & -\frac{r}{\ell} \end{bmatrix}, \quad B(t) = \begin{bmatrix} -\frac{1}{c} \\ 0 \end{bmatrix}, \quad C(t) = \begin{bmatrix} 0 \\ \frac{1}{\ell} \end{bmatrix}. \tag{9.90}$$

In this case, it is assumed that the initial conditions are perfectly known and equal to zero.

The parameters of the model in (9.88–9.90) were populated with the values in Table 9.1. Then, by using the formulae in (9.10–9.11) and (9.66–9.67) with $\eta(0) = \eta_1 = [1, 0]^\top$ and $\eta(0) = \eta_2 = [0, 1]^\top$, and (9.87), we obtained two bounding ellipsoids, $\mathcal{X}_{\eta_1}^+(t)$ and $\mathcal{X}_{\eta_2}^+(t)$, whose evolution with time is depicted in Fig. 9.2. Note that the reachable set at any time t must be contained in the intersection of the ellipsoids $\mathcal{X}_{\eta_1}^+(t)$ and $\mathcal{X}_{\eta_2}^+(t)$.

9.3 Continuous-Time Nonlinear Systems

Here we consider systems of the form

$$\frac{d}{dt}x(t) = f\big(t, x(t), w(t)\big), \tag{9.91}$$

where $x(t) \in \mathbb{R}^n$, $w(t) \in \mathbb{R}^m$, and $f \colon [0, \infty) \times \mathbb{R}^n \times \mathbb{R}^m \to \mathbb{R}^n$. Assume that $w(t) \in \mathcal{W}(t)$, $t \geq 0$, and $x(0) \in \mathcal{X}(0)$, where $\mathcal{W}(t)$ and $\mathcal{X}(0)$ are ellipsoids with center $\overline{w}(t) \in \mathbb{R}^m$ and $\overline{x}(0) \in \mathbb{R}^n$, respectively, and shape matrix $E(t) \in \mathbb{R}^{m \times m}$ and $F(0) \in \mathbb{R}^{n \times n}$, respectively; i.e.,

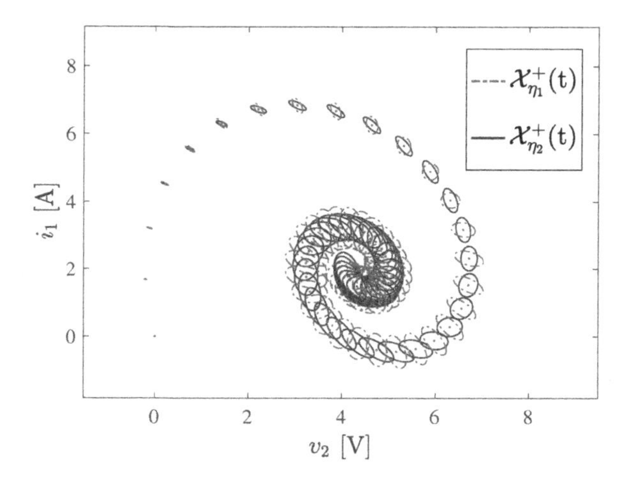

Figure 9.2 RLC circuit analysis results: evolution with time of two tight ellipsoids.

$$\mathcal{W}(t) = \{w \in \mathbb{R}^m\colon \big(w - \overline{w}(t)\big)^\top E^{-1}(t)\big(w - \overline{w}(t)\big) \leq 1\}, \tag{9.92}$$

$$\mathcal{X}(0) = \{x \in \mathbb{R}^n\colon \big(x - \overline{x}(0)\big)^\top F^{-1}(0)\big(x - \overline{x}(0)\big) \leq 1\}. \tag{9.93}$$

As we did earlier for the linear case, here we would like to find the reachable set of (9.91–9.93), which we denote by $\mathcal{X}(t)$. However, this is difficult in general; thus, instead, we will focus on approximately characterizing it by resorting to linearization. To this end, let $\overline{x}(t)$, $t \geq 0$, denote the trajectory followed by (9.91) for $w(t) = \overline{w}(t)$, $t \geq 0$, and $x(0) = \overline{x}(0)$. Let $\Delta x(t)$ denote a small variation of $x(t)$ around $\overline{x}(t)$ that results from a small variation of $w(t)$ around $\overline{w}(t)$, which we denote by $\Delta w(t)$. Thus, since $\Delta w(t) = w(t) - \overline{w}(t)$, $w(t) \in \mathcal{W}(t)$, and $\Delta x(0) = x(0) - \overline{x}(0)$, $x(0) \in \mathcal{X}(0)$, it follows that

$$\Delta w(t) \in \Delta\mathcal{W}(t) = \{\Delta w\colon (\Delta w)^\top E^{-1}(t)\Delta w \leq 1\}, \tag{9.94}$$

$$\Delta x(0) \in \Delta\mathcal{X}(0) = \{\Delta x\colon (\Delta x)^\top F^{-1}(0)\Delta x \leq 1\}. \tag{9.95}$$

Then, we have that

$$\frac{d}{dt}\big(\overline{x}(t) + \Delta x(t)\big) = f\big(t, \overline{x}(t) + \Delta x(t), \overline{w}(t) + \Delta w(t)\big), \tag{9.96}$$

with $\Delta w(t) \in \Delta\mathcal{W}(t)$ and $\Delta x(0) \in \Delta\mathcal{X}(0)$. Now, by linearizing $f(t, \cdot, \cdot)$ around $\overline{x}(t)$ and $\overline{w}(t)$, we can approximate $\Delta x(t)$ by some $\widetilde{\Delta x}(t)$ the evolution of which is governed by

$$\frac{d}{dt}\widetilde{\Delta x}(t) = A(t)\widetilde{\Delta x}(t) + B(t)\Delta w(t), \tag{9.97}$$

with $\Delta w(t) \in \Delta\mathcal{W}(t)$, $\Delta x(0) \in \Delta\mathcal{X}(0)$, and

$$A(t) = \left.\frac{\partial f(t,x,w)}{\partial x}\right|_{x=\overline{x}(t), w=\overline{w}(t)} \in \mathbb{R}^{n\times n},$$

$$B(t) = \left.\frac{\partial f(t,x,w)}{\partial w}\right|_{x=\overline{x}(t), w=\overline{w}(t)} \in \mathbb{R}^{n\times m}.$$

As discussed in Chapter 8 for the discrete-time nonlinear case, for the system in (9.97) to provide a good approximation, the deviation of $w(t)$ from $\overline{w}(t)$ needs to be sufficiently small. In the context of the system in (9.91), this translates into the ellipsoid $\mathcal{W}(t)$ being sufficiently small. In addition, the ellipsoid $\mathcal{X}(0)$ also needs to be small enough.

We can now use the techniques we developed for the continuous-time linear case (see Section 9.2) to find an ellipsoid $\widetilde{\Delta\mathcal{X}}^{+}(t)$ containing the reachable set of (9.97), denoted by $\widetilde{\Delta\mathcal{X}}(t)$. In this case because both $\Delta\mathcal{W}(t)$ and $\widetilde{\Delta\mathcal{X}}(0)$ are centered around zero, we have that

$$\widetilde{\Delta\mathcal{X}}^{+}(t) = \left\{\widetilde{\Delta x} \in \mathbb{R}^n : \widetilde{\Delta x}^{\top} \widetilde{F}^{-1}(t)\widetilde{\Delta x} \leq 1\right\},$$

with the evolution of $\widetilde{F}(t)$ governed by:

$$\frac{d}{dt}\widetilde{F}(t) = A(t)\widetilde{F}(t) + \widetilde{F}(t)A^{\top}(t) + \beta(t)\widetilde{F}(t) + \frac{1}{\beta(t)}B(t)E(t)B^{\top}(t), \tag{9.98}$$

with $\widetilde{F}(0) = F(0)$ and $\beta(t) > 0$ for all $t \geq 0$.

We can now set $\beta(t)$ to some positive constant β, or choose it so that $\widetilde{\Delta\mathcal{X}}^{+}(t)$ is optimal in some sense. For example, by choosing $\beta(t)$ as follows:

$$\beta(t) = \sqrt{\frac{\mathbf{tr}\left(\widetilde{F}^{-1}(t)B(t)E(t)B^{\top}(t)\right)}{n}}, \tag{9.99}$$

the volume of the resulting ellipsoid is minimum [among the class of ellipsoids with shape matrix as defined by (9.98)]. Also, by choosing $\beta(t)$ as follows:

$$\beta(t) = \sqrt{\frac{\mathbf{tr}\left(B(t)E(t)B^{\top}(t)\right)}{\mathbf{tr}\left(\widetilde{F}(t)\right)}}, \tag{9.100}$$

the sum of squares of the semi-axis lengths of the resulting ellipsoid is minimum [among the class of ellipsoids with shape matrix as defined by (9.98)]. Finally, by choosing $\beta(t)$ as follows:

$$\beta(t) = \sqrt{\frac{\eta^{\top}(t)B(t)E(t)B(t)^{\top}\eta(t)}{\eta^{\top}(t)\widetilde{F}(t)\eta(t)}}, \tag{9.101}$$

where $\eta(t)$ evolves according to

$$\frac{d}{dt}\eta(t) = -A^{\top}(t)\eta(t), \tag{9.102}$$

with $\eta(0)$ given, the resulting ellipsoid, $\widetilde{\Delta\mathcal{X}}^{+}(t)$, and the reachable set $\widetilde{\Delta\mathcal{X}}(t)$ will touch each other at the following point:

$$x^*(t) = \frac{1}{\sqrt{\eta^\top(t)\widetilde{F}(t)\eta(t)}}\widetilde{F}(t)\eta(t). \tag{9.103}$$

Then, since $x(t) \approx \overline{x}(t) + \widetilde{\Delta x}(t)$, we have that

$$\begin{aligned} x(t) \in \mathcal{X}(t) &\approx \widetilde{\mathcal{X}}(t) \\ &:= \{\overline{x}(t)\} + \widetilde{\Delta\mathcal{X}}(t) \\ &\subseteq \{\overline{x}(t)\} + \widetilde{\Delta\mathcal{X}}^{+}(t) \\ &=: \widetilde{\mathcal{X}}^{+}(t). \end{aligned} \tag{9.104}$$

Note that we cannot guarantee that $\mathcal{X}(t) \subseteq \widetilde{\mathcal{X}}^{+}(t)$ because the set $\widetilde{\mathcal{X}}(t)$ is not guaranteed to contain the reachable set $\mathcal{X}(t)$. However, in many practical applications, it will be the case that the set $\widetilde{\mathcal{X}}(t)$ will provide a good approximation to $\mathcal{X}(t)$.

9.4 Performance Requirements Verification

The computation of the reachable set allows us to determine whether or not the system violates certain performance requirements that impose bounds on the maximum excursions of certain system variables. In most applications, performance requirements generally consist of constraints in the form of interval ranges on variables of interest. These requirements can be described by a set $\mathcal{R}(t) \subseteq \mathcal{R}^n$, $t \geq 0$, as follows:

$$\mathcal{R}(t) = \mathcal{R}_1(t) \times \mathcal{R}_1(t) \times \cdots \times \mathcal{R}_n(t), \tag{9.105}$$

with $\mathcal{R}_i(t) = \{x_i \in \mathbb{R}\colon x_i^m(t) < x_i < x_i^M(t)\}$, where $x_i^m(t), x_i^M(t) \in \mathbb{R}$ respectively denote the minimum and maximum values that the ith entry of $x(t)$ can take. Thus, verifying that the system meets all performance requirements for any $w(t) \in \mathcal{W}(t)$, $0 \leq t \leq T$, is equivalent to checking whether or not $\mathcal{X}(t) \subseteq \mathcal{R}(t)$, for all $t \in [0, T]$. However, since in general we will not be able to exactly characterize $\mathcal{X}(t)$, we need to utilize the bounding ellipsoid techniques discussed earlier.

Linear Case. Consider (9.4) and assume we are given a finite number of bounding ellipsoids, $\mathcal{X}_1^+(t), \mathcal{X}_2^+(t), \ldots, \mathcal{X}_q^+(t)$, with centers $\overline{x}^{(j)}(t)$, $j = 1, 2, \ldots, q$, and shape matrices $F^{(j)}(t) \in \mathbb{R}^{n\times n}$, $j = 1, 2, \ldots, q$, respectively. Let $\hat{u}_i$ denote an n-dimensional vector whose i^{th} entry is equal to one and all others are equal to zero, and define:

$$M_i(t) = \min_j \left\{ \hat{u}_i^\top \overline{x}^{(j)}(t) + \sqrt{\hat{u}_i^\top F^{(j)}(t)\hat{u}_i} \right\},$$

$$m_i(t) = \max_j \left\{ \hat{u}_i^\top \overline{x}^{(j)}(t) - \sqrt{\hat{u}_i^\top F^{(j)}(t)\hat{u}_i} \right\}; \tag{9.106}$$

then, we have that $\mathcal{X}(t) \subset \mathcal{R}(t)$ if

$$M_i(t) < x_i^M(t),$$
$$m_i(t) > x_i^m(t), \tag{9.107}$$

for all $i = 1, 2, \ldots, n$. To establish this result, first note that, for any closed and convex set, we have that

$$S_{\mathcal{X}}(\hat{u}_i) = \max_{x \in \mathcal{X}} x_i,$$
$$S_{\mathcal{X}}(-\hat{u}_i) = \max_{x \in \mathcal{X}} -x_i, \tag{9.108}$$

i.e., $S_{\mathcal{X}}(\hat{u}_i)$ is the maximum value that the ith entry of x can take for any $x \in \mathcal{X}$, and $-S_{\mathcal{X}}(-\hat{u}_i)$ is the minimum value that the ith entry of x can take for any $x \in \mathcal{X}$. Now, since

$$S_{\mathcal{X}_j^+(t)}(\hat{u}_i) = \hat{u}_i^\top \overline{x}^{(j)}(t) + \sqrt{\hat{u}_i^\top F^{(j)}(t)\hat{u}_i},$$
$$S_{\mathcal{X}_j^+(t)}(-\hat{u}_i) = -\hat{u}_i^\top \overline{x}^{(j)}(t) + \sqrt{\hat{u}_i^\top F^{(j)}(t)\hat{u}_i}, \tag{9.109}$$

and the set $\cap_{i=1}^q \mathcal{X}_i^+(t)$ is convex, it follows that

$$\begin{aligned}
S_{\cap_{j=1}^q \mathcal{X}_j^+(t)}(\hat{u}_i) &= \max_{x \in \cap_{j=1}^q \mathcal{X}_j^+(t)} x_i \\
&= \min_j \underbrace{\max_{x \in \mathcal{X}_j^+(t)} x_i}_{= S_{\mathcal{X}_j^+(t)}(\hat{u}_i)} \\
&= \min_j \left\{ \hat{u}_i^\top \overline{x}^{(j)}(t) + \sqrt{\hat{u}_i^\top F^{(j)}(t)\hat{u}_i} \right\} \\
&= M_i(t), \\
S_{\cap_{j=1}^q \mathcal{X}_j^+(t)}(-\hat{u}_i) &= \max_{x \in \cap_{j=1}^q \mathcal{X}_j^+(t)} -x_i \\
&= \min_j \underbrace{\max_{x \in \mathcal{X}_j^+(t)} -x_i}_{= S_{\mathcal{X}_j^+(t)}(-\hat{u}_i)} \\
&= \min_j \left\{ -\hat{u}_i^\top \overline{x}^{(j)}(t) + \sqrt{\hat{u}_i^\top F^{(j)}(t)\hat{u}_i} \right\} \\
&= -\max_j \left\{ \hat{u}_i^\top \overline{x}^{(j)}(t) - \sqrt{\hat{u}_i^\top F^{(j)}(t)\hat{u}_i} \right\} \\
&= -m_i(t).
\end{aligned} \tag{9.110}$$

Then, since $\mathcal{X}(t) \subseteq \cap_{i=1}^{q} \mathcal{X}_i^+(t)$, it follows that $\mathcal{X}(t) \subseteq \mathcal{R}(t)$ if $M_i(t) < x_i^M(t)$ and $m_i(t) > x_i^m(t)$.

Nonlinear Case. Consider (9.91) and assume we are given a family of ellipsoids, $\widetilde{\mathcal{X}}_1^+(t), \widetilde{\mathcal{X}}_2^+(t), \ldots, \widetilde{\mathcal{X}}_q^+(t)$, obtained by linearization. As noted earlier, these ellipsoids are only guaranteed to upper bound the reachable set of the linearized system, but not necessarily the reachable set of the original nonlinear system in (9.91). Thus, even if the criteria in (9.107) are satisfied, we cannot say with certainty whether or not $\mathcal{X}(t) \subseteq \mathcal{R}(t)$, although in some practical applications this is likely to be the case.

Example 9.6 (Single machine infinite bus system) Consider the setting in Fig. 9.3, where the sinusoidal voltage source on the left together with the inductive reactance X_d' [p.u.] describe the terminal behavior of a synchronous generator, and where the sinusoidal voltage source on the right is an abstraction for the external system to which the generator is connected via a short transmission line with inductive reactance X_l [p.u.]. Assume the evolution of the phase angle $\delta(t)$ [rad] of the voltage source associated with the generator is described by the following state-space model:

$$\frac{d\delta(t)}{dt} = \omega(t) - \omega_0, \tag{9.111}$$

$$\frac{d\omega(t)}{dt} = -\frac{D}{M}\big(\omega(t) - \omega_0\big) + \frac{1}{M}P_m - \frac{V_0 v(t)}{M(X_d' + X_l)} \sin\big(\delta(t)\big), \tag{9.112}$$

where $\omega(t)$ [rad/s] denotes the generator electrical frequency at time t, ω_0 denotes the system nominal frequency, e.g., 60 Hz, D [s/rad] denotes the generator friction coefficient, M [s^2/rad] denotes the generator inertia constant, P_m [p.u.] denotes the mechanical power applied to the generator shaft, V_0 [p.u.] denotes the generator internal voltage magnitude, and $v(t)$ [p.u.] denotes the voltage magnitude of the voltage source describing the external system.

Assume that $v(t) \in [v^m, v^M]$, $t \geq 0$, where $0 < v^m < v^M$. Furthermore, assume that D, M, X_d', X_l, V_0, and P_m are positive constants. As we did in

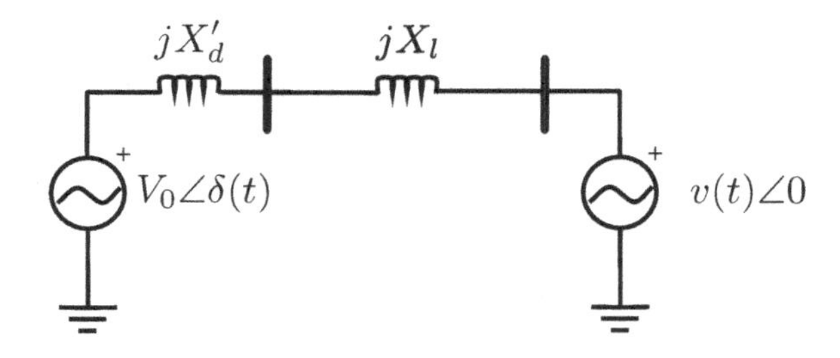

Figure 9.3 Single-machine infinite bus system.

Examples 9.1 and 9.5, the constraint $v(t) \in [v^m, v^M]$ can be rewritten as follows:

$$v(t) \in \left\{ v \in \mathbb{R} \colon E^{-1}(v - \overline{v}) \leq 1 \right\}, \tag{9.113}$$

where

$$E = \left(\frac{v^M - v^m}{2} \right)^2, \quad \overline{v} = \frac{v^M + v^m}{2}.$$

Also, assume that

$$P_m < \frac{V_0 \overline{v}}{X'_d + X_l},$$

and

$$\delta(0) = \arcsin \left(\frac{P_m}{\frac{V_0 \overline{v}}{X'_d + X_l}} \right) =: \delta_0, \quad \omega(0) = \omega_0. \tag{9.114}$$

Let $\overline{x}(t) = [\overline{\delta}(t), \overline{\omega}(t)]^\top$ denote the trajectory followed by (9.111–9.112) for $v(t) = \overline{v}$ and initial conditions as in (9.114); clearly

$$\overline{x}(t) = \left[\delta_0, \omega_0 \right]^\top, \tag{9.115}$$

for all $t \geq 0$. Define $\Delta x(t) = [\Delta \delta(t), \Delta \omega(t)]^\top$, where $\Delta \delta(t) = \delta(t) - \overline{\delta}(t)$, and $\Delta \omega(t) = \omega(t) - \overline{\omega}(t)$. Also define $\Delta v(t) = v(t) - \overline{v}$; then, by linearizing (9.111–9.112) around $\overline{x}(t)$, we obtain

$$\frac{d}{dt} \Delta x(t) = A \Delta x(t) + B \Delta v(t), \tag{9.116}$$

where

$$A = \begin{bmatrix} 0 & 1 \\ -\frac{V_0 \overline{v}}{M(X'_d + X_l)} \cos \delta_0 & -\frac{D}{M} \end{bmatrix}, \qquad B = \begin{bmatrix} 0 \\ -\frac{P_m}{M \overline{v}} \end{bmatrix}. \tag{9.117}$$

The parameters of the model above were populated using the values in Table 9.2. Then, by tailoring (9.98), (9.99), and (9.104) to the setting here, we obtained a minimum volume ellipsoid, $\widetilde{\mathcal{X}}_0^+(t)$; Fig. 9.4 depicts such ellipsoid for $t = 20$ s. Also, by using (9.98) with three arbitrarily chosen constant β's: $\beta_1 = 0.64$, $\beta_2 = 0.74$, and $\beta_3 = 0.84$, we obtained three bounding ellipsoids $\widetilde{\mathcal{X}}_{\beta_1}^+(t)$, $\widetilde{\mathcal{X}}_{\beta_2}^+(t)$, $\widetilde{\mathcal{X}}_{\beta_3}^+(t)$, which are also depicted in Fig. 9.4 for $t = 20$ s. Additionally, Fig. 9.4 shows a worst-case system trajectory, $x(t)$, $t \geq 0$, obtained by simulating the nonlinear model in (9.111–9.112), which verifies that the linearized model in (9.116) in this case provides a fairly accurate approximation.

Table 9.2 SMIB system model parameter values.

V_0 [p.u.]	X'_d [p.u.]	X_l [p.u.]	M [s^2/rad]	D [s/rad]	P_m [p.u.]	ω_0 [rad/s]	v^m [p.u.]	v^M [p.u.]
1	0.2	0.066	$\frac{1}{15\pi}$	0.08	1	120π	0.9	1.1

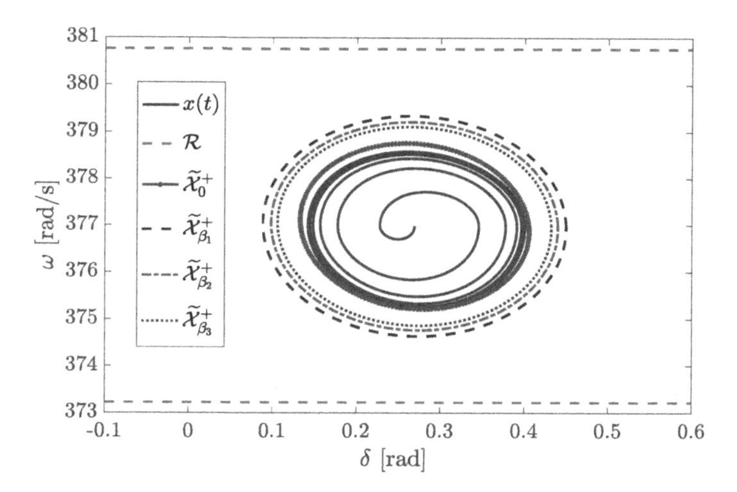

Figure 9.4 SMIB system reachable set computation results: worst-case trajectory, approximate minimum volume ellipsoid, and three approximate bounding ellipsoids computed for arbitrarily chosen β's: $\beta_1 = 0.64$, $\beta_2 = 0.74$, and $\beta_3 = 0.84$.

Assume the presence of an under/over-speed protection relay that will trip the generator offline when its speed is below 373.2 rad/s or above 380.6 rad/s; these performance requirements are depicted in Fig. 9.4 by the two horizontal dashed lines. One can check that all the ellipsoids in Fig 9.4 are fully contained within the region defined by the two horizontal dashed lines. Then, despite the variability in the infinite bus voltage magnitude, we can conclude that the under/over-speed protection relay will not unintentionally trip the generator offline.

9.5 Case Studies

In this section, we use the techniques developed earlier to analyze the performance of a buck DC-DC power converter. We also show how to use these techniques to assess the effect of variability associated with renewable-based electricity generation on power system dynamics, with a focus on time-scales involving electromechanical phenomena.

9.5.1 Buck Converter

Consider the circuit in Fig. 9.5, where the voltage $v_l(t)$ is regulated by stepping down the voltage $v_s(t)$ via high-frequency commutation of switches switch 1 and switch 2. Such a circuit is referred to as a buck DC-DC power converter. Assume that the values that $w(t) = \left[v_s(t), i_l(t)\right]^\top$ can take are known to belong to the following set:

$$\mathcal{W} = \left\{[v_s, i_l]^\top : |v_s - \overline{v}_s| \leq Q_v,\ |i_l - \overline{i}_l| \leq Q_i\right\}, \tag{9.118}$$

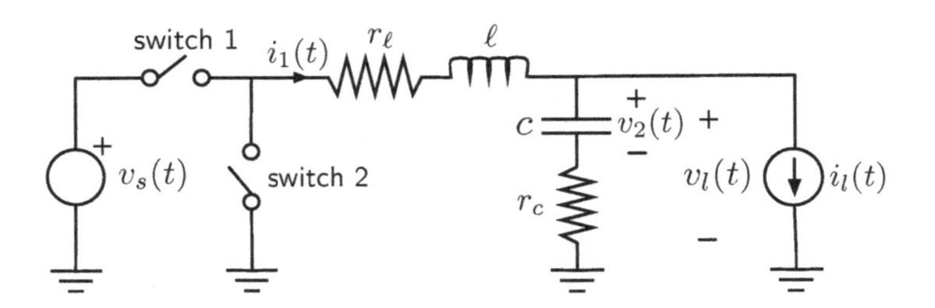

Figure 9.5 Buck converter schematic.

where $\overline{v}_s$, $\overline{i}_l$ represent the nominal values of $v_s(t)$ and $i_l(t)$, and $Q_v > 0$, $Q_i > 0$ describe the range of variation in $v_s(t)$ and $i_l(t)$ around $\overline{v}_s$ and $\overline{i}_l$, respectively. Let $q_1(\cdot)$ denote an indicator function that takes value 1 when switch 1 is closed and 0 otherwise. Similarly, let $q_2(\cdot)$ denote an indicator function that takes value 1 when switch 2 is closed and 0 otherwise. Assume that

$$q_1(t) = \begin{cases} 1, & kT \leq t < (k+D)T, \\ 0, & (k+D)T \leq t < (k+1)T, \end{cases}$$

$$q_2(t) = \begin{cases} 0, & kT \leq t < (k+D)T, \\ 1, & (k+D)T \leq t < (k+1)T, \end{cases}$$

where $0 \leq D \leq 1$ and $T > 0$ are fixed. In words, the switches are operated at fixed time instants, $t_0 = 0, t_1 = DT, t_2 = T, t_3 = (1+D)T, t_4 = 2T, \ldots$, so that their status is always opposite, i.e., if switch switch 1 is closed (open), then switch switch 2 is open (closed). Let $\sigma(t) = q_1(t) = 1 - q_2(t)$, then by using Kirchhoff's laws, one can check that the evolution of $x(t) = \left[v_2(t), i_1(t)\right]^\top$ is described by

$$\frac{d}{dt}x(t) = Ax(t) + Bw(t), \tag{9.119}$$

with

$$A = \begin{bmatrix} 0 & \frac{1}{c} \\ -\frac{1}{\ell} & -\frac{r_\ell + r_c}{\ell} \end{bmatrix}, \quad B = \begin{bmatrix} 0 & -\frac{1}{c} \\ \frac{\sigma(t)}{\ell} & \frac{r_c}{\ell} \end{bmatrix}, \tag{9.120}$$

where $w(t) \in \mathcal{W}$. In subsequent analysis, the parameters of the model in (9.118–9.120) are populated using the values in Table 9.3.

Reachable Set Computation

The techniques developed earlier in the chapter cannot be directly applied here because the input set $\mathcal{W}$ is not an ellipsoid. However, we can always bound $\mathcal{W}$ by one or more ellipsoids and then proceed with the analysis. For example, in Fig. 9.6(a), the set $\mathcal{W}$ is outer-bounded by an ellipsoid $\mathcal{W}^+$ with center $\overline{w} = [12, 4]^\top$, and shape matrix

Table 9.3 Open-loop controlled buck converter parameters.

Parameter	Value
r_ℓ [Ω]	0.1
r_c [Ω]	$5 \cdot 10^{-2}$
ℓ [H]	$12 \cdot 10^{-6}$
c [F]	$6 \cdot 10^{-4}$
T^{-1} [kHz]	250
D	0.45
$\overline{v}_s$ [V]	12
Q_v [V]	0.1
$\overline{i}_l$ [A]	4
Q_i [A]	1
λ	$7.5 \cdot 10^{-2}$
v_r [V]	5

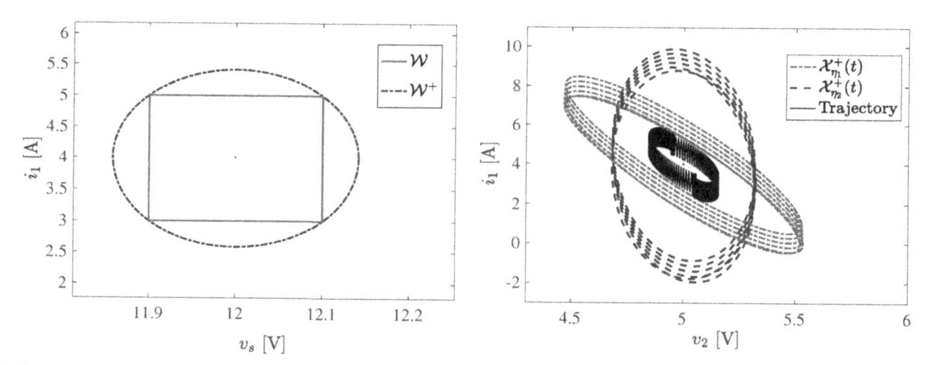

(a) Input space $\mathcal{W}$ and outer-bound ellipsoidal approximation $\mathcal{W}^+$

(b) Reachable set outer-bound ellipsoids $\mathcal{X}^+_{\eta_1}(t)$ and $\mathcal{X}^+_{\eta_2}(t)$, and trajectory

Figure 9.6 Outer-bound approximation for open-loop controlled buck converter: input space and reachable set approximation.

$$E = \begin{bmatrix} 0.02 & 0 \\ 0 & 1.99 \end{bmatrix}. \tag{9.121}$$

Then, by using (9.10–9.11) and (9.66–9.67) with $\eta(0) = \eta_1 = [1, 0]^\top$ and $\eta(0) = \eta_2 = [0, 1]^\top$, and (9.87), we obtained two bounding ellipsoids, $\mathcal{X}^+_{\eta_1}(t)$ and $\mathcal{X}^+_{\eta_2}(t)$. These ellipsoids are depicted in Fig. 9.6(b) for selected time instants after initial transients have died out. Figure 9.6(b) also shows a trajectory that results from a change in input from $w(t) = [11.9, 5]^\top$ to $w(t) = [12.1, 3]^\top$ and vice versa. As expected, $x(t)$ is contained within $\widetilde{\mathcal{X}}(t) = \mathcal{X}^+_{\eta_1}(t) \cap \mathcal{X}^+_{\eta_2}(t)$ because according to (9.82), the system reachable set, $\mathcal{X}(t)$, must be contained in $\widetilde{\mathcal{X}}(t)$.

Voltage Regulation Verification

Assume that design requirements specify that the voltage $v_l(t)$ must lie with an interval centered around a nominal value, v_r, i.e.,

$$|v_l(t) - v_r| \leq \lambda v_r, \quad t \geq 0, \tag{9.122}$$

where λ is a positive constant referred to as *voltage regulation*. Then, given $\widetilde{\mathcal{X}}(t) = \mathcal{X}^+_{\eta_1}(t) \cap \mathcal{X}^+_{\eta_2}(t)$ computed earlier, we would like to determine whether or not (9.122) will be satisfied. To this end, first, note that

$$v_l(t) = [1, r_c]^\top x(t) - r_c i_l(t). \tag{9.123}$$

Let $\mathcal{P}_{\widetilde{\mathcal{X}}(t)}$ denote the projection of the set $\widetilde{\mathcal{X}}(t)$ onto the direction defined by the unit vector $e = \frac{1}{\sqrt{1+r_c^2}}[1, r_c]^\top$, i.e.,

$$\mathcal{P}_{\widetilde{\mathcal{X}}(t)} = \left\{ z \in \mathbb{R} \colon z = \frac{1}{\sqrt{1+r_c^2}}[1, r_c]^\top x, \quad x \in \widetilde{\mathcal{X}}(t) \right\}. \tag{9.124}$$

Now since $i_l(t) \in \left[\bar{i}_l - Q_i, \bar{i}_l + Q_i\right]$, it follows that

$$v_l(t) \in \left\{ v_l : v_l = y - r_c i_l,\ y^m(t) \leq y \leq y^M(t),\ i_l \in \left[\bar{i}_l - Q_i, \bar{i}_l + Q_i\right] \right\}, \tag{9.125}$$

where $y^m(t)$ and $y^M(t)$ result from multiplying the endpoints of the real interval $\mathcal{P}_{\widetilde{\mathcal{X}}(t)}$ by $\sqrt{1+r_c^2}$. Finally, it follows from (9.123) and (9.125) that

$$y^m - r_c(\bar{i}_l + Q_i) \leq v_l \leq y^M - r_c(\bar{i}_l - Q_i). \tag{9.126}$$

Therefore, the regulation requirement in (9.122) is met if and only if

$$\begin{aligned} y^m &\geq r_c(\bar{i}_l + Q_i) + (1 - \lambda)v_r, \\ y^M &\leq r_c(\bar{i}_l - Q_i) + (1 + \lambda)v_r. \end{aligned} \tag{9.127}$$

Figure 9.7 shows the projection of $\widetilde{\mathcal{X}}(t)$ onto the subspace defined by

$$e = \frac{1}{\sqrt{1+r_c^2}}[1, r_c]^\top = [0.99, 0.05]^\top,$$

denoted by $\mathcal{P}_{\widetilde{\mathcal{X}}}$, for a fixed t after transients have died out. From the figure, we can conclude that $v_l(t)$ meets the voltage regulation requirement in (9.122) since in this case, $r_c(\bar{i}_l + Q_i) + (1 - \lambda)v_r = 4.875$ and $r_c(\bar{i}_l - Q_i) + (1 + \lambda)v_r = 5.525$; thus, the criteria in (9.127) are clearly met.

9.5.2 Three-Bus Power System

Consider the three-bus power system in Fig. 9.8 consisting of a synchronous generator connected to bus 1, a renewable-based generating resource connected to bus 2, and a load connected to bus 3. The synchronous generator is described by a third-order state-space model that includes the mechanical equations of motion and the governor. The electrical network is described by the standard

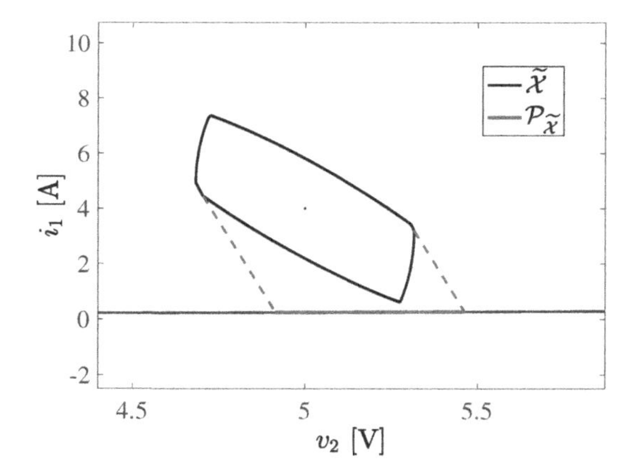

Figure 9.7 Intersection of outer-bounding ellipsoids, $\widetilde{\mathcal{X}}$, and its projection onto the direction defined by the unit vector $e = [0.99, 0.05]^\top$, denoted by $\mathcal{P}_{\widetilde{\mathcal{X}}}$, for a fixed time t when $\sigma(t) = 1$. Although in the figure the dotted parallel lines do not appear perpendicular to the line where the projection $\mathcal{P}_{\widetilde{\mathcal{X}}}$ is defined, in reality they are. This is just a skewing effect due to different axis scales.

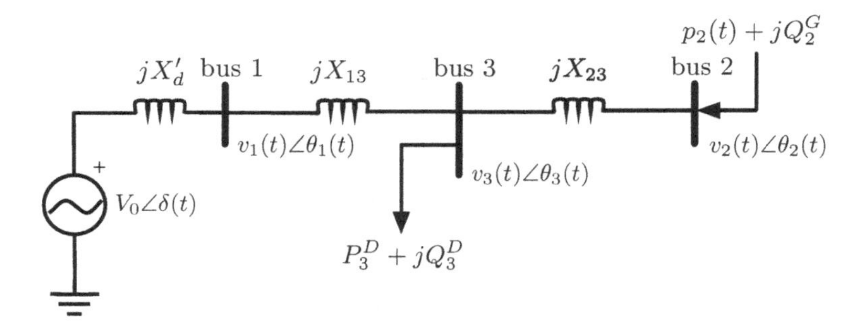

Figure 9.8 Three-bus power system.

power flow model (see Section B.3). The renewable-based generating resource is modeled as an unknown-but-bounded active positive power injection into bus 2 (the reactive power injection is assumed to be a known constant), whereas the active and reactive power consumed by the load at bus 3 are assumed to be known constants.

Synchronous Generator Model. As depicted in Fig. 9.8, and similar to what we did in Example 9.6, the synchronous generator is modeled as a reactance, X_d' [p.u.], in series with a sinusoidal voltage source whose voltage magnitude, V_0 [p.u.], is assumed to be constant and known, and whose phase angle, $\delta(t)$ [rad], which essentially describes generator rotor electrical angular position (relative to a synchronous reference frame rotating at the nominal system frequency, ω_0 [rad/s]), is described by

$$\frac{d\delta(t)}{dt} = \omega - \omega_0, \tag{9.128}$$

$$\frac{d\omega(t)}{dt} = \frac{1}{M}\Big(- D\big(\omega(t) - \omega_0\big) + p_m(t) - p_e(t)\Big), \tag{9.129}$$

$$\frac{dp_m(t)}{dt} = \frac{1}{\tau}\left(- p_m(t) - \frac{1}{R}\left(\frac{\omega(t)}{\omega_0} - 1\right) + P_c\right), \tag{9.130}$$

where $p_e(t)$ [p.u.] denotes the generator electrical power output, $p_m(t)$ [p.u.] denotes the mechanical power applied to its shaft, P_c [p.u.] denotes the generator power command, D [s/rad] denotes the generator friction coefficient, M [s^2/rad] denotes the generator inertia constant, τ [s] denotes the governor time constant, R [p.u.] is the slope of the speed-droop characteristic. In the remainder we assume that the constants ω_0, P_c, D, M, τ, and R are known.

Electrical network model. As depicted in Fig. 9.8, the transmission lines are modeled as purely imaginary series impedances. Let $v_i(t)$ and $\theta_i(t)$ respectively denote the voltage magnitude and phase angle of bus i, $i \in \mathcal{V} = \{1, 2, 3\}$, at time t. Then, the active and reactive power balance equations for bus 1 are

$$0 = \underbrace{\frac{V_0 v_1(t)}{X_d'} \sin\big(\theta_1(t) - \delta(t)\big)}_{= -p_e(t)} + \frac{v_1(t)v_3(t)}{X_{13}} \sin\big(\theta_1(t) - \theta_3(t)\big), \tag{9.131}$$

$$0 = \left(\frac{1}{X_d'} + \frac{1}{X_{13}}\right) v_1^2(t) - \frac{V_0 v_1(t)}{X_d'} \cos\big(\theta_1(t) - \delta(t)\big) - \frac{v_1(t)v_3(t)}{X_{13}} \cos\big(\theta_1(t) - \theta_3(t)\big), \tag{9.132}$$

where X_{13} denotes the reactance of the transmission line connecting bus 1 and bus 3. Let $p_2(t)$ and $q_2(t)$ respectively denote the active and reactive power injected into the network by the renewable-based generation resource at bus 2. Assume that $p_2(t)$ is uncertain but known to take values in some interval $[p_2^m, p_2^M]$, where $p_2^M \geq p_2^m > 0$. Also, assume that $q_2(t) = Q_2^G$ for all $t \geq 0$, where Q_2^G is known. Then, the active and reactive power balance equations for bus 2 are

$$p_2(t) = \frac{v_2(t)v_3(t)}{X_{23}} \sin\big(\theta_2(t) - \theta_3(t)\big), \tag{9.133}$$

$$Q_2^G = \frac{v_2^2(t)}{X_{23}} - \frac{v_2(t)v_3(t)}{X_{23}} \cos\big(\theta_2(t) - \theta_3(t)\big), \tag{9.134}$$

where X_{23} denotes the reactance of the transmission line connecting bus 2 and bus 3, and

$$p_2(t) \in \{p_2 \colon E^{-1}(p_2 - \overline{p}_2) \leq 1\}, \tag{9.135}$$

with

$$E = \left(\frac{p_2^M - p_2^m}{2}\right)^2, \quad \overline{p}_2 = \frac{p_2^M + p_2^m}{2}.$$

Assume that the active and reactive power consumed by the load connected to bus 3 are known constants denoted by P_3^D and Q_3^D, respectively; then the active and reactive power balance equations for bus 3 are

$$-P_3^D = \frac{v_3(t)v_1(t)}{X_{13}} \sin\big(\theta_3(t) - \theta_1(t)\big) + \frac{v_3(t)v_2(t)}{X_{23}} \sin\big(\theta_3(t) - \theta_2(t)\big), \quad (9.136)$$

$$-Q_3^D = \left(\frac{1}{X_{13}} + \frac{1}{X_{23}}\right)v_3^2(t) - \frac{v_3(t)v_1(t)}{X_{13}} \cos\big(\theta_3(t) - \theta_1(t)\big) - \frac{v_3(t)v_2(t)}{X_{23}} \cos\big(\theta_3(t) - \theta_2(t)\big). \quad (9.137)$$

Reachable Set Computation

Let $x(t) = \big[\delta(t), \omega(t), p_m(t)\big]^\top$, $y(t) = \big[\theta_1(t), \theta_2(t), \theta_3(t), V_1(t), V_2(t), V_3(t)\big]^\top$, and $w(t) = p_2(t)$. Then, the system in (9.128–9.137) can be compactly written as follows:

$$\frac{d}{dt}x(t) = f\big(x(t), y(t)\big), \quad (9.138)$$

$$0 = g\big(x(t), y(t), w(t)\big), \quad (9.139)$$

which is a differential-algebraic equation (DAE) model; thus, it does not match the form of the system in (9.91). However, we can use a similar approach based on linearization to compute an approximation to the set containing $x(t)$ for all $t \geq 0$. Assume that $w(0) = \overline{p}_2 =: w^{(0)}$, and $x(0) = x^{(0)}$ and $y(0) = y^{(0)}$ are such that

$$0 = f\big(x^{(0)}, y^{(0)}\big), \quad (9.140)$$

$$0 = g\big(x^{(0)}, y^{(0)}, w^{(0)}\big), \quad (9.141)$$

i.e., the system is initially in equilibrium. Let $\big(\overline{x}(t), \overline{y}(t)\big)$ denote the trajectory followed by (9.138–9.139) for $w(t) = \overline{w}(t) := \overline{p}_2$ and initial conditions $\big(x(0), y(0)\big) = \big(x^{(0)}, y^{(0)}\big)$; then clearly

$$\big(\overline{x}(t), \overline{y}(t)\big) = (x^{(0)}, y^{(0)}), \quad (9.142)$$

for all $t \geq 0$. Let $\big(\Delta x(t), \Delta y(t)\big)$ denote a small variation in $\big(x(t), y(t)\big)$ around $\big(\overline{x}(t), \overline{y}(t)\big) = \big(x^{(0)}, y^{(0)}\big)$ that results from a small variation of $w(t)$ around $\overline{w}(t) = w^{(0)}$, which we denote by $\Delta w(t)$. Therefore, since $\Delta w(t) = w(t) - \overline{w}(t)$, $w(t) \in \mathcal{W}(t)$, $\Delta x(0) = x(0) - \overline{x}(0) = \mathbf{0}_3$, and $\Delta y(0) = y(0) - \overline{y}(0) = \mathbf{0}_6$, it follows that

$$\Delta w(t) \in \Delta\mathcal{W}(t) = \big\{\Delta w\colon (\Delta w)^\top E^{-1}(t)\Delta w \leq 1\big\},$$

$$\Delta x(0) \in \Delta\mathcal{X}(0) = \{\mathbf{0}_3\},$$

$$\Delta y(0) \in \Delta\mathcal{Y}(0) = \{\mathbf{0}_6\}. \quad (9.143)$$

Then, we have that

$$\frac{d}{dt}\big(\overline{x}(t) + \Delta x(t)\big) = f\big(\overline{x}(t) + \Delta x(t), \overline{y}(t) + \Delta y(t)\big), \quad (9.144)$$

$$0 = g\big(\overline{x}(t) + \Delta x(t), \overline{y}(t) + \Delta y(t), \overline{w}(t) + \Delta w(t)\big), \quad (9.145)$$

where $\Delta w(t) \in \Delta\mathcal{W}(t)$, $\Delta x(0) \in \Delta\mathcal{X}(0)$, and $\Delta y(0) \in \Delta\mathcal{Y}(0)$. Assume that the functions $f(\cdot,\cdot)$ and $g(\cdot,\cdot,\cdot)$ in (9.138–9.139) are continuously differentiable with respect to their arguments in some neighborhood of $\big(x^{(0)}, y^{(0)}\big)$ and $\big(x^{(0)}, y^{(0)}, w^{(0)}\big)$, respectively. Then, we can approximate $\big(\Delta x(t), \Delta y(t)\big)$ by some $\big(\widetilde{\Delta x}(t), \widetilde{\Delta y}(t)\big)$ the evolution of which is governed by

$$\frac{d}{dt}\widetilde{\Delta x}(t) = A_1\widetilde{\Delta x}(t) + A_2\widetilde{\Delta y}(t), \quad (9.146)$$

$$0 = A_3\widetilde{\Delta x}(t) + A_4\widetilde{\Delta y}(t) + A_5\widetilde{\Delta w}(t), \quad (9.147)$$

with $\widetilde{\Delta w}(t) \in \Delta\mathcal{W}(t)$, $\widetilde{\Delta x}(0) \in \Delta\mathcal{X}(0)$, and $\widetilde{\Delta y}(0) \in \Delta\mathcal{Y}(0)$, and

$$A_1 = \left.\frac{\partial f(x,y)}{\partial x}\right|_{x=x^{(0)},y=y^{(0)}}, \quad A_2 = \left.\frac{\partial f(x,y)}{\partial y}\right|_{x=x^{(0)},y=y^{(0)}},$$

$$A_3 = \left.\frac{\partial g(x,y,w)}{\partial x}\right|_{x=x^{(0)},y=y^{(0)},w=w^{(0)}}, \quad A_4 = \left.\frac{\partial g(x,y,w)}{\partial y}\right|_{x=x^{(0)},y=y^{(0)},w=w^{(0)}},$$

$$A_5 = \left.\frac{\partial g(x,y,w)}{\partial w}\right|_{x=x^{(0)},y=y^{(0)},w=w^{(0)}}. \quad (9.148)$$

Now, by assuming A_4 is invertible, we can rewrite (9.147) as follows:

$$\widetilde{\Delta y}(t) = -A_4^{-1}\big(A_3\widetilde{\Delta x}(t) + A_5\widetilde{\Delta w}(t)\big), \quad (9.149)$$

which by plugging into (9.146) yields

$$\frac{d}{dt}\widetilde{\Delta x}(t) = A\widetilde{\Delta x}(t) + B\widetilde{\Delta w}(t), \quad (9.150)$$

with $\widetilde{\Delta w}(t) \in \Delta\mathcal{W}(t)$ and $\widetilde{\Delta x}(0) \in \Delta\mathcal{X}(0)$, and

$$A = A_1 - A_2A_4^{-1}A_3,$$
$$B = -A_2A_4^{-1}A_5.$$

Now we can apply the techniques in Section 9.2 to compute a bounding ellipsoid $\widetilde{\Delta\mathcal{X}}^+$ containing the reachable set of (9.150), denoted by $\widetilde{\Delta\mathcal{X}}$. Finally, since

$$x(t) = x^{(0)} + \Delta x(t)$$
$$\approx x^{(0)} + \widetilde{\Delta x}(t), \quad (9.151)$$

Table 9.4 Parameter values for generator model in three-bus system example.

V_0 [p.u.]	X'_d [p.u.]	M [s^2/rad]	D [s/rad]	P_c [p.u.]	ω_0 [rad/s]	τ [s]	R [p.u.]
1.13	0.2	$\frac{1}{15\pi}$	0.04	0.6	120π	0.2	0.05

Table 9.5 Parameter values for network model in three-bus system example.

X_{13} [p.u.]	X_{23} [p.u.]	p_2^m [p.u.]	p_2^M [p.u.]	Q_2^G [p.u.]	P_3^D [p.u.]	Q_3^D [p.u.]
0.1	0.15	0.08	0.72	0	1	0.5

we have that the set $\mathcal{X}(t)$ containing the possible values taken by $x(t)$ can be approximated as follows:

$$\begin{aligned}\mathcal{X}(t) &\approx \widetilde{\mathcal{X}}(t)\\ &:= \{x^{(0)}\} + \widetilde{\Delta\mathcal{X}}(t)\\ &\subseteq \{x^{(0)}\} + \widetilde{\Delta\mathcal{X}}^+(t)\\ &=: \widetilde{\mathcal{X}}^+(t).\end{aligned} \tag{9.152}$$

The parameters of the model above were populated using the values in Tables 9.4–9.5. In this case,

$$E = \left(\frac{p_2^M - p_2^m}{2}\right)^2 = 0.1024, \quad w^{(0)} = \overline{p}_2 = \frac{p_2^M + p_2^m}{2} = 0.4 \text{ p.u.};$$

thus, by plugging $w^{(0)}$ into (9.140–9.141) and solving for

$$x^{(0)} = \left[\delta^{(0)}, \omega^{(0)}, p_m^{(0)}\right]^\top, \quad y^{(0)} = \left[\theta_1^{(0)}, \theta_2^{(0)}, \theta_3^{(0)}, V_1^{(0)}, V_2^{(0)}, V_3^{(0)}\right]^\top,$$

we obtain $\omega^{(0)} = 120\pi$ rad/s, $p_m^{(0)} = 0.6$ pu, $V_1^{(0)} = 1$ pu, $V_2^{(0)} = 0.94$ pu, $V_3^{(0)} = 0.94$ pu, $\delta^{(0)} - \theta_1^{(0)} = 6.12°$, $\theta_1^{(0)} - \theta_3^{(0)} = 3.65°$, $\theta_2^{(0)} - \theta_3^{(0)} = 3.89°$. Next, by plugging these values into (9.148), we can obtain the values of the matrices A_1–A_5. Then, by using the techniques discussed earlier in the chapter, we computed a minimum volume ellipsoid, $\widetilde{\mathcal{X}}^+(t)$, for $0 \leq t \leq T$, where $T = 20$ s. Figure 9.9 displays the projection of this ellipsoid onto the subspace defined by the vectors corresponding to the ω and p_m axes for $t = 20$ s. In addition, Fig. 9.9 displays worst-case trajectories of the system nonlinear model and the corresponding linearized model obtained in both cases by switching between the two extreme values that the input can take; one can clearly see that the trajectory of the linearized model closely follows that of the nonlinear model. Performance requirements dictate that the generator speed, $\omega(t)$, remains within the interval $\mathcal{R} = [373.03, 380.57]$ rad/s, which is depicted in Fig. 9.9 by the vertical dashed traces. For the plotted trajectory, one can notice that $\omega(t)$ is not fully contained in the interval $\mathcal{R}$; thus performance requirements are not met. Such violation of performance requirements is also captured by the approximate reachable set obtained from the linearized model.

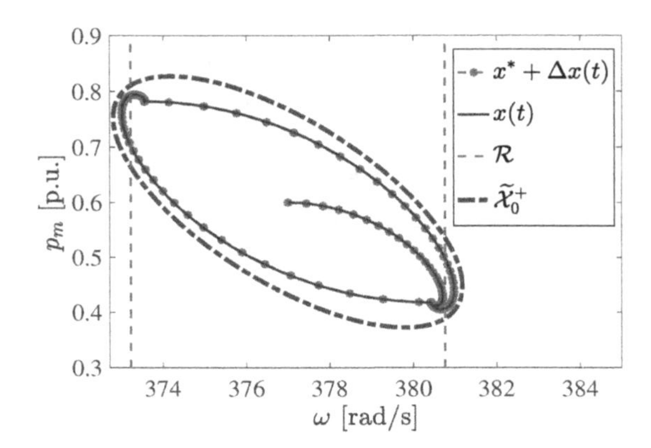

Figure 9.9 Three-bus power system reachable set computation results: exact and approximate worst-case trajectories and approximate minimum volume ellipsoid.

9.6 Notes and References

The derivation of the generic bounding ellipsoid derived from a discrete-time system by taking the limit as dt goes to zero is due to Schweppe [10]. The results on ellipsoids with minimum volume and minimum sum of squares of the semi-axis lengths are due to Chernousko [17]. The results on tight ellipsoids are due to Kurzhanski and Varaiya [50], although the derivation here is slightly different from that in [50]. The results in Example 9.5 and Section 9.5.1 appeared in [51] and were obtained using the Ellipsoidal Toolbox developed by Kurzhanskiy [19]. The results in Example 9.6 and Section 9.5.2 appeared in [52]. The reader is referred to [53] for detailed discussions on power-electronic converter modeling. The reader is referred to [54] for detailed discussions on power system electromechanical dynamics modeling.

Appendix A Mathematical Background

This appendix summarizes basic concepts from linear algebra and real analysis that are extensively used throughout the book. The treatment of each subject is far from being complete and is mostly included as a reference and to introduce basic notation used throughout the book.

A.1 Vectors

Basic Notions and Notation. An n-dimensional real vector x is an n-tuple whose components, denoted by $x_1, x_2, \ldots, x_n$, are real numbers. The sum of two n-dimensional real vectors x, y is another n-dimensional real vector, z, whose components are:

$$z_i = x_i + y_i,$$

$i = 1, 2, \ldots, n$. The multiplication of a real number c and an n-dimensional real vector, x, is another n-dimensional real vector, y, whose components are:

$$y_i = cx_i,$$

$i = 1, 2, \ldots, n$. We write $\mathbb{R}^n$ to denote the set of all n-dimensional real vectors equipped with the vector sum and the vector by scalar multiplication as defined above.

The components of a vector $x \in \mathbb{R}^n$ can be arranged in a column as follows:

$$\begin{bmatrix} x_1 \\ x_2 \\ \vdots \\ x_n \end{bmatrix}, \tag{A.1}$$

in which case we refer to x as a column vector, or in a row as follows:

$$[x_1, x_2, \ldots, x_n], \tag{A.2}$$

in which case we refer to x as a row vector. Unless otherwise stated, in this book, we will adopt the column arrangement convention; thus, when we write $x \in \mathbb{R}^n$, we mean the components of x are arranged as in (A.1). Also, for any column vector $x \in \mathbb{R}^n$, its transpose, denoted by $x^\top$, is defined as:

$$x^\top = [x_1, x_2, \ldots, x_n];$$

thus, $x^\top$ is an n-dimensional row vector. The transpose of a row vector $x^\top \in \mathbb{R}^n$ is a column vector, i.e., $(x^\top)^\top = x$.

We write $\mathbf{0}_n$ to denote an n-dimensional vector with all its entries equal to zero and refer to it as the n-dimensional all-zeros vector. We write $\mathbf{1}_n$ to denote an n-dimensional vector with all its entries equal to one and refer to it as the n-dimensional all-ones vector.

Linear Combination and Linear Dependence. Given $w_1, w_2, \ldots, w_m \in \mathbb{R}^n$, their linear combination is another vector $w \in \mathbb{R}^n$ defined as follows:

$$w = \alpha_1 w_1 + \alpha_2 w_2 + \cdots + \alpha_m w_m,$$

for given $\alpha_1, \alpha_2, \ldots, \alpha_m \in \mathbb{R}$. Vectors $w_1, w_2, \ldots, w_m \in \mathbb{R}^n$ are said to be linearly dependent if

$$\alpha_1 w_1 + \alpha_2 w_2 + \cdots + \alpha_m w_m = \mathbf{0}_n$$

for some $\alpha_1, \alpha_2, \ldots, \alpha_m \in \mathbb{R}$ such that at least one of them is non-zero. Vectors $w_1, w_2, \ldots, w_m \in \mathbb{R}^n$ are said to be linearly independent if they are not linearly dependent.

Inner Product and Norms. The inner product of two vectors $x, y \in \mathbb{R}^n$, denoted by $x^\top y$ (and sometimes by $<x, y>$), is defined as follows:

$$x^\top y = \sum_{i=1}^{n} x_i y_i.$$

We say two vectors $x, y \in \mathbb{R}^n$ are orthogonal if $x^\top y = 0$.

The Euclidean norm, or 2-norm, of a vector $x \in \mathbb{R}^n$, denoted by $\|x\|_2$, is defined as follows:

$$\|x\|_2 = \sqrt{x^\top x} = \sqrt{x_1^2 + x_2^2 + \cdots x_n^2}.$$

We refer to $\mathbb{R}^n$ equipped with the Euclidean norm, as the n-dimensional Euclidean space. Other norms include the ∞-norm, denoted by $\|x\|_\infty$, and defined as follows:

$$\|x\|_\infty = \max_i |x_i|,$$

and the 1-norm, denoted by $\|x\|_1$, and defined as follows:

$$\|x\|_1 = |x_1| + |x_2| + \cdots + |x_n|.$$

A.2 Matrices

Basic Notions and Notation. An $(n \times m)$-dimensional real matrix A is a rectangular array of the form

$$A = [a_{i,j}] = \begin{bmatrix} a_{1,1} & a_{1,2} & \cdots & a_{1,m} \\ a_{2,1} & a_{2,2} & \cdots & a_{2,m} \\ \vdots & \vdots & \ddots & \vdots \\ a_{n,1} & a_{n,2} & \cdots & a_{n,m} \end{bmatrix}, \tag{A.3}$$

where $a_{i,j}$, $i = 1, 2, \ldots, n$, $j = 1, 2, \ldots, m$, are real numbers referred to as the entries of A. In addition to using $a_{i,j}$ to denote the (i, j) entry of A, sometimes we also use $A(i, j)$ and $[A]_{i,j}$ as needed. We write $\mathbb{R}^{n\times m}$ to denote the set of all $(n\times m)$-dimensional real matrices. The transpose of a matrix $A = [a_{i,j}] \in \mathbb{R}^{n\times m}$, denoted by $A^\top$, is another matrix in $\mathbb{R}^{m\times n}$ defined as follows:

$$\left[A^\top\right]_{i,j} = a_{j,i},$$

$i = 1, 2, \ldots, n$, $j = 1, 2, \ldots, m$. Also, $\left(A^\top\right)^\top = A$ for any $A \in \mathbb{R}^{n\times m}$.

In light of (A.3), one can see the columns of an $n \times m$-dimensional real matrix as column vectors in $\mathbb{R}^n$, whereas the rows can be seen as row vectors in $\mathbb{R}^m$. For the case when $n > 1, m = 1$, the matrix that results has a single column and can be viewed as a single column vector. Conversely, a column vector in $\mathbb{R}^n$ can also be seen as an $(n \times 1)$-dimensional matrix. Similarly, for the case when $n = 1, m > 1$, the matrix that results has a single row and thus can be viewed as a single row vector. For the case when $m = n = 1$, the resulting matrix has a single entry and thus can be seen as just a scalar. Conversely, a row vector in $\mathbb{R}^m$ can also be seen as an $(1 \times m)$-dimensional matrix.

A matrix $A \in \mathbb{R}^{n\times m}$ can be partitioned into mutually exclusive smaller matrices, referred to as submatrices or blocks, such that each entry of the original matrix lies in one and only one block. For example, the matrix

$$A = \begin{bmatrix} a_{1,1} & a_{1,2} & a_{1,3} \\ a_{2,1} & a_{2,2} & a_{2,3} \\ a_{3,1} & a_{3,2} & a_{3,3} \end{bmatrix}$$

can be partitioned into four blocks as follows:

$$A_1 = \begin{bmatrix} a_{1,1} & a_{1,2} \\ a_{2,1} & a_{2,2} \end{bmatrix}, \quad A_2 = \begin{bmatrix} a_{1,3} \\ a_{2,3} \end{bmatrix}, \quad A_3 = \begin{bmatrix} a_{3,1} & a_{3,2} \end{bmatrix}, \quad A_4 = [a_{3,3}].$$

Then, we can rewrite the matrix A as follows:

$$A = \begin{bmatrix} A_1 & A_2 \\ A_3 & A_4 \end{bmatrix};$$

we refer to A as a block matrix. In light of the above discussion, given matrix of appropriate dimensions, we can always concatenate them to form another matrix. For example, given matrices $A_1 \in \mathbb{R}^{p,q}$, $A_2 \in \mathbb{R}^{p,m-q}$, $A_3 \in \mathbb{R}^{n-p,q}$, $A_4 \in \mathbb{R}^{n-p,m-q}$, we can form a matrix $A \in \mathbb{R}^{n\times m}$ as follows:

$$A = \begin{bmatrix} A_1 & A_2 \\ A_3 & A_4 \end{bmatrix}.$$

We write $\mathbf{0}_{n\times m}$ to denote an $(n \times m)$-dimensional matrix with all of its entries equal to zero and refer to it as the $(n \times m)$-dimensional zero matrix. We denote the dependence of the entries of a matrix $A \in \mathbb{R}^{n\times m}$ on some parameter p by writing $A(p)$ or A_p.

Operations with Matrices. Given two matrices $A, B \in \mathbb{R}^{n\times m}$, their sum, denoted by $A + B$, is another matrix in $\mathbb{R}^{n\times m}$ defined as follows:

$$[A + B]_{i,j} = a_{i,j} + b_{i,j},$$

$i = 1, 2, \ldots, n,\ j = 1, 2, \ldots, m$. The product of a real number c and a matrix $A \in \mathbb{R}^{n\times m}$, denoted by $c\,A$, is another matrix in $\mathbb{R}^{n\times m}$ whose entries are defined as follows:

$$[c\,A]_{i,j} = c\,a_{i,j},$$

$i = 1, 2, \ldots, n,\ j = 1, 2, \ldots, m$. The multiplication of a matrix $A \in \mathbb{R}^{n\times p}$ and $B \in \mathbb{R}^{p\times m}$, denoted by AB, is another matrix in $\mathbb{R}^{n\times m}$, defined as follows:

$$[AB]_{i,j} = \sum_{k=1}^{p} a_{i,k} b_{k,j},$$

$i = 1, 2, \ldots, n,\ j = 1, 2, \ldots, m$. The product of a matrix $A \in \mathbb{R}^{n\times m}$ and a vector $x \in \mathbb{R}^m$, denoted by Ax, is another vector $y \in \mathbb{R}^n$, defined as follows:

$$y_i = \sum_{k=1}^{m} a_{i,k} x_k,$$

$i = 1, 2, \ldots, n$; this definition is consistent with the definition of matrix multiplication above when the vector x is viewed as an $(m \times 1)$-dimensional matrix. The Kronecker product of $A = [a_{i,j}] \in \mathbb{R}^{n\times m}$ and $B = [b_{i,j}] \in \mathbb{R}^{p\times q}$, denoted by $A \otimes B \in \mathbb{R}^{np\times mq}$, is defined to be the following block matrix:

$$A \otimes B = \begin{bmatrix} a_{1,1}B & \ldots & a_{1,m}B \\ \vdots & \ddots & \vdots \\ a_{n,1}B & \ldots & a_{n,m}B \end{bmatrix}. \tag{A.4}$$

The derivative of a matrix $A(p) = [a_{i,j}(p)] \in \mathbb{R}^{n\times m}$ with respect to p, denoted by $\frac{d}{dp}A(p)$, is defined as follows:

$$\left[\frac{d}{dp}A(p)\right]_{i,j} = \frac{da_{i,j}(p)}{dp},$$

$i = 1, 2, \ldots, n$, $j = 1, 2, \ldots, m$. The integral of a matrix $A(p) = [a_{i,j}(p)] \in \mathbb{R}^{n\times m}$ with respect to p, denoted by $\int A(p)dp$, is defined as follows:

$$\left[\int A(p)dp\right]_{i,j} = \int a_{i,j}(p)dp,$$

$i = 1, 2, \ldots, n$, $j = 1, 2, \ldots, m$.

Square Matrices. A matrix A is said to be square when $n = m$. For a square matrix $A \in \mathbb{R}^{n\times n}$, entries of the form $a_{i,i}$, $i = 1, 2, \ldots, n$, are referred to as the diagonal entries of A, whereas entries of the form $a_{i,j}$, $i \neq j$, are referred to as the off-diagonal entries of A. We say a square matrix A is symmetric if $A^\top = A$, i.e., $a_{i,j} = a_{j,i}$ for all $i \neq j$. We say a square matrix A is diagonal if all of its off-diagonal entries are zero. We write $\mathbf{diag}(a_1, a_2, \ldots, a_n)$ to denote an $(n \times n)$-dimensional diagonal matrix whose diagonal entries are $a_1, a_2, \ldots, a_n$. A diagonal matrix in $\mathbb{R}^{n\times n}$ with all of its diagonal entries equal to one is referred to as the $(n \times n)$-dimensional identity matrix, and is denoted by I_n.

A square matrix $A \in \mathbb{R}^{n\times n}$ is said to be invertible, or nonsingular if there exists a matrix in $\mathbb{R}^{n\times n}$, denoted by A^{-1}, such that

$$AA^{-1} = I_n;$$

otherwise it is said to be singular. Properties of the inverse include:

I1. $AA^{-1} = A^{-1}A = I_n$ for any invertible matrix $A \in \mathbb{R}^n$.
I2. $(AB)^{-1} = B^{-1}A^{-1}$ for any invertible matrices $A, B \in \mathbb{R}^n$.
I3. $(kA)^{-1} = \frac{1}{k}A^{-1}$ for any non-zero $k \in \mathbb{R}$ and any invertible matrix $A \in \mathbb{R}^n$.
I4. $\left(A^\top\right)^{-1} = \left(A^{-1}\right)^\top$ for any invertible matrix $A \in \mathbb{R}^n$.

Given a square matrix $A \in \mathbb{R}^{n\times n}$, its trace, denoted by $\mathbf{tr}(A)$, is defined as the sum of its diagonal entries, i.e.,

$$\mathbf{tr}(A) = \sum_{i=1}^{n} a_{ii}.$$

Properties of the trace include:

T1. $\mathbf{tr}(A^\top) = \mathbf{tr}(A)$ for any $A \in \mathbb{R}^{n\times n}$.
T2. $\mathbf{tr}(A + B) = \mathbf{tr}(A) + \mathbf{tr}(B)$ for any $A, B \in \mathbb{R}^{n\times n}$.
T3. $\mathbf{tr}(kA) = k\,\mathbf{tr}(A)$ for any $k \in \mathbb{R}$ and any $A \in \mathbb{R}^{n\times n}$.
T4. $\mathbf{tr}(AB) = \mathbf{tr}(BA)$ for any $A, B \in \mathbb{R}^{n\times n}$.

In general, the trace of the product of two square matrices $A, B \in \mathbb{R}^{n\times n}$ is not equal to the product of the individual traces, i.e., $\mathbf{tr}(AB) \neq \mathbf{tr}(A)\mathbf{tr}(B)$.

For any $A(p) = [a_{i,j}(p)] \in \mathbb{R}^{n\times n}$, we have that

$$\frac{d}{dp}\mathbf{tr}(A(p)) = \mathbf{tr}\left(\frac{d}{dp}A(p)\right)$$

and

$$\int \mathbf{tr}\big(A(p)\big)dp = \mathbf{tr}\Big(\int A(p)dp\Big).$$

Given a square matrix $A \in \mathbb{R}^{n\times n}$, its determinant, denoted by $\mathbf{det}(A)$, can be recursively defined as follows:

$$\mathbf{det}(A) = \sum_{j=1}^{n} a_{i,j} \underbrace{(-1)^{i+j}\mathbf{det}(A_{i,j})}_{=:\, c_{i,j}} \tag{A.5}$$

for any $i = 1, 2, \ldots, n$, or equivalently

$$\mathbf{det}(A) = \sum_{i=1}^{n} a_{i,j}(-1)^{i+j}\mathbf{det}(A_{i,j}) \tag{A.6}$$

for any $j = 1, 2, \ldots, n$, with $A_{i,j}$ being an $\big((n-1)\times(n-1)\big)$-dimensional matrix that results from matrix A after deleting row i and column j, and where

$$c_{i,j} := (-1)^{i+j}\mathbf{det}(A_{i,j})$$

is referred to as the cofactor corresponding to $a_{i,j}$. Properties of the determinant include:

D1. $\mathbf{det}(A^\top) = \mathbf{det}(A)$ for any $A \in \mathbb{R}^{n\times n}$.
D2. $\mathbf{det}(A^{-1}) = \mathbf{det}(A)^{-1}$ for any $A \in \mathbb{R}^{n\times n}$ that is invertible.
D3. $\mathbf{det}(kA) = k^n\mathbf{det}(A)$ for any $k \in \mathbb{R}$ and any $A \in \mathbb{R}^{n\times n}$.
D4. $\mathbf{det}(AB) = \mathbf{det}(A)\mathbf{det}(B)$ for any $A, B \in \mathbb{R}^{n\times n}$.

In general, the determinant of the sum of two matrices $A, B \in \mathbb{R}^n$ is not equal to the sum of the determinants, i.e., $\mathbf{det}(A+B) \neq \mathbf{det}(A) + \mathbf{det}(B)$.

The adjugate (or classical adjoint) of matrix $A \in \mathbb{R}^{n\times n}$, denoted by $\mathbf{adj}(A)$, is another matrix in $\mathbb{R}^{n\times n}$ defined as follows:

$$\begin{aligned}\big[\mathbf{adj}(A)\big]_{i,j} &= c_{j,i}\\ &= (-1)^{i+j}\mathbf{det}(A_{j,i}),\end{aligned} \tag{A.7}$$

$i = 1, 2, \ldots, n,\ \ j = 1, 2, \ldots, n$. It can be shown that

$$A\,\mathbf{adj}(A) = \mathbf{adj}(A)A = \mathbf{det}(A)I_n; \tag{A.8}$$

thus, if $\mathbf{det}(A) \neq 0$, we can rewrite (A.8) as

$$\frac{1}{\mathbf{det}(A)}\mathbf{adj}(A)A = I_n, \tag{A.9}$$

then, the inverse of A can be computed as follows:

$$A^{-1} = \frac{1}{\mathbf{det}(A)}\mathbf{adj}(A); \tag{A.10}$$

therefore a matrix $A \in \mathbb{R}^{n\times n}$ is invertible if and only if $\mathbf{det}(A) \neq 0$.

For any $A(p) = \big[a_{i,j}(p)\big] \in \mathbb{R}^{n\times n}$, we have that

$$\frac{d}{dp}\mathbf{det}\big(A(p)\big) = \mathbf{tr}\Big(\mathbf{adj}\big(A(p)\big)\frac{d}{dp}A(p)\Big), \tag{A.11}$$

an expression that is referred to as Jacobi's formula. By using (A.10) if $A(p)$ is invertible, Jacobi's formula can be rewritten as

$$\frac{d}{dp}\mathbf{det}\big(A(p)\big) = \mathbf{det}\big(A\big)\mathbf{tr}\Big(A^{-1}(p)\frac{d}{dp}A(p)\Big). \tag{A.12}$$

Rank. Recall the interpretation of the columns and rows of a matrix $A \in \mathbb{R}^{n\times m}$ as n-dimensional column vectors and m-dimensional row vectors, respectively. Let c and r respectively denote the largest number of columns and rows of A that are linearly independent; it turns out $c = r$. The rank of the matrix A, denoted by $\mathbf{rank}(A)$, is then defined as follows:

$$\mathbf{rank}(A) = c = r;$$

thus, clearly $\mathbf{rank}(A) = \mathbf{rank}\big(A^\top\big)$ for any $A \in \mathbb{R}^{n\times m}$. Other properties of the rank include

- **R1.** $\mathbf{rank}(A) \leq \min\{n, m\}$ for any $A \in \mathbb{R}^{n\times m}$. A matrix $A \in \mathbb{R}^{n\times m}$ is said to be full rank if $\mathbf{rank}(A) = \min\{n, m\}$, otherwise it is said to be rank deficient.
- **R2.** $\mathbf{rank}\big(A + B\big) \leq \mathbf{rank}\big(A\big) + \mathbf{rank}\big(B\big)$ for any $A, B \in \mathbb{R}^{n\times m}$.
- **R3.** $\mathbf{rank}(A) = n$ for any $A \in \mathbb{R}^{n\times n}$ that is invertible; i.e., all the rows (columns) of the matrix A are linearly independent.
- **R4.** $\mathbf{rank}(A^\top A) = \mathbf{rank}(A)$ for any $A \in \mathbb{R}^n$.
- **R5.** If $A \in \mathbb{R}^{n\times n}$ and $C \in \mathbb{R}^{m\times m}$ are invertible, then, for any $B \in \mathbb{R}^{n\times m}$, we have that $\mathbf{rank}(AB) = \mathbf{rank}(B) = \mathbf{rank}(BC) = \mathbf{rank}(ABC)$.

Eigenvalues and Eigenvectors. Consider a matrix $A \in \mathbb{R}^{n\times n}$. We say λ is an eigenvalue of A if there exists some $v \in \mathbb{R}^n, v \neq \mathbf{0}_n$, such that

$$Av = \lambda v,$$

and refer to v as the (right) eigenvector of A associated with eigenvalue λ. Associated with each eigenvalue λ of A, there also exists some $w \in \mathbb{R}^n, w \neq \mathbf{0}_n$, such that

$$w^\top A = \lambda w^\top,$$

and we say w is a (left) eigenvector of A. The eigenvalues of a matrix $A \in \mathbb{R}^{n\times n}$ are the roots of its characteristic polynomial $p(\lambda)$, which is defined as follows:

$$p(\lambda) = \mathbf{det}\big(\lambda I_n - A\big). \tag{A.13}$$

The characteristic polynomial $p(\lambda)$ has n roots, denoted by $\lambda_1, \lambda_2, \ldots, \lambda_n$, not all of which are necessarily distinct, and some of which might be complex numbers. The set of eigenvalues of a matrix $A \in \mathbb{R}^{n\times n}$ is referred to as its spectrum and is denoted as $\sigma(A)$. The complex eigenvalues of a matrix $A \in \mathbb{R}^{n\times n}$ always occur in complex conjugate pairs, i.e., if $a+ib \in \sigma(A)$, then $a-ib \in \sigma(A)$. The algebraic multiplicity of an eigenvalue is the number of times that the eigenvalue appears as a root of the characteristic polynomial. The sum of the algebraic multiplicities of the eigenvalues of a matrix $A \in \mathbb{R}^{n\times n}$ is equal to n. The geometric multiplicity of an eigenvalue is the number of linearly independent eigenvectors associated with this eigenvalue. The geometric multiplicity of an eigenvector is always smaller than or equal to its algebraic multiplicity.

The maximum of the magnitude of the elements in $\sigma(A)$, denoted by $\rho(A)$, is called the spectral radius of A. For any matrix $A \in \mathbb{R}^{n\times n}$ with spectrum $\sigma(A) = \{\lambda_1, \lambda_2, \ldots, \lambda_n\}$, we have that

$$\mathbf{tr}(A) = \lambda_1 + \lambda_2 + \cdots + \lambda_n, \qquad \mathbf{det}(A) = \lambda_1\lambda_2\cdots\lambda_n.$$

As stated above, if A has complex eigenvalues, they will occur in complex conjugate pairs, which ensures that $\lambda_1 + \lambda_2 + \cdots + \lambda_n$ and $\lambda_1\lambda_2\cdots\lambda_n$ are real numbers; this is consistent with the fact that $\mathbf{tr}(A)$ and $\mathbf{det}(A)$ must be real numbers because the entries of A are real numbers.

Diagonalization. A matrix $A \in \mathbb{R}^{n\times n}$ is said to be diagonalizable if there exist an invertible matrix $P \in \mathbb{R}^{n\times n}$ and a diagonal matrix $D \in \mathbb{R}^{n\times n}$ such that $A = PDP^{-1}$. A matrix $A \in \mathbb{R}^{n\times n}$ is diagonalizable if and only if each of its eigenvalues has equal algebraic and geometric multiplicities.

A matrix $U \in \mathbb{R}^{n\times n}$ is called orthogonal if $UU^\top = U^\top U = I_n$, from where it follows that $U^{-1} = U^\top$. A symmetric matrix $A \in \mathbb{R}^{n\times n}$ is always diagonalizable by an orthogonal matrix, i.e., there exists an orthogonal matrix $U \in \mathbb{R}^{n\times n}$ such that

$$A = U\Lambda U^\top, \tag{A.14}$$

with $\Lambda = \mathbf{diag}(\lambda_1, \lambda_2, \ldots, \lambda_n)$, where $\lambda_1, \lambda_2, \ldots, \lambda_n$ are the eigenvalues of A.

Positive (Semi)Definiteness. A symmetric matrix $A \in \mathbb{R}^{n\times n}$ is said to be positive definite if

$$x^\top A x > 0 \tag{A.15}$$

for all $x \in \mathbb{R}^n$, $x \neq \mathbf{0}_n$. The eigenvalues of a positive definite matrix are real positive numbers. A symmetric matrix $A \in \mathbb{R}^{n\times n}$ is said to be positive semidefinite if, instead of the strict inequality in (A.15), we have that $x^\top A x \geq 0$ for all $x \in \mathbb{R}^n$. The eigenvalues of a positive semidefinite matrix are real nonnegative numbers. Properties of positive (semi)definite matrices include:

P1. $\mathbf{rank}(A) = n$ for any positive definite matrix.

P2. $\mathbf{rank}(A) < n$ for any positive semidefinite matrix.

P3. If $A \in \mathbb{R}^{n\times n}$ is positive definite and $B \in \mathbb{R}^{n\times n}$ is positive (semi)definite, then $A + B$ is positive definite.

P4. For any positive real number c and any positive (semi)definite $A \in \mathbb{R}^{n\times n}$, the matrix $cA \in \mathbb{R}^{n\times n}$ is positive (semi)definite.

Since a positive (semi)definite matrix $A \in \mathbb{R}^{n\times n}$ is symmetric, it is diagonalizable and there exists an orthogonal matrix $U \in \mathbb{R}^{n\times n}$ such that:

$$A = U\Lambda U^\top, \tag{A.16}$$

with $\Lambda = \mathbf{diag}(\lambda_1, \lambda_2, \dots, \lambda_n)$, where $\lambda_1, \lambda_2, \dots, \lambda_n$ are the eigenvalues of A. Given a positive (semi)definite matrix $A \in \mathbb{R}^{n\times n}$ decomposed as in (A.16), its square root, denoted by $\sqrt{A}$, is another matrix in $\mathbb{R}^{n\times n}$ defined as follows:

$$\sqrt{A} = U\sqrt{\Lambda}U^\top,$$

where $\sqrt{\Lambda} = \mathbf{diag}\big(\sqrt{\lambda_1}, \sqrt{\lambda_2}, \dots, \sqrt{\lambda_n}\big)$.

Consider a positive definite matrix $A \in \mathbb{R}^{n\times n}$ and a positive (semi)definite matrix $B \in \mathbb{R}^{n\times n}$. Then there exists a nonsingular matrix $W \in \mathbb{R}^{n\times n}$ such that

$$A = W\Gamma W^\top, \qquad B = W\Lambda W^\top,$$

where $\Gamma = \mathbf{diag}(\gamma_1, \gamma_2, \dots, \gamma_n)$ and $\Lambda = \mathbf{diag}(\lambda_1, \lambda_2, \dots, \lambda_n)$. Furthermore,

$$\psi_i = \lambda_i/\gamma_i, \quad i = 1, 2, \dots, n,$$

are the roots of the polynomial

$$p(\psi) = \mathbf{det}\big(B - \psi A\big).$$

Note that since A is positive definite, it follows that $\gamma_i > 0,\ i = 1, 2, \dots, n$; thus, we can define $\sqrt{\Gamma} = \mathbf{diag}\big(\sqrt{\gamma_1}, \sqrt{\gamma_2}, \dots, \sqrt{\gamma_n}\big)$, which is clearly invertible. Now, let $V = W\sqrt{\Gamma}$, then we have that

$$A = VV^\top, \qquad B = V\Psi V^\top,$$

where $\Psi = \mathbf{diag}(\psi_1, \psi_2, \dots, \psi_n)$.

A.3 Functions

Basic Notions and Notation. We write $f\colon \mathcal{X} \to \mathcal{Y}$ to denote a function mapping elements in the set $\mathcal{X}$ into elements in the set $\mathcal{Y}$. For example, we write $f\colon \mathbb{R}^n \to \mathbb{R}$ to denote a function mapping n-dimensional real vectors into real numbers. In this case we say f is a real-valued function and define its support, $\mathcal{S}_f$, as the set of vectors in $\mathbb{R}^n$ for which the function evaluates to a non-zero value, i.e.,

$$\mathcal{S}_f = \{x \in \mathbb{R}^n \colon f(x) \neq 0\}.$$

Similarly, we write $f\colon \mathbb{R}^n \to \mathbb{R}^m$ to denote a function mapping n-dimensional real vectors into m-dimensional real vectors; in this case we say f is a vector-valued function, and for any $x \in \mathbb{R}^n$, we have that

$$f(x) = \big[f_1(x), f_2(x), \dots, f_m(x)\big]^\top, \tag{A.17}$$

where $f_i\colon \mathbb{R}^n \to \mathbb{R}$. A real-valued function $f\colon \mathbb{R}^n \to \mathbb{R}$ is said to be continuous at $x \in \mathbb{R}^n$ if whenever a sequence $(x_k)_{k\in\mathbb{N}}$, where the x_k's are n-dimensional vectors, converges to some $x \in \mathbb{R}^n$, then the sequence $\big(f(x_k)\big)_{k\in\mathbb{N}}$ converges to $f(x)$. A vector-valued function $f\colon \mathbb{R}^n \to \mathbb{R}^m$ with components $f_1, f_2, \ldots, f_m$ is said to be continuous at $x \in \mathbb{R}^n$ if each f_i is continuous.

Differentiation. A real-valued function $f\colon \mathbb{R}^n \to \mathbb{R}$ is said to be differentiable at $x \in \mathbb{R}^n$ if the partial derivatives with respect to all the components of x exist. A function $f\colon \mathbb{R}^n \to \mathbb{R}$ is said to be differentiable if it is differentiable at every $x \in \mathbb{R}^n$. A function $f\colon \mathbb{R}^n \to \mathbb{R}$ is said to be continuously differentiable if it is differentiable and its partial derivatives are continuous functions. A vector-valued function $f\colon \mathbb{R}^n \to \mathbb{R}^m$ with components $f_1, f_2, \ldots, f_m$ is said to be differentiable at $x \in \mathbb{R}^n$ if each f_i is differentiable at $x \in \mathbb{R}^n$. A vector valued function $f\colon \mathbb{R}^n \to \mathbb{R}^m$ with components $f_1, f_2, \ldots, f_m$ is said to be (continuously) differentiable if each component f_i is (continuously) differentiable.

The gradient of a real-valued function $f\colon \mathbb{R}^n \to \mathbb{R}$ at $x \in \mathbb{R}^n$, denoted by $\nabla f(x)$, is defined as a column vector whose entries are the partial derivatives of f, i.e.,

$$\nabla f(x) = \left[\frac{\partial f(x)}{\partial x_1}, \frac{\partial f(x)}{\partial x_2}, \ldots, \frac{\partial f(x)}{\partial x_n}\right]^\top. \tag{A.18}$$

The Jacobian of a vector-valued function $f\colon \mathbb{R}^n \to \mathbb{R}^m$, with components $f_1, f_2, \ldots, f_m$, is an $(m \times n)$-dimensional matrix, denoted by $\frac{\partial f(x)}{\partial x}$ (and sometimes by $J_f(x)\big)$, defined as follows:

$$\frac{\partial f(x)}{\partial x} = \begin{bmatrix} \frac{\partial f_1(x)}{\partial x_1} & \frac{\partial f_1(x)}{\partial x_2} & \cdots & \frac{\partial f_1(x)}{\partial x_n} \\ \frac{\partial f_2(x)}{\partial x_1} & \frac{\partial f_2(x)}{\partial x_2} & \cdots & \frac{\partial f_2(x)}{\partial x_n} \\ \vdots & \vdots & \ddots & \vdots \\ \frac{\partial f_m(x)}{\partial x_1} & \frac{\partial f_m(x)}{\partial x_2} & \cdots & \frac{\partial f_m(x)}{\partial x_n} \end{bmatrix}, \tag{A.19}$$

and we write $\left.\frac{\partial f(x)}{\partial x}\right|_{x=x_0}$ to denote the Jacobian of f evaluated at $x = x_0$. By using (A.18), we can rewrite (A.19) as follows:

$$\frac{\partial f(x)}{\partial x} = \begin{bmatrix} \big(\nabla f_1(x)\big)^\top \\ \big(\nabla f_2(x)\big)^\top \\ \vdots \\ \big(\nabla f_n(x)\big)^\top \end{bmatrix}. \tag{A.20}$$

Inverse Functions. Consider a vector-valued function $f\colon \mathbb{R}^n \to \mathbb{R}^n$ and a set $\mathcal{X} \in \mathbb{R}^n$. Define

$$\mathcal{Y} = \big\{y \in \mathbb{R}^n \colon y = f(x),\ x \in \mathcal{X}\big\}.$$

We say the function f is invertible on $\mathcal{X}$ if, for every $y \in \mathcal{Y}$, there is a unique $x \in \mathcal{X}$ such that $y = f(x)$. If the function f is invertible on $\mathcal{X}$, there exists a function $f^{-1}\colon \mathcal{Y} \to \mathcal{X}$ defined as follows: for every $y \in \mathcal{Y}$, $x = f^{-1}(y)$ is the unique element in $\mathcal{X}$ for which $y = f(x)$.

THEOREM A.1 *(Inverse Function Theorem) Consider a continuously differentiable vector-valued function $f\colon \mathbb{R}^n \to \mathbb{R}^n$ whose Jacobian at $x_0 \in \mathbb{R}^n$, $J_f(x_0)$, is full rank (and therefore invertible). Define*

$$\mathcal{X} = \{x \in \mathbb{R}^n \colon \|x - x_0\|_2 < r\},$$

where r is some positive constant, and

$$\mathcal{Y} = \{y \in \mathbb{R}^n \colon y = f(x),\ x \in \mathcal{X}\}.$$

Then, for r small enough, f is invertible on $\mathcal{X}$ and there exists an inverse function $f^{-1}\colon \mathcal{Y} \to \mathcal{X}$ that is continuously differentiable. Moreover, for $y_0 = f(x_0)$, we have that

$$J_{f^{-1}}(y_0) = J_f^{-1}(x_0).$$

Linear Approximation. A real-valued function $f\colon \mathbb{R} \to \mathbb{R}$ that is smooth at any point $x \in \mathbb{R}$, i.e., $f^{(k)}(x) = \frac{d^k f(x)}{dx^k}$ exists for all $k \in \mathbb{N}$, can be described by its Taylor series as follows:

$$f(x) = \sum_{k=0}^{\infty} \frac{f^{(k)}(a)}{k!}(x - a)^k, \tag{A.21}$$

for any $a \in \mathbb{R}$ and all $x \in \mathbb{R}$, where $f^{(0)}(x) = f(x)$. Let $a = x_0$ and $x = x_0 + \Delta x$, then it follows from (A.21) that

$$f(x_0 + \Delta x) = f(x_0) + \frac{df(x_0)}{dx}\Delta x + \text{h.o.t.}(\Delta x), \tag{A.22}$$

where $\text{h.o.t.}(\Delta x)$ denotes the higher-order terms of Δx. For Δx small enough, $\text{h.o.t.}(\Delta x)$ can be neglected; thus,

$$f(x_0 + \Delta x) \approx f(x_0) + \frac{df(x_0)}{dx}\Delta x. \tag{A.23}$$

Next, consider a differentiable real-valued function $f\colon \mathbb{R}^n \to \mathbb{R}$; then for any $x_0 \in \mathbb{R}^n$ and sufficiently small $\Delta x_1, \Delta x_2, \ldots, \Delta x_n$, the result in (A.23) can be extended as follows:

$$\begin{aligned} f(x_0 + \Delta x) &\approx f(x_0) + \sum_{i=1}^{n} \frac{\partial f(x_0)}{\partial x_i}\Delta x_i, \\ &\approx f(x_0) + \left(\nabla f(x_0)\right)^\top \Delta x, \end{aligned} \tag{A.24}$$

where $\Delta x = [\Delta x_1, \Delta x_2, \ldots, \Delta x_n]^\top$. Finally, consider a vector-valued function $f\colon \mathbb{R}^n \to \mathbb{R}^m$ whose components $f_1, f_2, \ldots, f_m$ are differentiable. Then, for any $x_0 \in \mathbb{R}^n$ and sufficiently small $\Delta x_1, \Delta x_2, \ldots, \Delta x_n$, it follows from (A.24) that

$$f(x_0 + \Delta x) = \begin{bmatrix} f_1(x_0 + \Delta x) \\ f_2(x_0 + \Delta x) \\ \vdots \\ f_n(x_0 + \Delta x) \end{bmatrix}$$

$$= \begin{bmatrix} f_1(x_0) \\ f_2(x_0) \\ \vdots \\ f_n(x_0) \end{bmatrix} + \begin{bmatrix} (\nabla f_1(x_0))^\top \\ (\nabla f_2(x_0))^\top \\ \vdots \\ (\nabla f_n(x_0))^\top \end{bmatrix} \Delta x$$

$$= f(x_0) + \left. \frac{\partial f(x)}{\partial x} \right|_{x=x_0} \Delta x, \tag{A.25}$$

where the last equality follows from (A.20).

Integration by Substitution. Consider a continuous function $f \colon \mathbb{R} \to \mathbb{R}$ and a differentiable function $g \colon \mathbb{R} \to \mathbb{R}$ with continuous derivative, $g' \colon \mathbb{R} \to \mathbb{R}$. Then, we have that

$$\int_a^b f\big(g(x)\big) g'(x) dx = \int_\alpha^\beta f(y) dy, \tag{A.26}$$

where $\alpha = g(a)$ and $\beta = g(b)$. This result can be extended to the multi-dimensional case as follows. Consider a real-valued function $f \colon \mathbb{R}^n \to \mathbb{R}$ and a one-to-one continuously differentiable vector-valued function $\varphi \colon \mathbb{R}^n \to \mathbb{R}^n$. Then, for any closed and bounded set $\mathcal{X} \in \mathbb{R}^n$ such that the Jacobian of $\varphi(\cdot)$ is non-zero for every $x \in \mathcal{X}$, i.e., $J_\varphi(x) \neq 0, \ \forall x \in \mathcal{X}$, we have that

$$\int_{\mathcal{X}} f\big(\varphi(x)\big) \big|\mathbf{det}\big(J_\varphi(x)\big)\big| dx = \int_{\mathcal{Y}} f(y) dy, \tag{A.27}$$

where $dx = dx_1 dx_2 \dots dx_n$, $dy = dy_1 dy_2 \dots dy_n$, and

$$\mathcal{Y} = \{y \colon y = \varphi(x), \ x \in \mathcal{X}\}.$$

Integration by Parts. Consider two differentiable functions $u \colon \mathbb{R} \to \mathbb{R}$ and $v \colon \mathbb{R} \to \mathbb{R}$ with derivatives $u' \colon \mathbb{R} \to \mathbb{R}$ and $v' \colon \mathbb{R} \to \mathbb{R}$, respectively; then, we have that

$$\int_a^b u(x) v'(x) dx = u(b)v(b) - u(a)v(a) - \int_a^b u'(x) v(x) dx. \tag{A.28}$$

By using Green's first identity, the result above can be extended to vector-valued functions as follows. Consider a real-valued function $\varphi \colon \mathbb{R}^n \to \mathbb{R}$ supported on some open-bounded set $\mathcal{X} \in \mathbb{R}^n$ and a vector-valued function $v \colon \mathbb{R}^n \to \mathbb{R}^n$ with components $v_1, v_2, \dots, v_n$. Assume φ and v_i, $i = 1, 2, \dots,$ are continuously differentiable in the closure of $\mathcal{X}$, i.e., in the set $\mathbf{cl}(\mathcal{X}) = \mathcal{X} \cup \partial\mathcal{X}$, where $\partial\mathcal{X}$ denotes the boundary of $\mathcal{X}$. Then, we have that

$$\int_{\mathcal{X}} \langle \nabla\varphi(x), v(x) \rangle dV = \int_{\partial\mathcal{X}} \langle \varphi(x) v(x), n(x) \rangle dS - \int_{\mathcal{X}} \langle \varphi(x) \nabla, v(x) \rangle dV, \tag{A.29}$$

where $\nabla = \left[\frac{\partial}{\partial x_1}, \frac{\partial}{\partial x_2}, \ldots, \frac{\partial}{\partial x_n}\right]^\top$, $dV = dx_1 dx_2 \ldots dx_n$, $n(x)$ is a unit vector perpendicular to the surface arising from $\partial\mathcal{X}$, and pointing outward from the set $\mathcal{X}$, and dS is the surface differential element.

A.4 Notes and References

The material on vectors and matrix analysis is standard and can be found in [41, 55, 56]. The material on differentiation of functions can be found in standard textbooks such as [57].

Appendix B Power Flow Modeling

Here, we provide some background on power flow modeling. In subsequent developments, we consider a three-phase power system comprising n buses ($n > 1$) and l transmission lines $(n-1 \leq l \leq n(n-1)/2)$ and assume the following hold:

A1. The power system is balanced and operating in sinusoidal regime.
A2. There is at most one transmission line connecting each pair of buses.
A3. There is a single voltage level in the system; thus, there are no transformers.

B.1 Transmission Line Lumped-Parameter Model

The per-phase terminal behavior of a transmission line connecting buses i and k can be described by the lumped-parameter circuit model in Fig. B.1, typically referred to as the Π-equivalent circuit model. By using Kirchhoff's laws, one can check that

$$\begin{aligned}\overline{I}_{ik} &= \frac{\overline{Y}_{ik}}{2}\overline{V}_i + \frac{1}{\overline{Z}_{ik}}(\overline{V}_i - \overline{V}_k) \\ &= \left(\frac{\overline{Y}_{ik}}{2} + \frac{1}{\overline{Z}_{ik}}\right)\overline{V}_i - \frac{1}{\overline{Z}_{ik}}\overline{V}_k, \end{aligned} \tag{B.1}$$

$$\begin{aligned}\overline{I}_{ki} &= \frac{\overline{Y}_{ik}}{2}\overline{V}_k + \frac{1}{\overline{Z}_{ik}}(\overline{V}_k - \overline{V}_i) \\ &= -\frac{1}{\overline{Z}_{ik}}\overline{V}_i + \left(\frac{\overline{Y}_{ik}}{2} + \frac{1}{\overline{Z}_{ik}}\right)\overline{V}_k, \end{aligned} \tag{B.2}$$

where the shunt admittance element, $\overline{Y}_{ik}$, and the series impedance element, $\overline{Z}_{ik}$, are functions of the line length, conductor arrangement and spacing. For short and lossless transmission lines, we have that $\overline{Y}_{ik} \approx 0$ and $\overline{Z}_{ik} \approx jX_{ik}$, where $X_{ik} > 0$ and $j = \sqrt{-1}$.

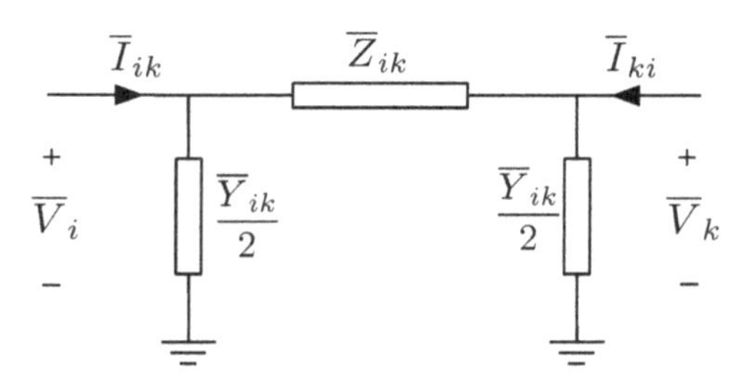

Figure B.1 Transmission line Π-equivalent circuit model.

B.2 Network Admittance Matrix

Let $\overline{I}_i$ denote the phasor associated with the (sinusoidal) current injection at bus i; then, by using (B.1), it follows that

$$\begin{aligned}\overline{I}_i &= \sum_{k=1}^{n} \overline{I}_{ik} \\ &= \sum_{k=1}^{n} \overline{y}_{i,k} \overline{V}_k, \qquad i = 1, 2, \ldots, n,\end{aligned} \tag{B.3}$$

where

$$\overline{y}_{i,k} = \begin{cases} \displaystyle\sum_{l=1}^{n} \left(\frac{\overline{Y}_{il}}{2} + \frac{1}{\overline{Z}_{il}} \right), & \text{if } i = k, \\ \displaystyle -\frac{1}{\overline{Z}_{ik}}, & \text{if } i \neq k. \end{cases} \tag{B.4}$$

Define $\overline{I} = [\overline{I}_1, \overline{I}_2, \ldots, \overline{I}_n]^\top$, $\overline{V} = [\overline{V}_1, \overline{V}_2, \ldots, \overline{V}_n]^\top$, and $\overline{Y} = [\overline{y}_{i,k}] \in \mathbb{R}^{n \times n}$; then, (B.3) can be rewritten in matrix form as follows:

$$\begin{aligned}\overline{I} &= \overline{Y}\,\overline{V} \\ &= (G + jB)\overline{V},\end{aligned} \tag{B.5}$$

where $G = [g_{i,k}] \in \mathbb{R}^{n \times n}$, with $g_{i,k} = \mathbf{real}(\overline{y}_{i,k})$, and $B = [b_{i,k}] \in \mathbb{R}^{n \times n}$, with $b_{i,k} = \mathbf{imag}(\overline{y}_{i,k})$.

B.3 Power Flow Equations

Let p_i and q_i respectively denote the active and reactive power injected into the network via bus i. Also, let v_i and θ_i denote the magnitude and angle of the phasor $\overline{V}_i$. Then, by using (B.3), it follows that the complex power injected into the network via bus i, which we denote by $\overline{S}_i$, is given by

$$\begin{aligned}
\overline{S}_i &= p_i + jq_i \\
&= \overline{V}_i \overline{I}_i^* \\
&= \overline{V}_i \left(\sum_{k=1}^{n} \overline{y}_{i,k} \overline{V}_k \right)^* \\
&= v_i e^{j\theta_i} \left(\sum_{k=1}^{n} (g_{i,k} + jb_{i,k}) v_k e^{j\theta_k} \right)^* \\
&= \sum_{k=1}^{n} v_i v_k (g_{i,k} - jb_{i,k}) e^{j(\theta_i - \theta_k)} \\
&= \sum_{k=1}^{n} v_i v_k (g_{i,k} - jb_{i,k}) \big(\cos(\theta_i - \theta_k) + j \sin(\theta_i - \theta_k) \big) \\
&= \sum_{k=1}^{n} v_i v_k \big(g_{i,k} \cos(\theta_i - \theta_k) + b_{i,k} \sin(\theta_i - \theta_k) \big) \\
&\quad + j \sum_{k=1}^{n} v_i v_k \big(g_{i,k} \sin(\theta_i - \theta_k) - b_{i,k} \cos(\theta_i - \theta_k) \big).
\end{aligned} \tag{B.6}$$

Then, by equating real and imaginary parts on both sides of (B.6), we obtain the active and reactive power balance equations:

$$p_i = \sum_{k=1}^{n} v_i v_k \big(g_{i,k} \cos(\theta_i - \theta_k) + b_{i,k} \sin(\theta_i - \theta_k) \big), \quad i = 1, 2, \ldots, n, \tag{B.7}$$

$$q_i = \sum_{k=1}^{n} v_i v_k \big(g_{i,k} \sin(\theta_i - \theta_k) - b_{i,k} \cos(\theta_i - \theta_k) \big), \quad i = 1, 2, \ldots, n, \tag{B.8}$$

also referred to as the power flow equations.

B.4 Network Flow Model

Here we are interested in obtaining a model describing the relation between the flow of active power across transmission lines and the power injections into the network. In order to simplify such model, we will make several assumptions about the system network parameters and operating conditions. This simplified model is heavily featured in Chapters 3 and 7.

Assume that all transmission lines are lossless and there are no shunt elements, then $\overline{Y}_{ik} = 0$ for all i, k, and $\overline{Z}_{ik} = jX_{ik}$; thus

$$g_{i,k} = 0, \tag{B.9}$$

for all i, k, and

$$b_{i,k} = \begin{cases} -\sum_{l \neq i} \frac{1}{X_{il}}, & \text{if } i = k, \\ \frac{1}{X_{ik}}, & \text{if } i \neq k, \end{cases} \tag{B.10}$$

for all i. Therefore, (B.7–B.8) can be simplified as follows:

$$\begin{aligned} p_i &= \sum_{k=1}^{n} v_i v_k b_{i,k} \sin(\theta_i - \theta_k) \\ &= \sum_{k=1}^{n} \frac{v_i v_k}{X_{ik}} \sin(\theta_i - \theta_k), \quad i = 1, 2, \ldots, n, \end{aligned} \tag{B.11}$$

$$\begin{aligned} q_i &= -\sum_{k=1}^{n} v_i v_k b_{i,k} \cos(\theta_i - \theta_k) \\ &= \left(\sum_{k \neq i} \frac{1}{X_{ik}} \right) v_i^2 - \sum_{k \neq i} \frac{v_i v_k}{X_{ik}} \cos(\theta_i - \theta_k), \quad i = 1, 2, \ldots, n. \end{aligned} \tag{B.12}$$

Note that because the system is lossless, there is one additional constraint that the p_i's need to satisfy:

$$\sum_{i=1}^{n} p_i = 0; \tag{B.13}$$

this can be obtained by summing (B.11) over all $i = 1, 2, \ldots, n$. Now, assuming the $b_{i,k}$'s are known, there is a total of $4n$ variables in (B.11–B.12) as follows:

- bus voltage magnitudes, $v_1, v_2, \ldots, v_n$,
- bus voltage phase angles, $\theta_1, \theta_2, \ldots, \theta_n$,
- bus active power injections, $p_1, p_2, \ldots, p_n$,
- bus reactive power injections, $q_1, q_2, \ldots, q_n$.

However, we only have $2n$ equations (n active power balance equations and n reactive power balance equations); therefore (B.11–B.12) do not form a closed system of equations. Thus, we need to fix $2n$ variables as follows.

First, we assume that at each bus i, $i = 1, 2, \ldots, n$, there is some control mechanism that ensures the voltage magnitude remains fixed, i.e.,

$$v_i = V_i, \quad i = 1, 2, \ldots, n, \tag{B.14}$$

where V_i is some positive scalar. Next, assume that each bus i, $i = 1, 2, \ldots, m$, $1 \leq m \leq n - 1$, corresponds to a generator or a load with the active power it injects into the network, denoted by ξ_i, being extraneously set (taking a positive value if bus i corresponds to a generator and taking a negative value if bus i corresponds to a load), i.e.,

$$p_i = \xi_i, \quad i = 1, 2, \ldots, m. \tag{B.15}$$

Similarly, assume that each bus $i = m+1, m+2, \ldots, n$, corresponds to a generator or a load with the active power it injects into the network, denoted by u_i, being controllable, i.e.,

$$p_i = u_i, \quad i = m+1, m+2, \ldots, n. \tag{B.16}$$

Next, since we do not have control over the values that ξ_i, $i = 1, 2, \ldots, m$ take, in order for (B.13) to be satisfied, we need to specify how the total amount of extraneous power injected into the network, $\sum_{i=1}^{m} \xi_i$, is allocated among the controlled injections. In this regard, we assume that the controlled power injection into bus i, $i = m+1, m+2, \ldots, n$, is equal to some nominal (setpoint) value, denoted by r_i, plus a term that is proportional to the sum of the extraneous power injection values and the controlled power injection setpoints, i.e.,

$$u_i = r_i + \alpha_i \left(\sum_{j=1}^{m} \xi_j + \sum_{j=m+1}^{n} r_j \right), \quad i = m+1, m+2, \ldots, n, \tag{B.17}$$

where α_i is a nonpositive scalar such that $\sum_{i=m+1}^{n} \alpha_i = -1$. Thus, (B.11–B.12) together with (B.14–B.17) form a system with $2n$ equations and $2n$ unknowns: $\theta_1, \theta_2, \ldots, \theta_n$, and $q_1, q_2, \ldots, q_n$. Furthermore, (B.11) together with (B.14–B.17) form a closed system of n equations involving only $\theta_1, \theta_2, \ldots \theta_n$; thus, since we are only interested in establishing a model for the flow of active power across transmission lines, in the remainder we only consider these equations. However, note that out of these n equations, only $n-1$ are linearly independent; thus, if we were interested in solving for θ_i, $i = 1, 2, \ldots, n$, we would have to fix the value of one of them a priori. This is not an issue as the equations depend on the difference between pairs of angles.

Assign an arbitrary orientation for the positive flow on each transmission line; then, the power system network topology together with the chosen orientation can be described by a directed graph, $\mathcal{G} = \{\mathcal{V}, \mathcal{E}\}$, where each element in the set $\mathcal{V} = \{1, 2, \ldots, n\}$ corresponds to a bus, and $\mathcal{E} \subset \mathcal{V} \times \mathcal{V} \setminus \{(i,i) \colon i \in \mathcal{V}\}$ so that $(i,j) \in \mathcal{E}$ if there is a transmission line connecting bus i and bus j with the flow of power on this line assigned to be positive from bus i to bus j. Let $\mathbb{I}$ denote a one-to-one mapping that assigns each element $(i,j) \in \mathcal{E}$ to an element $e \in \mathcal{L} = \{1, 2, \ldots, l\}$. Then, we can define the node-to-edge incidence matrix of $\mathcal{G}$, denoted by $M = [m_{i,e}] \in \mathbb{R}^{n \times l}$, as follows:

$$m_{i,e} = \begin{cases} 1, & \text{if } e = \mathbb{I}\big((i,j)\big),\ (i,j) \in \mathcal{E}, \\ -1, & \text{if } e = \mathbb{I}\big((l,i)\big),\ (l,i) \in \mathcal{E}, \\ 0, & \text{otherwise.} \end{cases} \tag{B.18}$$

Define $\theta = [\theta_1, \theta_2, \ldots, \theta_n]^\top$, $\xi = [\xi_1, \xi_2, \ldots, \xi_m]^\top$, $r = [r_{m+1}, r_{m+2}, \ldots, r_n]^\top$, and $\alpha = [\alpha_{m+1}, \alpha_{m+2}, \ldots, \alpha_n]^\top$. Then, by using the node-to-edge incidence matrix of the graph $\mathcal{G}$, as defined in (B.18), we can rewrite (B.11) together with (B.14–B.17) in matrix form as follows:

$$\begin{bmatrix} \xi \\ u \end{bmatrix} = M\Gamma\mathbf{sin}(M^\top\theta), \tag{B.19}$$

$$u = r + \alpha\left(\mathbf{1}_m^\top\xi + \mathbf{1}_{n-m}^\top r\right), \tag{B.20}$$

where $\Gamma = \mathbf{diag}(\gamma_1, \gamma_2, \ldots, \gamma_l)$, with

$$\gamma_e = \frac{V_iV_j}{X_{ij}}, \quad e = \mathbb{I}\big((i,j)\big),\ (i,j) \in \mathcal{E},$$

and where for $x := M^\top\theta \in \mathbb{R}^l$, $\mathbf{sin}(x)$ is defined as:

$$\mathbf{sin}(x) = \left[\sin(x_1), \sin(x_2), \ldots, \sin(x_l)\right]^\top. \tag{B.21}$$

Then, by plugging (B.20) into (B.19), we obtain that

$$\begin{bmatrix} I_m \\ \alpha\mathbf{1}_m^\top \end{bmatrix}\xi + \begin{bmatrix} \mathbf{0}_{m\times(n-m)} \\ I_{n-m} + \alpha\mathbf{1}_{n-m}^\top \end{bmatrix} r = M\Gamma\mathbf{sin}(M^\top\theta). \tag{B.22}$$

Under normal operating conditions for some ξ and r, it is desirable that the corresponding $\theta = [\theta_1, \theta_2, \ldots, \theta_n]^\top$ is such that

$$|\theta_i - \theta_j| < \pi/2,\ \forall(i,j) \in \mathcal{E}; \tag{B.23}$$

this avoids power flow circulations around the loops in the network, while ensuring that the system is stable around the operating point (ξ, θ). Then, assuming (B.23) holds, it is possible to rewrite the power balance relation in (B.22) as a relation between the extraneous power injection vector, ξ, and the active power flows along the transmission lines instead of the angles. To this end, define the active power flow vector, $\phi = [\phi_1, \phi_2, \ldots, \phi_e]^\top$, where ϕ_e, $e = \mathbb{I}\big((i,j)\big)$, $(i,j) \in \mathcal{E}$, is defined to be positive in the direction corresponding to the orientation chosen when defining the node-to-edge incidence matrix, M, and negative otherwise. Then, we have that

$$\begin{bmatrix} I_m \\ \alpha\mathbf{1}_m^\top \end{bmatrix}\xi + \begin{bmatrix} \mathbf{0}_{m\times(n-m)} \\ I_{n-m} + \alpha\mathbf{1}_{n-m}^\top \end{bmatrix} r = M\phi, \tag{B.24}$$

$$0 = N\mathbf{arcsin}(\Gamma^{-1}\phi), \tag{B.25}$$

where the rows of $N \in \mathbb{R}^{(l-n+1)\times l}$ form a basis for the null space of M, i.e., N is a full rank matrix satisfying $MN^\top = \mathbf{0}_n$; and where for $y \in \mathbb{R}^l$, $\mathbf{arcsin}(y)$ is defined as:

$$\mathbf{arcsin}(y) = [\arcsin(y_1), \arcsin(y_2), \ldots, \arcsin(y_l)]^\top. \tag{B.26}$$

Essentially, (B.25) is enforcing that the angle differences across transmission lines are consistent in the sense that they add up to zero along loops in the networks. Note that for a given ξ, there are $l+1$ equations in (B.24–B.25), while the dimension of the unknown vector, ϕ, is l. Thus, when solving for ϕ, we can remove one of the equations in (B.24), which is linearly dependent on the others as the rank of M is $n-1$ when $\mathcal{G}$ is connected.

B.5 Per-Unit Normalization

The basic premise in per-unit normalization is to choose *base values* for quantities of interest, e.g., voltages, currents, impedances, and power, and to define each particular quantity in per unit as follows:

$$\text{quantity in per-unit} = \frac{\text{actual quantity}}{\text{base value quantity}}.$$

The chosen base quantities must be such that they satisfy the same relationships as those satisfied by the actual variables they are associated with. For example, given a complex impedance $\overline{Z}$, the relation between the complex voltage across its terminals, denoted by $\overline{V}$, and the current flowing through it, denoted by $\overline{I}$, is

$$\overline{V} = \overline{Z}\,\overline{I}; \tag{B.27}$$

then, the base quantities corresponding to $\overline{V}$, $\overline{I}$, and $\overline{Z}$, denoted by V_B, I_B, and Z_B, respectively, must satisfy

$$V_B = Z_B I_B. \tag{B.28}$$

Dividing (B.27) by (B.28) we obtain

$$\overline{V}^{\text{p.u.}} = \overline{Z}^{\text{p.u.}}\,\overline{I}^{\text{p.u.}},$$

where

$$\begin{aligned} \overline{V}^{\text{p.u.}} &= \frac{\overline{V}}{V_B}, \\ \overline{Z}^{\text{p.u.}} &= \frac{\overline{Z}}{Z_B}, \\ \overline{I}^{\text{p.u.}} &= \frac{\overline{I}}{I_B}, \end{aligned}$$

which is of the same form as $\overline{V} = \overline{Z}\,\overline{I}$; thus, we can do analysis with the quantities in per unit (p.u.).

Example B.1 (Simple DC circuit) Consider the circuit in Fig. B.2 (left). By using Ohm's law, we have that

$$\begin{aligned} I &= \frac{V_s}{R_1 + R_2} \\ &= \frac{10}{3}\ \text{A}. \end{aligned} \tag{B.29}$$

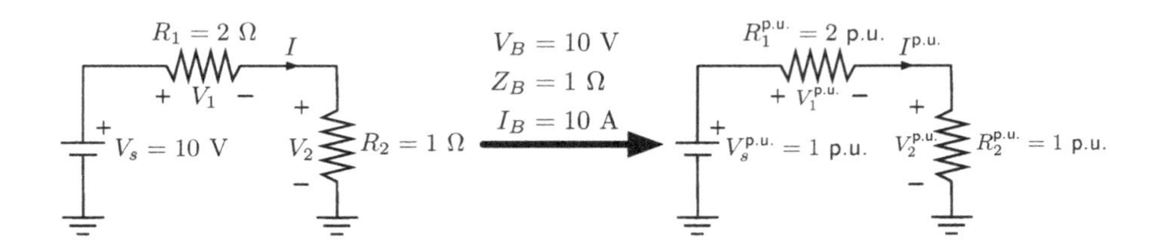

Figure B.2 DC circuit diagrams with actual quantities (left) and per-unit quantities (right).

Now, choose base quantities as follows: $V_B = 10$ V, $Z_B = 1\ \Omega$, and $I_B = 10$ A. Then, source per-unit voltage, $V_s^{\text{p.u.}}$, is obtained as follows:

$$\begin{aligned} V_s^{\text{p.u.}} &= \frac{V_s}{V_B} \\ &= 1 \text{ p.u.} \end{aligned} \tag{B.30}$$

Similarly, the per-unit values of R_1 and R_2, denoted by $R_1^{\text{p.u.}}$ and $R_2^{\text{p.u.}}$, respectively, are given by

$$\begin{aligned} R_1^{\text{p.u.}} &= \frac{R_1}{Z_B} \\ &= 2 \text{ p.u.}, \end{aligned} \tag{B.31}$$

$$\begin{aligned} R_2^{\text{p.u.}} &= \frac{R_2}{Z_B} \\ &= 1 \text{ p.u.} \end{aligned} \tag{B.32}$$

We can also define the per-unit values of I, V_1, and V_2, denoted by $I^{\text{p.u.}}$, $V_1^{\text{p.u.}}$, and $V_2^{\text{p.u.}}$, respectively, as follows:

$$I^{\text{p.u.}} = \frac{I}{I_B}, \quad V_1^{\text{p.u.}} = \frac{V_1}{V_B}, \quad V_2^{\text{p.u.}} = \frac{V_2}{V_B}.$$

As captured by the circuit in Fig. B.2 (right), the per-unit quantities defined above must satisfy the same relations as their corresponding quantities. Thus, we can solve for $I^{\text{p.u.}}$ as follows:

$$\begin{aligned} I^{\text{p.u.}} &= \frac{V_s^{\text{p.u.}}}{R_1^{\text{p.u.}} + R_2^{\text{p.u.}}} \\ &= \frac{1}{3} \text{ p.u.} \end{aligned} \tag{B.33}$$

Then, we have that $I = I^{\text{p.u.}} I_B = \frac{10}{3}$ A, which coincides with the result in (B.29) obtained using the actual variables directly.

Next, consider the expression for complex power:

$$\overline{S} = \overline{V}\,\overline{I}^*; \tag{B.34}$$

then, the base quantities corresponding to $\overline{S}$, $\overline{V}$, and $\overline{I}$, denoted by S_B, V_B, and I_B, respectively, must be such that

$$S_B = V_B I_B. \tag{B.35}$$

Thus,

$$\overline{S}^{\text{pu}} = \overline{V}^{\text{pu}}\left(\overline{I}^{\text{pu}}\right)^*, \tag{B.36}$$

where

$$\begin{aligned} \overline{S}^{\text{p.u.}} &= \frac{\overline{S}}{S_B}, \\ \overline{V}^{\text{p.u.}} &= \frac{\overline{V}}{V_B}, \\ \overline{I}^{\text{p.u.}} &= \frac{\overline{I}}{I_B}. \end{aligned}$$

Note that if we simultaneously consider $V_B = Z_B I_B$ and $S_B = V_B I_B$, we have four variables and two equations. Thus, if we choose the bases of two variables, the other two bases are automatically determined. For example, pick S_B and V_B, then:

$$Z_B = \frac{V_B^2}{S_B}, \tag{B.37}$$

$$I_B = \frac{S_B}{V_B}. \tag{B.38}$$

For a three-phase balanced system, three-phase complex power, denoted by $\overline{S}_{3\phi}$, is equal to the sum of the individual phase powers, denoted by $\overline{S}_a, \overline{S}_b, \overline{S}_c$, which happen to be all equal, i.e.,

$$\overline{S}_{3\phi} = 3\overline{S}_{\phi}, \tag{B.39}$$

where $\overline{S}_{\phi} = \overline{S}_a = \overline{S}_b = \overline{S}_c$. Also, recall that for a three-phase balanced system, the line-to-line voltage, $V_{\ell-\ell}$, and the phase voltage, V_{ϕ}, are related as follows:

$$V_{\ell-\ell} = \sqrt{3}V_{\phi}. \tag{B.40}$$

Let $S_{B,\phi}$, $V_{B,\phi}$, $I_{B,\phi}$, $Z_{B,\phi}$ denote phase base quantities chosen so as to satisfy (B.35) and (B.37–B.38). Then, in light of (B.39–B.40), it is natural to define the three-phase power base, $S_{B,3\phi}$, and line-to-line voltage base, $V_{B,\ell-\ell}$ as follows:

$$S_{B,3\phi} = 3S_{B,\phi}, \tag{B.41}$$

$$V_{B,\ell-\ell} = \sqrt{3}V_{B,\phi}. \tag{B.42}$$

Then,

$$\begin{aligned}
\overline{S}_{3\phi}^{\text{p.u.}} &= \frac{\overline{S}_{3\phi}}{S_{B,3\phi}} \\
&= \frac{3\overline{S}_{\phi}}{3S_{B,\phi}} \\
&= \overline{S}_{\phi}^{\text{p.u.}},
\end{aligned}$$

and

$$\begin{aligned}
V_{\ell-\ell}^{\text{p.u.}} &= \frac{V_{\ell-\ell}}{V_{B,\ell-\ell}} \\
&= \frac{\sqrt{3}V_{\phi}}{\sqrt{3}V_{B,\phi}} \\
&= V_{\phi}^{\text{p.u.}}.
\end{aligned}$$

Thus, when represented in per unit, it is not necessary in general to distinguish between per-phase quantities and their three-phase counterparts as their value is the same.

B.6 Notes and References

The material on the Π-equivalent circuit model and the power flow formulation in Section B.3 can be found in standard power system textbooks [58, 59]. The model in Section B.4 was inspired by [23] and has been utilized in [24]. The material on per-unit normalization is also standard and follows the developments in [58].

References

[1] C. A. Desoer and L. A. Zadeh, *Linear System Theory*. New York, NY: McGraw-Hill, 1963.

[2] M. Athans and P. Falb, *Optimal Control: An Introduction to the Theory and Its Application*. New York, NY: McGraw-Hill, 1966.

[3] R. E. Kalman, "Analysis and design principles of second and higher order saturating servomechanisms," *Transactions of the American Institute of Electrical Engineers, Part II: Applications and Industry*, vol. 74, no. 5, pp. 294–310, 1955.

[4] R. E. Kalman, "Canonical structure of linear dynamical systems," *Proceedings of the National Academy of Sciences of the United States of America*, vol. 48, no. 4, pp. 596–600, April 1962.

[5] D. Luenberger, *Introduction to Dynamic Systems: Theory, Models, and Applications*. New York, NY: John Wiley & Sons, 1979.

[6] K. Ogata, *Modern Control Engineering*, 4th ed. Upper Saddle River, NJ: Prentice Hall, 2001.

[7] H. K. Khalil, *Nonlinear Systems*, 3rd ed. Upper Saddle River, NJ: Prentice Hall, 2002.

[8] J. P. Hespanha, *Linear Systems Theory*. Princeton, NJ: Princeton University Press, 2018.

[9] G. F. Franklin, D. J. Powell, and A. Emami-Naeini, *Feedback Control of Dynamic Systems*, 8th ed. Upper Saddle River, NJ: Prentice Hall, 2019.

[10] F. Schweppe, *Uncertain Dynamic Systems*. Upper Saddle River, NJ: Prentice Hall, 1973.

[11] G. Grimmett and D. Stirzaker, *Probability and Random Processes*, 3rd ed. Oxford: Oxford University Press, 2001.

[12] D. P. Bertsekas and J. N. Tsitsiklis, *Introduction to Probability*, 2nd ed. Nashua, NH: Athena Scientific, 2008.

[13] A. Oppenheim and G. Verghese, *Signals, Systems & Inference*. London: Pearson, 2016.

[14] B. Hajek, *Probability with Engineering Applications: ECE 313 Course Notes*. [Online] https://courses.grainger.illinois.edu/ece313/fa2016/probabilityJan17.pdf,2017.

[15] B. Hajek, *Random Processes for Engineers*. Cambridge: Cambridge University Press, 2015.

[16] C. Hadjicostis, *Estimation and Inference in Discrete Event Systems.* London: Springer Nature, 2020.

[17] F. L. Chernousko, *State Estimation for Dynamic Systems.* Boca Raton, FL: CRC Press, 1994.

[18] A. Kurzhanski and I. Vályi, *Ellipsoidal Calculus for Estimation and Control.* Boston, MA: Birkhauser, 1997.

[19] A. A. Kurzhanskiy and P. Varaiya, "Ellipsoidal toolbox," EECS Department, University of California, Berkeley, Tech. Rep., May 2006.

[20] S. Boyd and L. Vandenberghe, *Convex Optimization.* Cambridge: Cambridge University Press, 2004.

[21] M. Kendall, *A Course in the Geometry of n Dimensions.* New York, NY: Hafner, 1961.

[22] P. McMullen, "Volumes of projections of unit cubes," *Bulletin of the London Mathematical Society*, vol. 16, no. 3, pp. 278–280, May 1984.

[23] F. Dörfler, M. Chertkov, and F. Bullo, "Synchronization in complex oscillator networks and smart grids," *Proceedings of the National Academy of Sciences USA*, vol. 110, no. 6, pp. 2005–2010, 2013.

[24] M. Zholbaryssov and A. D. Domínguez-García, "Convex relaxations of the network flow problem under cycle constraints," *IEEE Transactions on Control of Network Systems*, pp. 1–1, 2019.

[25] B. Borkowska, "Probabilistic load flow," *IEEE Transactions on Power Apparatus and Systems*, vol. 93, no. 3, pp. 752–759, Aug. 1974.

[26] P. Sauer, "Generalized stochastic power flow algorithm," PhD dissertation, Purdue University, West Lafayette, IN, 1977.

[27] J. Kemeny and J. Snell, *Finite Markov Chains.* New York, NY: Springer-Verlag, 1983.

[28] J. Norris, *Markov Chains.*

[29] P. Brémaud, *Markov chains: Gibbs fields, Monte Carlo simulation, and queues.*

[30] R. Barlow and F. Proschan, *Mathematical Theory of Reliability.* New York, NY: John Wiley & Sons, 1965.

[31] J. Endrenyi, *Reliability Modeling in Electric Power Systems.* Chichester: John Wiley & Sons, 1978.

[32] M. L. Shooman, *Probabilistic Reliability: An Engineering Approach.* Malabar, FL: Robert E. Krieger Publishing Company, 1990.

[33] M. Rausand and A. Høyland, *System Reliability Theory.* Hoboken, NJ: Wiley Interscience, 2004.

[34] W. Bouricius, W. Carter, and P. Schneider, "Reliability modeling techniques for self-repairing computer systems," in *Proceedings of the 24th National ACM Conference.* New York, NY: ACM Press, 1969.

[35] T. Arnold, "The concept of coverage and its effect on the reliability model a repairable system," *IEEE Transactions on Computers*, vol. C-22, no. 3, pp. 251–254, March 1973.

[36] P. Babcock, "An introduction to reliability modeling of fault-tolerant systems," The Charles Stark Draper Laboratory, Cambridge, MA, Tech. Rep. CSDL-R-1899, 1986.

[37] P. Babcock, G. Rosch, and J. J. Zinchuk, "An automated environment for optimizing fault-tolerant systems designs," in *Proc. Annual Reliability and Maintainability Symposium*, pp. 360–367, 1991.

[38] P. R. Kumar and P. Varaiya, *Stochastic Systems: Estimation, Identification and Adaptive Control.* Upper Saddle River, NJ: Prentice Hall, 1986.

[39] S. T. Cady, M. Zholbaryssov, A. D. Domínguez-García, and C. N. Hadjicostis, "A distributed frequency regulation architecture for islanded inertialess AC microgrids," *IEEE Transactions on Control Systems Technology*, vol. 25, no. 6, pp. 1961–1977, 2017.

[40] K. J. Åström, *Introduction to Stochastic Control.* Mineola, NY: Dover, 2006.

[41] R. A. Horn and C. R. Johnson, *Topics in Matrix Analysis.* New York, NY: Cambridge University Press, 1991.

[42] L. C. Evans, *An Introduction to Stochastic Differential Equations.* Providence, RI: American Mathematical Society, 2013.

[43] B. Øskendal, *Stochastic Differential Equations: An Introduction with Applications*, 6th ed. Berlin: Springer-Verlag, 2013.

[44] R. Brockett, *Lecture Notes on Stochastic Control.* [Online] www.eeci-institute.eu/pdf/M15/M15/RogersStochastic.pdf, 2009.

[45] A. Girard, "Reachability of uncertain linear systems using zonotopes," *Hybrid Systems: Computation and Control*, pp. 291–305, 2005.

[46] Yu Christine Chen, Xichen Jiang, and A. D. Domínguez-García, "Impact of power generation uncertainty on power system static performance," in *North American Power Symposium*, pp. 1–5, 2011.

[47] X. Jiang and A. D. Domínguez-García, "A zonotope-based method for capturing the effect of variable generation on the power flow," in *North American Power Symposium*, pp. 1–6, 2014.

[48] Z. Wang and F. Alvarado, "Interval arithmetic in power flow analysis," *IEEE Transaction on Power Systems*, vol. 7, no. 3, pp. 1341–1349, Aug. 1992.

[49] A. Kurzhanskiy and P. Varaiya, "Ellipsoidal techniques for reachability analysis of discrete-time linear systems," *IEEE Transactions on Automatic Control*, vol. 52, no. 1, pp. 26–38, Jan. 2007.

[50] A. Kurzhanski and P. Varaiya, "Ellipsoidal techniques for reachability analysis. Parts I & II," *Optimization Methods and Software*, vol. 17, pp. 177–237, Feb. 2002.

[51] E. Hope, X. Jiang, and A. Domínguez-García, "A reachability-based method for large-signal behavior verification of DC-DC converters," *IEEE Transactions on Circuits and Systems I: Regular Papers*, vol. 58, no. 12, pp. 2944–2955, Dec. 2011.

[52] Y. Chen and A. Domínguez-García, "A method to study the effect of renewable resource variability on power system dynamics," *IEEE Transactions on Power Systems*, vol. 27, no. 4, pp. 1978–1989, Nov. 2012.

[53] P. T. Krein, *Elements of Power Electronics.* New York, NY: Oxford University Press, 1998.

[54] P. Sauer and M. Pai, *Power System Dynamics and Stability.* Upper Saddle River, NJ: Prentice Hall, Inc., 1998.

[55] R. A. Horn and C. R. Johnson, *Matrix Analysis.* New York, NY: Cambridge University Press, 1985.

[56] G. H. Golub and C. F. Van Loan, *Matrix Computations*, 2nd ed. Baltimore, MD: Johns Hopkins University Press, 1989.

[57] W. Rudin, *Principles of Mathematical Analysis.*

[58] R. B. Bergen and V. Vittal, *Power systems analysis.* Upper Saddle River, NJ: Prentice Hall, 2000.

[59] J. Glover, M. Sarma, and T. Overbye, *Power System Analysis and Design.* Boston, MA: Cengage Learning, 2011.

Index

Π-equivalent circuit model, 78, 151, 320

active power, 4, 9, 13, 14, 60, 82, 151, 193, 216, 227
additive disturbance, 195
algebraic equations, 5
attainable failure sequence, 106, 113, 115, 117, 120, 121
availability model, 119, 120, 123, 124
average frequency error, 154, 195, 264

bathtub curve, 111
Bayes' formula, 17
Bernoulli distribution, 19
Bernoulli process, 30
  moments, 32
Binomial distribution, 19
buck DC-DC power converter, 297
buck dc-dc power converter
  voltage regulation, 300
  voltage regulation verification, 300

Cartesian product, 36
Chapman–Kolmogorov equations, 94
Chapman-Kolmogorov equations, 98, 126–128
characteristic function, 142
Chebyshev's inequality, 75
closed set, 36
common cause failures, 12, 104
component failure sequence, 101, 102
conditional expectation, 24
conditional pdf, 23
conditional probability, 16
convex set, 38
  support function, 239
correlation, 25
covariance, 25
covariance matrix, 29, 132, 146

deterministic inputs, 58, 68, 137, 179, 220, 258, 289
differential-algebraic equation (DAE) model, 303
  linearization, 303
Dirac delta function, 67, 167
directed graph, 78, 151, 193, 227, 263, 324
  node-to-edge incidence matrix, 78, 151, 227, 264, 324
disjoint sets, 36
droop coefficient, 152, 194, 264
dynamical systems, 8
  continuous-time, 5, 13
  discrete-time, 5, 13
Dynkin's formula, 184, 187, 189

ellipsoid, 8, 14, 40, 203, 204, 238, 273
  center, 40, 203, 204, 239, 259, 275
  minimum sum of squares of the semi-axis lengths, 268
  minimum volume, 268
  Minkowski sum, 42
  shape matrix, 40, 203, 204, 239, 259, 275
  support function, 40, 204, 240, 275
  tight, 268
  volume, 42
Euler's gamma function, 42, 246, 282
event, 16
event algebra, 16
events
  independent, 17
  mutually exclusive, 17
expectation, 18, 20
exponential distribution, 21
extraneous disturbance, 156
extraneous power injection vector, 85, 228, 233

failed mode, 11, 101
failure, 2
failure coverage, 117
failure rate, 109, 116
first moment, 18, 20
frequency regulation, 11

frequency-droop control, 10
  continuous-time, 194
  discrete-time, 152, 263

Gaussian distribution, 21
Gaussian systems, 140, 142, 145, 181
Generator matrix, 99
Geometric distribution, 19
gradient, 41, 206, 316
Green's first identity, 318

Hurwitz matrix, 279
hyperplane, 209

i.i.d. random variables, 30
i.i.d. random vectors, 138, 141, 150
independence, 16
inner product, 308
input, 1
input uncertainty
  probabilistic, 7, 8, 12, 13
  set-theoretic, 7, 8, 14, 273
input vector, 6, 55, 65, 76, 273
input-to-state mapping, 6, 8, 13, 203
input-to-state-mapping, 62
integration by parts, 185, 186, 190, 191, 318
integration by substitution, 318
inverse function theorem, 69, 317
Itô's chain rule, 187
Itô's lemma, 183

Jacobi's formula, 49, 282
Jacobian matrix, 68, 70, 72, 212, 221, 316
joint pdf, 28
jointly distributed random variables, 21

Kirchhoff's laws, 3, 54, 298, 320
Kirchhoff's voltage law, 66, 215
Kronecker product, 135, 177, 196, 279, 310

Lagrange multiplier, 41, 206
law of total probability, 17, 94, 95
Leibniz's integral differentiation rule, 51
linear combination, 138, 219, 308
linear dependence, 308
linear dynamical systems, 16, 46, 130
  asymptotic stability, 48, 52
  continuous-time, 46, 166, 168, 273
  discrete-time, 46, 131, 237, 238
  equilibrium point, 48, 52
  input, 46, 49
  Lyapunov stability, 48, 52
  state, 46, 49
  state-transition matrix, 46, 49
  time-invariant, 47, 135, 177, 243, 244
  trajectory, 46, 49

linearization, 13, 14, 59, 144, 146, 212, 259, 291
loopy networks, 85, 231, 235

Markov availability model, 119
Markov chain, 8, 13, 91, 92
  continuous-time, 94
  discrete-time, 94
  homogeneous, 95
  transition matrix, 95
Markov process, 8, 93
Markov property, 93, 97
Markov reliability model, 112
matrix
  adjugate, 312
  characteristic polynomial, 313
  classical adjoint, 312
  determinant, 312
  eigenvalue, 313
  eigenvector, 313
  partition, 309
  positive (semi)definite, 314
  rank, 313
  spectrum, 313
  square, 311
  trace, 311
mean, 28
mean time to failure (MTTF), 110, 116
microgrids, 3, 13, 14, 151, 192, 263
Minkowski difference, 37
Minkowski sum, 37, 239
  support function, 37, 239
moments
  first and second, 7, 12, 131

network admittance matrix, 216
network topology, 9, 11, 193, 263
node-to-edge incidence matrix, 193
nominal mode, 11, 101
non-repairable component, 109
non-repairable system, 11, 109, 116
nonlinear dynamical systems
  continuous-time, 187, 290
  discrete-time, 144, 259

open set, 36
ordinary differential equations (ODEs), 5
outage, 2
outer-bounding ellipsoid, 278
  minimum sum of squares of the semi-axis lengths, 251, 278, 284
  minimum volume, 246, 278, 281, 282
  tight, 254, 285, 286

per-unit normalization, 61, 326
  base quantities, 326

performance characterization, 74, 100
performance requirements, 55, 225, 238, 261, 274, 293
performance requirements verification, 225, 261, 293
phasor
  magnitude, 78, 151, 193, 216, 263
  phase angle, 151, 193, 216, 263
Poisson distribution, 19
polytope, 64
power, 1
power electronics, 3, 151
power flow analysis, 77, 226
power flow model, 78, 216, 227, 301, 321
  active power balance equations, 323
  reactive power balance equations, 323
power flow problem, 9
power injection uncertainty
  probabilistic, 151
  set-theoretic, 263
power injections, 9, 79
power system
  bulk, 3, 15
  buses, 9, 78, 151, 320
  electric, 1
  electromechanical phenomena, 297
  generator bus, 79, 324
  load bus, 79, 324
  network, 9
  network admittance matrix, 321
  network topology, 78, 82
  transmission lines, 9, 78, 151, 320
probabilistic power flow, 90
probability, 16
probability density function (pdf), 7, 20
  conditional, 23
  joint, 21
  marginal, 22
probability mass function (pmf), 18
  conditional, 28
  joint, 28, 30
  marginal, 27
probability measure, 16
probability space, 16
probability theory, 12

random variable, 17
  continuous, 20
  cumulative distribution function (cdf), 18, 20
  discrete, 18
  expectation, 18
  Gaussian, 140, 182
  probability density function (pdf), 20
  probability mass function (pmf), 18
  realization, 17
  second moment, 19
  standard deviation, 19
  variance, 19, 20
random variables
  independent, 23
  joint cdf, 21
  joint pdf, 21
  marginal pdf, 22
  marginal pmf, 27
  uncorrelated, 25
random vector, 55, 92
  characteristic function, 142
  continuous, 27
  correlation matrix, 29, 56, 58, 59
  covariance matrix, 29, 56, 58, 59
  cumulative distribution function (cdf), 27
  discrete, 27
  expectation, 28
  first and second moments, 55
  Gaussian, 142, 182
  mean, 56, 58, 59
random vectors, 27
  conditional pmf, 28
  independent, 28
  joint cdf, 28–30
  joint pdf, 55
  joint pmf, 28
  marginal pmf, 28
  uncorrelated, 157, 172
reachable set, 238, 273–275
  approximate characterization, 257, 288
  exact characterization, 257, 288
  outer-bounding ellipsoid, 243
reactive power, 60, 216
reduced-order Markov models, 124
  aggregation, 126
  truncation, 124
reliability function, 110, 115
reliability model, 119, 123
repair rate, 119, 123
repairable system, 118

sample space, 16
second moment, 19
set, 35
  boundary, 36
  cardinality, 35
  closure, 36
  difference, 36
  interior, 36
  interior point, 36
  intersection, 36
  singleton, 35, 229
  union, 36
set theory, 16, 35, 273
single machine infinite bus system, 295

singleton, 35, 229
spectrum, 135, 177, 243, 279
standard deviation, 19
standard Wiener vector process, 181, 187
state, 5
state vector, 6, 55, 65, 76, 273
state-space model, 13, 14, 46
  continuous-time, 166, 273
  discrete-time, 130, 237
static systems, 5, 14
  linear, 56, 62, 65, 202, 204, 219
  nonlinear, 58, 221
  performance characterization, 74
  performance requirements verification, 225
  reliability analysis, 11, 92, 101
stationarity, 32
stochastic process, 8, 12, 16, 29, 130, 131
  continuous-time, 30
  correlation function, 31, 131, 135, 138, 144, 146, 176
  covariance function, 8, 31, 131, 132, 138, 144–146, 157, 166, 173
  discrete time, 29
  first and second moments, 145, 171
  independent increments, 168, 182
  mean function, 8, 30, 131, 132, 138, 144–146, 157, 166, 172
  sample paths, 33, 182
  strict sense stationary, 32
  wide sense stationary, 32
  zero mean, 168
subset, 35
support function, 37, 205
supporting hyperplane, 39
supporting hyperplane theorem, 38
synchronous generator, 295, 300
  electrical angular position, 301
  electrical frequency, 295
  electrical power output, 302
  friction coefficient, 302
  governor constant, 302
  inertia constant, 295, 302
  internal voltage, 295
  mechanical power, 295, 302
  power command, 302
  speed-droop characteristic slope, 302
system
  behavior, 1
  model parameters, 3
  nominal response, 1
  response, 1
  structure, 1
  variables, 3
system performance, 54
system status, 104
  nonoperational, 12, 105
  operational, 12, 105

Taylor series expansion, 59, 85, 217, 221, 317
three-bus power system, 300
three-phase voltage source inverter, 152, 193
tree network, 80, 228, 234

uncertainty
  input, 7
  structural, 7
uncertainty models, 1
  probabilistic, 7
  set-theoretic, 7
unconstrained optimization problem, 41, 206
  maximizer, 207
uncorrelated, 133
uniform distribution, 20
union bound, 74
unitary ball, 38

variance, 19
vector norm, 308
  1-norm, 308
  ∞-norm, 308
  Euclidean norm, 308

Weibull distribution, 111
white noise process, 8, 14, 167, 168
  Gaussian, 14
Wiener process, 33, 182
  continuous sample paths, 33, 182
  correlation function, 34
  covariance function, 34
  standard, 34

zonotope, 8, 14, 43, 219
  center, 43, 203, 219
  generators, 43, 203, 219
  Minkowski sum, 43
  support function, 43, 219
  volume, 44

Printed in the United States
by Baker & Taylor Publisher Services